Lecture Notes in Mathematics

Edited by A. Dold, B. Eckmann and F. Takens

1442

L. Accardi W. von Waldenfels (Eds.)

Quantum Probability and Applications V

Proceedings of the Fourth Workshop, held in Heidelberg, FRG, Sept. 26–30, 1988

Springer-Verlag

Berlin Heidelberg New York London

Paris Tokyo Hong Kong Barcelona

Editors

Luigi Accardi
Dipartimento di Matematica, Università di Roma II
Via Orazio Raimondo, 00173 Rome, Italy

Wilhelm von Waldenfels
Institut für Angewandte Mathematik
Universität Heidelberg
Im Neuenheimer Feld 294
6900 Heidelberg, Federal Republic of Germany

Mathematics Subject Classification (1980): 46LXX, 47DXX, 58F11, 60FXX, 60GXX, 60HXX, 60JXX, 82A15

ISBN 3-540-53026-6 Springer-Verlag Berlin Heidelberg New York
ISBN 0-387-53026-6 Springer-Verlag New York Berlin Heidelberg

Printed in Germany

Printing and binding: Druckhaus Beltz, Hemsbach/Bergstr.
2146/3140-543210 – Printed on acid-free paper

INTRODUCTION

This volume, the fifth one of the quantum probability series, contains the proceedings of the Fourth Workshop on Quantum Probability, held in Heidelberg, September 26-30, 1988.

The workshop was made possible by the support of the Deutsche Forschungsgemeinschaft (Sonderforschungsbereich 123, University of Heidelberg) and of the Centro Volterra, University of Rome II.

We are glad to thank all the participants for their contributions to the present volume and for their contributions in discussions and conversations.

Luigi Accardi
Wilhelm von Waldenfels

TABLE OF CONTENTS

L. ACCARDI, A. FRIGERIO, L.Y. GANG, Quantum Langevin Equation in the Weak Coupling Limit 1

L. ACCARDI, L.Y. GANG, On the Low Density Limit of Boson Models 17

L. ACCARDI, R.L. HUDSON, Quantum Stochastic Flows and Non Abelian Cohomology 54

D. APPELBAUM, Quantum Diffusions on Involutive Algebras 70

A. BARCHIELLI, Some Markov Semigroups in Quantum Probability 86

V. BELAVKIN, A Quantum Stochastic Calculus in Fock Space of Input and Output Nondemolition Processes 99

C. CECCHINI, B. KÜMMERER, Stochastic Transitions on Preduals of von Neumann Algebras 126

F. FAGNOLA, Quantum Stochastic Calculus and a Boson Levy Theorem 131

K.-H. FICHTNER, U. SCHREITER, Locally Independent Boson Systems 145

A. FRIGERIO, Time-Inhomogeneous and Nonlinear Quantum Evolutions 162

D. GODERIS, A. VERBEURE, P. VETS, Quantum Central Limit and Coarse Graining 178

G.C. HEGERFELDT, An Open Problem in Quantum Shot Noise 194

E. HENSZ, A Method of Operator Estimation and a Strong Law of Large Numbers in von Neumann Algebras 204

A.S. HOLEVO, An Analog of the Ito Decomposition for Multiplicative Processes with Values in a Lie Group 211

R.L. HUDSON, P. SHEPPERSON, Stochastic Dilations of Quantum Dynamical Semigroups Using One-dimensional Quantum Stochastic Calculus 216

G.-L. INGOLD, H. GRABERT, Sluggish Decay of Preparation Effects in Low Temperature Quantum Systems 219

R. JAJTE, Almost Sure Convergence of Iterates of Contractions in Noncommutative L_2-Spaces 231

J.M. LINDSAY, H. MAASSEN, Duality Transform as *-Algebraic Isomorphism 247

J.M. LINDSAY, K.R. PARTHASARATHY, Rigidity of the Poisson Convolution 251

H. MAASSEN, A Discrete Entropic Uncertainty Relation 263

B. NACHTERGAELE, Working with Quantum Markov States and their Classical Analogues 267

H. NARNHOFER, Dynamical Entropy, Quantum K-Systems and Clustering 286

K.R. PARTHASARATHY, A Continuous Time Version of Stinespring's Theorem on Completely Positive Maps 296

A. PASZKIEWICZ, The Topology of the Convergence in Probability in a W^*-Algebra is Normal 301

D. PETZ, First Steps Towards a Donsker and Varadhan Theory in Operator Algebras 311

S. PULMANNOVÁ, Quantum Conditional Probability Spaces 320

P. ROBINSON, Quantum Diffusions on the Rotation Algebras and the Quantum Hall Effect 326

J.-L. SAUVAGEOT, Quantum Dirichlet Forms, Differential Calculus and Semigroups 334

M. SCHÜRMANN, Gaussian States on Bialgebras 347

G.L. SEWELL, Quantum Macrostatistics and Irreversible Thermodynamics 368

A.G. SHUHOV, Y.M. SUHOV, Correction to the Hydrodynamical Approximation for Groups of Bogoljubov Transformations 384

S.J. SUMMERS, Bell's Inequalities and Quantum Field Theory 393

QUANTUM LANGEVIN EQUATION IN THE WEAK COUPLING LIMIT

Luigi Accardi [*], Alberto Frigerio [**], and Lu Yun-Gang [*,***]

[*] Centro Matematico V. Volterra,
Dipartimento di Matematica, II Università di Roma,
00173 La Romanina, Roma, Italy.

[**] Dipartimento di Matematica e Informatica,
Università di Udine, 33100 Udine, Italy.

[***] On leave from Department of Mathematics,
Beijing Normal University, Beijing, P.R. China.

0. Introduction

This work is part of a series of papers [2,3,4] where, expanding some heuristic ideas in [11], we develop the theory of the weak coupling limit for open quantum systems. It has been known for fifteen years that the reduced dynamics of an open quantum system converges to a quantum dynamical semigroup in the weak coupling limit (coupling constant $\lambda \to 0$, microscopic time $s \to \infty$, with macroscopic time $t = \lambda^2 s$ held constant) [18,6]. We investigate whether and in which sense the full time evolution of an open quantum system converges, in the weak coupling limit, to an evolution driven by quantum Brownian motion [16,14,15]. In [2] we obtained rigorous results for the time evolution operator $U^{(\lambda)}_{t/\lambda^2}$; in [3] we studied the time evolved observables $j^{(\lambda)}_{t/\lambda^2}(X) = U^{(\lambda)+}_{t/\lambda^2}(X \otimes 1)U^{(\lambda)}_{t/\lambda^2}$, and we proved that $j^{(\lambda)}_{t/\lambda^2}$ converges to a quantum diffusion j_t [10] governed by a quantum Langevin equation.

Here we wish to present the results of [3] in a self-contained way. To this end we adopt a method of proof which is simpler than the one of [3], by reducing the problem to a time-dependent generalization of the derivation of a quantum dynamical semigroup in the weak coupling limit [18,6]. However, the price to be paid for this simplification is that the present method is specific for boson reservoirs with linear coupling to the system of interest, whereas the method of [3] is suitable for generalization to the fermion case [4], to more general interactions, and to the low density limit [9,5].

1. Notations and Preliminaries

We consider a quantum system S, with associated Hilbert space $\mathcal{H}$ and Hamiltonian H, coupled to another quantum system S', with Hilbert space $\mathcal{H}'$ and Hamiltonian H', by an interaction λV, where the coupling constant λ is assumed to be "small". All Hilbert spaces in this paper will be understood to be separable. S is spatially confined, meaning that the Hamiltonian H is self-adjoint in $\mathcal{H}$, bounded from below, and such that $\exp[-\beta H]$ is trace class for all positive β. S' is an infinitely extended quasi-free boson system, meaning that $\mathcal{H}' = \Gamma(\mathcal{H}_1)$ is the symmetric Fock space over the one-particle Hilbert space $\mathcal{H}_1$, and $H' = d\Gamma(H_1)$ is the differential second quantization of the one-particle Hamiltonian H_1, a non-negative self-adjoint operator in $\mathcal{H}_1$ with Lebesgue spectrum. We shall also assume that there exists a nonzero (nonclosed) linear subspace K_1 of $\mathcal{H}_1$ such that, for all $f,g \in K_1$, we have

$$\int_{-\infty}^{+\infty} |\langle f,\exp[iH_1 t]g\rangle|\, dt \;<\; +\infty . \tag{1.1}$$

The annihilation and creation operators in $\mathcal{H}'$ corresponding to the test functions f in $\mathcal{H}_1$ will be denoted by $a(f)$, $a^+(f)$; they satisfy the CCR $[a(f),a^+(g)] = \langle f,g\rangle$ and $a(f)\Phi_o = 0$, where Φ_o is the Fock vacuum vector.

By doubling the space $\mathcal{H}_1$ to $\mathcal{H}_1 \oplus \bar{\mathcal{H}}_1$, $\bar{\mathcal{H}}_1$ denoting the conjugate space to $\mathcal{H}_1$, this formalism allows us to consider also a representation of the CCR algebra determined by a gauge-invariant quasi-free state w_Q which is stationary under the time evolution determined by H'. Specifically, let Q be a positive self-adjoint operator in $\mathcal{H}_1$, commuting strongly with H_1 and satisfying $Q \geq 1$, and let w_Q be determined by

$$w_Q(a(f)a^+(g)) = \langle f, \tfrac{1}{2}(Q+1)g\rangle , \quad w_Q(a^+(g)a(f)) = \langle f, \tfrac{1}{2}(Q-1)g\rangle .$$

Then $a(f)$ is represented by $\pi_Q(a(f)) = a(Q_+ f \oplus 0) + a^+(0 \oplus \overline{Q_- f})$, where

$$Q_+ = [\tfrac{1}{2}(Q+1)]^{1/2} , \quad Q_- = [\tfrac{1}{2}(Q-1)]^{1/2} ; \tag{1.2}$$

we have indeed $\langle \Phi_\circ, \pi_Q(a(f))\pi_Q(a^+(g))\Phi_\circ\rangle = w_Q(a(f)a^+(g)) : f,g \in H_1$.

We identify $\mathcal{H}_1$ and $\bar{\mathcal{H}}_1$ as sets, assuming that $\mathcal{H}_1$ is mapped onto $\bar{\mathcal{H}}_1$ by a conjugation commuting with H_1 and with Q, so that

$$Q_+ \exp[iH_1t]f \oplus \overline{Q_- \exp[iH_1t]f} = \exp[iH_1t]Q_+f \oplus \exp[-iH_1t]\overline{Q_- f} .$$

The interaction V between S and S' is assumed to be of the form

$$V = i \sum_{j=1}^{n} [B_j \otimes a^+(g_j) - B_j^+ \otimes a(g_j)] , \tag{1.3}$$

where $B_1,\ldots,B_n$ are bounded operators on H, $g_1,\ldots,g_n \in K_1$, and where

$$[H,B_j] = -w_j B_j , \quad w_j > 0 : j = 1,\ldots,n \tag{1.4}$$

$$\langle g_j, \exp[iH_1t]g_k\rangle = 0 \text{ for all } t \text{ if } j \neq k . \tag{1.5}$$

Such interactions arise in the so-called rotating wave approximation.

In order to derive a reduced dynamics for the observables of S, we assume that the initial state of S' is a gauge-invariant quasi-free state w_Q. Again, by doubling $\mathcal{H}_1$ to $\mathcal{H}_1 \oplus \bar{\mathcal{H}}_1$ and the set of indices $(1,\ldots,n)$ to $(1,\ldots,2n)$, we can reduce this to the case where the initial state of S' is the Fock vacuum: the new expression of the interaction V becomes

$$V = i \sum_{j=1}^{2n} [B_j \otimes a^{+}(\tilde{g}_j) - B_j^{+} \otimes a(\tilde{g}_j)] \quad , \tag{1.6}$$

where $\tilde{g}_j = Q g_{+j} \oplus 0$, $\tilde{g}_{n+j} = 0 \oplus \overline{Q g_{-j}}$, and where $B_{n+j} = - B_j^{+}$: $j = 1,\ldots,n$. In this and in the following two Sections, tildes will be dropped and 2n will be called n again.

Let $H_\lambda = H + H' + \lambda V$ $(= H \otimes 1 + 1 \otimes H' + \lambda V)$ be the total Hamiltonian for the composite system S + S', and let $H_\circ = H_{\lambda=0}$. Let also

$$U_t^{(\lambda)} = \exp[iH_\circ t] \exp[-iH_\lambda t] \quad : t \geq 0 \tag{1.7}$$

be the time evolution operators, in the interaction picture, for state vectors in $\mathcal{H} \otimes \mathcal{H}'$. Then it has been shown [18,6] that, for all u , v in $\mathcal{H}$ and for all X in $B(\mathcal{H})$, one has

$$\lim_{\lambda \to 0} \langle u \otimes \Phi_\circ , U_{t/\lambda^2}^{(\lambda)+} (X \otimes 1) U_{t/\lambda^2}^{(\lambda)} v \otimes \Phi_\circ \rangle = \langle u, T_t(X) v \rangle , \tag{1.8}$$

where $(T_t : t \geq 0)$ is a quantum dynamical semigroup (in the sense of [12,17], or a quantum Markovian semigroup in the sense of [1]) whose infinitesimal generator L is given by

$$L(X) = (\sum_{j=1}^{n} c_j(w_j) B_j^{+} X B_j) + K^{+} X + X K , \tag{1.9}$$

where

$$c_j(w_j) = \frac{1}{2} \int_{-\infty}^{+\infty} \langle g_j , \exp[i(H_1 - w_j)t] g_j \rangle \, dt \quad (\geq 0) \tag{1.10}$$

and where

$$K = -\sum_{j=1}^{n} \int_{-\infty}^{0} \langle g_j, \exp[i(H_1 - w_j)t] g_j \rangle \, dt \; B_j^{+} B_j . \qquad (1.11)$$

2. Statement of main result

In our previous paper [2] we have shown that there is a precise sense in which $U^{(\lambda)}_{t/\lambda^2}$ (defined by (1.7)) converges as $\lambda \to 0$ to the solution $U(t)$ of a quantum stochastic differential equation (QSDE; in the sense of [16]) of the form

$$dU(t) = \Big(\sum_{j=1}^{n} [B_j \, dA_j^{+}(t) - B_j^{+} \, dA_j(t)] + K \, dt \Big) U(t) \qquad (2.1)$$

with $U(0) = 1$, where $\{A_j(t), A_j^{+}(t) : t \geq 0, \; j = 1,\dots,n\}$ are mutually independent Fock quantum Brownian motions satisfying

$$dA_j(t) \, dA_k^{+}(t) = \delta_{jk} \, c_j(w_j) \, dt \qquad (2.2)$$

and where K is given by (1.11) (actually, the proof in [2] is just for the case $n = 1$, but an extension to arbitrary n can be easily given). In the present paper, like in [3], we shall prove that there is a precise sense in which

$$j^{(\lambda)}_{t/\lambda^2}(X) := U^{(\lambda)+}_{t/\lambda^2} (X \otimes 1) \, U^{(\lambda)}_{t/\lambda^2} \quad : \quad X \in B(\mathcal{H}) \qquad (2.3)$$

converges as $\lambda \to 0$ to

$$j_t(X) := U^{+}(t) (X \otimes 1) U(t) \quad : \quad X \in B(\mathcal{H}), \qquad (2.4)$$

where $U(t)$ is the solution of the QSDE (2.1): j_t is a quantum diffusion (in the sense of [10]) satisfying the following quantum Langevin equation:

$$dj_t(X) = \sum_{j=1}^{n} \{ j_t([B_j^+,X])\, dA_j(t) - j_t([B_j,X])\, dA_j^+(t) \}$$

$$+ j_t(L(X))\, dt \quad , \tag{2.5}$$

where $dA_j(t)$, $dA_j^+(t)$ are as in (2.1) and where L is given by (1.9).

We introduce the same definitions as in [2,3]. Because of (1.5), we can define a strongly continuous group $\{S_t : t \in R\}$ of unitary operators on H_1 such that

$$S_t g_j = \exp[i(H_1 - w_j)t]\, g_j \quad : \quad t \in R \ , \ j = 1,\ldots,n \ . \tag{2.6}$$

We introduce a scalar product $(.|.)$ on K_1 by

$$(f|g) = \int_{-\infty}^{+\infty} \langle f, S_t g\rangle\, dt \quad : \quad f \ , \ g \in K_1 \ , \tag{2.7}$$

and we denote by K the completion of $K_1/\ker(.|.)$ with respect to the norm got from $(.|.)$. We also introduce collective Weyl operators on H' as $W(h *_\lambda f)$: $h \in L^2(R)$, $f \in K_1$, where, as usual, $W(g) = \exp[a^+(g) - a(g)]$, and where

$$h *_\lambda f = \lambda \int_{-\infty}^{+\infty} h(\lambda^2 t)\, S_t f\, dt \quad : \quad h \in L^2(R) \ , \ f \in K_1 \tag{2.8}$$

(in [2,3], h has been always taken to be the indicator function of an interval).

In addition to the Fock space $\mathcal{H}' = \Gamma(\mathcal{H}_1)$ we shall also need the Fock space $\mathcal{H}'' = \Gamma(L^2(R) \otimes K)$ over $L^2(R) \otimes K$. Weyl operators on either Fock space will be always denoted by $W(.)$, and the Fock vacuum vector of either space will be always denoted by Φ_o .

Let also $\{A_j(t), A_j^+(t) : t \geq 0 \ , \ j = 1,\ldots,n\}$ be the annihilation and creation operators in $\mathcal{H}''$ corresponding to the test functions $\chi_{[0,t]} \otimes g_j : t \geq 0$, $j = 1,\ldots,n$, g_j being regarded as elements of K; they are mutually independent Fock quantum Brownian motions satisfying the quantum Ito table (2.2), as shown in [2].

With the use of all the definitions and notations given above, the main result of this paper can be stated as follows:

Theorem 2.1. For all u, v in $\mathcal{H}$, h in $L^2(R)$, f in K_1, and for all X in $B(\mathcal{H})$ we have

$$\lim_{\lambda\to 0} \langle u \otimes W(h * f_\lambda)\Phi_\circ , j^{(\lambda)}_{t/\lambda^2}(X)\, v \otimes W(h * f_\lambda)\Phi_\circ \rangle_{\mathcal{H}\otimes\mathcal{H}'}$$

$$= \langle u \otimes W(h \otimes f)\Phi_\circ , j_t(X)\, v \otimes W(h \otimes f)\Phi_\circ \rangle_{\mathcal{H}\otimes\mathcal{H}''} , \qquad (2.9)$$

where, in the r.h.s. of (2.9), f is regarded as an element of K.

The proof will be given in the following Section, through a number of Lemmas. It should be noted that the interest of a result like Theorem 2.1 lies in the fact that, in general, linear combinations of coherent states $\langle W(f)\Phi_\circ , \cdot\, W(f)\Phi_\circ \rangle$ are dense in the set of all normal states. So the result here is only slightly less general than the one of [3], where it has been shown that, for all u, v in $\mathcal{H}$, h_1, h_2 in $L^2(R)$, f_1, f_2 in K_1, X in $B(\mathcal{H})$, one has

$$\lim_{\lambda\to 0} \langle u \otimes W(h_1 * f_{\lambda 1})\Phi_\circ , j^{(\lambda)}_{t/\lambda^2}(X)\, v \otimes W(h_2 * f_{\lambda 2})\Phi_\circ \rangle_{\mathcal{H}\otimes\mathcal{H}'}$$

$$= \langle u \otimes W(h_1 \otimes f_1)\Phi_\circ , j_t(X)\, v \otimes W(h_2 \otimes f_2)\Phi_\circ \rangle_{\mathcal{H}\otimes\mathcal{H}''} .$$

3. Proofs

Lemma 3.1. The left-hand side of (2.9) can be rewritten as follows:

$$\langle u \otimes W(h * _\lambda f)\Phi_\circ , j^{(\lambda)}_{t/\lambda^2}(X)\, v \otimes W(h *_\lambda f)\Phi_\circ \rangle_{\mathcal{H} \otimes \mathcal{H}'}$$

$$= \langle u \otimes \Phi_\circ , j^{(\lambda,h,f)}_{t/\lambda^2}(X)\, v \otimes \Phi_\circ \rangle_{\mathcal{H} \otimes \mathcal{H}'} , \tag{3.1}$$

where

$$j^{(\lambda,h,f)}_{t/\lambda^2}(X) = U^{(\lambda,h,f)+}_{t/\lambda^2}(X \otimes 1)\, U^{(\lambda,h,f)}_{t/\lambda^2} \tag{3.2}$$

and where

$$U^{(\lambda,h,f)}_{t/\lambda^2} = [1 \otimes W(h *_\lambda f)]^{-1}\, U^{(\lambda)}_{t/\lambda^2}\, [1 \otimes W(h *_\lambda f)] . \tag{3.3}$$

<ins>Proof</ins>. A straightforward manipulation. ∎

From (1.3) -- (1.5) we have

$$\frac{d}{dt} U^{(\lambda)}_{t/\lambda^2} = - \frac{i}{\lambda} V(t/\lambda^2)\, U^{(\lambda)}_{t/\lambda^2} , \tag{3.4}$$

where

$$V(t/\lambda^2) = i \sum_{j=1}^{n} [B_j \otimes a^+(S_{t/\lambda^2} g_j) - B^+_j \otimes a(S_{t/\lambda^2} g_j)] . \tag{3.5}$$

Now we derive a differential equation for $U^{(\lambda,h,f)}_{t/\lambda^2}$.

<ins>Lemma 3.2.</ins> We have

$$\frac{d}{dt} U^{(\lambda,h,f)}_{t/\lambda^2} = - \frac{i}{\lambda} V^{(\lambda,h,f)}(t/\lambda^2)\, U^{(\lambda)}_{t/\lambda^2} , \tag{3.6}$$

where

$$V^{(\lambda,h,f)}(t/\lambda^2) = [1 \otimes W(h *_\lambda f)]^{-1}\, V(t/\lambda^2)\, [1 \otimes W(h *_\lambda f)]$$

$$= V(t/\lambda^2) + \Delta H(\lambda,h,f,t) \tag{3.7}$$

and where

$$\Delta H(\lambda,h,f,t) = i \sum_{j=1}^{n} [\, \langle h *_\lambda f, S_{t/\lambda^2} g\rangle B_j - \langle S_{t/\lambda^2} g_j , h *_\lambda f\rangle B^+_j] . \tag{3.8}$$

Proof. Note that the map $X \longmapsto [1 \otimes W(h * _\lambda f)]^{-1} X [1 \otimes W(h * _\lambda f)]$ ($X \in B(\mathcal{H} \otimes \mathcal{H}')$) is an automorphism; then use the commutation relations $[a(f),W(g)] = \langle f,g\rangle W(g)$. ∎

Lemma 3.3. The right-hand side of (2.9) can be rewritten as follows:

$$\langle u \otimes W(h \otimes f)\Phi_\circ , j_t(X) v \otimes W(h \otimes f)\Phi_\circ \rangle_{\mathcal{H} \otimes \mathcal{H}''}$$

$$= \langle u \otimes \Phi_\circ , j_t^{(h,f)}(X) v \otimes \Phi_\circ \rangle_{\mathcal{H} \otimes \mathcal{H}''} , \tag{3.9}$$

where

$$j_t^{(h,f)}(X) = U^{(h,f)+}(t) (X \otimes 1) U^{(h,f)}(t) \tag{3.10}$$

and where

$$U^{(h,f)}(t) = [1 \otimes W(h \otimes f)]^{-1} U(t) [1 \otimes W(h \otimes f)] . \tag{3.11}$$

Proof. A straightforward manipulation. ∎

Lemma 3.4. $U^{(h,f)}(t)$ satisfies the following QSDE (with $U^{(h,f)}(0) = 1$):

$$dU^{(h,f)}(t) = \left(\sum_{j=1}^{n} [B_j dA_j^+(t) - B_j^+ dA_j(t)] + [K + \Delta K(h,f,t)]dt\right)U^{(h,f)}(t) \tag{3.12}$$

where, for each t such that $h(.)$ is continuous at t,

$$\Delta K(h,f,t)dt = [1 \otimes W(h \otimes f)]^{-1} \left[\sum_{j=1}^{n} [B_j dA_j^+(t) - B_j^+ dA_j(t)], [1 \otimes W(h \otimes f)] \right]$$

$$= \sum_{j=1}^{n} [\bar{h}(t) (f|g_j) B_j - h(t) (g_j|f) B_j^+] dt . \tag{3.13}$$

Proof. Note that the map $X \longmapsto [1 \otimes W(h \otimes f)]^{-1} X [1 \otimes W(h \otimes f)]$

($X \in B(\mathcal{H} \otimes \mathcal{H}'')$) is an automorphism; then use the commutation relations $[a(f),W(g)] = \langle f,g\rangle W(g)$, recalling that $dA_j(t) = a(\chi_{[t,t+dt]} \otimes g_j)$. ∎

Note that $\Delta K(h,f,t)$ is a bounded, skew-adjoint operator.

Lemma 3.5. Let h be a continuous function of compact support. Then

$$\lim_{\lambda\to 0} \left\| -\frac{i}{\lambda} \Delta H(\lambda,h,f,t) - \Delta K(h,f,t) \right\| = 0 , \tag{3.14}$$

uniformly on compact intervals in t .

Proof. It suffices to prove that, for all $f , g \in K_1$ and for all continuous h of compact support, one has

$$\lim_{\lambda\to 0} \frac{1}{\lambda} \langle h *_\lambda f, S_{t/\lambda^2} g\rangle = \bar{h}(t) \, (f|g) \tag{3.15}$$

uniformly on compact intervals in t . Indeed, we have

$$\frac{1}{\lambda} \langle h *_\lambda f, S_{t/\lambda^2} g\rangle = \int_{-\infty}^{+\infty} \bar{h}(\lambda^2 s) \langle S_{t/\lambda^2} f, S_s g\rangle \, ds$$

$$= \int_{-\infty}^{+\infty} \bar{h}(\lambda^2 s) \langle f, S_{s-t/\lambda^2} g\rangle \, ds$$

(with the change of variable $s-t/\lambda^2 = u$)

$$= \int_{-\infty}^{+\infty} \bar{h}(t + \lambda^2 u) \langle f, S_u g\rangle \, du \xrightarrow[\lambda\to 0]{} \bar{h}(t) \int_{-\infty}^{+\infty} \langle f, S_u g\rangle \, du$$

by the dominated convergence theorem. Recalling the definition (2.7) of $(f|g)$ and taking into account that a continuous function of compact support is uniformly continuous, the claim follows. ∎

Lemma 3.6. Let h be continuous of compact support, and let

$$\tilde{j}^{(\lambda,h,f)}_{t/\lambda^2}(X) = \tilde{U}^{(\lambda,h,f)+}_{t/\lambda^2} (X \otimes 1) \, \tilde{U}^{(\lambda,h,f)}_{t/\lambda^2} \quad : X \in B(\mathcal{H}) , \tag{3.16}$$

where $\tilde{U}^{(\lambda,h,f)}_{t/\lambda^2}$ is the (unitary) solution of

$$\frac{d}{dt} \tilde{U}^{(\lambda,h,f)}_{t/\lambda^2} = [-\frac{i}{\lambda} V(t/\lambda^2) + \Delta K(h,f,t)] \, \tilde{U}^{(\lambda,h,f)}_{t/\lambda^2} , \tag{3.17}$$

with $\tilde{U}^{(\lambda,h,f)}_{0} = 1$. Then, for all $X \in B(\mathcal{H})$, we have

$$\lim_{\lambda\to 0} \| j^{(\lambda,h,f)}_{t/\lambda^2}(X) - \tilde{j}^{(\lambda,h,f)}_{t/\lambda^2}(X)\| = 0 , \tag{3.18}$$

uniformly on compact intervals in t.

Proof. It is a consequence of Lemma 3.5. Indeed, we can write

$$U^{(\lambda,h,f)}_{t/\lambda^2} - \tilde{U}^{(\lambda,h,f)}_{t/\lambda^2} = U^{(\lambda,h,f)}_{t/\lambda^2} Z^{(\lambda,h,f)}_{t/\lambda^2} ,$$

where $Z^{(\lambda,h,f)}_{t/\lambda^2} = 1 - U^{(\lambda,h,f)+}_{t/\lambda^2} \tilde{U}^{(\lambda,h,f)}_{t/\lambda^2}$ satisfies $Z^{(\lambda,h,f)}_{0} = 0$ and

$$\frac{d}{dt} Z^{(\lambda,h,f)}_{t/\lambda^2} = U^{(\lambda,h,f)+}_{t/\lambda^2} \left(-\frac{i}{\lambda}\Delta H(\lambda,h,f,t) - \Delta K(\lambda,h,f,t)\right) \tilde{U}^{(\lambda,h,f)}_{t/\lambda^2} .$$

Since $U^{(\lambda,h,f)}_{t/\lambda^2}$ and $\tilde{U}^{(\lambda,h,f)}_{t/\lambda^2}$ are unitary, we have

$$\| Z^{(\lambda,h,f)}_{t/\lambda^2} \| \le t \left\| -\frac{i}{\lambda}\Delta H(\lambda,h,f,t) - \Delta K(\lambda,h,f,t)\right\| ,$$

which tends to 0 as $\lambda \to 0$, uniformly on compact intervals in t, by Lemma 3.5. The claim follows. ∎

Lemma 3.7. Let h be continuous of compact support, and let $(T^{(h,f)}_t : t \ge 0)$ be the family of completely positive maps of $B(\mathcal{H})$ such that, for all $X \in B(\mathcal{H})$,

$$\frac{d}{dt} T^{(h,f)}_t(X) = L(X) + \Delta K(h,f,t)^+ X + X \Delta K(h,f,t), \tag{3.19}$$

with $T^{(h,f)}_0(X) = X$. Then, for all $u, v \in \mathcal{H}$, $X \in B(\mathcal{H})$, we have

$$\lim_{\lambda\to 0} \left| \langle u \otimes \Phi_\circ , \tilde{j}^{(\lambda,h,f)}_{t/\lambda^2}(X)\, v \otimes \Phi_\circ \rangle_{\mathcal{H}\otimes\mathcal{H}'} - \langle u, T^{(h,f)}_t(X) v\rangle_{\mathcal{H}} \right| = 0, \tag{3.20}$$

uniformly on compact intervals in t.

Proof (Sketch). The same arguments as in [18,6] can be used, since the time dependence of $\Delta K(h,f,t)$, being on the "macroscopic time scale", does not give particular problems (see instead [7] for the case of time dependence on the microscopic time scale). ∎

Lemma 3.8. Let h be continuous of compact support. Then, for all $u, v \in \mathcal{H}$ and for all $X \in B(\mathcal{H})$, we have

$$\langle u \otimes \Phi_\circ , j_t^{(h,f)}(X)\, v \otimes \Phi_\circ \rangle_{\mathcal{H} \otimes \mathcal{H}''} = \langle u, T_t^{(h,f)}(X) v \rangle_{\mathcal{H}} . \quad (3.21)$$

Proof (Sketch). This follows from the QSDE (3.12), with the use of the quantum Ito table (2.2), as in [16]; the time dependence of $\Delta K(h,f,t)$ does not give any particular problem. ∎

Proof of Theorem 2.1. The Lemmas above give the proof for the special case where h is continuous of compact support. The general case may be obtained through approximation of $h \in L^2(R)$ with continuous functions of compact support. ∎

Remark. The present method of proof could be easily generalized to obtain the convergence of multi-time correlation functions, through a time-dependent generalization of the techniques of [8,12]. However, here we do not wish to investigate this matter further.

4. Finite-temperature quantum Brownian motion

We now consider explicitly the situation in which w_Q is a state describing thermal equilibrium at inverse temperature β. Let $Q = \coth(\beta H_1/2)$ (we recall that $H_1 \geq 0$ with Lebesgue spectrum, the chemical potential μ has been put equal to 0 since the interaction (1.3) does not conserve the particle number). Then, for each $f \in H_1$ we replace the Weyl operator $W(f)$ by $W(Q_+ f \oplus \overline{Q_- f})$, where

$$Q_+ = [\tfrac{1}{2}(Q+1)]^{\frac{1}{2}} = (1 - \exp[-\beta H_1])^{-\frac{1}{2}},$$

$$Q_- = [\tfrac{1}{2}(Q-1)]^{\frac{1}{2}} = (\exp[\beta H_1] - 1)^{-\frac{1}{2}} = \exp[-\beta H_1/2]\, Q_+ , \tag{4.1}$$

thus reducing ourselves from the consideration of the state w_Q on the C^*-algebra generated by $\{W(f) : f \in \mathrm{Dom}(Q)\}$ to the Fock vacuum state on the C^*-algebra generated by $\{W(Q_+ f \oplus \overline{Q_- f}) : f \in \mathrm{Dom}(Q)\}$ (see Section 2). Then the interaction V between S and S' becomes

$$V = i \sum_{j=1}^{2n} [B_j \otimes a^+(\tilde{g}_j) - B_j^+ \otimes a(\tilde{g}_j)] , \tag{4.2}$$

where $\tilde{g}_j = Q_+ g_j \oplus 0$, $\tilde{g}_{n+j} = 0 \oplus \overline{Q_- g_j}$, and where $B_{n+j} = - B_j^+$.

Here we shall be very careful in distinguishing $\tilde{g}_j$ from g_j.

Then the reduced dynamics is given by

$$\lim_{\lambda \to 0} \langle u \otimes \Phi_\circ , U^{(\lambda)+}_{t/\lambda^2} (X \otimes 1) U^{(\lambda)}_{t/\lambda^2} v \otimes \Phi_\circ \rangle = \langle u, T_t(X) v\rangle , \tag{4.3}$$

where $T_t = \exp[Lt]$, with

$$L(X) = \left(\sum_{j=1}^{2n} \tilde{c}_j(w_j) B_j^+ X B_j\right) + K^+ X + X K , \tag{4.4}$$

$$\tilde{c}_j(w_j) = \frac{1}{2} \int_{-\infty}^{+\infty} \langle \tilde{g}_j, \exp[i(\tilde{H}_1 - w_j)t] \tilde{g}_j \rangle \, dt , \tag{4.5}$$

$$K = - \sum_{j=1}^{2n} \int_{-\infty}^{0} \langle \tilde{g}_j, \exp[i(\tilde{H}_1 - w_j)t] \tilde{g}_j \rangle \, dt \; B_j^+ B_j , \tag{4.6}$$

where $\tilde{H}_1 = H_1 \oplus (-H_1)$, and where $w_{n+j} = - w_j$. If we write

$$c_j(w_j) = \frac{1}{2} \int_{-\infty}^{+\infty} \langle g_j, \exp[i(H_1 - w_j)t] g_j \rangle \, dt , \tag{4.7}$$

we obtain, for $j = 1,\dots,n$,

$$\tilde{c}_j(w_j) = (1 - \exp[-\beta w_j])^{-\frac{1}{2}} c_j(w_j) \ ,$$
$$\tilde{c}_{n+j}(w_{n+j}) = (\exp[\beta w_j] - 1)^{-\frac{1}{2}} c_j(w_j) \ . \tag{4.8}$$

Let also $\{\tilde{A}_j(t), \tilde{A}_j^{+}(t) : t \geq 0 \ , \ j = 1,\ldots,2n\}$ be the annihilation and creation operators in $\mathcal{H}''$ corresponding to the test functions $\chi_{[0,t]} \otimes \tilde{g}_j : t \geq 0 \ , \ j = 1,\ldots,n,$ g_j being regarded as elements of K ; they are mutually independent Fock quantum Brownian motions satisfying the quantum Ito table (2.2). We can define mutually independent finite-temperature (non-Fock) quantum Brownian motions [14,15] $\{A_j^{(\beta)}(t), A_j^{(\beta)+}(t) : t \geq 0 \ , \ j = 1,\ldots,n\}$ by

$$A_j^{(\beta)}(t) = \tilde{A}_j(t) + \tilde{A}_{n+j}^{+}(t) \ : \ j = 1,\ldots,n \ . \tag{4.9}$$

They satisfy the quantum Ito table

$$dA_j^{(\beta)}(t)\, dA_k^{(\beta)+}(t) = \delta_{jk} (1 - \exp[-\beta w_j])^{-\frac{1}{2}} c_j(w_j)\, dt \ ,$$
$$dA_j^{(\beta)+}(t)\, dA_k^{(\beta)}(t) = \delta_{jk} (\exp[\beta w_j] - 1)^{-\frac{1}{2}} c_j(w_j)\, dt \ . \tag{4.10}$$

With this notation, and recalling that $B_{n+j} = - B_j^{+} : j = 1,\ldots,n,$ we can rewrite the quantum stochastic differential equations (2.1) and (2.5) for the case of a thermal reservoir in the following way:

$$dU(t) = \left(\sum_{j=1}^{2n} [B_j\, d\tilde{A}_j^{+}(t) - B_j^{+}\, d\tilde{A}_j(t)] + K\, dt \right) U(t)$$

$$= \left(\sum_{j=1}^{n} [B_j\, dA_j^{(\beta)+}(t) - B_j^{+}\, dA_j^{(\beta)}(t)] + K\, dt \right) U(t) \ , \tag{4.11}$$

$$
\begin{aligned}
dj_t(X) &= \sum_{j=1}^{2n} \left(j_t([B_j^+, X])\, d\tilde{A}_j(t) - j_t([B_j, X])\, d\tilde{A}_j^+(t) \right) \\
&\quad + j_t(L(X))\, dt \quad , \\
&= \sum_{j=1}^{n} \left(j_t([B_j^+, X])\, dA_j^{(\beta)}(t) - j_t([B_j, X])\, dA_j^{(\beta)+}(t) \right) \\
&\quad + j_t(L(X))\, dt \quad . \qquad (4.12)
\end{aligned}
$$

This being a mere change of notation, Theorem 2.1 remains true.

We wish to comment on the physical significance of our results. When the weak coupling limit is applicable, starting from a description of the heat bath where the one-particle energy is positive and the initial state is (a local perturbation of) a state satisfying the KMS condition with respect to the time evolution, we have rigorously obtained the two typical features of finite-temperature quantum Brownian motion in the sense of [14,15], namely:
(i) the one-particle energy (generator of time translations) is unbounded from below as well as from above;
(ii) the reference state of the "heat bath" satisfies the KMS condition not for time evolution, but for a much more trivial automorphism group, consisting in multiplying the annihilation operators $A_j^{(\beta)}(t)$ by a phase factor $\exp[-i\omega_j t]$.

References
----

1. Accardi, L.: Quantum Markov chains. In: Proceedings of the School of Mathematical Physics, University of Camerino (1974).

2. Accardi, L., Frigerio, A., and Lu Y.-G.: The weak coupling limit as a quantum functional central limit theorem. To appear in Commun. Math. Phys.

3. Accardi, L., Frigerio, A., and Lu Y.-G.: The quantum weak coupling limit (II). Langevin equation and finite temperature case. Preprint University of Rome II (1989).

4. Accardi, L., Frigerio, A., and Lu Y.-G.: On the weak coupling limit for fermions. Preprint University of Rome II (1989).

5. Accardi, L., and Lu Y.-G.: The low density limit: the boson Fock case. Preprint Centro Matematico Vito Volterra, Rome (1989).

6. Davies, E.B.: Markovian master equations. Commun. Math. Phys. 39 (1974) 91-110.

7. Davies, E.B., and Spohn, H.: J. Stat. Phys. 19 (1978) 511- .

8. Dümcke, R.: Convergence of multi-time correlation functions in the weak and singular coupling limits. J. Math.Phys. 24 (1983) 311-315.

9. Dümcke, R.: The low density limit for an N-level system interacting with a free Bose or Fermi gas. Commun.Math.Phys. 97 (1985) 331-359.

10. Evans, M., and Hudson, R.L.: Multidimensional quantum diffusions. In: Accardi, L., and von Waldenfels, W. (Eds.): Quantum Probability and Applications III (Proceedings, Oberwolfach 1987). Lecture Notes in Mathematics vol. 1303. Springer-Verlag, Berlin Heidelberg New York Tokyo (1988) pp. 69-88.

11. Frigerio, A.: Quantum Poisson processes: physical motivations and applications. In: Accardi, L., and von Waldenfels, W. (Eds.): Quantum Probability and Applications III (Proceedings, Oberwolfach 1987). Lecture Notes in Mathematics vol. 1303. Springer-Verlag, Berlin Heidelberg New York Tokyo (1988) pp. 107-127.

12. Frigerio, A., and Gorini, V.: Markov dilations and quantum detailed balance. Commun. Math. Phys. 93 (1984) 517-532.

13. Gorini, V., Kossakowski, A., and Sudarshan, E.C.G.: Completely positive dynamical semigroups of N-level systems. J. Math. Phys. 17 (1976) 821-825.

14. Hudson, R.L., and Lindsay, J.M.: A non-commutative martingale representation theorem for non-Fock quantum Brownian motion. J. Funct. Anal. 61 (1985) 202-221.

15. Hudson, R.L., and Lindsay, J.M.: Uses of non-Fock quantum Brownian motion and a martingale representation theorem. In: Accardi, L., and von Waldenfels, W. (Eds.): Quantum Probability and Applications II (Proceedings, Heidelberg 1984)). Lecture Notes in Mathematics vol. 1136. Springer-Verlag, Berlin Heidelberg New York Tokyo (1985) pp. 276-305.

16. Hudson, R.L., and Parthasarathy, K.R.: Quantum Ito's formula and stochastic evolutions. Commun. Math. Phys. 91 (1984) 301-323.

17. Lindblad, G.: On the generators of quantum dynamical semigroups. Commun. Math. Phys. 48 (1976) 119-130.

18. Pulè, J.V.: The Bloch equations. Commun. Math. Phys. 38 (1974) 241-256.

ON THE LOW DENSITY LIMIT OF BOSON MODELS

L. Accardi **Lu Yun Gang ***

Centro Matematico V.Volterra
Dipartimento di Matematica
Universita' di Roma II

ABSTRACT

We study a nonrelativistic quantum system coupled, via a quadratic interaction (cf. formula (1.10) below), to a free Boson gas in the Fock state. We prove that, in the low density limit (z = fugacity $\to 0$), the family of processes given by the collective Weyl operators and the collective coherent vectors , converge to the Fock quantum Brownian motion over $L^2(\mathbf{R}, dt; K)$, where K is an appropriate Hilbert space (cf. Section (1.)) . Moreover we prove that the matrix elements of the wave operator of the system at time t/z^2 in the collective coherent vectors converge to the matrix elements, in suitable coherent vectors of the quantum Brownian motion process, of a unitary Markovian cocycle satisfying a quantum stochastic differential equation ruled by some pure number process (i.e. no quantum diffusion part and only the quantum analogue of the purely discontinuous, or jump, processes).

§1 . Introduction

In the theory of quantum open systems the analogy between the weak coupling and the low density limit is well known and has been systematically studied by Palmer [9].

In spite of this analogy, the results available in the low density case are much poorer than the corresponding ones for the weak coupling case, for example the existence of the Markovian semi-group has been proved only for a short time interval in a very special Fermion model (Dümcke [11], [12]).

In the last two years our understanding of the weak coupling limit has drastically increased since it has been possible to acquire control of the limit not only of the reduced dynamics, but of the full quantum dynamics, both in the Schrödinger and in the Heisenberg form [3], [4], [5], [6], [7]. The basic new idea in these papers was the introduction of some classes of "collective states"(collective coherent states or collective number states) and the study of the limit of the matrix elements of the Schrödinger as well as of the Heisenberg dynamics with respect to these collective states. The previous investigations were limited

---

* On leave of absence from Beijing Normal Uniersity

to the vacuum expectation value of the Heisenberg evolute of an observable of the system which, in the limit, yielded the reduced (Markovian) dynamics. The above mentioned results allowed, in particular, to give a precise mathematical meaning to the statement that the Fock (resp. finite temperature) heat bath converges in the weak coupling limit, both in the Boson and in the Fermion case to a Fock (resp, finite temperature) quantum Brownian motion over a Hilbert space canonically associated to the one-particle dynamics (cf. Section (6.) of [3]). The analogy between the weak coupling and the low density limit naturally suggests to try and realize a similar program in the latter case. Frigerio and Maassen [15] and later Frigerio and Alicki [9] conjectured that, while in the weak coupling limit the Schrödinger dynamics (in interaction picture) converges (in the sense of [3]) to a unitary Markovian cocycle satisfying a quantum stochastic differential equation of diffusion type (i.e. driven only by the A, A^+ noises), in the low density limit the same dynamics should convergence to a unitary Markovian cocycle satisfying a quantum stochastic differential equation of pure jump type (i.e. driven only by the "number" noises). The goal of the present paper is to outline a solution of the above mentioned problem. Here we shall outline the strategy and prove the fundamental estimates. The full details of the proof can be found in [8] of which the present paper can be considered a preliminary version.

As in [3], we state the problem for a general quasi free state and we prove the convergence of the kinematical process of the collective coherent vectors to quantum Brownian motion in the general case. Starting from Section (3.) we restrict our attention to the Fock case. Here we shall only discuss the Boson case.

Let H_0 denote the system Hilbert space and H_1 the one particle resevoir Hilbert space . Let

$$W(H_1) := \{W(f) : f \in H_1\} \tag{1.1}$$

be the Weyl C^*-algebra on H_1 ; let H be a self-adjoint bounded below operator on H_1 and z, β positive real numbers interpreted respectively as density of the resevoir particles and inverse temperature. Define

$$Q_z := \left(1 + z^2 e^{-\beta H}\right) \cdot \left(1 - z^2 e^{-\beta H}\right)^{-1} = \coth(\beta H + \mu) \tag{1.2}$$

(with $z^2 = e^{\mu}$) and suppose that, for each z in a an interval $[0, Z]$, Q_z is a self-adjoint operator on a domain D, independent on z Denote φ_{Q_z} the mean zero gauge invariant qausi-free state on $W(H_1)$ with covariance operator Q_z, i.e.

$$\varphi_{Q_z}(W(f)) = \exp\left(-\frac{1}{2} < f, Q_z f >\right) \quad , \quad \forall f \in H_1 \tag{1.3}$$

and let $\{\mathcal{H}_{Q_z}, \pi_{Q_z}, \Phi_{Q_z}\}$ be the GNS - triple of $\{W(H_1), \varphi_{Q_z}\}$, so that

$$< \Phi_{Q_z}, \pi_{Q_z}(W(f))\Phi_{Q_z} > = \varphi_{Q_z}(W(f)) \tag{1.4}$$

We shall write W_{Q_z} for $\pi_{Q_z} \circ W$. The Fock represenation corresponds to the case $Q_z = 1$, i.e. $z = 0$. In this case the GNS representation will be simply denoted $\{\mathcal{H}, \pi, \Phi\}$. Let S_t be a unitary group on $B(H_1)$ (the one particle free evolution of the resevoir) and suppose that

$$S_t \cdot Q_z = Q_z \cdot S_t \quad , \quad \forall t \geq 0 \tag{1.5}$$

where the equality is meant on D. This implies that the second quantization of S_t, denoted $W(S_t)$, leaves φ_{Q_z} invariant hence it is implemented, in the GNS representation, by a unitary 1-parameter group $V_t^{(z)}$ whose generator $H^{(z)} =: H_R$ is called the free Hamiltonian of the resevoir. As in [3] we assume that there exists a non zero subspace K of H_1 (in all the examples it is a dense subspace) such that

$$\int_{\mathbf{R}} |< f, S_t g >| dt < \infty \quad , \quad \forall f, g \in K \tag{1.6}$$

Moreover , we suppose that $Q_z K \subseteq K$ and $e^{-tH_R} K \subseteq K$ for all $t \geq 0$.

Let be given a self-adjoint operator H_S on the system space H_o, called the system Hamiltonian. The total free Hamiltonian is defined to be

$$H_0 \otimes 1 + 1 \otimes H_1 \tag{1.7}$$

We define the interaction Hamiltonian V as in [11] and [19] i.e., we fix two functions $g_1, g_o \in K$ and define

$$\begin{aligned} V &:= i\left(D \otimes A^+(g_0) \cdot A(g_1) \; - \; D^+ \otimes A^+(g_1) \cdot A(g_0)\right) \\ &= i \sum_{\varepsilon \in \{0,1\}} D_\varepsilon \otimes A^+(g_\varepsilon) \cdot A(g_{1-\varepsilon}) \end{aligned} \tag{1.8}$$

with the notations

$$D_0 = D \quad , \quad D_1 = -D^+ \tag{1.9}$$

and where D is a bounded operator on H_o satisfying

$$\exp(-itH_S) \cdot D \cdot \exp(itH_S) = D \tag{1.10}$$

Moreover we assume that g_0 and g_1 have disjoint energy spectra, i.e.

$$< g_0, S_t g_1 > = 0 \quad , \quad \forall t \in \mathbf{R} \tag{1.11}$$

a condition that is natural because the condition (2.5) of [14] and has already been used in the literature on the weak coupling limit (cf. [10] , §3). As we shall prove in subsequent papers the assumptions (1.10), (1.12) are not essential, however they simplify several computations. With these notations, the total Hamiltonian is

$$H_{total} := H_S \otimes 1 + 1 \otimes H_R + V \tag{1.12}$$

and the wave operator at time t is defined by

$$U(t) := \exp\left(-itH^{(0)}\right) \cdot \exp\left(itH_{total}\right) \tag{1.13}$$

with

$$H^{(0)} := H_S \otimes 1 + 1 \otimes H_R \tag{1.14}$$

Therefore

$$\frac{d}{dt}U(t) = \frac{1}{i}V(t)U(t) \quad ; \quad U(0) = 1 \tag{1.15}$$

where ,

$$V(t) := \exp(-itH_S \otimes 1)\ V\ \exp(itH_S \otimes 1) = i \sum_{\varepsilon \in \{0,1\}} D_\varepsilon \otimes A^+(S_t g_\varepsilon) A(S_t g_{1-\varepsilon}) \tag{1.16}$$

Moreover the solution of (1.15) is given by the iterated series

$$U(t) = \sum_{n=0}^{\infty} \int_0^t dt_1 \cdots \int_0^{t_{n-1}} dt_n V(t_1) \cdots V(t_n) \tag{1.17}$$

which is weakly convergent on the domain of the vectors of the form

$$u \otimes \Phi_{Q_z}(z \int_{S/z^2}^{T/z^2} S_u f du) := u \otimes W_{Q_z}(z \int_{S/z^2}^{T/z^2} S_u f du)\Phi_{Q_z} \tag{1.17a}$$

where $u \in H_o$, $f \in K$, $S, T \in \mathbf{R}$. From Lemma (3.2) of [3], we know that the assumption (1.7) implies that the sesquilinear form $(\cdot|\cdot) : K \times K \longrightarrow C$ defined by

$$(f|g) := \int_{\mathbf{R}} < f, S_t g > dt \quad , \quad f, g \in K \tag{1.18}$$

defines a pre-scalar product on K. We denote $\{K, (\cdot|\cdot)\}$, or simply K, the completion of the quotient of K by the zero $(\cdot|\cdot)$ -norm elements .

Definition(1.1) Let $\mathcal{K}$ be a Hilbert space, T an interval in $\mathbf{R}$, $Q \geq 1$ be a self-adjoint operator on $\mathcal{K}$ and let

$$\{\mathcal{H}_Q, \pi_Q, \Phi_Q\} \tag{1.19}$$

denote the GNS representation of the CCR over $L^2(T, dt; \mathcal{K})$ with respect to the quasi-free state φ_Q on $W(L^2(T, dt; \mathcal{K}))$ characterized by

$$\varphi_Q(W(\xi)) = e^{-\frac{1}{2}<\xi, 1 \otimes Q\xi>} \quad ; \quad \xi \in L^2(T, dt; \mathcal{K}) \tag{2.20}$$

Denote $\mathcal{D}$ the set of all vectors of the form $\pi(W(\xi))\Phi_Q = W_Q(\xi) \cdot \Phi_Q$ with $\xi \in L^2(T, dt; \mathcal{K})$. The stochastic process

$$\{\mathcal{H}_Q\ ,\ \mathcal{D}\ ,\ W_A(\chi_{(s,t]} \otimes f)\ ;\ (s,t] \subseteq T\ ,\ f \in \mathcal{K}\} \tag{2.21}$$

is called the Q**–quantum Brownian motion on** $L^2(T, dt, \mathcal{K})$.

If $Q = 1$ we speak of the **Fock Brownian Motion** on $L^2(T, dt, \mathcal{K})$; if Q is the multiplication by a constant $(\beta \geq 1)$, then we speak of the **finite temperature quantum Brownian Motion** , in the terminology of [18] or of the **universal invariant quantum Brownian Motion** in the terminology of [17].

Sometimes, when no confusion can arise, we call quantum Brownian motion also the process

$$\{\mathcal{H}_Q, \mathcal{D}, A(\chi_{(s,t]} \otimes f), A^+(\chi_{(s,t]} \otimes f) \; ; \; s,t \in T, \; f \in K\} \tag{2.22}$$

where $A(\cdot)$, $A^+(\cdot)$ denote respectively the annihilation and creation fields in the representation (2.25). For the normalized coherent vectors we use the notation:

$$W_Q(\chi_{[s,t]} \otimes f) \cdot \Phi_Q = \Phi_Q(\chi_{[s,t]} \otimes f).$$

Our main result in this paper is to prove that, the limit

$$\lim_{z\to 0} < u \otimes W(z \int_{S/z^2}^{T/z^2} S_u f du)\Phi, U(t/z^2)v \otimes W(z \int_{S'/z^2}^{T'/z^2} S_u f' du)\Phi > \tag{1.23}$$

exists and is equal to

$$< u \otimes W(\chi_{[S,T]} \otimes f)\Psi, U(t) \cdot v \otimes W(\chi_{[S',T']} \otimes f')\Psi > \tag{1.24}$$

where $\{\mathcal{H}, W, \Psi\}$ is the Fock Brownian motion on $L^2(\mathbf{R}, dt; K)$ and $U(t)$ satisfies a quantum stochastic differential equation driven by purely discontinuous noises in the sense of [15] and [18], whose form is given by (6.1).

ACKNOWLEDGEMENTS L. Accardi acknowledges support from Grant AFOSR 870249 and ONR N00014-86-K-0538 through the Center for Mathematical System Theory, University of Florida.

§2 Convergence of the collective process

In this Section we show that, at a purely kinematical level, i.e. with $t = 0$ in (1.23), the low density limit coincides with the weak coupling limit and the limiting process is the Fock Brownian motion on $L^2(\mathbf{R}, dt; K)$ where K is equipped with the scalar product (1.18).

First recall from [3] (Lemma (3.2)) that for each $f, f' \in K, S, S', T, T' \in \mathbf{R}$, one has

$$\lim_{z\to 0} < z \int_{S/z^2}^{T/z^2} S_u f du, z \int_{S'/z^2}^{T'/z^2} S_u f' du > = < \chi_{[S,T]}, \chi_{[S',T']} >_{L^2(\mathbf{R})} \cdot \int_{\mathbf{R}} < f, S_t f' > dt$$

$$= < \chi_{[S,T]} \otimes f, \chi_{[S',T']} \otimes f' >_{L^2(\mathbf{R},dt;K)} \tag{2.1}$$

LEMMA (2.1) For each $n \in \mathbf{N}, \{f_k\}_{k=1}^n \subset K$, $\{S_k, T_k\}_{k=1}^n \subset \mathbf{R}$, $\{x_k\}_{k=1}^n \subset \mathbf{R}$,

$$\lim_{z\to 0} < \Phi_{Q_z}, W(x_1 z \int_{S_1/z^2}^{T_1/z^2} S_u f_1 du) \cdots W(x_n z \int_{S_n/z^2}^{T_n/z^2} S_u f_n du)\Phi_{Q_z} >$$

$$= < \Psi, W(x_1 \chi_{[S_1,T_1]} \otimes f_1) \cdots W(x_n \chi_{[S_n,T_n]} \otimes f_n)\Psi > \tag{2.2}$$

and the convergence is uniform for $\{x_k\}_{k=1}^n, \{S_k, T_k\}_{k=1}^n$ in a bounded set of $\mathbf{R}$.

PROOF. In the above notations one has

$$< \Phi_{Q_z}, W(x_1 z \int_{S_1/z^2}^{T_1/z^2} S_u f_1 du) \cdots W(x_n z \int_{S_n/z^2}^{T_n/z^2} S_u f_n du)\Phi_{Q_z} >$$

$$= \exp\left(-Im \sum_{1\le j<k\le n} x_j x_k z^2 \int_{S_j/z^2}^{T_j/z^2}\int_{S_k/z^2}^{T_k/z^2} dudv < S_u f_j, S_v f_k >\right)\cdot$$

$$\cdot < \Phi_{Q_z}, W(\sum_{k=1}^{n} x_k z \int_{S_k/z^2}^{T_k/z^2} S_u f_k du)\Phi_{Q_z} >=$$

$$= \exp\left(-Im \sum_{1\le j<k\le n} x_j x_k z^2 \int_{S_j/z^2}^{T_j/z^2}\int_{S_k/z^2}^{T_k/z^2} dudv < S_u f_j, S_v f_k >\right)\cdot$$

$$\cdot \exp\left(-\frac{1}{2} < \sum_{k=1}^{n} x_k z \int_{S_k/z^2}^{T_k/z^2} S_u f_k du, \sum_{k=1}^{n} x_k z \int_{S_k/z^2}^{T_k/z^2} S_u Q_z f_k du >\right) \tag{2.3}$$

and the only difference with the situation considered in Theorem (3.4) of [3] is the presence of the term Q_z (instead of a z-independent Q). By definition , one has

$$Q_z = (1 + z^2 e^{-\beta H})\cdot(1 - z^2 e^{-\beta H})^{-1} = 1 + 2\sum_{n=1}^{\infty} z^{2n} e^{-\beta n H} \tag{2.4}$$

and the convergence is in norm since H is bounded below. So , by (2.1) , for each $1 \le j < k \le n$,

$$z^2 < \int_{S_k/z^2}^{T_k/z^2} S_u f_k du, \int_{S_j/z^2}^{T_j/z^2} S_u Q_z f_j du > \tag{2.5}$$

$$= z^2 < \int_{S_k/z^2}^{T_k/z^2} S_u f_k du, \int_{S_j/z^2}^{T_j/z^2} S_u f_j du > + z^2 O(1) \longrightarrow < \chi_{[S_k,T_k]} \otimes f_k, \chi_{[S_j,T_j]} \otimes f_j >$$

Hence ,

$$\lim_{z\to 0} < \Phi_{Q_z}, W(x_1 z \int_{S_1/z^2}^{T_1/z^2} S_u f_1 du) \cdots W(x_n z \int_{S_n/z^2}^{T_n/z^2} S_u f_n du)\Phi_{Q_z} >$$

$$= \exp\left(-\frac{1}{2} \sum_{1\le j,k\le n} x_k x_j < \chi_{[S_k,T_k]} \otimes f_k, \chi_{[S_j,T_j]} \otimes f_j >\right)\cdot$$

$$\cdot \exp\left(-Im \sum_{1\le j<k\le n} x_k x_j < \chi_{[S_k,T_k]} \otimes f_k, \chi_{[S_j,T_j]} \otimes f_j >\right)$$

$$=< \Psi, W(x_1 \chi_{[S_1,T_1]} \otimes f_1) \cdots W(x_n \chi_{[S_n,T_n]} \otimes f_n)\Psi > \tag{2.6}$$

The uniformity of the convergence is proved as in Theorem (3.4) of [3].

REMARK Notice that we obtain the Fock Brownian motion even if we start from a finite temperature resevoir. This is due to the choice of the normalization. The normalization leading to the finite temperature Brownian motion (for $z \neq 0$) shall be discussed elsewhere.

§3 . The basic estimates in the Fock case

As in the weak coupling limit, our starting point will be the analysis of the quantities

$$\Delta_n^\varepsilon(z,t) := \int_0^{t/z^2} dt_1 \int_0^{t_1} dt_2 \cdots \int_0^{t_{n-1}} dt_n < W(z\int_{S/z^2}^{T/z^2} S_u f du)\Phi, \tag{3.1}$$

$$A^+(S_{t_1} g_{\varepsilon(1)}) A(S_{t_1} g_{1-\varepsilon(1)}) \cdots A^+(S_{t_1} g_{\varepsilon(1)}) A(S_{t_n} g_{1-\varepsilon(n)}) \cdot W(z\int_{S'/z^2}^{T'/z^2} S_u f' du)\Phi >$$

LEMMA (3.1) For each $n \in \mathbf{N}$ and $\varepsilon \in \{0,1\}^n$

$$\Delta_n^\varepsilon(z,t) = \tag{3.2}$$

$$= \sum_{d=0}^{n} \sum_{2\le q_1<\cdots<q_d\le n} \sum_{\substack{1\le p_1,\cdots,p_d\le n \\ |\{p_h\}_{h=1}^d|=d,\, p_h<q_h,\, h=1,\cdots,d}} z^{2n-2d} \int_0^{t/z^2} dt_1 \int_0^{t_1} dt_2 \cdots \int_0^{t_{n-1}} dt_n$$

$$< W(z\int_{S/z^2}^{T/z^2} S_u f du)\Phi, W(z\int_{S'/z^2}^{T'/z^2} S_u f' du)\Phi >$$

$$\prod_{h=1}^{d} < S_{t_{p_h}} g_{1-\varepsilon(p_h)}, S_{t_{q_h}} g_{\varepsilon(q_h)} > \prod_{\alpha\in\{1,\cdots,n\}\setminus\{q_h\}_{h=1}^d} \int_{S/z^2}^{T/z^2} < S_u f, S_{t_\alpha} g_{\varepsilon(\alpha)} > du$$

$$\prod_{\alpha\in\{1,\cdots,n\}\setminus\{p_h\}_{h=1}^d} \int_{S'/z^2}^{T'/z^2} < S_{t_\alpha} g_{1-\varepsilon(\alpha)}, S_u f' > du$$

with the convention that if $d = 0$, then the sums over $q_1 < \cdots < q_d$ and $p_1, \cdots, p_d$ and the product over h, in (3.2) are equal to 1.

PROOF For each $n \in \mathbf{N}$, $\varepsilon \in \{0,1\}^n$, we want to bring the product

$$A^+(S_{t_1} g_{\varepsilon(1)}) A(S_{t_1} g_{1-\varepsilon(1)}) \cdots A^+(S_{t_n} g_{\varepsilon(n)}) A(S_{t_n} g_{1-\varepsilon(n)})$$

in Wick ordered form , i.e.

$$A^+(S_{t_1} g_{\varepsilon(1)}) A(S_{t_1} g_{1-\varepsilon(1)}) \cdots A^+(S_{t_n} g_{\varepsilon(n)}) A(S_{t_n} g_{1-\varepsilon(n)})$$

$$= \sum_{d=0}^{n} C(g,\varepsilon,d)(A^+ \cdots)^{n-d} \cdot (A \cdots)^{n-d} \tag{3.3}$$

where , $C(g,\varepsilon,d)$ is a product of some scalar products of the g_j and d is the number of scalar products arosen in the exchange of a creator with an annihilator at its left. Notice that, in the left hand side of (3.5), the operator in the extreme right is an annihilator and the one on the extreme left a creator. Therefore $n-d \geq 1$, i.e. $d \leq n-1$. For each d , we can choose d creators among

$$A^+(S_{t_2}g_{\varepsilon(2)}),\cdots,A^+(S_{t_n}g_{\varepsilon(n)})$$

which are used to produce scalar products with annihilators . Each choice determines a unique subset of $\{2,\cdots,n\}$, denoted $\{q_1,\cdots,q_d\}$. Clearly , we can suppose that

$$q_1 < \cdots < q_d \tag{3.4}$$

If the operators $A^+(S_{t_\alpha}g_{\varepsilon(\alpha)})$ with $\alpha \in \{1,\cdots,n\}\setminus\{q_h\}_{h=1}^d$ have been moved to the left of the annihilation operators, it means that for each fixed q_h, there exists a $p_h < q_h$ such that the operator $A(S_{t_{p_h}}g_{1-\varepsilon(p_h)})$ has been used to produce the scalar product $< S_{t_{p_h}}g_{1-\varepsilon(p_h)}, S_{t_{q_h}}g_{\varepsilon(q_h)} >$. Therefore the remaining set of annihilators is $\{A(S_{t_\alpha}g_{1-\varepsilon(\alpha)});\ \alpha\in\{1,\cdots,n\}\setminus\{p_h\}_{h=1}^d\}$. Thus the right hand side of (3.3) is equal to, with the same convention as in the statement of the Lemma,

$$\sum_{d=0}^{n}\sum_{2\leq q_1<\cdots<q_d\leq n}\ \sum_{\substack{1\leq p_1,\cdots,p_d\leq n\\ |\{p_h\}_{h=1}^d|=d,\,p_h<q_h,\,h=1,\cdots,d}}\ \prod_{h=1}^{d} < S_{t_{p_h}}g_{1-\varepsilon(p_h)}, S_{t_{q_h}}g_{\varepsilon(q_h)} >$$

$$\prod_{\alpha\in\{1,\cdots,n\}\setminus\{q_h\}_{h=1}^d} A^+(S_{t_\alpha}g_{\varepsilon(\alpha)})\cdot \prod_{\alpha\in\{1,\cdots,n\}\setminus\{p_h\}_{h=1}^d} A(S_{t_\alpha}g_{1-\varepsilon(\alpha)}) \tag{3.5}$$

From (3.5), one deduces (3.2) by taking the matrix elements in the collective coherent vectors.

Notice that, because of the assumption (1.8), for each $h=1,\cdots,d$, if $1-\varepsilon(p_h)\neq\varepsilon(q_h)$, then , $< S_{t_{p_h}}g_{1-\varepsilon(p_h)}, S_{t_{q_h}}g_{\varepsilon(q_h)} >= 0$, Therefore many of the products in (3.5) are zero. The following Lemma introduces a notation in order to keep track of this fact.

LEMMA (3.2) For each $n\in\mathbb{N}$, $\varepsilon\in\{0,1\}^n$, denote $\{i_1,\cdots,i_k\}$ the unique subset of $\{1,\cdots,n\}$, such that

$$\{i_1,\cdots,i_k\} = \{i\ :\ \varepsilon(i)=1\} \tag{3.6}$$

and

$$i_1 < \cdots < i_k \tag{3.7}$$

With this notation, the quantity $\Delta_n^\varepsilon(z,t)$, defined by (3.3) is equal to

$$\sum_{m=0}^{k\wedge(n-k)}\ \sum_{\substack{2\leq q_1<\cdots<q_m\leq n\\ \{q_h\}_{h=1}^m\subset\{i_h\}_{h=1}^k}}\ \sum_{m'=0}^{k\wedge(n-k)}\ \sum_{\substack{2\leq q'_1<\cdots<q'_{m'}\leq n\\ \{q'_h\}_{h=1}^{m'}\subset\{1,\cdots,n\}\setminus\{i_h\}_{h=1}^k}} \tag{3.8}$$

$$\sum_{(p_1,\cdots,p_m)}' \sum_{(p'_1,\cdots,p'_{m'})} z^{2n-2(m+m')} \int_0^{t/z^2} dt_1 \int_0^{t_1} dt_2 \cdots \int_0^{t_{n-1}} dt_n$$

$$< W(z \int_{S/z^2}^{T/z^2} S_u f du)\Phi, W(z \int_{S'/z^2}^{T'/z^2} S_u f' du)\Phi >$$

$$\prod_{h=1}^{m} < S_{t_{p_h}} g_1, S_{t_{q_h}} g_1 > \cdot \prod_{h=1}^{m'} < S_{t_{p'_h}} g_0, S_{t_{q'_h}} g_0 >$$

$$\prod_{\alpha \in \{1,\cdots,n\} \setminus (\{q_h\}_{h=1}^m \cup \{q'_h\}_{h=1}^{m'})} \int_{S/z^2}^{T/z^2} < S_u f, S_{t_\alpha} g_{\varepsilon(\alpha)} > du$$

$$\prod_{\alpha \in \{1,\cdots,n\} \setminus (\{p_h\}_{h=1}^m \cup \{p'_h\}_{h=1}^{m'})} \int_{S'/z^2}^{T'/z^2} < S_{t_\alpha} g_{1-\varepsilon(\alpha)}, S_u f' > du$$

where , $\sum'_{(p_1,\cdots,p_m)}$ means the sum over all $\{p_h\}_{h=1}^m \subset \{1,\cdots,n\} \setminus \{i_h\}_{h=1}^k$ with $p_h < q_h$, $h = 1,\cdots,m$ and $\sum'_{(p'_1,\cdots,p'_{m'})}$ means the sum over all $\{p'_h\}_{h=1}^{m'} \subset \{i_h\}_{h=1}^k$ with $p'_h < q'_h$, $h = 1,\cdots,m'$.

PROOF In the right hand side of (3.2) , for each $d = 0,1,\cdots,n-1$, $2 \le q_1 < \cdots < q_d \le n$, there exists a unique subset $\{q''_h\}_{h=1}^m \subset \{q_h\}_{h=1}^d$ such that $\varepsilon(q''_h) = 1$, $h = 1,\cdots,m$, and $\varepsilon(q) = 0$ if $q \in \{q_h\}_{h=1}^d \setminus \{q''_h\}_{h=1}$. Clearly $m \le k$, $d-m \le n-k$. Denote $\{q'_h\}_{h=1}^{d-m} = \{q_h\}_{h=1}^d \setminus \{q''_h\}_{h=1}^m$, without loss of generality , one can suppose that

$$q''_1 < \cdots < q''_m \quad ; \quad q'_1 < \cdots < q'_{d-m}$$

Similarly $\{p_h\}_{h=1}^d$ is split into $\{p''_h\}_{h=1}^m \cup \{p'_h\}_{h=1}^{d-m}$ where p'_h (resp. p''_h) is the index paired with q'_h (resp. q''_h) . Since by definition $\varepsilon(q''_h) = 1$ ($h = 1,\cdots,m$) and $\varepsilon(q'_h) = 0$ ($h = 1,\cdots,d-m$), the only non zero products will arise from those p_j such that $\varepsilon(p''_h) = 0$, for $h = 1,\cdots,m$ and $\varepsilon(p'_h) = 1$ for $h = 1,\cdots,d-m$, i.e. $\{p''_h\}_{h=1}^m \subset \{1,\cdots,n\} \setminus \{i_h\}_{h=1}^k$ and $\{p'_h\}_{h=1}^{d-m} \subset \{i_h\}_{h=1}^k$. So , $m \le n-k$, and $d-m \le k$. From this (3.6) is obtained by putting $d-m = m'$, $q''_h = q_h$ for $h = 1,\cdots,m$; $p''_h = p_h$ for $h = 1,\cdots,m$.

Now, proceeding as in [3], for each $n \in \mathbf{N}$ and $\varepsilon \in \{0,1\}^n$, in the expression $\Delta_n^\varepsilon(z,t)$, defined by (3.1), we separate the terms which will be relevant in the limit $z \to 0$ (type I terms) from those which will vanish in the limit (type II terms). More precisely we define

$$\Delta_n^\varepsilon(z,t) =: I_n^\varepsilon(z,t) + II_n^\varepsilon(z,t) \tag{3.9}$$

where

$$I_n^\varepsilon(z,t) := \sum_{m=0}^{k\wedge(n-k)} \sum_{(q_1,\cdots,q_m)}' \sum_{m'=0}^{k\wedge(n-k)} \sum_{(q'_1,\cdots,q'_{m'})} \tag{3.10}$$

$$z^{2n-2(m+m')}\int_0^{t/z^2}dt_1\int_0^{t_1}dt_2\cdots\int_0^{t_{n-1}}dt_n$$

$$<W(z\int_{S/z^2}^{T/z^2}S_uf du)\Phi, W(z\int_{S'/z^2}^{T'/z^2}S_uf'du)\Phi>$$

$$\prod_{h=1}^{m}<g_1,S_{(t_{q_h}-t_{q_h-1})}g_1>\cdot\prod_{h=1}^{m'}<g_0,S_{(t_{q'_h}-t_{q'_h-1})}g_0>$$

$$\prod_{\alpha\in\{1,\cdots,n\}\setminus(\{q_h\}_{h=1}^m\cup\{q'_h\}_{h=1}^{m'})}\int_{S/z^2}^{T/z^2}<S_uf,S_{t_\alpha}g_{\varepsilon(\alpha)}>du$$

$$\prod_{\alpha\in\{1,\cdots,n\}\setminus(\{q_h-1\}_{h=1}^m\cup\{q'_h-1\}_{h=1}^{m'})}\int_{S'/z^2}^{T'/z^2}<S_{t_\alpha}g_{1-\varepsilon(\alpha)},S_uf'>du$$

and

$$II_n^\varepsilon(z,t)=\sum_{m=0}^{n-1}\sum_{2\leq\bar{q}_1<\cdots<\bar{q}_m\leq n}\sum'_{(\bar{q}_1,\bar{p}_1,\cdots,\bar{q}_m,\bar{p}_m)}$$

$$<W(z\int_{S/z^2}^{T/z^2}S_uf du)\Phi, W(z\int_{S'/z^2}^{T'/z^2}S_uf'du)\Phi>$$

$$z^{2n-2m}\int_0^{t/z^2}dt_1\int_0^{t_1}dt_2\cdots\int_0^{t_{n-1}}dt_n\prod_{h=1}^{m}<S_{t_{\bar{p}_h}}g_{1-\varepsilon(\bar{p}_h)},S_{t_{\bar{q}_h}}g_{\varepsilon(\bar{q}_h)}>\cdot$$

$$\prod_{\alpha\in\{1,\cdots,n\}\setminus\{\bar{q}_h\}_{h=1}^m}\int_{S/z^2}^{T/z^2}<S_uf,S_{t_\alpha}g_{\varepsilon(\alpha)}>du$$

$$\prod_{\alpha\in\{1,\cdots,n\}\setminus\{\bar{p}_h\}_{h=1}^m}\int_{S'/z^2}^{T'/z^2}<S_{t_\alpha}g_{1-\varepsilon(\alpha)},S_uf'>du$$

$$=:\sum_{m=0}^{n-1}\sum_{2\leq\bar{q}_1<\cdots<\bar{q}_m\leq n}\sum'_{(\bar{q}_1,\bar{p}_1,\cdots,\bar{q}_m,\bar{p}_m)}\Delta_n^\varepsilon(z,t,\{\bar{q}_h,\bar{p}_h\}_{h=1}^m)\tag{3.11}$$

where , $\sum'_{(\bar{q}_1,\bar{p}_1,\cdots,\bar{q}_m,\bar{p}_m)}$ means the sum for all $1\leq\bar{p}_1,\cdots,\bar{p}_m\leq n$ with $|\{\bar{p}_h\}_{h=1}^m|=m$, $\bar{p}_h<\bar{q}_h$, $h=1,\cdots,m$ and $\bar{p}_h<\bar{q}_h-1$, for some $h=1,\cdots,m$.

LEMMA (3.3) For each $n\in\mathbf{N}$, $\varepsilon\in\{0,1\}^n$,

$$\lim_{z\to0}II_n^\varepsilon(z,t)=0\tag{3.12}$$

PROOF For each $n\in\mathbf{N}$, $\varepsilon\in\{0,1\}^n$, since the coherent vectors are unit vectors, we have from (3.12)

$$|II_n^\varepsilon(z,t)| \le \sum_{m=0}^{n-1} \sum_{2\le \bar q_1 < \cdots < \bar q_m \le n} \sum'_{(\bar q_1, \bar p_1, \cdots, \bar q_m, \bar p_m)}$$

$$C_1^n \cdot z^{2n-2m} \int_0^{t/z^2} dt_1 \int_0^{t_1} dt_2 \cdots \int_0^{t_{n-1}} dt_n \prod_{h=1}^{m} |< S_{t_{p_h}} g, S_{t_{q_h}} g >| \tag{3.13}$$

where , $g = g_0$ or g_1 and

$$C_1 := \max_{F=f,f',G=g_0,g_1} \int_{-\infty}^{\infty} |< G, S_t F >| dt \tag{3.14}$$

Now let

$$h_0 := \min\{h : \ \bar q_h - \bar p_h \ge 2 \ , \ h = 1, \cdots, m\}$$

then ,

$$|\Delta_n^\varepsilon(z,t)| \le C_1^n \cdot z^{-2m} \int_0^{t} dt_1 \int_0^{t_1} dt_2 \cdots \int_0^{t_{n-1}} dt_n \prod_{h=1}^{m} |< g, S_{(t_{q_h} - t_{p_h})/z^2} g >|$$

$$\le C_1^n \cdot z^{-2(m-1)} \int_0^{t} dt_1 \int_0^{t_1} dt_2 \cdots \int_0^{t_{q_1-2}} dt_{\bar q_1 - 1} \int_{-t_{p_1/z^2}}^{(t_{q_1-1} - t_{p_1})/z^2} dt'_{\bar q_1} |< g, S_{t'_{q_1}} g >|$$

$$\int_0^{t_{p_1} + z^2 t'_{q_1}} dt_{\bar q_1 + 1} \cdots \int_0^{t_{n-1}} dt_n \prod_{h=2}^{m} |< g, S_{(t_{q_h} - t_{p_h})/z^2} g >| \ , \tag{3.15}$$

where , if for some $h = 2, \cdots, m$, $\bar p_h = \bar q_1$, then , $t_{\bar p_h}$ means $t_{\bar p_1} + z^2 t'_{\bar q_1}$. Since $t'_{\bar q_1} \in [-t_{\bar p_1/z^2}, (t_{\bar q_1 - 1} - t_{\bar p_1})/z^2))$, then $t_{\bar p_1} + z^2 t'_{\bar q_1} \in [0, t_{\bar q_1 - 1})$. So, by extending the $dt_{\bar q_1 + 1}$-integral to the interval $[0, t_{\bar q_1 - 1}]$, we majorize the right hand side of (3.15) by

$$|\Delta_n^\varepsilon(z,t)| \le$$

$$C_1^n \cdot z^{-2(m-1)} \int_0^{t} dt_1 \int_0^{t_1} dt_2 \cdots \int_0^{t_{q_1-2}} dt_{\bar q_1 - 1} \int_{-t_{p_1/z^2}}^{(t_{q_1-1} - t_{p_1})/z^2} dt'_{\bar q_1} |< g, S_{t'_{q_1}} g >|$$

$$\int_0^{t_{q_1-1}} dt_{\bar q_1 + 1} \cdots \int_0^{t_{n-1}} dt_n \prod_{h=2}^{m} |< g, S_{(t_{q_h} - t_{p_h})/z^2} g >| \tag{3.16}$$

Iterating the procedure, we obtain

$$|\Delta_n^\varepsilon(z,t)| \le$$

$$\le C_1^n \cdot C_1' \cdot \int_0^{t} dt_1 \int_0^{t_1} dt_2 \cdots \int_0^{t_{q_1-2}} dt_{\bar q_1 - 1} \int_{-t_{p_1/z^2}}^{(t_{q_1-1} - t_{p_1})/z^2} dt'_{\bar q_1} |< g, S_{t'_{q_1}} g >|$$

$$\int_0^{t_{q_1-1}} dt_{\bar q_1 + 1} \cdots \int_0^{t_{q_2-2}} dt_{\bar q_2 - 1} \int_{-t_{\bar p_2/z^2}}^{(t_{\bar q_2-1} - t_{\bar p_2})/z^2} dt'_{\bar q_2} |< g, S_{t'_{q_2}} g >|$$

$$\int_0^{t_{q_2-1}} dt_{\bar{q}_2+1} \cdots \int_0^{t_{q_m-2}} dt_{\bar{q}_m-1} \int_{-t_{\bar{p}_m}^{-}/z^2}^{(t_{q_m-1}^{-}-t_{\bar{p}_m}^{-})/z^2} dt'_{\bar{q}_m} | < g, S_{t'_{q_m}} g > |$$

$$\int_0^{t_{q_m-1}} dt_{\bar{q}_m+1} \cdots \int_0^{t_{n-1}} dt_n \tag{3.17}$$

with

$$t_{\bar{q}_h-1}^{-} = \begin{cases} t_{\bar{q}_h-1} , & \text{if } \bar{q}_h > \bar{q}_{h-1}+1 ; \\ t_{\bar{q}_2-\alpha-1} , & \text{if } \bar{q}_h = \bar{q}_{h-1}+1 = \bar{q}_{h-2}+2 = \cdots = \bar{q}_{h-\alpha}+\alpha > \bar{q}_{h-(\alpha-1)}+(\alpha+1) \end{cases} \tag{3.18}$$

$$t_{\bar{p}_h}^{-} = \begin{cases} t'_{\bar{q}_\alpha} , & \text{if } \bar{p}_h = \bar{q}_\alpha ; \\ t_{\bar{p}_h} , & \text{if } \bar{p}_h \notin \{\bar{q}_h\}_{h=1}^m \end{cases} \tag{3.19}$$

So , if $\bar{q}_{h_0} - \bar{q}_{h_0-1} = 1$, then there exists a $\alpha \in \mathbf{N}$, such that $t_{q_{h_0}-1}^{-} = t_{q_{h_0}-\alpha-1}$. By the definition of h_0 , it follows that $\bar{p}_{h_0} < \bar{q}_{h_0} - \alpha - 1$, hence , $t_{q_{h_0}-1}^{-} - t_{p_{h_0}}^{-} < 0$ almost everywhere . So ,

$$|\Delta_n^\varepsilon(z,t)| \le$$

$$\le C_2^n \cdot \int_0^t dt_1 \int_0^{t_1} dt_2 \cdots \int_0^{t_{q_1-2}} dt_{\bar{q}_1-1} \int_0^{t_{q_1-1}} dt_{\bar{q}_1+1} \cdots$$

$$\int_0^{t_{q_{h_0}-2}} dt_{\bar{q}_{h_0}-1} \int_{-t_{p_{h_0}}^{-}/z^2}^{(t_{q_{h_0}-1}^{-}-t_{p_{h_0}}^{-})/z^2} dt'_{\bar{q}_{h_0}} | < g, S_{t'_{q_{h_0}}} g > |$$

$$\cdots \int_0^{t_{q_m-2}} dt_{\bar{q}_m-1} \int_0^{t_{q_m-1}} dt_{\bar{q}_m+1} \cdots \int_0^{t_{n-1}} dt_n$$

$$\longrightarrow 0 \tag{3.20}$$

Where , C_2 is also a constant .

Having proved that the type II terms tend to zero, we shall now compute the limit of the type I terms .

<u>*LEMMA* (3.4)</u> For each $n \in \mathbf{N}$, $\varepsilon \in \{0,1\}^n$,

$$\lim_{z\to 0} I_n^\varepsilon(z,t) := \sum_{m=0}^{k\wedge(n-k)} \sum_{(q_1,\cdots,q_m)}' \sum_{m'=0}^{k\wedge(n-k)} \sum_{(q'_1,\cdots,q'_{m'})} (g_1|g_1)_-^m \cdot (g_o|g_o)_-^{m'} \tag{3.21}$$

$$\int_{0\le t_n\le\cdots\le \widehat{t_{q'_{m'}}}\le\cdots\le\widehat{t_{q_m}}\le\cdots\le\widehat{t_{q'_1}}\le\cdots\le\widehat{t_{q_1}}\le\cdots\le t_1\le t} dt_1\cdots\widehat{dt_{q_1}}\cdots\widehat{dt_{q'_1}}\cdots\widehat{dt_{q_m}}\cdots\widehat{dt_{q_{m'}}}\cdots dt_n$$

$$\prod_{\alpha\in\{1,\cdots,n\}\setminus(\{q_h\}_{h=1}^m\cup\{q'_h\}_{h=1}^{m'})} (f|g_{\varepsilon(\alpha)}) \cdot \chi_{[S,T]}(t_\alpha)$$

$$\prod_{\alpha\in\{1,\cdots,n\}\setminus(\{q_h-1\}_{h=1}^m\cup\{q'_h-1\}_{h=1}^{m'})\cap\{1,\cdots,\bar q_1-2\}} (g_{1-\varepsilon(\alpha)}|f')\cdot\chi_{[S',T']}(t_\alpha)$$

$$\cdot(g_{1-\varepsilon(\bar q_{r_1})}|f')\cdot\chi_{[S',T']}(t_{\bar q_1-1})\cdot$$

$$\prod_{\alpha\in\{1,\cdots,n\}\setminus(\{q_h-1\}_{h=1}^m\cup\{q'_h-1\}_{h=1}^{m'})\cap\{\bar q_{r_1}+1,\cdots,\bar q_{r_1+1}-2\}} (g_{1-\varepsilon(\alpha)}|f')\cdot\chi_{[S',T']}(t_\alpha)$$

$$\cdot(g_{1-\varepsilon(\bar q_{r_2})}|f')\cdot\chi_{[S',T']}(t_{\bar q_1+1-1})\cdot$$

$$\cdots\cdots$$

$$\prod_{\alpha\in\{1,\cdots,n\}\setminus(\{q_h-1\}_{h=1}^m\cup\{q'_h-1\}_{h=1}^{m'})\cap\{\bar q_{r_{x-1}}+1,\cdots,\bar q_{r_{x-1}+1}-2\}} (g_{1-\varepsilon(\alpha)}|f')\cdot\chi_{[S',T']}(t_\alpha)$$

$$\cdot(g_{1-\varepsilon(\bar q_{r_x})}|f')\cdot\chi_{[S',T']}(t_{\bar q_{x-1}-1})\cdot$$

$$\prod_{\alpha\in\{1,\cdots,n\}\setminus(\{q_h-1\}_{h=1}^m\cup\{q'_h-1\}_{h=1}^{m'})\cap\{\bar q_{r_x}+1,\cdots,n\}} (g_{1-\varepsilon(\alpha)}|f')\cdot\chi_{[S',T']}(t_\alpha)$$

$$< W(\chi_{[S,T]}\otimes f)\Psi, W(\chi_{[S',T']}\otimes f')\Psi >$$

where $\widehat{t}$ means that the t is absent and

$$\{\bar q_h\}_{h=1}^{m+m'} := \{q_h\}_{h=1}^m \cup \{q'_h\}_{h=1}^{m'}$$

with

$$\bar q_1 < \cdots < \bar q_{m+m'} \tag{3.22}$$

x and $\{r_h\}_{h=1}^x$ are uniquely determined by the prescription

$$\{\bar q_h\}_{h=1}^{m+m'} = \{\bar q_h\}_{h=1}^{r_1} \cup \{\bar q_h\}_{h=r_1+1}^{r_2} \cup\cdots\cup \{\bar q_h\}_{h=r_{x-1}+1}^{r_x} \tag{3.23}$$

and

$$\bar q_{r_y+1} = \bar q_{r_y+2}-1 = \bar q_{r_y+3}-2 = \cdots = \bar q_{r_{y+1}} - (r_{y+1}-r_y-1) < \bar q_{r_{y+1}+1} - (r_{y+1}-r_y) \tag{3.24}$$

with

$$y = 0, 1, \cdots, x-1, r_0 = 0$$

<u>*REMARK*</u> The conditions (3.23), (3.24) mean that we split the ordered sequence (3.22) into ordered subsequences which are maximal with respect to the property that $\bar q_{h+1} = \bar q_h$ for all elements in the subsequence (a subsequence may contain a single element).

<u>*PROOF*</u> . For each m, m' , there exist unique x and $\{r_h\}_{h=1}^x$ such that $\{\bar q_h\}_{h=1}^{m+m'}$ satisfies (3.23) and (3.24). If, in the expression of $I_n^\varepsilon(z,t)$, we put

$$t'_\alpha = \begin{cases} t_\alpha, & \text{if } \alpha\in\{1,\cdots,n\}\setminus\{\bar q_h\}_{h=1}^{m+m'} ; \\ (t_{\bar q_h} - t_{\bar q_h-1})/z^2, & \text{if } \alpha\in\{\bar q_h\}_{h=1}^{m+m'} . \end{cases} \tag{3.25}$$

then , $t_{\bar{q}_h} = z^2 t'_{\bar{q}_h} + t'_{\bar{q}_h - 1}$, $h = 1, \cdots, m + m'$. So that

$$I_n^\varepsilon(z,t) = \sum_{m=0}^{k\wedge(n-k)} \sum_{(q_1,\cdots,q_m)}' \sum_{m'=0}^{k\wedge(n-k)} \sum_{(q'_1,\cdots,q'_{m'})} \tag{3.26}$$

$$< W(z \int_{S/z^2}^{T/z^2} S_u f du)\Phi, W(z \int_{S'/z^2}^{T'/z^2} S_u f' du)\Phi >$$

$$\int_0^t dt'_1 \cdots \int_0^{t'_{q_1-2}} dt'_{\bar{q}_1 - 1} \int_{-t'_{q_1-1}/z^2}^0 dt'_{\bar{q}_1} \int_{-t'_{q_1-1}/z^2 - t'_{q_1}}^0 dt'_{\bar{q}_2} \cdots$$

$$\cdots \int_{-t'_{q_1-1}/z^2 - t'_{q_1} - \cdots - t'_{q_{r_1}-1}}^0 dt'_{\bar{q}_{r_1}} \int_0^{t'_{q_1-1} + z^2(t'_{q_1} + \cdots + t'_{q_{r_1}})} dt'_{\bar{q}_{r_1}+1} \cdots$$

$$\cdots \int_0^{t'_{q_{r_1+1}-2}} dt'_{\bar{q}_{r_1+1}-1} \int_{-t'_{q_{r_1+1}-1}/z^2}^0 dt'_{\bar{q}_{r_1+1}} \int_{-t'_{q_{r_1+1}-1}/z^2 - t'_{q_{r_1+1}}}^0 dt'_{\bar{q}_{r_1+2}} \cdots$$

$$\cdots \int_{-t'_{q_{r_1+1}-1}/z^2 - t'_{q_{r_1+1}} - \cdots - t'_{q_{r_2}-1}}^0 dt'_{\bar{q}_{r_2}} \int_0^{t'_{q_{r_1+1}-1} + z^2(t'_{q_{r_1+1}} + \cdots + t'_{q_{r_2}})} dt'_{\bar{q}_{r_2+1}} \cdots$$

$$\int_0^{t'_{q_{r_{x-1}+1}-2}} dt'_{\bar{q}_{r_{x-1}+1}-1} \int_{-t'_{q_{r_{x-1}+1}-1}/z^2}^0 dt'_{\bar{q}_{r_{x-1}+1}} \int_{-t'_{q_{r_{x-1}+1}-1}/z^2 - t'_{q_{r_{x-1}+1}}}^0 dt'_{\bar{q}_{r_{x-1}+2}} \cdots$$

$$\int_{-t'_{q_{r_{x-1}+1}-1}/z^2 - t'_{q_{r_{x-1}+1}} - \cdots - t'_{q_{r_x}-1}}^0 dt'_{\bar{q}_{r_x}} \int_0^{t'_{q_{r_{x-1}+1}-1} + z^2(t'_{q_{r_{x-1}+1}} + \cdots + t'_{q_{r_x}})} dt'_{\bar{q}_{r_x+1}} \cdots \int_0^{t_{n-1}} dt$$

$$\prod_{h=1}^m < g_1, S_{t'_{q_h}} g_1 > \cdot \prod_{h=1}^{m'} < g_0, S_{t'_{q'_h}} g_0 >$$

$$\prod_{\alpha \in \{1,\cdots,n\} \setminus (\{q_h\}_{h=1}^m \cup \{q'_h\}_{h=1}^{m'})} \int_{(S-t'_\alpha)/z^2}^{(T-t'_\alpha)/z^2} < S_u f, g_{\varepsilon(\alpha)} > du$$

$$\prod_{\alpha \in \{1,\cdots,\bar{q}_1 - 2\}} \int_{(S'-t'_\alpha)/z^2}^{(T'-t'_\alpha)/z^2} < g_{1-\varepsilon(\alpha)}, S_u f' > du$$

$$\int_{(S'-t'_{q_1-1})/z^2 - t'_{q_1} - \cdots - t'_{q_{r_1}-1}}^{(T'-t'_{q_1-1})/z^2 - t'_{q_1} - \cdots - t'_{q_{r_1}-1}} < g_{1-\varepsilon(\bar{q}_{r_1})}, S_u f' > du$$

$$\prod_{\alpha \in \{\bar{q}_{r_1}+1,\cdots,\bar{q}_{r_1+1}-2\}} \int_{(S'-t'_\alpha)/z^2}^{(T'-t'_\alpha)/z^2} < g_{1-\varepsilon(\alpha)}, S_u f' > du$$

$$\cdots\cdots$$

$$\int_{(S'-t'_{q_{r_x-1}+1}-1)/z^2-t'_{q_{r_x-1}+1}-\cdots-t'_{q_{r_x}-1}}^{(T'-t'_{q_{r_x-1}+1}-1)/z^2-t'_{q_{r_x-1}+1}-\cdots-t'_{q_{r_x}-1}} < g_{1-\varepsilon(\bar{q}_{r_x})}, S_u f' > du$$

$$\prod_{\alpha\in\{\bar{q}_{r_x}+1,\cdots,n\}} \int_{(S'-t'_\alpha)/z^2}^{(T'-t'_\alpha)/z^2} < g_{1-\varepsilon(\alpha)}, S_u f' > du$$

In the limit $z \to 0$ the integrals of the form

$$\int_{(S'-t'_\alpha)/z^2}^{(T'-t'_\alpha)/z^2} < F, S_u G > du$$

converge to

$$\chi_{[S',T']}(t'_\alpha)(F \mid G)$$

the integrals of the form

$$\int_0^{t'_{q_\alpha}+z^2(\ldots)}$$

converge to

$$\int_0^{t'_{q_\alpha}}$$

and the integrals

$$\int_{-t_{q_\alpha}/z^2-(\ldots)}^{0}$$

give rise to the factors $(g_j \mid g_j)_-$ and the corresponding integration variable disappears. The limiting procedure is justified by dominated convergence.

§4 The uniform estimate

In §3, for each $n \in \mathbf{N}$, we have computed the limit, as $z \to 0$, of the n-th term of the iterated series. In this section , we shall prove some uniform estimates on these terms, which will allow to take the term by term limit of the series.

In the following , we shall use the notation

$$\|g_\varepsilon\|_-^2 := \int_{-\infty}^0 |< g_\varepsilon, S_u g_\varepsilon >|du \quad , \quad \varepsilon = 0,1 \tag{4.1}$$

<u>*LEMMA* (4.1)</u> For each $n \in \mathbf{N}$,

$$\sum_{\varepsilon\in\{0,1\}^n} |I_n^\varepsilon(z,t)| \le \sum_{k=0}^{n} \binom{n}{k} \sum_{m=0}^{k\wedge(n-k)} \sum_{m'=0}^{k\wedge(n-k)} \binom{k}{m}\binom{n-k}{m'}$$

$$\left[\max_{F=f,f';G=g_0,g_1}\int_{-\infty}^{\infty}|<F,S_uG>|du\right]^{2(n-m-m')}\cdot$$

$$\cdot\left(\|g_0\|_-\bigvee\|g_1\|_-\right)^{2(m+m')}\cdot\frac{t^{n-m-m'}}{(n-m-m')!} \tag{4.2}$$

PROOF Notice that the expression (3.8) is exactly of the same type as the expression $I_g(n,\lambda)$, estimated in Lemma (5.2) of [3]. Therefore, from that Lemma we obtain

$$\sum_{\epsilon\in\{0,1\}^n}|I_n^\epsilon(z,t)|\leq\sum_{k=0}^{n}\sum_{1\leq i_1<\cdots<i_k\leq n}\sum_{m=0}^{k\wedge(n-k)}\sum_{m'=0}^{k\wedge(n-k)} \tag{4.3}$$

$$\sum_{(q_1,\cdots,q_m)}'\sum_{(q'_1,\cdots,q'_{m'})}\max_{\epsilon\in\{0,1\}}\left[\int_{-\infty}^{\infty}|<S_uf,g_\epsilon>|du\right]^{n-m-m'}$$

$$\max_{\epsilon\in\{0,1\}}\left[\int_{-\infty}^{\infty}|<S_uf,g_\epsilon>|du\right]^{n-m-m'}\cdot\|g_0\|_-^{2m}\cdot\|g_1\|_-^{2m'}$$

$$\int_{0\leq t_n\leq\cdots\leq\widehat{t_{q'_{m'}}}\leq\cdots\leq\widehat{t_{q_m}}\leq\cdots\leq\widehat{t_{q'_1}}\leq\cdots\leq\widehat{t_{q_1}}\leq\cdots\leq t_1\leq t}dt_1\cdots\widehat{dt_{q_1}}\cdots\widehat{dt_{q'_1}}\cdots\widehat{dt_{q_m}}\cdots\widehat{dt_{q_{m'}}}\cdots dt_n$$

The sum $\sum_{1\leq i_1<\cdots<i_k\leq n}$ gives us the factor $\binom{n}{k}$; the sum $\sum'_{(q_1,\cdots,q_m)}$ gives us the factor $\binom{k}{m}$; the sum $\sum_{(q'_1,\cdots,q'_{m'})}$ gives us the factor $\binom{n-k}{m'}$ and the integral

$$\int_{0\leq t_n\leq\cdots\leq\widehat{t_{q'_{m'}}}\leq\cdots\leq\widehat{t_{q_m}}\leq\cdots\leq\widehat{t_{q'_1}}\leq\cdots\leq\widehat{t_{q_1}}\leq\cdots\leq t_1\leq t}dt_1\cdots\widehat{dt_{q_1}}\cdots\widehat{dt_{q'_1}}\cdots\widehat{dt_{q_m}}\cdots\widehat{dt_{q_{m'}}}\cdots dt_n$$

$$=\frac{t^{n-m-m'}}{(n-m-m')!} \tag{4.4}$$

From this (4.2) follows easily .

COROLLARY (4.2) For each $n\in\mathbf{N}$,

$$\sum_{\epsilon\in\{0,1\}^n}|I_n^\epsilon(z,t)|\leq \tag{4.5}$$

$$4^n\cdot\max_{0\leq m\leq n}\left(\frac{t^{n-m}}{(n-m)!}\cdot\left(\|g_0\|_-\bigvee\|g_1\|_-\right)^{2m}\cdot\left[\max_{F=f,f';G=g_0,g_1}\int_{-\infty}^{\infty}|<F,S_uG>|du\right]^{2(n-m)}\right)$$

In particular, for each $D\in B(H_0)$, if

$$\left(\|g_0\|\bigvee\|g_1\|\right)^2<\frac{1}{4\|D\|} \tag{4.6}$$

then , the series

$$\sum_{n=0}^{\infty} \sum_{\varepsilon\in\{0,1\}^n} < u, D_{\varepsilon(1)} \cdots D_{\varepsilon(n)} v > I_n^{\varepsilon}(z,t) \tag{4.7}$$

converges absolutely and unformly for $z > 0$, and is continuous in $u, v \in K$.

PROOF Clear from Lemma (4.1).

Notice the difference between this estimate and the uniform estimate of Lemma (5.2) in [3]. This difference is not related to the differences between weak coupling and low density, but to the fact that the here the interaction is quadratic and there it was linear.

The following Lemma is a further (with respect to [3]) generalization of the Pule' inequality (i.e. Lemma (3.2) in [20]).

LEMMA (4.3) Let $f : \mathbf{R} \longrightarrow \mathbf{R}_+$ be a positive integrable symmetric function. Then, for each $n \in \mathbf{N}$, $1 \le m \le n-1$, $2 \le q_1 < \cdots < q_m \le n$, $1 \le p_1 < \cdots < p_m \le n-1$, if $\{p_h, q_h\}_{h=1}^m = \{1, \cdots, n\}$ and $p_h < q_h$, $h = 1, \cdots, m$, one has

$$\lambda^{-2m} \int_0^t dt_1 \cdots \int_0^{t_{n-1}} dt_n \sum_{\sigma\in S_m^o} \prod_{h=1}^m f\left(\frac{t_{p_{\sigma(h)}} - t_{q_h}}{\lambda^2}\right) \le \frac{t^{n-m}}{(n-m)!} \left[\int_{-\infty}^0 f(t)dt\right]^m \tag{4.8}$$

PROOF Let , for each $\sigma \in S_m^o$,

$$\varepsilon_\sigma(j) = \begin{cases} q_h , & \text{if } j = 2h ; \\ p_{\sigma(h)} , & \text{if } j = 2h-1 . \end{cases} \tag{4.9}$$

then , $\varepsilon \in S_{2m,n}^o$, where ,

$$S_{2m,n}^o := \{\varepsilon : \{1, \cdots, 2m\} \longrightarrow \{1, \cdots, n\} , \varepsilon(2) < \cdots < \varepsilon(2m) \text{ and } \varepsilon(2h-1) < \varepsilon(2h)\}$$

Notice that , for $\varepsilon \in S_{2m,n}^o$, there exists a subset $\{r_1^\varepsilon, \cdots, r_{2m-n}^\varepsilon\} \subset \{1, \cdots, n\}$, such that

$$\{r_1^\varepsilon, \cdots, r_{2m-n}^\varepsilon\} \subset \{\varepsilon(2h-1)\}_{h=1}^m \cap \{\varepsilon(2h)\}_{h=1}^m .$$

For $\sigma \neq \sigma'$, one has $\varepsilon_\sigma \neq \varepsilon'_\sigma$, so

$$\lambda^{-2m} \int_0^t dt_1 \cdots \int_0^{t_{n-1}} dt_n \sum_{\sigma\in S_m^o} \prod_{h=1}^m f\left(\frac{t_{p_{\sigma(h)}} - t_{q_h}}{\lambda^2}\right)$$

$$\le \lambda^{-2m} \int_0^t dt_1 \cdots \int_0^{t_{n-1}} dt_n \sum_{\varepsilon\in S_{2m,n}^o} \prod_{h=1}^m f\left(\frac{t_{\varepsilon(2h-1)} - t_{\varepsilon(2h)}}{\lambda^2}\right) \tag{4.10}$$

Put

$$S_t^{(n)} := \{(t_1, \cdots, t_n) , t \ge t_1 \ge \cdots \ge t_n \ge 0\} \tag{4.11}$$

$$x_h^{(\varepsilon)} := \frac{t_{\varepsilon(2h-1)} - t_{\varepsilon(2h)}}{\lambda^2} , \quad h = 1, \cdots, m. \tag{4.12}$$

$$y_h^{(\varepsilon)} := t_{\varepsilon(\delta_h^\varepsilon)}, \quad h = 1, \cdots, n-m. \tag{4.13}$$

where , $\delta_1^\varepsilon < \cdots < \delta_{n-m}^\varepsilon$ and $\{\varepsilon(\delta_h^\varepsilon)\}_{h=1}^{n-m} \cap \{\varepsilon(2h-1)\}_{h=1}^m = \emptyset$. So , one has

$$0 \le x_h^{(\varepsilon)} < \infty \ , \ \left(y_1^{(\varepsilon)}, \cdots, y_{n-m}^{(\varepsilon)}\right) \in \mathcal{S}_t^{(n-m)} \tag{4.14}$$

Therfore ,

$$\lambda^{-2m} \int_0^t dt_1 \cdots \int_0^{t_{n-1}} dt_n \sum_{\sigma \in \mathcal{S}_m^o} \prod_{h=1}^m f\left(\frac{t_{p_{\sigma(h)}} - t_{q_h}}{\lambda^2}\right)$$

$$\le \int_{\mathcal{S}_t^{(n-m)}} \int_{\mathbf{R}_+^m} \prod_{h=1}^m f(x_h) dy_1 \cdots dy_{n-m} dx_1 \cdots dx_m$$

$$\frac{t^{n-m}}{(n-m)!} \left[\int_{-\infty}^0 f(t)dt\right]^m \tag{4.15}$$

<u>*COROLLARY*</u> (4.4) Let $f : \mathbf{R} \longrightarrow \mathbf{R}_+$ be a positive integrable symmetric function , for each $n \in \mathbf{N}$, $1 \le m \le n-1$, $2 \le q_1 < \cdots < q_m \le n$, $1 \le p_1 < \cdots < p_m \le n-1$, $p_h < q_h$, $h = 1, \cdots, m$, one has

$$\lambda^{-2m} \int_0^t dt_1 \cdots \int_0^{t_{n-1}} dt_n \sum_{\sigma \in \mathcal{S}_m^o} \prod_{h=1}^m f\left(\frac{t_{p_{\sigma(h)}} - t_{q_h}}{\lambda^2}\right)$$

$$\le \frac{t^{n-m}}{(n-\bar{m})! \cdot (\bar{m}-m)!} \left[\int_{-\infty}^0 f(t)dt\right]^m \tag{4.16}$$

where , $\bar{m} := |\{p_h, q_h\}_{h=1}^m|$.

<u>*PROOF*</u> Denote , for each $\sigma \in \mathcal{S}_n^0$,

$$\{\alpha_h\}_{h=1}^{\bar{m}} := \{p_{\sigma(h)}, q_h\}_{h=1}^m \quad , \quad \{\beta_h\}_{h=1}^{n-\bar{m}} := \{1, \cdots, n\} \setminus \{\alpha_h\}_{h=1}^{\bar{m}}$$

with

$$\alpha_1 < \alpha_2 < \cdots < \alpha_{\bar{m}} \quad , \quad \beta_1 < \beta_2 < \cdots < \beta_{n-\bar{m}}$$

Enlarging $t_{\beta_1 - 1}$ to t , $t_{\beta_2 - 1}$ to t_{β_1} , ... , $t_{\beta_{n-m}-1}$ to $t_{\beta_{n-m-1}}$ and $t_{\alpha_1 - 1}$ to t ,$t_{\alpha_2 - 1}$ to t_{α_1} , ... , $t_{\alpha_m - 1}$ to $t_{\alpha_{m-1}}$, one has

$$\lambda^{-2m} \int_0^t dt_1 \cdots \int_0^{t_{n-1}} dt_n \sum_{\sigma \in \mathcal{S}_m^o} \prod_{h=1}^m f\left(\frac{t_{p_{\sigma(h)}} - t_{q_h}}{\lambda^2}\right) =$$

$$= \lambda^{-2m} \sum_{\sigma \in \mathcal{S}_m^o} \int_0^t dt_1 \cdots \int_0^{t_{\beta_1 - 1}} dt_{\beta_1} \cdots \int_0^{t_{\beta_2 - 1}} dt_{\beta_2}$$

$$\cdots \int_0^{t_{\beta_{n-m}-1}} dt_{\beta_{n-m}} \int_0^{t_{n-1}} dt_n \prod_{h=1}^m f\left(\frac{t_{p_{\sigma(h)}} - t_{q_h}}{\lambda^2}\right) \le$$

$$\leq \lambda^{-2m} \sum_{\sigma \in S_m^o} \int_0^t dt_1 \cdots\cdots \int_0^{t_{m-1}} dt_{\beta m} \prod_{h=1}^m f(\frac{t_{p_{\sigma(h)}} - t_{q_h}}{\lambda^2})$$

$$\int_0^t dt_{\beta_1} \int_0^{t_{\beta_1}} dt_{\beta_2} \cdots \int_0^{t_{\beta_{n-m-1}}} dt_{\beta_{n-m}}$$

$$= \lambda^{-2m} \sum_{\sigma \in S_m^o} \int_0^t dt_1 \cdots\cdots \int_0^{t_{m-1}} dt_{\beta m} \prod_{h=1}^m f(\frac{t_{p_{\sigma(h)}} - t_{q_h}}{\lambda^2})$$

$$t^{n-\bar{m}} \cdot \frac{1}{(n-\bar{m})!}$$

Notice that , where the $\{p_h, q_h\}_{h=1}^m$ is not same as that in the left hand side of (4.16) but the order-relations is remained . So , using the Lemma (4.3) , one gets (4.16) .

<u>*LEMMA* (4.5)</u> For each $n \in \mathbf{N}$,

$$\sum_{\varepsilon \in \{0,1\}^n} |II_n^\varepsilon(z,t)| \leq \tag{4.17}$$

$$16^n \cdot \max_{0 \leq m \leq n-1} \frac{t^{n-m}}{(n-m)!} \cdot \left(\|g_0\|_- \bigvee \|g_1\|_-\right)^{2m} \left[\max_{F=f,f';G=g_0,g_1} \int_{-\infty}^{\infty} |< F, S_t G > |dt\right]^{2(n-m)}$$

<u>*PROOF*</u> . By (3.16), one has

$$\sum_{\varepsilon \in \{0,1\}^n} |II_n^\varepsilon(z,t)| \leq \sum_{k=0}^n \sum_{1 \leq i_1 < \cdots < i_k \leq n} \sum_{m=0}^{n-1} \sum_{2 \leq \bar{q}_1 < \cdots < \bar{q}_m \leq n} \sum_{(\bar{q}_1, \bar{p}_1, \cdots, \bar{q}_m, \bar{p}_m)}'$$

$$\left[\max_{F=f,f';G=g_0,g_1} \int_{-\infty}^{\infty} |< F, S_t G > |dt\right]^{2(n-m)}$$

$$\cdot z^{2n-2m} \int_0^{t/z^2} dt_1 \int_0^{t_1} dt_2 \cdots \int_0^{t_{n-1}} dt_n \prod_{h=1}^m |< S_{t_{\bar{p}_h}} g, S_{t_{\bar{q}_h}} g > |$$

$$\leq \sum_{k=0}^n \sum_{1 \leq i_1 < \cdots < i_k \leq n} \sum_{m=0}^{n-1} \sum_{2 \leq \bar{q}_1 < \cdots < \bar{q}_m \leq n} \sum_{1 \leq p_1 < \cdots < p_m \leq n-1} \sum_{\sigma \in S_m^o}$$

$$\left[\max_{F=f,f';G=g_0,g_1} \int_{-\infty}^{\infty} |< F, S_t G > |dt\right]^{2(n-m)}$$

$$\cdot z^{2n-2m} \int_0^{t/z^2} dt_1 \int_0^{t_1} dt_2 \cdots \int_0^{t_{n-1}} dt_n \prod_{h=1}^m F\left(t_{q_h} - t_{p_h}\right) \tag{4.18}$$

where ,

$$F(u) := \max_{\varepsilon \in \{0,1\}} |< g_\varepsilon, S_t g_{1-\varepsilon} > | \tag{4.19}$$

So, by Lemma (4.3) and Corollary (4.4),

$$\sum_{\varepsilon\in\{0,1\}^n} |II_n^\varepsilon(z,t)| \leq \sum_{k=0}^{n} \sum_{1\leq i_1<\cdots i_k\leq n} \sum_{m=0}^{n-1} \binom{n}{m}^2 \cdot$$

$$\left[\max_{F=f,f';G=g_0,g_1} \int_{-\infty}^{\infty} |< F, S_t G >|dt\right]^{2(n-m)}$$

$$\max_{m\leq n_0\leq n} \frac{t^{n-m}}{(n-n_0)!(n_0-m)!} \cdot \left(\|g_0\|_- \vee \|g_1\|_-\right)^{2m}$$

$$\leq \sum_{k=0}^{n} \sum_{1\leq i_1<\cdots<i_k\leq n} 8^n \max_{0\leq m\leq n-1} \left(\frac{t^{n-m}}{(n-m)!} \cdot \left(\|g_0\|_- \vee \|g_1\|_-\right)^{2m} \cdot\right.$$

$$\left.\cdot\left[\max_{F=f,f';G=g_0,g_1} \int_{-\infty}^{\infty} |< F, S_t G >|dt\right]^{2(n-m)}\right) \tag{4.20}$$

which easily implies (4.17) .

<u>*COROLLARY* (4.5)</u> For each $D \in B(H_0)$ if

$$\left(\|g_0\|_- \vee \|g_1\|_-\right)^2 < \frac{1}{16\|D\|} \tag{4.21}$$

then , the series

$$\sum_{n=0}^{\infty} \sum_{\varepsilon\in\{0,1\}^n} II_n^\varepsilon(z,t) \tag{4.22}$$

converges absolutely and uniformly for $z > 0$.

§5 The low desity limit and its derivative

In this section, using the results of §3 and §4 we derive the existence of the low density limit (1.23) and its explicit form. Using the latter, we deduce an integral equation for this limit.

<u>*THEOREM* (5.1)</u> For each $u, v \in H_0$ and $D \in B(H_0)$ satisfying (1.10) and for each $g_0, g_1 \in K$ satisfying (1.11) and

$$\left(\|g_0\|_- \vee \|g_1\|_-\right)^2 < \frac{1}{16\|D\|} \tag{5.0b}$$

the low density limit

$$\lim_{z\to 0} < u \otimes W(z \int_{S/z^2}^{T/z^2} S_u f du)\Phi, U(t/z^2)v \otimes W(z \int_{S'/z^2}^{T'/z^2} S_u f' du)\Phi > \tag{5.1}$$

exists and is equal to

$$\sum_{n=0}^{\infty}\sum_{k=0}^{n}\sum_{1\le i_1<\cdots<i_k\le n} < u, D_{\varepsilon(1)}\cdots D_{\varepsilon(n)}v > \cdot$$

$$\sum_{m=0}^{k\wedge(n-k)}\sum_{(q_1,\cdots,q_m)}^{\prime}\sum_{m'=0}^{k\wedge(n-k)}\sum_{(q'_1,\cdots,q'_{m'})}(g_1|g_1)_-^m\cdot(g_0|g_0)_-^{m'}$$

$$\int_{0\le t_n\le\cdots\le\widehat{t_{q'_{m'}}}\le\cdots\le\widehat{t_{q_m}}\le\cdots\le\widehat{t_{q'_1}}\le\cdots\le\widehat{t_{q_1}}\le\cdots\le t_1\le t} dt_1\cdots\widehat{dt_{q_1}}\cdots\widehat{dt_{q'_1}}\cdots\widehat{dt_{q_m}}\cdots\widehat{dt_{q_{m'}}}\cdots dt_n$$

$$\prod_{\alpha\in\{1,\cdots,n\}\backslash(\{q_h\}_{h=1}^m\cup\{q'_h\}_{h=1}^{m'})}(f|g_{\varepsilon(\alpha)})\cdot\chi_{[S,T]}(t_\alpha)$$

$$\prod_{\alpha\in\{1,\cdots,n\}\backslash(\{q_h-1\}_{h=1}^m\cup\{q'_h-1\}_{h=1}^{m'})\cap\{1,\cdots,\bar q_1-2\}}(g_{1-\varepsilon(\alpha)}|f')\cdot\chi_{[S',T']}(t_\alpha)$$

$$\cdot(g_{1-\varepsilon(\bar q_{r_1})}|f')\cdot\chi_{[S',T']}(t_{\bar q_1-1})\cdot$$

$$\prod_{\alpha\in\{1,\cdots,n\}\backslash(\{q_h-1\}_{h=1}^m\cup\{q'_h-1\}_{h=1}^{m'})\cap\{\bar q_{r_1}+1,\cdots,\bar q_{r_1+1}-2\}}(g_{1-\varepsilon(\alpha)}|f')\cdot\chi_{[S',T']}(t_\alpha)$$

$$\cdot(g_{1-\varepsilon(\bar q_{r_2})}|f')\cdot\chi_{[S',T']}(t_{\bar q_1+1-1})\cdot$$

$$\cdots\cdots$$

$$\prod_{\alpha\in\{1,\cdots,n\}\backslash(\{q_h-1\}_{h=1}^m\cup\{q'_h-1\}_{h=1}^{m'})\cap\{\bar q_{r_{x-1}}+1,\cdots,\bar q_{r_{x-1}+1}-2\}}(g_{1-\varepsilon(\alpha)}|f')\cdot\chi_{[S',T']}(t_\alpha)$$

$$\cdot(g_{1-\varepsilon(\bar q_{r_x})}|f')\cdot\chi_{[S',T']}(t_{\bar q_{x-1}-1})\cdot$$

$$\prod_{\alpha\in\{1,\cdots,n\}\backslash(\{q_h-1\}_{h=1}^m\cup\{q'_h-1\}_{h=1}^{m'})\cap\{\bar q_{r_x}+1,\cdots,n\}}(g_{1-\varepsilon(\alpha)}|f')\cdot\chi_{[S',T']}(t_\alpha)$$

$$< W(\chi_{[S,T]}\otimes f)\Psi, W(\chi_{[S',T']}\otimes f')\Psi > \tag{5.2}$$

PROOF

$$< u\otimes W(z\int_{S/z^2}^{T/z^2}S_u f du)\Phi, U^{(z)}(t/z^2)v\otimes W(z\int_{S'/z^2}^{T'/z^2}S_u f' du)\Phi >$$

$$=\sum_{n=0}^{\infty}(-i)^n\int_0^{t/z^2}dt_1\int_0^{t_1}dt_2\cdots\int_0^{t_{n-1}}dt_n < u\otimes W(z\int_{S/z^2}^{T/z^2}S_u f du)\Phi,$$

$$V(t_1)\cdots V(t_n)v\otimes W(z\int_{S'/z^2}^{T'/z^2}S_u f' du)\Phi >$$

$$= \sum_{n=0}^{\infty}\sum_{k=0}^{n}\sum_{1\le i_1<\cdots<i_k\le n} < u, D_n(i_1,\cdots,i_k)u > (I_n^\varepsilon(z,t) + II_n^\varepsilon(z,t)) \tag{5.3}$$

Because of the uniform estimate one has

$$\lim_{z\to 0} < u\otimes W(z\int_{S/z^2}^{T/z^2} S_u f du)\Phi, U^{(z)}(t/z^2)v\otimes W(z\int_{S'/z^2}^{T'/z^2} S_u f' du)\Phi >$$

$$= \sum_{n=0}^{\infty}\sum_{k=0}^{n}\sum_{1\le i_1<\cdots<i_k\le n} < u, D_n(i_1,\cdots,i_k)u > \lim_{z\to 0}(I_n^\varepsilon(z,t) + II_n^\varepsilon(z,t)) \tag{5.4}$$

and the result follows from Lemma (3.3) and Lemma (3.4) .

In order to find an integral equation satisfied by the limit (5.1), we proceed in analogy with the similar problem dealt with in [2] and, for each $n \in \mathbf{N}$ we use the explicit form (1.16) of $V(t_1)$ in the expression

$$(-i)^n \int_0^{t/z^2} dt_1 \int_0^{t_1} dt_2 \cdots \int_0^{t_{n-1}} dt_n < u\otimes W(z\int_{S/z^2}^{T/z^2} S_u f du)\Phi,$$

$$V(t_1)\cdots V(t_n)v\otimes W(z\int_{S'/z^2}^{T'/z^2} S_u f' du)\Phi >=$$

$$= \frac{1}{z^2}(-i)^{n-1}\int_0^t dt_1 \int_0^{t_1/z^2} dt_2 \int_0^{t_2} dt_3 \cdots \int_0^{t_{n-1}} dt_n < u\otimes W(z\int_{S/z^2}^{T/z^2} S_u f du)\Phi \tag{5.5a}$$

Now we write (5.5a) as the sum of four terms in two of which $(I_n(1,z,t), I_n(2,z,t))$ the A, A^+ have acted on the coherent vectors giving rise to scalar products; while the remaining two $(II_n(0,z,t), II_n(1,z,t))$ contain the commutators needed to produce the scalar products. Explicitly, (5.5a) is equal to:

$$(D\otimes A^+(S_{t_1/z^2}g_0)A(S_{t_1/z^2}g_1) - D^+\otimes A^+(S_{t_1/z^2}g_1)A(S_{t_1/z^2}g_0))\cdot$$

$$\cdot V(t_2)\cdots V(t_n)v\otimes W(z\int_{S'/z^2}^{T'/z^2} S_u f' du)\Phi >$$

$$= \int_0^t dt_1 \int_{S/z^2}^{T/z^2} < S_u f, S_{t_1/z^2}g_0 > du\cdot \int_{S'/z^2}^{T'/z^2} < S_{t_1/z^2}g_1, S_u f' > du\cdot$$

$$\cdot(-i)^{n-1}\int_0^{t_1/z^2} dt_2 \int_0^{t_2} dt_3 \cdots \int_0^{t_{n-1}} dt_n < D^+u\otimes W(z\int_{S/z^2}^{T/z^2} S_u f du)\Phi,$$

$$\cdot V(t_2)\cdots V(t_n)v\otimes W(z\int_{S'/z^2}^{T'/z^2} S_u f' du)\Phi >$$

$$+ \int_0^t dt_1 \int_{S/z^2}^{T/z^2} < S_u f, S_{t_1/z^2}g_1 > du\cdot \int_{S'/z^2}^{T'/z^2} < S_{t_1/z^2}g_0, S_u f' > du\cdot$$

$$\cdot(-i)^{n-1}\int_0^t dt_1\int_0^{t_1/z^2}dt_2\int_0^{t_2}dt_3\cdots\int_0^{t_{n-1}}dt_n < -Du\otimes W(z\int_{S/z^2}^{T/z^2}S_u f du)\Phi,$$

$$\cdot V(t_2)\cdots V(t_n)v\otimes W(z\int_{S'/z^2}^{T'/z^2}S_u f' du)\Phi >$$

$$+\int_0^t dt_1\int_{S/z^2}^{T/z^2} < S_u f, S_{t_1/z^2}g_0 > du\cdot(-i)^{n-1}\frac{1}{z}$$

$$\int_0^{t_1/z^2}dt_2\int_0^{t_2}dt_3\cdots\int_0^{t_{n-1}}dt_n < D^+u\otimes W(z\int_{S/z^2}^{T/z^2}S_u f du)\Phi,$$

$$\cdot\left[1\otimes A(S_{t_1/z^2}g_1), V(t_2)\cdots V(t_n)\right]v\otimes W(z\int_{S'/z^2}^{T'/z^2}S_u f' du)\Phi >$$

$$+\int_0^t dt_1\int_{S/z^2}^{T/z^2} < S_u f, S_{t_1/z^2}g_1 > du\cdot(-i)^{n-1}\frac{1}{z}$$

$$\int_0^{t_1/z^2}dt_2\int_0^{t_2}dt_3\cdots\int_0^{t_{n-1}}dt_n < -Du\otimes W(z\int_{S/z^2}^{T/z^2}S_u f du)\Phi,$$

$$\cdot\left[1\otimes A(S_{t_1/z^2}g_0), V(t_2)\cdots V(t_n)\right]v\otimes W(z\int_{S'/z^2}^{T'/z^2}S_u f' du)\Phi >$$

$$:= I_n(1,z,t)+I_n(2,z,t)+II_n(0,z,t)+II_n(1,z,t) \tag{5.5}$$

Now we analyse separately the contributions of the four type of terms (5.5) to the limit (5.1). From §4 , we know that

$$|\int_0^{t_1/z^2}dt_2\int_0^{t_2}dt_3\cdots\int_0^{t_{n-1}}dt_n < D^+u\otimes W(z\int_{S/z^2}^{T/z^2}S_u f du)\Phi,$$

$$\cdot V(t_2)\cdots V(t_n)v\otimes W(z\int_{S'/z^2}^{T'/z^2}S_u f' du)\Phi >|$$

$$\le C\Big(16\|D\|\Big)^n\cdot\|u\|\cdot\|v\|\cdot\max_{0\le m\le n-1}\frac{t_1^{n-m}}{(n-m)!}\Big(\|g_0\|_-\vee\|g_1\|_-\Big)^{2m} \tag{5.6}$$

with a constant C . So , if $\Big(\|g_0\|_-\vee\|g_1\|_-\Big)^2<\frac{1}{16\|D\|}$, one has , by dominated convergence

$$\lim_{z\to0}\sum_{n=1}^{\infty}I_n(1,z,t)=\sum_{n=1}^{\infty}\lim_{z\to0}\int_0^t dt_1\int_{S/z^2}^{T/z^2} < S_u f, S_{t_1/z^2}g_0 > du\cdot$$

$$\cdot\int_{S'/z^2}^{T'/z^2} < S_{t_1/z^2}g_1, S_u f' > du\cdot$$

$$\cdot(-i)^{n-1}\int_0^{t_1/z^2}dt_2\int_0^{t_2}dt_3\cdots\int_0^{t_{n-1}}dt_n < D^+u\otimes W(z\int_{S/z^2}^{T/z^2}S_u f du)\Phi,$$

$$\cdot V(t_2)\cdots V(t_n)v\otimes W(z\int_{S'/z^2}^{T'/z^2} S_uf'du)\Phi>$$

$$=\sum_{n=1}^{\infty}\int_0^t dt_1 \lim_{z\to 0}\int_{S/z^2}^{T/z^2} <S_uf, S_{t_1/z^2}g_0> du\cdot\int_{S'/z^2}^{T'/z^2} <S_{t_1/z^2}g_1, S_uf'> du\cdot$$

$$\cdot(-i)^{n-1}\int_0^{t_1/z^2} dt_2\int_0^{t_2} dt_3\cdots\int_0^{t_{n-1}} dt_n <D^+u\otimes W(z\int_{S/z^2}^{T/z^2} S_ufdu)\Phi,$$

$$\cdot V(t_2)\cdots V(t_n)v\otimes W(z\int_{S'/z^2}^{T'/z^2} S_uf'du)\Phi> \tag{5.7}$$

Now, if $\left(\|g_0\|_- \vee \|g_1\|_-\right)^2 < \frac{1}{16\|D\|}$, then the limit (5.1) is continuous for $u, v\in H_0$. So we can write this limit as

$$<u, G(t)> \tag{5.8a}$$

with $G(t)\in H_0$ for each $t\geq 0$. With this notation (5.7) becomes

$$\lim_{z\to 0}\sum_{n=1}^{\infty} I_n(1,z,t) = \int_0^t dt_1(f|g_0)\chi_{[S,T]}(t_1)\cdot(g_1|f')\chi_{[S',T']}(t_1) <D^+u, G(t_1)> \tag{5.8}$$

Similarly , we obtain

$$\lim_{z\to 0}\sum_{n=1}^{\infty} I_n(2,z,t) = \int_0^t dt_1(f|g_1)\chi_{[S,T]}(t_1)\cdot(g_0|f')\chi_{[S',T']}(t_1) <-Du, G(t_1)> \tag{5.9}$$

Now, we investigate $II_n(0,z,t)$. By definition for $n\geq 2$

$$II_n(0,z,t) = \int_0^t dt_1\int_{S/z^2}^{T/z^2} <S_uf, S_{t_1/z^2}g_0> du\cdot(-i)^{n-1}\frac{1}{z}$$

$$\int_0^{t_1/z^2} dt_2\int_0^{t_2} dt_3\cdots\int_0^{t_{n-1}} dt_n <D^+u\otimes W(z\int_{S/z^2}^{T/z^2} S_ufdu)\Phi,$$

$$\cdot\left[1\otimes A(S_{t_1/z^2}g_1), V(t_2)\cdots V(t_n)\right]v\otimes W(z\int_{S'/z^2}^{T'/z^2} S_uf'du)\Phi> \tag{5.10}$$

and we want to consider the limit

$$\lim_{z\to 0} II_{n+1}(0,z,t) \tag{5.11}$$

Put $s=t_1$, $s_1=t_2$, $\cdots$, $s_n=t_{n+1}$,

$$II_{n+1}(0,z,t) = \int_0^t ds\int_{(S-s)/z^2}^{(T-s)/z^2} <S_uf, g_0> du\cdot(-i)^n\frac{1}{z}$$

$$\int_0^{s/z^2} ds_1 \int_0^{s_1} ds_2 \cdots \int_0^{s_{n-1}} ds_n < D^+ u \otimes W(z \int_{S/z^2}^{T/z^2} S_u f du)\Phi,$$

$$\cdot [1 \otimes A(S_{s/z^2} g_1), V(s_1) \cdots V(s_n)] v \otimes W(z \int_{S'/z^2}^{T'/z^2} S_u f' du)\Phi > \tag{5.12}$$

By Lemma (3.2) and in the notations of formula (3.11) , one has

$$[1 \otimes A(S_{s/z^2} g_1), V(s_1) \cdots V(s_n)] = \sum_{\varepsilon \in \{0,1\}^n} [1 \otimes A(S_{s/z^2} g_1), D_{\varepsilon(1)} \cdots D_{\varepsilon(n)} \otimes (I_n^\varepsilon + II_n^\varepsilon)] \tag{5.13}$$

The main idea of the estimates which follow is that the matrix elements of type II (cf. §3) will vanish in the limit $z \to 0$. In order to control the limit of the remaining ones, we need their explicit form. Reducing the commutator (5.13) to the normal rodered form produces the following (unfortunately rather long expression):

$$= \sum_{\varepsilon \in \{0,1\}^n} D_{\varepsilon(1)} \cdots D_{\varepsilon(n)} \otimes \Big(\sum_{m=0}^{k \wedge (n-k)} \sum_{(q_1,\cdots,q_m)}' \sum_{m'=0}^{k \wedge (n-k)} \sum_{(q'_1,\cdots,q'_{m'})}$$

$$\prod_{h=1}^{m} < g_1, S_{(s_{q_h} - s_{q_h - 1})} g_1 > \cdot \prod_{h=1}^{m'} < g_0, S_{(s_{q'_h} - s_{q'_h - 1})} g_0 >$$

$$[A(S_{s/z^2} g_1), \prod_{\alpha \in \{1,\cdots,n\} \setminus (\{q_h\}_{h=1}^m \cup \{q'_h\}_{h=1}^{m'})} A^+(S_{s_\alpha} g_{\varepsilon(\alpha)})]$$

$$\prod_{\alpha \in \{1,\cdots,n\} \setminus (\{q_h - 1\}_{h=1}^m \cup \{q'_h - 1\}_{h=1}^{m'})} A(S_{s_\alpha} g_{1-\varepsilon(\alpha)})$$

$$+ \sum_{m=0}^{k \wedge (n-k)} \sum_{\substack{2 \le q_1 < \cdots < q_m \le n \\ \{q_h\}_{h=1}^m \subset \{i_h\}_{h=1}^k}} \sum_{m'=0}^{k \wedge (n-k)} \sum_{\substack{2 \le q'_1 < \cdots < q'_{m'} \le n \\ \{q'_h\}_{h=1}^{m'} \subset \{1,\cdots,n\} \setminus \{i_h\}_{h=1}^k}}$$

$$\sum_{(q_1,p_1,\cdots,q_m,p_m)}' \sum_{(q'_1,p'_1,\cdots,q'_{m'},p'_{m'})} \prod_{h=1}^{m} < S_{s_{p_h}} g_1, S_{s_{q_h}} g_1 > \cdot \prod_{h=1}^{m'} < S_{s_{p'_h}} g_0, S_{s_{q'_h}} g_0 >$$

$$[A(S_{s/z^2} g_1), \prod_{\alpha \in \{1,\cdots,n\} \setminus (\{q_h\}_{h=1}^m \cup \{q'_h\}_{h=1}^{m'})} A^+(S_{s_\alpha} g_{\varepsilon(\alpha)})]$$

$$\prod_{\alpha \in \{1,\cdots,n\} \setminus (\{p_h\}_{h=1}^m \cup \{p'_h\}_{h=1}^{m'})} A(S_{s_\alpha} g_{1-\varepsilon(\alpha)} 0) +$$

$$+ \sum_{m=0}^{k \wedge (n-k)} \sum_{(q_1,\cdots,q_m)}' \sum_{m'=0}^{k \wedge (n-k)} \sum_{\substack{2 \le q'_1 < \cdots < q'_{m'} \le n \\ \{q'_h\}_{h=1}^{m'} \subset \{1,\cdots,n\} \setminus \{i_h\}_{h=1}^k}} \sum_{(q'_1,p'_1,\cdots,q'_{m'},p'_{m'})}$$

$$\prod_{h=1}^{m} < S_{s_{q_h-1}} g_1, S_{s_{q_h}} g_1 > \cdot \prod_{h=1}^{m'} < S_{s_{p'_h}} g_0, S_{s_{q'_h}} g_0 >$$

$$[A(S_{s/z^2} g_1), \prod_{\alpha\in\{1,\cdots,n\}\setminus(\{q_h\}_{h=1}^{m}\cup\{q'_h\}_{h=1}^{m'})} A^+(S_{s_\alpha} g_{\varepsilon(\alpha)})]$$

$$\prod_{\alpha\in\{1,\cdots,n\}\setminus(\{q_h-1\}_{h=1}^{m}\cup\{p'_h\}_{h=1}^{m'})} A(S_{s_\alpha} g_{1-\varepsilon(\alpha)})+$$

$$+\sum_{m=0}^{k\wedge(n-k)} \sum_{\substack{2\le q_1<\cdots<q_m\le n \\ \{q_h\}_{h=1}^{m}\subset\{i_h\}_{h=1}^{k}}} \sum_{(q_1,p_1,\cdots,q_m,p_m)}' \sum_{m'=0}^{k\wedge(n-k)} \sum_{(q'_1,\cdots,q'_{m'})}$$

$$\prod_{h=1}^{m} < S_{s_{p_h}} g_1, S_{s_{q_h}} g_1 > \cdot \prod_{h=1}^{m'} < S_{s_{q'_h-1}} g_0, S_{s_{q'_h}} g_0 >$$

$$[A(S_{s/z^2} g_1), \prod_{\alpha\in\{1,\cdots,n\}\setminus(\{q_h\}_{h=1}^{m}\cup\{q'_h\}_{h=1}^{m'})} A^+(S_{s_\alpha} g_{\varepsilon(\alpha)})]$$

$$\prod_{\alpha\in\{1,\cdots,n\}\setminus(\{p_h\}_{h=1}^{m}\cup\{q'_h-1\}_{h=1}^{m'})} A(S_{s_\alpha} g_{1-\varepsilon(\alpha)})\Big)$$

$$= \sum_{\varepsilon\in\{0,1\}^n} D_{\varepsilon(1)}\cdots D_{\varepsilon(n)} \otimes \Big(\sum_{m=0}^{k\wedge(n-k)} \sum_{\substack{(q_0,\cdots,q_m) \\ q_0=1}}' \sum_{m'=0}^{k\wedge(n-k)} \sum_{(q'_1,\cdots,q'_{m'})}$$

$$\prod_{h=0}^{m} < g_1, S_{(s_{q_h}-s_{q_h-1})} g_1 > \cdot \prod_{h=1}^{m'} < g_0, S_{(s_{q'_h}-s_{q'_h-1})} g_0 >$$

$$\prod_{\alpha\in\{2,\cdots,n\}\setminus(\{q_h\}_{h=1}^{m}\cup\{q'_h\}_{h=1}^{m'})} A^+(S_{s_\alpha} g_{\varepsilon(\alpha)})$$

$$\prod_{\alpha\in\{1,\cdots,n\}\setminus(\{q_h-1\}_{h=1}^{m}\cup\{q'_h-1\}_{h=1}^{m'})} A(S_{s_\alpha} g_{1-\varepsilon(\alpha)})$$

$$+\sum_{m=0}^{(n-1)} \sum_{1=\bar q_0<\bar q_1<\cdots<\bar q_m\le n} \sum_{\substack{(\bar q_0,\bar p_0,\cdots,\bar q_m,\bar p_m) \\ \bar q_0=1,\bar p_0=0,}}'$$

$$\prod_{h=0}^{m} < S_{s_{\bar p_h}} g_{1-\varepsilon(\bar p_h)}, S_{s_{\bar q_h}} g_{\varepsilon(\bar q_h)} > \cdot$$

$$\prod_{\alpha\in\{1,\cdots,n\}\setminus\{\bar q_h\}_{h=0}^{m}} A^+(S_{s_\alpha} g_{\varepsilon(\alpha)}) \prod_{\alpha\in\{1,\cdots,n\}\setminus\{\bar p_h\}_{h=1}^{m}} A(S_{s_\alpha} g_{1-\varepsilon(\alpha)})\Big) \tag{5.14}$$

Seaprating, in the above formula, the contridutions of the type I terms from that of the (irrelevent in the limit) type II, one can write

$$II_{n+1}(0,z,t) := \sum_{\varepsilon\in\{0,1\}^n} \Big(\Delta_1(n,z,\varepsilon,t) + \Delta_2(n,z,\varepsilon,t)\Big) \tag{5.15}$$

with

$$\Delta_1(n,z,\varepsilon,t) = < u, D_{\varepsilon(1)} \cdots D_{\varepsilon(n)} v > \sum_{m=0}^{k\wedge(n-k)} \sum_{\substack{(q_0,\cdots,q_m)\\ q_0=1}}^{\prime} \sum_{m'=0}^{k\wedge(n-k)} \sum_{(q'_1,\cdots,q'_{m'})}$$

$$\int_0^t ds \int_{(S-s)/z^2}^{(T-s)/z^2} < S_u f, g_0 > du \cdot (-i)^n \frac{1}{z} \int_0^{s/z^2} ds_1 \int_0^{s_1} ds_2 \cdots \int_0^{s_{n-1}} ds_n$$

$$\prod_{h=0}^{m} < g_1, S_{(s_{q_h} - s_{q_h-1})} g_1 > \cdot \prod_{h=1}^{m'} < g_0, S_{(s_{q'_h} - s_{q'_h-1})} g_0 >$$

$$\prod_{\alpha\in\{2,\cdots,n\}\setminus(\{q_h\}_{h=1}^m \cup \{q'_h\}_{h=1}^{m'})} A^+(S_{s_\alpha} g_{\varepsilon(\alpha)})$$

$$\prod_{\alpha\in\{1,\cdots,n\}\setminus(\{q_h-1\}_{h=1}^m \cup \{q'_h-1\}_{h=1}^{m'})} A(S_{s_\alpha} g_{1-\varepsilon(\alpha)}) \tag{5.15a}$$

and

$$\Delta_2(n,z,\varepsilon,t) = < u, D_{\varepsilon(1)} \cdots D_{\varepsilon(n)} v >$$

$$\sum_{m=0}^{(n-1)} \sum_{1=\bar q_0 < \bar q_1 < \cdots < \bar q_m \le n} \sum_{\substack{(\bar q_0,\bar p_0,\cdots,\bar q_m,\bar p_m)\\ \bar q_0=1,\bar p_0=0}}^{\prime}$$

$$\int_0^t ds \int_{(S-s)/z^2}^{(T-s)/z^2} < S_u f, g_0 > du \cdot (-i)^n \frac{1}{z} \int_0^{s/z^2} ds_1 \int_0^{s_1} ds_2 \cdots \int_0^{s_{n-1}} ds_n$$

$$\prod_{h=0}^{m} < S_{s_{\bar p_h}} g_{1-\varepsilon(\bar p_h)}, S_{s_{\bar q_h}} g_{\varepsilon(\bar q_h)} > \cdot$$

$$\prod_{\alpha\in\{1,\cdots,n\}\setminus\{\bar q_h\}_{h=0}^m} A^+(S_{s_\alpha} g_{\varepsilon(\alpha)}) \prod_{\alpha\in\{1,\cdots,n\}\setminus\{\bar p_h\}_{h=1}^m}^{\bullet} A(S_{s_\alpha} g_{1-\varepsilon(\alpha)})\Big) \tag{5.15b}$$

Applying the same arguments used in the proofs of the results in §3 and §4 to (5.15a) and (5.15b) , we obtain the following estimates:

<u>*LEMMA*</u> (5.1) There exists a constant C_1 such that for each $n \in \mathbf{N}$, $\varepsilon = 0, 1$,

$$|II_{n+1}(\varepsilon,z,t)| \le \Big(16\|D\|\Big)^n \|u\| \cdot \|v\| C_1 \cdot \max_{0\le m\le n} \frac{t^{n+1-m}}{(n+1-m)!} \Big(\|g_0\|_- \bigvee \|g_1\|_-\Big)^{2m} \tag{5.16}$$

<u>*PROOF*</u> (5.16) is an immediate consequence of the uniform estimate of Lemma (4.5).

<u>*LEMMA* (5.2)</u> For each $n \in \mathbf{N}$,

$$\lim_{z \to 0} \Delta_2(n, z, \varepsilon, t) = 0 \tag{5.17}$$

<u>*PROOF*</u> The proof is the same as the proof that the terms of type II tend to zero as $z \to 0$.

<u>*LEMMA* (5.3)</u> For each $n \in \mathbf{N}$, the limit

$$\lim_{z \to 0} \Delta_1(n, z, \varepsilon, t) \tag{5.18}$$

exists .

<u>*PROOF*</u> The proof is the same as that used to prove that the terms of type I tend to a limit as $z \to 0$.

<u>*COROLLARY*</u> For each $n \in \mathbf{N}$, $\varepsilon = 0, 1$ the limit

$$\lim_{z \to 0} II_{n+1}(\varepsilon, z, t) \tag{5.19}$$

exists .

<u>*PROOF*</u> Clear from (5.15) and Lemmata (5.2), (5.3).

Now, for each $n \in \mathbf{N}$, consider the commutator

$$\big[1 \otimes A(S_{s/z^2} g_1), V(s_1) \cdots V(s_n)\big] = \big[1 \otimes A(S_{s/z^2} g_1), V(s_1)\big] V(s_2) \cdots V(s_n)$$

$$+ \sum_{j=2}^{n} V(s_1) \cdots V(s_{j-1}) \big[1 \otimes A(S_{s/z^2} g_1), V(s_j)\big] V(s_{j+1}) \cdots V(s_n) \tag{5.20}$$

The following lemma proves that the matrix elements of the right hand side of (5.20) tend to zero rapidly as $z \to 0$.

<u>*LEMMA* (5.4)</u> The limit

$$\lim_{z \to 0} \int_0^t ds \int_{(S-s)/z^2}^{(T-s)/z^2} < S_u f, g_0 > du \cdot (-i)^n \frac{1}{z} \int_0^{s/z^2} ds_1 \int_0^{s_1} ds_2 \cdots \int_0^{s_{n-1}} ds_n$$

$$< D^+ u \otimes W(z \int_{S/z^2}^{T/z^2} S_u f du)\Phi, \sum_{j=2}^{n} V(s_1) \cdots V(s_{j-1})$$

$$\big[1 \otimes A(S_{s/z^2} g_1), V(s_j)\big] V(s_{j+1}) \cdots V(s_n) v \otimes W(z \int_{S'/z^2}^{T'/z^2} S_u f' du)\Phi > \tag{5.21}$$

exists and is equal to zero.

PROOF

$$\int_0^t ds \int_{(S-s)/z^2}^{(T-s)/z^2} < S_u f, g_0 > du \cdot (-i)^n \frac{1}{z} \int_0^{s/z^2} ds_1 \int_0^{s_1} ds_2 \cdots \int_0^{s_{n-1}} ds_n$$

$$< D^+ u \otimes W(z \int_{S/z^2}^{T/z^2} S_u f du)\Phi, \sum_{j=2}^{n} V(s_1) \cdots V(s_{j-1})$$

$$[1 \otimes A(S_{s/z^2} g_1), V(s_j)] V(s_{j+1}) \cdots V(s_n) u \otimes W(z \int_{S'/z^2}^{T'/z^2} S_u f' du)\Phi >$$

$$= z \int_0^{t/z^2} ds \int_{S/z^2-s}^{T/z^2-s} < S_u f, g_0 > du \cdot (-i)^n \int_0^{s} ds_1 \int_0^{s_1} ds_2 \cdots \int_0^{s_{n-1}} ds_n$$

$$< D^+ u \otimes W(z \int_{S/z^2}^{T/z^2} S_u f du)\Phi, \sum_{j=2}^{n} V(s_1) \cdots V(s_{j-1})$$

$$[1 \otimes A(S_s g_1), V(s_j)] V(s_{j+1}) \cdots V(s_n) v \otimes W(z \int_{S'/z^2}^{T'/z^2} S_u f' du)\Phi > \tag{5.22}$$

Notice that each term of the right hand side of (5.22) is in $\Delta_2(n, z, \varepsilon, t)$, so , by (5.17) , one gets (5.21).

Similarly , we get the following ,

LEMMA (5.5) For each $n \in \mathbf{N}$, the limit

$$\lim_{z \to 0} II_{n+1}(\varepsilon, z, t) \tag{5.23}$$

exists and is equal to

$$\lim_{z \to 0} z \int_0^{t/z^2} ds \int_{S/z^2-s}^{T/z^2-s} < S_u f, g_\varepsilon > du \cdot (-i)^n \int_0^{s} ds_1 \int_0^{s_1} ds_2 \cdots \int_0^{s_{n-1}} ds_n$$

$$< D_\varepsilon^+ u \otimes W(z \int_{S/z^2}^{T/z^2} S_u f du)\Phi, [1 \otimes A(S_s g_{1-\varepsilon}), V(s_1)]$$

$$V(s_2) \cdots V(s_n) v \otimes W(z \int_{S'/z^2}^{T'/z^2} S_u f' du)\Phi > \tag{5.24}$$

Let us introduce, for simplicity the notation

$$\Phi_z(f, S, T) := W(z \int_{S/z^2}^{T/z^2} S_u f du)\Phi \tag{5.25}$$

Then, using Lemma (5.4) and the uniform estimate of Lemma (5.1) we obtain that the limit

$$\lim_{z\to 0}\sum_{n=0}^{\infty} II_{n+1}(\epsilon,z,t)$$

exists and is equal to

$$\lim_{z\to 0} z\int_0^{t/z^2} ds\int_{S/z^2-s}^{T/z^2-s} < S_u f, g_\varepsilon > du\cdot\sum_{n=1}^{\infty}(-i)^n\int_0^s ds_1\int_0^{s_1} ds_2\cdots\int_0^{s_{n-1}} ds_n$$

$$< D_\varepsilon^+ u\otimes\Phi_z(f,S,T),[1\otimes A(S_s f_{1-\varepsilon}),V(s_1)\cdots V(s_n)v\otimes\Phi_z(f',S',T') > \tag{5.26}$$

Summing the iterated series inside the commutator and with the change of variables $sz^2 = r$, we find that the limit (5.26) exists and is equal to

$$\lim_{z\to 0}\int_0^t dr\chi_{[S,T]}(r)(f,g_\varepsilon)du\cdot < D_\varepsilon^+ u\otimes\Phi_z(f,S,T),$$

$$\frac{1}{z}[1\otimes A(S_{r/z^2}f_{1-\varepsilon}),U(r/z^2)v\otimes\Phi_z(f',S',T') > \tag{5.27}$$

Clearly this limit is continuous in u (and in $D_\varepsilon^+ u$), hence it will have the form

$$< D_\varepsilon^+ u, K_\varepsilon(t) >=< u, D_\varepsilon K_\varepsilon(t) > \tag{5.28}$$

for some $K_\varepsilon(t)\in H_0$. Our next step shall be to deduce an equation for $D_\varepsilon K_\varepsilon(t)$.

<u>*LEMMA* (5.6)</u> Let $K_\varepsilon(t)$ and $G(t)$ be defined respectively by equations (5.27), (5.28) and (5.8a). Then $K_\varepsilon(t)$ satisfies the equation.

$$\int_0^t < D_\varepsilon^+ u, K_\varepsilon(s) > (f|g_\varepsilon)\chi_{[S,T]}(s)ds = \tag{5.29}$$

$$= \int_0^t ds(f|g_\varepsilon)\chi_{[S,T]}(s)(g_{1-\varepsilon}|g_{1-\varepsilon}) - < D_{1-\varepsilon}^+ D_\varepsilon^+ u, K_{1-\varepsilon}(s) >$$

$$+\int_0^t ds(f|g_\varepsilon)\chi_{[S,T]}(s)(g_{1-\varepsilon}|g_{1-\varepsilon}) - \chi_{[S',T']}(s)(g_\varepsilon|f') < D_{1-\varepsilon}^+ D_\varepsilon^+ u, +G(s) >$$

<u>*PROOF*</u> The left hand side of (5.29) can be written, using dominated convergence, (5.27) and Lemma (5.4), as:

$$\lim_{z\to 0}\int_0^t ds(f|g_\varepsilon)\chi_{[S,T]}(s)ds \quad \sum_{n=1}^{\infty}(-i)^n\int_0^{1/z^2} ds_1\cdots\int_0^{s_{n-1}} ds_n \tag{5.30}$$

$$< D_\varepsilon^+ u\otimes\Phi_z(f,S,T),\frac{1}{z}[1\otimes A(S_{1/z^2}g_{1-\varepsilon}),V(s_1)]\cdot V(s_2)\cdots V(s_n)v\otimes\Phi_z(f',S',T') >$$

Now notice that

$$[1 \otimes A(S_{s/z^2} g_{1-\varepsilon}), V(s_1)] = i(D_{1-\varepsilon} \otimes 1) < S_{s/z^2} g_{1-\varepsilon}, S_{s_1} g_{1-\varepsilon} > A(S_{s_1} g_\varepsilon)$$

therefore, bringing $A(S_{s_1} g_\varepsilon)$ to the right of the product $V(s_2) \cdot \ldots \cdot V(s_n)$, we obtain a commutator term plus a term coming from the action of $A(S_{s_1} g_\varepsilon)$ on the coherent vector $\Phi_z(f', S', T')$. Thus the limit (5.30) is equal to

$$\lim_{z \to 0} \int_0^t ds (f|g_\varepsilon) \chi_{[S,T]}(s) \sum_{n=1}^{\infty} (-i)^{n-1} \int_0^{s/z^2} ds \cdots \int_0^{s_{n-1}} ds_n \tag{5.31}$$

$$< D^+_{1-\varepsilon} D^+_\varepsilon u \otimes \Phi_z(f, S, T), \frac{1}{z} [1 \otimes A(S_{s_1} g_\varepsilon), V(s_2)] \cdot$$

$$\cdot V(s_3) \cdots V(s_n) v \otimes \Phi_z(f', S', T') >< S_{s/z^2} g_{1-\varepsilon}, S_{s_1} g_{1-\varepsilon} >< S_{s/z^2} g_{1-\varepsilon}, S_{s_1} g_{1-\varepsilon} > +$$

$$< D^+_{1-\varepsilon} D^+_\varepsilon$$

$$u \otimes \Phi_z(f, S, T), V(s_2) \cdots V(s_n) v \otimes \Phi_z(f', S', T') > \int_{S'/z^2}^{T'/z^2} du < S_1, g_\varepsilon, S_u f' > du$$

with the change of variable in the first term:

$$s_1 z^2 = r_1 \tag{5.32}$$

we see that the limit (5.31) is equal to

$$\lim_z \int_0^t ds (f|g_\varepsilon) \chi_{[S,T]}(s) \sum_{n=1}^{\infty} (-i)^{n-1} \int_0^s dr_1 \int_0^{r_1/z^2} \int_0^{s_2} ds_3 \cdots \int_0^{s_{n-1}} ds_n$$

$$\Big\{ \frac{1}{z^2} < g_{1-\varepsilon}, S_{(r_1-s)/z^2} g_{1-\varepsilon} > \cdot < D^+_{1-\varepsilon} D^+_\varepsilon u \otimes \Phi_z(f, S, T),$$

$$\frac{1}{z} \Big[1 \otimes A(S_{r_1/z^2} g_\varepsilon), V(s_2) \Big] V(s_3) \cdots V(s_n) \cdot v \otimes \Phi_z(f', S', T') > +$$

$$+ < D^+_{1-\varepsilon} D^+_\varepsilon u \otimes \Phi_z(f, S, T), V(s_2) \cdots V(s_n) v \otimes \Phi_z(f', S', T') >$$

$$\frac{1}{z^2} < g_{1-\varepsilon}, S_{(r_1-s)/z^2} g_{1-\varepsilon} > \int_{S'-r_1)/z^2}^{(r'-r_1)/z^2} dv < g_\varepsilon, S_v f' > \Big\} \tag{5.33}$$

The limit (5.33) is the sum of two terms. Concerning the first one, we notice that, because of uniform convergence we can exchange the series with the dr_1 integral and, with the change of variables we obtain the integral

$$\int_0^t ds (f|g_\varepsilon) \chi_{[S,T]}(s) \int_{-s/z^2}^0 ds_1 < g_{1-\varepsilon}, S_{s_1} g_{1-\varepsilon} >$$

$$\sum_{n=1}^{\infty} (-i)^{n-1} \int_0^{s_1+s/z^2} ds_2 \int_0^{s_2} ds_3 \cdots \int_0^{s_{n-1}} ds_n$$

$$< D^+_{1-\varepsilon} D^+_\varepsilon n \otimes \Phi_z(f,S,T), \frac{1}{z}\left[1 \otimes A(S_{s_1+s/z^2} g_\varepsilon), V(s_2)\right] V(s_3) \cdots V(s_2) v \otimes \Phi_z(f',S',T') > \tag{5.34}$$

From Lemma (6.3) of [3] one immediately deduces that, as $z \to 0$, the limit of the expression (5.34) is the same as

$$\lim_{z\to 0} \int_0^t ds (f|g_\varepsilon) \chi_{[S,T]}(s)(g_{1-\varepsilon}|g_{1-\varepsilon}) - \sum_{n=1}^{\infty} (-i)^{n-1} \int_0^{s/z^2} ds_1 \int_0^{s_1} ds_2 \cdots \int_0^{s_{n-2}} ds_{n-1}$$

$$< D^+_{1-\varepsilon} D^+_\varepsilon u \otimes \Phi_z(f,S,T), \frac{1}{z}\left[1 \otimes A(S_{s/z^2} g_\varepsilon), V(s_1)\right] V(s_2) \cdots V(s_{n-1}) v \otimes \Phi_z(f',S',T') > \tag{5.35}$$

By Lemma (5.4), the limit (5.35) is equal to

$$\lim_{z\to 0} \int_0^t ds (f|g_\varepsilon) \chi_{[S,T]}(s)(g_{1-\varepsilon}|g_{1-\varepsilon}) - \sum_{n=1}^{\infty} (-i)^{n-1}$$

$$\int_0^{s/z^2} ds_1 \int_0^{s_1} ds_2 \cdots \int_0^{s_{n-2}} ds_{n-1} < D^+_{1-\varepsilon} D^+_\varepsilon u \otimes \Phi_z(f,S,T),$$

$$\frac{1}{z}\left[1 \otimes A(S_{s/z^2} g_\varepsilon), V(s_1) V(s_2) \cdots V(s_{n-1})\right] v \otimes \Phi_z(f',S',T') > \tag{5.36}$$

Thus we can resume the iterated series inside the commutator obtaining that the limit of the expression (5.36) is equal to

$$\int_0^t ds (f|g_\varepsilon) \chi_{[S,T]}(s)(g_{1-\varepsilon}|g_{1-\varepsilon}) - \cdot$$

$$\cdot \lim_{z\to 0} < D^+_{1-\varepsilon} D^+_\varepsilon u \otimes \Phi_z(f,S,T), \frac{1}{z}\left[1 \otimes A(S_{s/z^2} g_\varepsilon), U_{s/z^2}\right] v \otimes \Phi_z(f',S',T') >$$

$$= \int_0^t ds (f|g_\varepsilon) \chi_{[S,T]}(s)(g_{1-\varepsilon}|g_{1-\varepsilon}) - < D^+_{1-\varepsilon} D^+_\varepsilon u, K_{1-\varepsilon}(s) >$$

Concerning the second term we notice that, with the same change of variable (5.32) and again using uniform convergence, it can be written in the form:

$$\int_0^t ds (f|g_\varepsilon) \chi_{[S,T]}(s) \int_{-s/z^2}^0 ds_1 < g_{1-\varepsilon}, S_{s_1} g_{1-\varepsilon} >$$

$$\sum_{n=1}^{\infty} (-i)^{n-1} \int_0^{s_1+s/z^2} ds_2 \int_0^{s_2} ds_3 \cdots \int_0^{s_{n-1}} ds_n$$

$$< D^+_{1-\varepsilon} D^+_\varepsilon u \otimes \Phi_z(f,S,T), V(s_2) \cdots V(s_n)$$

$$v \otimes \Phi_z(f',S',T') > \int_{(S'-z^2 s_1 - s)/z^2}^{(T'-z^2 s_1 - s)/z^2} dv < g_\varepsilon, S_v f' >$$

Again by Lemma (6.3) of [3], the limit of the above expression is equal to

$$\lim_{z\to 0}\int_0^t ds(f|g_\varepsilon)\chi_{[S,T]}(s)(g_{1-\varepsilon}|g_{1-\varepsilon})_- \sum_{n=1}^{\infty}(-i)^{n-1}\int_0^{s/z^2} ds_1\int_0^{s_1} ds_2\cdots\int_0^{s_{n-1}} ds_n$$

$$\cdot\chi_{[S',T']}(s)(g_\varepsilon|f') < D^+_{1-\varepsilon}D^+_\varepsilon u\otimes\Phi_z(f,S,T), V(s_1)\cdots V(s_n)v\otimes\Phi_z(f',S',T') >=$$

$$= \lim_{z\to 0}\int_0^t ds(f|g_\varepsilon)\chi_{[S,T]}(s)(g_{1-\varepsilon}|g_{1-\varepsilon})_-\cdot\chi_{[S',T']}(s)(g_\varepsilon|f')$$

$$< D^+_{1-\varepsilon}D^+_\varepsilon u\otimes\Phi_z(f,S,T), U_{s/z^2}v\otimes\Phi_z(f',S',T') > \tag{5.38}$$

and, because of the definition (5.8a) of $G(t)$ and of dominated convergence, this limit is equal to

$$\int_0^t ds(f|g_\varepsilon)\chi_{[S,T]}(s)(g_{1-\varepsilon}|g_{1-\varepsilon})_-\cdot\chi_{[S',T']}(s)\cdot(g_\varepsilon|f')\cdot < D^+_{1-\varepsilon}D^+_\varepsilon u, G(s) > \tag{5.39}$$

In conclusion from (5.33), (5.37), (5.38), we obtain that equation (5.29).

Now, by solving equation (5.29), we find the explicit form of $D_\varepsilon K_\varepsilon(t)$.

<u>*LEMMA* (5.7)</u> In the above notations, denoting for each $\varepsilon\in\{0,1\}$

$$D_g(\varepsilon) := (1-(g_\varepsilon|g_\varepsilon)_-(g_{1-\varepsilon}|g_{1-\varepsilon})_- D_\varepsilon D_{1-\varepsilon})^{-1}$$

$$= \sum_{n=0}^{\infty}(g_\varepsilon|g_\varepsilon)^n_-(g_{1-\varepsilon}|g_{1-\varepsilon})^n_-(D_\varepsilon D_{1-\varepsilon})^n \tag{5.39}$$

one has:

$$D_\varepsilon K_\varepsilon(t) = \chi_{[S',T']}(t)(g_{1-\varepsilon}|f')(g_\varepsilon|g_\varepsilon)_-(g_{1-\varepsilon}|g_{1-\varepsilon})_- D_g(\varepsilon)D_\varepsilon D_{1-\varepsilon}D_\varepsilon G(t)+$$

$$+\chi_{[S',T']}(t)(g_\varepsilon|f')(g_{1-\varepsilon}|g_{1-\varepsilon})_- D_g(\varepsilon)D_\varepsilon D_{1-\varepsilon}D_\varepsilon G(t) \tag{5.40}$$

<u>*PROOF*</u> First notice that, since f, S, T, u are arbitrary, then equation (5.29) is equivalent to

$$D_\varepsilon K_\varepsilon(t) = (g_{1-\varepsilon}|g_{1-\varepsilon})_- D_\varepsilon D_{1-\varepsilon}K_{1-\varepsilon}(t)+$$

$$+(g_{1-\varepsilon}|g_{1-\varepsilon})_-(g_\varepsilon|f')\chi_{[S',T']}(t)D_\varepsilon D_{1-\varepsilon}G(t) \tag{5.41}$$

Replacing $K_{1-\varepsilon}(t)$ by its expression (5.27) and with the notation

$$a_\varepsilon := (g_\varepsilon|g_\varepsilon)_- \tag{5.42}$$

we obtain

$$D_\varepsilon K_\varepsilon(t) = (a_\varepsilon a_{1-\varepsilon}D_\varepsilon D_{1-\varepsilon})D_\varepsilon K_\varepsilon(t)+$$

$$\Big\{(g_{1-\varepsilon}|f')\chi_{[S',T']}(s)(a_\varepsilon a_{1-\varepsilon}D_\varepsilon D_{1-\varepsilon})D_\varepsilon G(t)+$$

$$+(g_\varepsilon|f')\chi_{[S',T']}(s)a_{1-\varepsilon}D_\varepsilon D_{1-\varepsilon})D_\varepsilon G(t)\Big\} \tag{5.42}$$

Thus $D_\varepsilon K_\varepsilon(t)$ satisfies the operator equation

$$(1-T_\varepsilon)D_\varepsilon K_\varepsilon(t)=G_\varepsilon(t) \tag{5.43}$$

where, $G_\varepsilon(t)$ is the term in braces in (5.42) and

$$T\varepsilon := a_\varepsilon a_{1-\varepsilon} D_\varepsilon D_{1-\varepsilon} \tag{5.44}$$

Notice that

$$\|T_\varepsilon\| \leq \|g_0\|^2 \cdot \|g_1\|^2 \cdot \|D\|^2$$

which is less than (not equal to) 1 because of our assumption (5.0b). This implies that

$$D_\varepsilon K_\varepsilon(t)=\sum_{n=0}^{\infty} T_\varepsilon^n G_\varepsilon(t) \tag{5.45}$$

From this and (5.44), (5.40) immediately follows.

Up to now, one can get the following

<u>*THEOREM* (5.8)</u> The low density limit

$$<u, G(t)>:=\lim_{z\to 0} <u\otimes W(z\int_{S/z^2}^{T/z^2} S_u f du)\Phi, U_z(t/z^2)v\otimes W(z\int_{S'/z^2}^{T'/z^2} S_u f' du)\Phi> \tag{5.46}$$

satisfies the integral equation

$$<u,G(t)>=<u,G(0)>+\int_0^t ds \sum_{\varepsilon\in\{0,1\}} \Big(\sum_{n=1}^{\infty}(f|g_\varepsilon)\chi_{[S,T]}(s)\cdot(g_\varepsilon|f')\chi_{[S',T']}(s)$$
$$\cdot(g_{1-\varepsilon}|g_{1-\varepsilon})_-^n\cdot(g_\varepsilon|g_\varepsilon)_-^{n-1} <(D_{1-\varepsilon}^+D_\varepsilon^+)^n u, G(s)> +$$
$$+\sum_{n=1}^{\infty}(f|g_\varepsilon)\chi_{[S,T]}(s)\cdot(g_{1-\varepsilon}|f')\chi_{[S',T']}(s)$$
$$\cdot(g_{1-\varepsilon}|g_{1-\varepsilon})_-^{n-1}\cdot(g_\varepsilon|g_\varepsilon)_-^{n-1} <D_\varepsilon^+(D_{1-\varepsilon}^+D_\varepsilon^+)^{n-1}u, G(s)>\Big) \tag{5.47}$$

In the following , we shall use the notations

$$D_1(\varepsilon):=D_g(\varepsilon)D_\varepsilon \tag{5.48}$$

and

$$D_2(\varepsilon):=(g_{1-\varepsilon}|g_{1-\varepsilon})_-\cdot D_\varepsilon D_{1-\varepsilon}D_g(\varepsilon) \tag{5.49}$$

Then , (5.47) can be written as

$$<u,G(t)>=<u,G(0)>+\int_0^t ds \sum_{\varepsilon\in\{0,1\}} \Big((f|g_\varepsilon)\chi_{[S,T]}(s)\cdot(g_\varepsilon|f')\chi_{[S',T']}(s)$$
$$<D_2^+(\varepsilon)u, G(s)> +(f|g_\varepsilon)\chi_{[S,T]}(s)\cdot(g_{1-\varepsilon}|f')\chi_{[S',T']}(s) <D_1^+(\varepsilon)u, G(s)>\Big) \tag{5.50}$$

§6 . The Quantum Stochastic Differential Equation

In this Section we identify the integral equation (5.50), satisfied by the low density limit (5.46), with the weak form of the quantum stochastic differential equation

$$U(t) = 1 + \int_0^t \sum_{\varepsilon\in\{0,1\}} \Big[D_1(\varepsilon)\otimes dN_s(g_\varepsilon, g_{1-\varepsilon}) + D_2(\varepsilon)\otimes dN_s(g_\varepsilon, g_\varepsilon)\Big]U(s) \tag{6.1}$$

on $H_0 \otimes \Gamma(L^2(\mathbf{R})\otimes(K,(\cdot|\cdot)))$, where , N is number process and for each $g, g' \in K$,

$$N(g,g') := N_s(\chi_{[0,s]}\otimes |g><g'|) \tag{6.2}$$

Throughout the section we shall use freely the notations, definitions and results of the Hudson-Parthasarathy paper [18], with the only exception that we call "number process" the process called "gauge " in [18] and we denote it N rather than Λ.

Since D is bounded, it follows that the q.s.d.e. (6.1) has a unique solution $U(t)$ which is given by the iterated series [18]. Moreover, we have the following

THEOREM (6.1) The solution of q.s.d.e. (6.1) is unitary .

PROOF The proof consists in a lengthy direct calculation showing that the coefficients $D_j(\epsilon)$ whose explicit form is given by (5.34), (5.35), (5.36), satisfy the unitarity conditions of [18] (cf. [8] for the details).

We sum up our results in the following theorem

THEOREM (6.2) For each $f, f', g_0, g_1 \in K$, $u, v, \in H_0$, $D \in B(H_0)$ satisfying (1.11) , $S,T,S',T' \in \mathbf{R}$, $t \geq 0$, if $\|g_0\|_- \vee \|g_1\|_- < \frac{1}{16\|D\|}$ and $<g_0, S_t g_1> = 0$, $\forall t \geq 0$, then the low density limit

$$\lim_{z\to 0} < u\otimes W(z\int_{S/z^2}^{T/z^2} S_u f du)\Phi, U(t/z^2)v\otimes W(z\int_{S'/z^2}^{T'/z^2} S_u f' du)\Phi > \tag{6.3}$$

exists , where ,

$$\frac{d}{dt}U(t) = -iV(t)U(t)$$

and

$$V(t) = i\Big(D\otimes A^+(S_t g_0)A(S_t g_1) - D^+\otimes A^+(S_t g_1)A(S_t g_0)\Big)$$

.

Moreover , the low density limit (6.3) is equal to

$$< u\otimes W(\chi_{[S,T]}\otimes f)\Psi, U(t)v\otimes W(\chi_{[S',T']}\otimes f')\Psi > \tag{6.4}$$

where , $U(t)$ is the solution of q.s.d.e. (6.1) on $H_0\otimes\Gamma(L^2(\mathbf{R})\otimes(K,(\cdot|\cdot)))$ where

(i) Ψ is the vacuum of $\Gamma(L^2(\mathbf{R})\otimes(K,(\cdot|\cdot)))$;

(ii) for each $g, g' \in K$,$N_s(g,g')$ is the number process

$$N_s(g,g') := N(\chi_{[0,s]} \otimes |g><g'|)$$

(iii) $D_1(\varepsilon)$, $D_2(\varepsilon)$ are given by (5.48) ,(5.49) respectively .

PROOF The theorem (5.1) has shown that the low density limit (6.3) exists . Now , we shall prove that it is equal to (6.4) . Clearly , (6.4) is continuous or $u, v \in H_0$, so , one can write (6.4) into

$$< u, F(t) >$$

where , $F(t) \in H_0$. Hence , we have

$$< u, F(0) > = < u \otimes W(\chi_{[S,T]} \otimes f)\Psi, v \otimes W(\chi_{[S',T']} \otimes f')\Psi >$$

$$= < u, G(0) > \tag{6.5}$$

Moreover ,

$$< u, F(t) > = < u, F(0) > + \int_0^t \sum_{\varepsilon \in \{0,1\}} < u \otimes W(\chi_{[S,T]} \otimes f)\Psi,$$

$$\Big[D_1(\varepsilon) \otimes dN_s(g_\varepsilon, g_{1-\varepsilon}) + D_2(\varepsilon) \otimes dN_s(g_\varepsilon, g_\varepsilon)\Big] v \otimes W(\chi_{[S',T']} \otimes f')\Psi > U(s) \tag{6.6}$$

Apply the theorem (4.3) of [17] to (6.6) , one obtains

$$< u, F(t) > = < u, F(0) > + \int_0^t ds \sum_{\varepsilon \in \{0,1\}} \Big((f|g_\varepsilon)\chi_{[S,T]}(s) \cdot (g_\varepsilon|f')\chi_{[S',T']}(s)$$

$$< D_2^+(\varepsilon)u, F(s) > +$$

$$+ (f|g_\varepsilon)\chi_{[S,T]}(s) \cdot (g_{1-\varepsilon}|f')\chi_{[S',T']}(s) < D_1^+(\varepsilon)u, F(s) > \Big) \tag{6.7}$$

Since (6.1) has unique solution , one knows that (6.7) has a unique solution . Therefore

$$< u, F(t) > = < u, G(t) > , t \geq 0$$

REFERENCES

[1] Accardi L. On the quantum Feynmann-Kac formula. Rendiconti del seminario Matematico e Fisico, Milano 48(1978) 135-180.

[2] Accardi L. Noise and dissipation in quantum theory , submitted to Reviews of Modern Phys.

[3] L.Accardi , A.Frigerio , Lu YunGang : The weak coupling limit as the quantum functional cenrtal limit. to appear in Commu. Math. Phys.

[4] L.Accardi , A.Frigerio , Lu YunGang : On the weak coupling limit (II) , Preprint February, 1989

[5] L.Accardi , A.Frigerio , Lu YunGang : The weak coupling limit for Fermion case, Preprint March 1989

[6] L.Accardi , Lu YunGang : On the weak coupling limit (IV) , Preprint May 1989

[7] L.Accardi , Lu YunGang : On the weak coupling limit (V) , Preprint May 1989

[8] L.Accardi , Lu YunGang : The low density for Boson case , Preprint May 1989

[9] Alicki R., Frigerio A. Quantum Poisson noise and linear Boltzmann equation. preprint, March 1989

[10]Davies E.B., Marfovian master equation. Commun. Math. Phys. 39, 91-110 (1974)

[11] R.Dümcke : The low density limit for n - level systems . Springer LNM. 1136 .

[12] R.Dümcke : The low density limit for N- level system interacting with a Free Boson or Fermi gas. to appear in Commu. Math. Phys.

[13] F. Fagnola: A martingale characterization of quantum Poisson process. preprint February, 1989

[14] Frigerio, A. Quantum Poisson processes: physical motivations and applications. Lect. Math. vol. 1303(1988), 107-127

[15] A.Frigerio , H.Maassen : Quantum Poisson processes and dilations of dynamical semigroups . to appear .

[16] H.Grad : Principles of the kinetic theory of Gases . Handbuch der Physik , vol . 12 , Springer (1958).

[17] R.L.Hudson , M.Lindsay : Uses of non-Fockquantum Borwnian motion and a quantum martingale representation theorem. 276 - 305 . 1136 .

[18] R.L.Hudson , K.R.Parthasarathy : Quantum Ito's formula and quantum stachastic evolutions. Commu . Math . Phys . (93) , 301 - 323 (1984) .

[19] P.F.Palmer : Ph. D. Thesis , Oxford University .

[20] Pulé J.V., The Bloch equation. Commu. Math. Phys. 38, 241-256 (1974)

QUANTUM STOCHASTIC FLOWS AND NON ABELIAN COHOMOLOGY

L. Accardi
Dipartimento di Matematica
Centro Matematico V.Volterra
Universita' di Roma II, Roma, Italy

R.L.Hudson
Department of Mathematics
University of Nottingham
Nottingham NG72RO, UK

Abstract

The notion of (quantum) stochastic flow is introduced. The analysis of their infinitesimal generators leads to the introduction of the notion of stochastic derivation. The Lue non-abelian first cohomology space for stochastic derivations plays the analogue role of the first Hochshild cohomology group for usual derivations (i.e. it gives an idea of how large the space of inner (stochastic) derivations is with respect to the space of all derivations) . We prove an addition theorem for stochastic derivations and an existence theorem for stochastic flows with bounded coefficients.

(1.) Introduction

In [10], [11], [6], [7],[8], [9] the classical notion of diffusion on a manifold was generalized to a quantum probabilistic framework and it was proved that certain cohomological obstructions arise for a quantum diffusion to be inner. In [1], [5] a generalization of the usual quantum dynamics was proposed, based on the assumption that the scattering operator of a quantum system at each time t admits a forward derivative, rather than a derivative in the usual sense. In the present paper, the two programmes are merged by considering scattering automorphisms admitting a forward derivative and showing that this leads to a natural extension of the cohomological relations of [10]. In Section (3.) we introduce the notion of **stochastic derivation** on a $*$-algebra and a (nonlinear) composition law among stochastic derivations. Then we show that, just as the first Hochschild cohomology group describes the obstructions for a usual derivation to be inner, the first non abelian cohomology space in the sense of Lue [12], [13] describes the obstructions preventing a stochastic derivation from being inner.

In the following, $\mathcal{A}$ denotes a topological $*$-algebra of operators acting on a Hilbert space $\mathcal{H}$ and defined on a common invariant dense domain $D \subseteq \mathcal{H}$. By a derivation (respectively, automorphism) of $\mathcal{A}$ we mean a $*$-derivation (respectively, $*$-automorphism). The set of linear maps $\mathcal{A} \to \mathcal{A}$ will be denoted $\mathcal{L}(\mathcal{A})$ and composition between two elements of $\mathcal{L}(\mathcal{A})$ will be denoted $\circ$. With this operation $\mathcal{L}(\mathcal{A})$ is an associative algebra. We shall consider $\mathcal{L}(\mathcal{A})$–valued finitely additive measures on $\mathbf{R}$ which we shall simply call $\mathcal{L}(\mathcal{A})$–valued measures. If L is such a measure, j_s is a 1-parameter family in $\mathcal{L}(\mathcal{A})$ and $(x_s), (y_s)$ 1-parameter families of operators in $\mathcal{A}$, expressions like

$$\int_0^t j_s(dL_s(x_s)), \int_0^t < j_s(dL_s(x_s))\xi, y_s\eta >$$

($\xi, \eta \in \mathcal{H}$) will be meant as limits of the corresponding Riemann sums. In the present paper we shall assume that all the integrals we are dealing with exist and we shall neglect domain problems. Such questions, as well as the problem giving a meaning to the stochastic calculus in a representation free context, are dealt with in [3], [4] . We also refer to these papers for all the notations concerning covariant projective families of conditional expectations and the associated notion of forward derivative.

If L, M are $\mathcal{L}(\mathcal{A})$–valued measures their bracket (or mutual quadratic variation) $[[L, M]]$ is the $\mathcal{L}(\mathcal{A})$–valued measure defined as in [4] with respect to the multiplication in $\mathcal{L}(\mathcal{A})$, i.e.

$$[[L, M]](s,t) = \lim_{|P(s,t)|\to 0} \sum_{(t_k,t_{k+1})\in P(s,t)} L_{(t_k,t_{k+1})} \circ M_{(t_k,t_{k+1})}$$

where the limit is meant in the pointwise topology on $\mathcal{A}, P(s,t)$ denotes a finite partition of the interval $[s,t]$ and $|P(s,t)|$ is the maximum lenght of the intervals in the partition. Frequently we shall use the notation

$$[[L, M]](t, t+dt) = dL_t \circ dM_t$$

$(\mathcal{A}_{t]})$ will denote a past filtration in $(\mathcal{A})$ equipped with a projective family of conditional expectations $E_{t]}) : (\mathcal{A}) \to (\mathcal{A}_{t]})$. A process is a map $F : \mathbf{R} \to \mathcal{A}$ such that, for each $t \in \mathbf{R}$, $F_t \in \mathcal{A}_{t]}$. Thus, in our notation, process means $(\mathcal{A}_{t]})$–adapted process.

An $\mathcal{A}$–valued (resp. $\mathcal{L}(\mathcal{A})$–valued) function $F : \mathbf{R}^2 \to \mathcal{A}$ (resp. $F : \mathbf{R}^2 \to \mathcal{L}(\mathcal{A})$) will be said to be $o(dt)$ if

$$\lim_{|P(s,t)|\to 0} \sum_{(t_k,t_{k+1})\in P(s,t)} F(t_k, t_{k+1} - t_k) = 0$$

in the topology of $\mathcal{A}$ (resp. $\mathcal{L}(\mathcal{A})$ where the latter is equipped with the topology of pointwise convergence. In this case we write

$$F \equiv 0$$

If F, G are two such functions, we write $F \equiv G$ if $F - G \equiv 0$.

__*ACKNOWLEDGMENTS*__ The authors are grateful to prof. P.A.Meyer for the invitation to Strasbourg, where this paper was begun in the Autumn of 1987. The first named author also acknowledges support from grant AFOSR 870249 and ONR N00014–86–K–0538 through the Center for Mathematical System theory, University of Flórida.

(2.) Deterministic flows

Let us first state, in the deterministic case, the problems we are going to generalize, the next two Sections, to the stochastic case. The following considerations are valid for the classical as well as for the quantum case.

In the notations of Section (1.), let u_t, u_t^o be 1-parameter automorphisms groups of $\mathcal{A}$ and let

$$u_t = e^{t\delta} \quad ; \quad u_t^o = e^{t\delta_0} \tag{2.1}$$

The equations (2.1) mean that, for each $x \in \mathcal{A}$

$$\frac{d}{dt}u_t(x) = u_t(\delta(x)) \ , \ u_o(x) = x \quad ; \quad \frac{d}{dt}u_t^o(x) = u_t^o(\delta_0(x)) \ , \ u_o^o(x) = x \tag{2.2}$$

and, following the notation introduced in Section (1.), we write these equations in the form

$$du_t(x) \equiv u_t(\delta(x))dt \quad ; \quad du_t^o(x) \equiv u_t^o(\delta_0(x))dt \tag{2.3}$$

Notice that, introducing the $\mathcal{A}$-valued measures on $\mathbf{R}$

$$[s,t) \to u_t(x) - u_s(x) = u_{(s,t)}(x) \in \mathcal{A} \quad ; \quad x \in \mathcal{A} \tag{2.4}$$

(and similarly for $u_{(s,t)}^o$), the identities (2.3) are equivalent to identities among $\mathcal{A}$-valued measures, i.e.

$$u_{(s,t)}(x) = \int_s^t u_r(\delta(x))dr \quad ; \quad u_{(s,t)}^o(x) = \int_s^t u_r^o(\delta_0(x))dr \tag{2.5}$$

The 1-parameter family

$$j_t := u_t \circ u_{-t}^o \tag{2.6}$$

satisfies the cocycle equation

$$j_{s+t} = j_s \circ u_s^o \circ j_t \circ u_{-s}^o \tag{2.7}$$

and, by analogy with the terminology used in the physical literature, is called an **interaction cocycle** or a family of **wave automorphisms** t (with respect to the **free evolution** u_t^o). Conversely, if j_t satisfies the cocycle equation (2.7), then u_t, defined by

$$u_t = j_t \circ u_t^o \tag{2.8}$$

is a 1-parameter group of automorphisms of $\mathcal{A}$.

Because of the cocycle equation (2.7), the two-parameter family $j_{(s,t)}$ defined by:

$$j(0,s) = j_s \quad ; \quad j(s, s+t) = u_s^o \circ j_t \circ u_{-s}^o \tag{2.9}$$

satisfies the evolution equation

$$j(r,s) \circ j(s,t) = j(r,t) \quad ; \quad r < s < t \tag{2.9a}$$

By analogy with the terminology used in the probabilistic literature, the family $j(s,t)$ is called a **homogeneous stochastic flow** or an **automorphism valued multiplicative functional** (with respect to the time shift u_t°). Here the term homogeneous refers to the property

$$u_r^\circ \circ j(s,t) \circ u_{-r}^\circ = j(s+r, t+r) \tag{2.10}$$

A simple algebraic computation leads to the identity

$$dj_t(x) \equiv j_t(l_t(x))dt = j_t \circ u_{-t}^\circ \circ [l_0 dt] \circ u_t^\circ(x) \tag{2.11}$$

with

$$l_t(x) := u_t^\circ \circ (\delta - \delta_0) \circ u_{-t}^\circ(x) =: (\delta_t - \delta_0)(x) \tag{2.12}$$

or, in terms of $\mathcal{A}$-valued measures,

$$j_{(s,t)}(x) = \int_s^t j_r(l_r(x))dr \tag{2.13}$$

Notice that l_t is a self-adjoint derivation on $\mathcal{A}$ for each t. Moreover

$$u_s^\circ \circ l_t \circ u_{-s}^\circ = l_{t+s} \tag{2.14}$$

Conversely, if (l_t) is a 1-parameter family of derivations on $\mathcal{A}$ satisfying (2.14) then, if $\mathcal{A}$ is a C^*-algebra and l_0 is bounded in norm on $\mathcal{A}$, one has, for each $t \in \mathbf{R}$

$$\| l_t \| \leq \| l_0 \| \tag{2.15}$$

Theorem (2.1). Let A be a C^*-algebra and let (l_t) be a norm continuous 1-parameter family of bounded derivations on $\mathcal{A}$ satisfying (2.14) and (2.15). Then, for each $x \varepsilon \mathcal{A}$, the equation

$$j_t(x) = x + \int_0^t j_s(l_s(x))ds \quad ; \quad x \in \mathcal{A} \tag{2.16}$$

has a unique solution. Moreover each j_t is an automorphism of A.

Proof. Under our assumptions, the iterated series

$$j_t(x) = x + \sum_{n=1}^{\infty} \int_0^t dt_1 \int_0^{t_1} dt_2 \ldots \int_0^{t_{n-1}} dt_n l_{t_n} \circ l_{t_{n-1}} \circ \ldots \circ l_{t_1}(x) \tag{2.17}$$

converges uniformly in the pair (t,x) if both t and $\| x \|$ vary in a bounded set. Moreover one has the estimate

$$\| j_s(x) \| \leq C_1 \| x \| \quad ; \quad s \leq t \tag{2.18}$$

$$\| j_{dt}(x) - (x + \int_0^{dt} l_s(x)ds) \| \leq C_2 \| x \| dt^2. \tag{2.19}$$

From this it follows that

$$\|j_{dt}(x)j_{dt}(y) - j_{dt}(xy)\| \le$$

$$C_3\|x\|\|y\|dt^2 + \left\|(x + \int_0^{dt} l_s(x)ds)(y + \int_0^{dt} l_s(y)ds) - j_{dt}(xy)\right\| \le$$

$$\le C_4\|x\|\|y\|dt^2 + \left\|xy + \int_0^{dt} l_s(xy)ds - j_{dt}(xy)\right\| \le C_5\|x\|\|y\|dt^2 \tag{2.20}$$

The convergence of the series (2.17) implies that the equation

$$X_t(x) - x = \int_0^t X_r(l_r(x))dr$$

has a unique solution. In particular, for each $s \in \mathbf{R}$, the map $t \to j_{t+s}(x)$ is the unique solution of

$$X_t(x) - j_s(x) = \int_s^t dt_1 X_{t_1}(l_{t_1+s}(x))$$

Now given $s, t \in$ and denoting, for $x \in \mathcal{A}$:

$$Y_t(x) := j_s \circ u_s^\circ \circ j_t \circ u_s^\circ(x)$$

one has, due to (2.15) and (2.14)

$$dY_t(x) \equiv j_s \circ u_s^\circ \circ j_t(l_t(u_s^\circ(x)))dt = j_s \circ u_s^\circ \circ j_t \circ u_s^\circ(l_{t+s}(x))dt = X_t(l_{t+s}(x))dt$$

Uniqueness then implies

$$X_t(x) = j_{s+t}(x) = Y_t(x) = j_s \circ u_s^\circ \circ j_t \circ u_s^\circ(x)$$

Thus j_t is a u_t° cocycle.

Now let $T < +\infty$ be fixed and let $dt = T/N$ (N - a natural integer). For $n = 0, 1, 2, \ldots, N$ denote

$$T_n(x) = u_{ndt}^\circ \circ j_{dt} \circ u_{-ndt}^\circ(x) \quad ; \quad x \in \mathcal{A} \tag{2.21}$$

The cocycle property implies that

$$\begin{aligned} T_{n-1} \circ T_n(x) &= u_{-(n-1)dt}^\circ \circ j_{dt} \circ (u_{-dt}^\circ \circ j_{dt} \circ u_{dt}^\circ) \circ u_{(n-1)dt}^\circ(x) \\ &= u_{-(n-1)dt}^\circ \circ j_{2dt} \circ u_{(n-1)dt}^\circ(x) \end{aligned}$$

Thus

$$T_0 \circ T_1 \circ T_2 \ldots \circ T_{N-1}(x) = j_{Ndt}(x) = j_T(x) \tag{2.22}$$

From the cocycle property it also follows that:

$$j_{Ndt}(xy) = j_{(N-1)dt}\left[u_{-(N-1)dt}^\circ \circ j_{dt} \circ u_{(N-1)dt}^\circ(xy)\right]$$

therefore, from the estimate (2.20)

$$j_{Ndt}(xy) = j_{(N-1)dt}(T_{N-1}(x) \cdot T_{N-1}(y)) + O(dt^2)$$

Iterating this identity we find

$$\begin{aligned} j_T(xy) = j_{Ndt}(xy) &= [T_0 \circ T_1 \ldots \circ T_{N-1}(x)] \cdot [T_0 \circ T_1 \circ \ldots \circ T_{N-1}(y)] + NO(dt^2) \\ &= j_T(x) \circ j_T(y) + NO(T^2/N^2) \end{aligned}$$

Letting $N \to \infty$, we obtain $j_T(xy) = j_T(x)j_T(y)$. The property $j_t(x^*) = j_t(x)^*$ follows immediately from the corresponding property for $l_s(x)$ and the convergence of the iteration series.

Remark If j_t, j_t^o are u_t^o-cocycles then in general $j_t \circ j_t^o$ will not be a u_t^o-cocycle. However, if j_t^o is a u_t^o-cocycle and j_t is a u_t-cocycle, with $u_t = j_t^o \circ u_t^o$, then $j_t \circ j_t^o$ is a u_t^o-cocycle. Moreover, if (l_t^o), is the generator of j_t^o and (l_t) the generator of j_t, the generator of $j_t \circ j_t^o$ is

$$\kappa_t = \tilde{l}_t + l_t^o \quad ; \quad \tilde{l}_t = (j_t^o)^{-1} \circ l_t \circ j_t \tag{2.23}$$

Denoting in (2.12)

$$\delta_I := l_0 = \delta - \delta_0$$

one immediately checks that u_t, defined by (2.8), satisfies:

$$du_t(x) \equiv u_t([\delta_0 + \delta_I](x))dt \tag{2.24}$$

so that the identity (2.8) is the integrated version of the addition formula for the generators

$$\delta = \delta_0 + \delta_I$$

We shall see in Section (4) that the corresponding addition formula for stochastic evolutions takes the same integrated form, but involves a quadratic formula at the level of generators.

(3.) Stochastic derivations

Let $u_t^o : \mathcal{A} \to \mathcal{A} (t \in \Re)$ be a 1-parameter group of automorphisms and let $j_t : \mathcal{A} \to \mathcal{A}$ be an u_t^o-cocycle; thus each j_t is an automorphism and

$$j_{s+t} = j_s \circ u_s^o \circ j_t \circ u_{-s}^o \tag{3.1}$$

Assume that the **forward derivative**

$$\lim_{\varepsilon \to 0+} E_{t]}\left(\frac{j_{t+\varepsilon}(x) - j_t(x)}{\varepsilon}\right) = b_t(x) \tag{3.2}$$

exists in $\mathcal{A}$ for each $t \in \mathbf{R}_+$ and $x \in \mathcal{A}$. Then from [1] one knows that

$$M_{(st)}(x) = j_t(x) - j_s(x) - \int_s^t b_r(x)dr \tag{3.3}$$

is a difference martingale. Clearly, for $r < s < t$

$$M_{(r,s)} + M_{(s,t)} = M_{(r,t)} \tag{3.3a}$$

Let us assume that $E_{t]}$ is u_t^o-covariant, that is

$$u_s^o \cdot E_{t]} = E_{t+s]} \cdot u_s^o$$

Therefore the cocycle property and the fact that j_t is adapted, imply:

$$b_t(x) = \lim_{\varepsilon \to 0^+} E_{t]}(j_t \circ u_t^o \left[\frac{j_\varepsilon(u_{-t}^o x) - u_{-t}^o x}{\varepsilon}\right]) = \lim_{\varepsilon \to 0^+} j_t(E_{t]} \circ u_t^o \left[\frac{j_\varepsilon(u_{-t}^o x) - u_{-t}^o x}{\varepsilon}\right])$$

$$= \lim_{\varepsilon\to 0+} j_t \circ u_t^o \circ E_{0]}\left(\frac{j_\varepsilon(u_{-t}^o x) - u_{-t}^o x}{\varepsilon}\right) = j_t \circ u_t^o \circ b^o(u_{-t}^o x) = j_t(b_t^o(x)) \tag{3.4}$$

where we have introduced the notations:

$$b^o(y) = \lim_{\varepsilon\to 0+} E_{0]}\left(\frac{j_\varepsilon(y) - y}{\varepsilon}\right) \tag{3.5}$$

$$b_t^o(x) = u_t^o \circ b^o \circ u_{-t}^o(x) \tag{3.6}$$

With these notations (3.3) becomes:

$$M_{(s,t)}(x) = j_t(x) - j_s(x) - \int_s^t j_r(b_r^o(x))dr \tag{3.7}$$

On the other hand

$$j_t(x) - j_s(x) = j_s \circ u_s^o \circ j_{t-s} \circ u_{-s}^o(x) - j_s(x) = j_s \circ u_s^o \left[j_{t-s}(u_{-s}^o x) - u_{-s}^o x\right] =$$

$$= j_s \circ u_s^o \left[\int_0^{t-s} j_r(b_r^o u_{-s}^o x)dr = M_{(o,t-s)}(u_{-s}^o x)\right] =$$

$$= \int_0^{t-s} j_s \circ u_s^o \circ j_r \circ u_r^o \circ b^o \circ u_{-(r+s)}^o(x)dr + j_s \circ u_s^o \circ M_{(o,t-s)}(u_{-s}^o x) =$$

$$= \int_0^{t-s} j_{s+r}(b_{s+r}^o(x))dr + j_s \circ u_s^o \circ M_{(o,t-s)} \circ u_{-s}^o(x)$$

from which one deduces that

$$M_{(s,t)}(x) = j_s \circ u_s^o \circ M(0,t-s) \circ u_{-s}^o(x) \tag{3.8}$$

and therefore

$$j_t(x) - j_s(x) = \int_s^t j_r(b_r^o(x))dr + j_s \circ u_s^o \circ M_{(o,t-s)} \circ u_s^o(x) \tag{3.9}$$

From (3.7) and (3.8) it follows that the limit

$$\lim_{|P(s,t)|\to 0} \sum_{t_k\in P(s,t)} j_{tk} \circ u_{t_k}^o \circ M_{(o,t_{k+1}-t_k)} \circ u_{-t_k}^o(x) \tag{3.10}$$

exists in a trivial way and is equal to $M(x)(s,t)$. Denoting

$$\int_s^t j_r(dM_r(x)) \tag{3.10a}$$

this limit, which corresponds to the symbolic notation

$$dM_t(x) = u_t^o \circ M_{(o,dt)} \circ u_{-t}^o(x) \quad ; \quad t, dt > 0 \tag{3.11}$$

equation (3.9) can be written

$$j_t(x) - j_s(x) = \int_s^t j_r\left(b_r^\circ(x)dr + dM_r(x)\right) \tag{3.12}$$

The notation (3.10a) is symbolic because $dM_t(x)$ need not be an $\mathcal{A}$-valued measure, only

$$j_t(dM_t(x)) = M_{(t,t,+dt)}(x) \tag{3.13}$$

is such.

<u>*EXAMPLES*</u> In all the examples known up to now $dM_t(x)$ is a measure or at least differs from a measure by terms of order $o(dt)$. The typical situation is the following:

$$M_{(o,t)}(x) \equiv l^\alpha(x)M_\alpha(o,t) \tag{3.14}$$

where $\alpha = q,\dots,d \in \mathbf{N}$, summation on dummy indices is understood in (3.14), the $l^\alpha : \mathcal{A} \to \mathcal{A}$ are linear maps satisfying

$$l^\alpha(\mathcal{A}_{0]}) \subseteq \mathcal{A}_{0]} \tag{3.15}$$

and each $M_\alpha(s,t)$ is a difference martingale in $\mathcal{A}$ commuting with the past, i.e.

$$[\mathcal{A}_{s]}, M_\alpha(s,t)] = 0 \tag{3.16}$$

and such that (localization):

$$j_s(M_\alpha(s,t)) = M_\alpha(s,t) \quad ; \quad \forall\alpha\ ,\ \forall s < t \tag{3.17}$$

(if j_s is inner, then (3.17) is a consequence of (3.16)). Moreover (shift-covariance)

$$u_r^\circ(M_\alpha(s,t)) = M_\alpha(s+r,t+r) \tag{3.18}$$

Under these conditions, and with the notation

$$l_t^\alpha(x) = u_t^\circ \circ l^\alpha \circ u_{-t}^\circ(x) \tag{3.19}$$

we obtain

$$\begin{aligned} dM_t(x) &= \left[u_t^\circ \circ l^\alpha \circ u_{-t}^\circ(x)\right] M_\alpha(t,t+dt) = l_t^\alpha(x)M_\alpha(t,t+dt) \\ &\equiv \int_t^{t+dt} l_r^\alpha(x)dM_\alpha(r) \end{aligned} \tag{3.20}$$

which is a measure and

$$j_t(dM_t(x)) \equiv j_t(l_t^\alpha(x))M_\alpha(t,t+dt) \tag{3.21}$$

By continuity (3.21) implies that

$$dj_t(x) \equiv j_t(dL_t(x)) \tag{3.22}$$

where

$$dL_t(x) = u_t^\circ \circ L_{(0,dt)} \circ u_{-t}^\circ(x) = L_{(t,t+dt)}(x) \tag{3.23}$$

$$L_{(0,dt)}(x) = b^\circ(x)dt + M_{(0,dt)}(x) \tag{3.24}$$

or equivalently:

$$j_t(x) - j_s(x) \equiv \int_s^t j_r(dL_r(x)) \tag{3.25}$$

Now, since each j_t is an automorphism of $\mathcal{A}$, it follows that, for each $x, y \in \mathcal{A}$, one has

$$j_t(x)^* = j_t(x^*) \quad ; \quad j_t(xy) = j_t(x)j_t(y) \tag{3.26}$$

The first of these relations implies

$$j_t(dL_t(x^*)) \equiv dj_t(x^*) = (dj_t(x))^* \equiv (j_t(dL_t(x)))^* = j_t(dL_t(x)^*)$$

and, since j_t is invertible, this implies that

$$dL_t(x^*) \equiv dL_t(x)^* \tag{3,27}$$

or equivalently:

$$L_{(0,dt)}(x)^* \equiv L_{(0,dt)}(x^*) \quad ; \quad \forall dt \geq 0 \tag{3.28a}$$

The second relation of (3.26) implies that

$$\begin{aligned} j_t(dL_t(xy)) &\equiv dj_t(xy) \equiv (dj_t(x))j_t(y) + j_t(x)dj_t(y) + dj_t(x)dj_t(y) \\ &\equiv j_t(dL_t(x)y) + j_t(xdL_t(y)) + j_t(dL_t(x)dL_t(y)) \end{aligned} \tag{3.28b}$$

so that

$$dL_t(xy) \equiv dL_t(x)y + xdL_t(y) + dL_t(x)dL_t(y) \tag{3.29}$$

or, using (3.23) and the arbitrariness of $t, dt \geq 0, x, y \in \mathcal{A}$

$$L_{[0,dt]}(xy) \equiv L_{[0,dt]}(x)y + xL_{[0,dt]}(y) + L_{[0,dt]}(x)L_{[0,dt]}(y) \tag{3.30}$$

Definition (3.1). Let $\mathcal{A}$ be a topological $*$-algebra and denote $\mathcal{L}(\mathcal{A})$ the family of linear continuous maps of $\mathcal{A}$ into itself. A continuous stochastic $*$-derivation on $\mathcal{A}$ is an $\mathcal{L}(\mathcal{A})$-valued measure on $\mathbf{R}$:

$$L : (s,t) \subseteq \mathbf{R} \longrightarrow L_{(s,t)} \in \mathcal{L}(\mathcal{A})$$

satisfying the conditions (3.28), (3.30).

Remark. Notice that both conditions (3.28), (3.30) concern the germ of the measure L at zero, rather than the whole measure L. It is this germ that should be considered the true analogue of the deterministic generator $\delta - \delta_o$ (cf. (2.12)) . By analogy with the deterministic case, we call this germ, the (infinitesimal) generator of j_t. In the following, when speaking of the brackets of two stochastic derivations, we shall always mean the germ of these brackets at zero.

Summing the relation (3.29) over the intervals of a partition of (s,t) we obtain that or each $(s,t) \subseteq \mathbf{R}$, the limit

$$\lim_{|P(s,t,)|\to 0} \sum_{dt\in P(s,t,)} dL_t(x)dL_t(y) = \lim_{|P(s,t)|\to 0} \sum_{t_j\in P(s,t)} dL_t(x)dL_t(y) = \\ =: [[L(x), L(y)]]\,(s,t) \tag{3.31}$$

exists and is equal to

$$\int_s^t dL_r(xy) - \int_s^t dL_r(x)y - \int_s^t x\,dL_r(y) \tag{3.32}$$

Now let us exploit the specific form (3.24) of the $\mathcal{A}$-valued measure $L_{(s,t)}$. Using this equation (3.29) becomes

$$b^o(xy)dt + M_{(0,dt)}(xy) \equiv [b^o(x)y + xb^o(y)]\,dt + \\ + xM_{(0,dt)}(y) + M_{(0,dt)}(x)y + M_{(0,dt)}(x)M_{(0,dt)}(y) \tag{3.33}$$

Since $M_{(s,t)}(z)$ is a difference martingale, this implies

$$b^o(xy) - b^o(x)y - xb^o(y) \equiv E_{0]}\Big(\frac{M_{(0,dt)}(x)M_{(0,dt)}(y)}{dt}\Big) \tag{3.34}$$

and since the left hand side of (3.34) is independent of dt, it follows that the limit

$$\sigma(x,y) := \lim_{dt\to 0} E_{0]}\Big(\frac{M_{(0,dt)}(x)M_{(0,dt)}(y)}{dt}\Big) \tag{3.35}$$

exists and the equality

$$b^o(x,y) - b^o(x)y - xb^o(y) = \sigma(x,y) \tag{3.36}$$

holds. Using this equality in (3.34), we obtain

$$M_{(0,dt)}(xy) - M_{(0,dt)}(x)y - xM_{(0,dt)}(y) \equiv M_{(0,dt)}(x)M_{(0,dt)}(y) - \sigma(x,y)dt \tag{3.37}$$

Going back to (3.29) and with the notations

$$dM_{t_j}(z) = u^o_{t_j} \circ M_{(0,dt)} \circ u^o_{-t_j}(z) \tag{3.38}$$

for any partition $s = t_1 < t_2 < \ldots < t_n = t$ of the interval $(0,t)$

$$\sigma_t(x,y) = u^o_t \circ \sigma(u^o_{-t}(x), u^o_{-t}(y)) \tag{3.39}$$

we conclude, using (3.36), that the limit

$$\lim_{|P(s,t)|\to 0} \sum_{t_j\in P(s,t)} dM_{t_j}(x)dM_{t_j}(y) =: [[M(x), M(y)]](s,t) \tag{3.40}$$

exists and is equal to

$$\int_s^t dM_r(x,y) - \int_s^t dM_r(x)y - \int_s^t x dM_r(y) - \int_s^t \sigma_r(x,y)dr \tag{3.41}$$

(4.) The addition formula

The sum of two stochastic derivations is not, in general, a stochastic derivative. As shown by the following theorem, the correct composition law between two stochastic derivations is quadratic in the components.

Theorem (4.1) (Addition formula) Let L, M be stochastic derivations on $\mathcal{A}$. Then, if the bracket $[[L, M]]$ is defined using composition in $\mathcal{L}(\mathcal{A})$ (and not pointwise in $\mathcal{A}$),

$$K := L + M + [[L, M]] \tag{4.1}$$

is a stochastic derivation on $\mathcal{A}$.

Proof Let $x, y \in \mathcal{A}$. Notice that in the dL-notation, (4.1) becomes

$$dK = dL + dM + dL \circ dM \tag{4.2}$$

Using this notation and expanding $dK(x \cdot y)$ we find

$$\begin{aligned} dK(x\cdot y) = {} & dL(x)y + xdL(y) + dL(x)\cdot dL(y) + dM(x)y + xdM(y) + dM(x)\cdot dM(y) + \\ & +(dL\circ dM)(x)\cdot y + dM(x)\cdot dM(y) + (dL\circ dM)(x)\cdot dL(y) + \\ & +dL(x)\cdot dM(y) + x\cdot(dL\circ dM)(y) + dL(x)\cdot(dL\circ dM)(y) + \\ & +dM(x)\cdot dM(y) + dM(x)\cdot(dL\circ dM)(y) + (dL\circ dM)(x)\cdot(dL\circ dM)(y) \end{aligned} \tag{4.3}$$

On the other hand, expanding the expression

$$dK(x)y + xdK(y) + dK(x)\cdot dK(y)$$

using the right hand side of (4.2), we find

$$\begin{aligned} & dL(x)y + dM(x)y + (dL\circ dM)(x)y + \\ & +xdL(y) + xdM(y) + x(dL\circ dM)(y) + \\ & +dL(x)dL(y) + dM(x)dM(y) + dM(x)(dL\circ dM)(y) + \\ & +dL(x)dM(y) + dM(x)dL(y) + dL(x)(dL\circ dM)(y) + \\ & + (dL\circ dM)(x)dL(y) + (dL\circ dM)(x)dM(y) + (dL\circ dM)(x)(dL\circ dM)(y) \end{aligned} \tag{4.4}$$

Comparing one by one the fifteen terms on the right hand side of (4.3) with those in the expression (4.4), we see that equality holds.

Finally, K commutes with the involution in $\mathcal{A}$ because both L and M do.

Remark. The origin of the addition formula (4.1) can be understood in terms of perturbations of perturbations of stochastic evolutions as follows: let j_t^o be an u_t^o - cocycle satisfying the equation

$$dj_t^o(x) \equiv j_t^o(dL_t(x)) \quad ; \quad x \in \mathcal{A} \tag{4.5}$$

Let u_t be the 1 - parameter automorphism group defined by

$$u_t = j_t^o \cdot u_t^o \tag{4.6}$$

and let j_t be an u_t - cocycle satisfying the equation

$$dj_t(x) \equiv j_t(dM_t(x)) \quad ; \quad x \in \mathcal{A} \tag{4.7}$$

Then $j_t \cdot j_t^o =: \kappa_t$ is an u_t^o - cocycle satisfying the equation

$$d\kappa_t(x) \equiv \kappa_t(dL_t(x) + d\widetilde{M}_t(x) + (d\widetilde{M}_t \circ dL_t)(x)) \tag{4.8}$$

with

$$d\widetilde{M}_t = (j_t^o)^{-1} \cdot dM_t \cdot j_t^o \tag{4.9}$$

The equations (4.8), (4.9) are the stochastic generalizations of (2.23). Looking at these equations in a neighborhood of $t = 0$, one finds (4.1) .

(5.) Existence and multiplicativity of j_t

Now we consider the following particular situation: the algebra $\mathcal{A}$ acts on a Hilbert space . Let there be given: a natural integer d, an Ito algebra $\mathfrak{I}$ of $\mathcal{A}$ valued measures on $\mathbf{R}$ with generators $M_1, \ldots, M_d$ and with constant structure coefficients (in the sense of [acqu] i.e. for each $h, k = 1, \ldots, d$ and $(s,t) \subseteq \mathbf{R}$ one has

$$[[M_h^+, M_k]](s,t) = c_{hk}^n M_n(s,t) \tag{5.1}$$

for some constants $c_{hk}^n \in \mathbf{C}$. Throughout this Section summation over repeated indices will be undestood. Under these assumption $\mathfrak{I}$ is the complex linear span of the $M_1, \ldots, M_d$. We assume the covariance of the M_k, i.e.

$$u_r^o(M_k(s,t)) = M_k(s+r, t+r) \tag{5.2}$$

commutativity with the past, i.e.

$$[\mathcal{A}_{s]}, M_k(s,t)] = 0 \quad ; \quad t > s \tag{5.3}$$

and that they are adapted, i.e.

$$M_k(s,t) \in \mathcal{A}_{t]} \quad ; \quad k = 1, \ldots, d \quad ; \quad t > s \tag{5.4}$$

We also assume that the family $M_k(s,t)$ is self-adjoint, in the sense that, for each $k = 1, \ldots, d$, there is an index, denoted $k^+ = 1, \ldots, d$, such that

$$M_k(s,t)^+ = M_{k^+}(s,t) \tag{5.5}$$

we also require that, for each adapted $\mathcal{A}$-valud process a, for each $s < t$, and for each $\xi \in D, M_k(s,t)\xi$ is in the domain of a_s.

The main analytical assumption on the M_k is that there exists a total subset $D \subseteq \mathcal{H}$ such that, for each $\xi, \eta \in D$, for each $k = 1, \ldots, d$ and for each quadruple a, b, c, d of $\mathcal{A}$-valued adapted processes, one has:

$$\left| \int_0^t < a_s dM_k(s)\xi, b_s\eta > - \int_0^t < c_s dM_k(s)\xi, d_s\eta > \right| \tag{5.5a}$$

$$\leq \int_0^t dm_{\xi,\eta}(s) \, |< a_{s]}\xi, b_{s]}\eta > - < c_{s]}\xi, d_{s]}\eta >|$$

for some Radon measure $m_{\xi,\eta}$ on $\mathbf{R}$. The basic integrators of the Fock and universally invariant Brownian motions satisfy this condition .

An estimate of the type (5.5a) will be called a weak **semimartingale estimate** (cf. [3]).

The topology on $\mathcal{A}$ will be the one specified by the weak convergence on D. Notice that from (5.1) and (5.2) one deduces that:

$$\begin{aligned} M_h(t, t+dt) \cdot M_k(t, t+dt) &= u_t^o(M_n(0, dt) \cdot M_k(0, dt)) \equiv \\ \equiv u_t^o(c_{hk}^n M_n(0, dt)) &= c_{hk}^n M_n(t, t+dt) \end{aligned} \tag{5.6}$$

Definition (5.1). A family of maps $l^k : \mathcal{A} \to \mathcal{A} (k = 1, \ldots, d)$ is called a family of $\mathfrak{S}$-derivation if for each $n = 1, \ldots, d$ and for each $x, y \in \mathcal{A}_{0]}$

$$l^k(\mathcal{A}_{0]}) \subseteq \mathcal{A}_{0]} \tag{5.7}$$

$$l^k(x)^* = l^{(k^+)}(x^*) \tag{5.8}$$

$$l^n(xy) = l^n(x)y + xl^n(y) + l^h(x)l^k(y)c_{hk}^n \tag{5.9}$$

Remark. The term $\mathfrak{S}$-derivation is due to the fact that the structure constants c_{hk}^n in (5.1) and in (5.9) are the same.

In the following we shall also assume that the l^k are bounded, i.e.

$$\| l^k(x) \|_\infty \leq l \cdot \| x \| \quad ; \quad l \in \mathbf{R}_+ \quad ; \quad x \in \mathcal{A}_{0]} \tag{5.10}$$

Under these assumptions it is clear that, denoting

$$l_t^k(x) = u_t^o \circ l^k \circ u_{-t}^o(x) \tag{5.11}$$

$$dL_t(x) = l_t^k(x) dM_k(t) \tag{5.12}$$

the identities (3.28),(3.30) , defining a stochastic derivation, are satisfied. In fact

$$dL_t(xy) = l_t^n(xy) dM_n(t) =$$

$$= l_t^n(x) y dM_n(t) + x l_t^n(y) dM_n(t) + l_t^h(x) l_t^k(y) c_{hk}^n dM_n(t)$$

$$\equiv dL_t(x)y + xdL_t(y) + dL_t(x)dL_t(y)$$

Moreover, because of (5.5) and (5.8)

$$dL_t(x^*) = dL_t(x)^* \quad ; \quad x \in \mathcal{A}_{0]} \tag{5.13}$$

The semimartingale estimate (5.5a), together with (5.10) and (5.11), implies that the iteration series

$$j_t(x_t) := x_0 + \sum_{n=1}^{\infty} \int_0^t dM_{k_1}(t_1) \int_0^{t_1} dM_{k_2}(t_2) \ldots$$

$$+ \ldots \int_0^{t_{n-1}} dM_{k_n}(t_n) l_{t_n}^{k_n} \circ l_{t_{n-1}}^{k_{n-1}} \circ \ldots \circ l_{t_1}^{k_1}(x_{t_1})$$

converges in $\mathcal{A}$ for each continuous curve $t \in \mathbf{R} \to x_t \in \mathcal{A}_{0]}$. Moreover for any $\xi, \eta \in D$

$$|< j_t(x_t)\xi, \eta >| \leq c_t(\sup_{s \in [0,t]} \| x_s \|) \| \xi \| \cdot \| \eta \| \tag{5.14}$$

and, with $dL_t(x_t)$ given by (5.12), one has

$$j_t(x_t) - x_t = \int_0^t j_s(dL_j(x_s)) = \int_0^t j_s(l_s^k(x_s)) dM_k(s) \tag{5.15}$$

Finally, if $s \to z_s$, is a continuous $\mathcal{A}$-valued adapted process, then for each $\xi, \eta \in D$, the integral

$$\int_0^t < j_s\Big(l_s^h(x_s)\Big) dM_h(s)\xi, z_s\eta > \tag{5.16}$$

in the sense of Section (1.), is well defined.

The fact that each j_t is an automorphism now follows easily from the following lemma in whose proof we adapt, to our context, a technique first developed by M. Evans [7].

Lemma (5.2).

Under the above notations and assumptions on the l^k, M_k , for every $x, y \in \mathcal{A}_{0]}$, $\xi, \eta \in D$ and $t \in \mathbf{R}_+$ one has:

$$< j_t(x \cdot y)\xi, \eta > = < j_t(y)\xi, j_t(x^*)\eta > \tag{5.17}$$

Proof. Fix $t > 0$ and let $x_s, y_s (s \in [0, t])$ be continuous in $\mathcal{A}_{0]}$.

Notice that

$$d < j_t(y_t)\xi, j_t(x_t^*)\eta > \equiv < j_t(l_t^h(y_t)) dM_h(t)\xi, j_t(x_t^*)\eta > +$$

$$+ < j_t(y_t)\xi, j_t(l_t^k(x_t^*)) dM_k(t)\eta > + c_{hk}^n < j_t(l_t^h(y_t))\xi, dM_n(t) j_t(l_t^k(x^*))\eta >$$

From this one deduces

$$|< j_t(x_t y_t)\xi, \eta > - < j_t(y_t)\xi, j_t(x_t^*)\eta >| = \tag{5.18}$$

$$= \left| \int_0^t < j_1(l_s^n(x_s y_s)) dM_n(s)\xi, \eta > - \int_0^t < j_1(l_1^h(y_s)) dM_h(s)\xi, j_s(x_s^*)\eta > \right.$$

$$- \int_0^t < j_s(y_s)\xi, j_s(l_s^h(x_s^*)) dM_h(s)\eta >$$

$$\left. - \int_0^t c_{hk}^n < j_s(l_s^h(y_s)) dM_n(s)\xi, j_s(l_s^k(x_s^*))\eta > \right|$$

Since the family $(l_t^n)(n = 1, \ldots, d)$ is an $\mathfrak{J}$ - derivation on $\mathcal{A}_{0]}$, the right hand side of (5.18) is majorized by

$$\left| \int_0^t < j_s(l_s^n(x_s) y_s) dM_n(s)\xi, \eta > - \int_0^t < j_s(y_s)\xi, j_s(l_s^n(x_s)^*) dM_n(s)\eta > \right|$$

$$+ \left| \int_0^t < j_s(x_s l_s^n(y_s)) dM_n(s)\xi, \eta > - \int_0^t < j_s(l_s^n(y_s)) dM_n(s)\xi, j_s(x_s^*)\eta > \right|$$

$$= | c_{hk}^n | \cdot \left| \int_0^t < j_s(l_s^h(x_s) \cdot l_s^k(y_s)) dM_n(s)\xi, \eta > \right.$$

$$\left. - \int_0^t < j_s(l_s^h(y_s)) dM_n(s)\xi m j_s(l_s^k(x_s)^*)\eta_s > \right| \tag{5.19}$$

The semimartingale estimate (5.5a) implies that (5.19) is majorized by:

$$c \cdot (3d + d^2) \max_{\alpha, \beta = 0, \ldots, d} \tag{5.20}$$

$$\int_0^t m_{\xi,\eta}(s) ds \, |< j_s(x_s^{(\alpha)} \cdot y_s^{(\beta)})\xi, \eta > - < j_s(y_s^{(\beta)})\xi, j_s(x_s^{(\alpha)*})\eta >|$$

where

$$c = \max_{h^o, k, \eta = 1, \ldots, d} | c_{h,k}^n | \tag{5.21}$$

$$x_s^{(0)} = x_s \quad ; \quad x_s^{(\alpha)} = l_s^\alpha(x_s) \quad ; \quad \alpha = 1, \ldots, d \tag{5.22}$$

and similarly for $y_s^{(\beta)}$. Iterating N times the inequality (5.20), we obtain that, for each natural integer N (and in obvious notations)

$$|< j_t(x_t y_t)\xi, \eta > - < j_t(y_t)\xi, j_t(x_t^*)\eta >| \leq \tag{5.23}$$

$$\leq c^N (3d + d^2)^N \max_{\alpha_1, \ldots, \alpha_N, \beta_1, \ldots, \beta_N = 0, \ldots, d}$$

$$\int_0^t m_{\xi,\eta}(t_1) dt_1 \int_0^{t_1} m_{\xi,\eta}(t_2) dt_2 \cdot \ldots \cdot \int_0^{t_{N-1}} m_{\xi,\eta}(t_N) dt_N$$

$$|< j_{1_N}(x_{t_N}^{(\alpha_1, \ldots, \alpha_N)} y_{t_N}^{(\beta_1, \ldots, \beta_N)})\xi, \eta > - < j_{1_N}(y_{t_N}^{(\beta_1, \ldots, \beta_N)})\xi, j_{t_N}(x_{t_N}^{(\alpha_1, \ldots, \alpha_N)^*})\eta >|$$

The boundedness of the l^α (cf. (5.10) then implies that

$$\sup_{s\in\mathbb{R}} \max_{\alpha_j,\beta_j} \left\{ \|x_s^{(\alpha_1,\dots\alpha_N)}\|, \|y_s^{(\beta_1,\dots,\beta_N)}\| \right\} \leq l^n \max\{\|x\|, \|y\|\} \tag{5.24}$$

Hence, majorizing the modulus of the difference in the right hand side of (5.23) with the sum of the moduli and using the estimates (5.10) and (5.11) we obtain that for each $N \in \mathbb{N}$ and for some (easily estimated) constant c, depending on $t, d, \| l \|$, one has :

$$|< j_t(xy)\xi, \eta > - < j_t(y)\xi, j_t(x^*)\eta >| \leq$$
$$\leq c^N \|x\|^2 \cdot \|y\|^2 \cdot (\int_0^t m_{\xi,\eta}(s)ds)^N \cdot \frac{1}{N!} \leq d_1^N/N!$$

Since N is arbitrary, this ends the proof.

BIBLIOGRAPHY

[1] Accardi L. Mathematical theory of quantum noise . Proceedings of the 1-st World Conference of the Bernoulli Society. Tashkent 1986 ; VNU Science Press 1987, pg.427-444

[2] Accardi L. , Fagnola F. Stochastic Integration in: Quantum Probability and Applications III Springer LNM n. 1303 (1988),6-19

[3] Accardi L. , Quaegebeur J. Ito algebras of Gaussian quantum fields. Journ.Funct.Anal. (1989)

[4] Accardi L. , Fagnola F., Quaegebeur Representation free Quantum Stochastic Calculus preprint 1989

[5] Accardi L. Noise and dissipation in quantum theory. preprint 1989

[6] Hudson R.L., P.Robinson Quantum Diffusion and the Noncommutative Torus. Lett.Math.Phys. 15(1988) 47-53.

[7] Evans M. Existence of quantum diffusions Probability Theory and related fields (1989)

[8] Evans M., R.L.Hudson Multidimensional quantum diffusions. In Quantum Probability and Applications III, L.Accardi, W.von Waldenfels (eds.) Springer LNM 1303, 68-89.

[8a] Evans M., R.L.Hudson Perturbations of quantum diffusions. to appear in Proc. London Math. Soc.

[9] Hudson R.L. Quantum diffusions on the algebra of all bounded operators on a Hilbert space. IV, to appear.

[10] Hudson R.L. Algebraic theory of quantum diffusions Proceedings Swansea (1986), A. Truman (ed.),to appear.

[11] Hudson R.L. Quantum diffusions and Cohomology of algebras. Proceedings First World Congress of Bernoulli Society, vol. 1., Yu.Prohorov, V.V. Sazonov (eds.) (1987), 479-485.

[12] Lue A.S.T. Non abelian cohomology of associative algebras. Quart.J.Math.Oxford (2),19(1968), 159-180.

[13] Lue A.S.T. Cohomology of groups relative to a variety. Journ. of Alg. 69 (1981), 155-174.

QUANTUM DIFFUSIONS ON INVOLUTIVE ALGEBRAS

BY

DAVID APPLEBAUM
Department of Mathematics, Statistics and Operational Research
Trent Polytechnic
Burton Street
NOTTINGHAM
NG1 4BU
ENGLAND

Abstract

We introduce a refinement of the notion of quantum diffusion which has the advantage over the earlier definition that the algebraic conditions which are necessary for existence are always satisfied. In this framework every quantum stochastic parallel transport process induces horizontal lifts. Applications are given to group respresentations and a factorisation property is obtained for parallel transport processes driven by classical Brownian motion on the $d \times d$ matrices.

Introduction

Let h_0 be a complex Hilbert space. H be symmetric Fock space over $L^2(\mathbb{R}^+)$ and $h = h_0 \otimes H$. R. L. Hudson [5] has defined a quantum diffusion on a unital *-subalgebra A of $B(h_0)$ to be a family of injective *-homomorphisms $(j_t,\ t \geq 0)$ from A into $B(h)$ satisfying the stochastic differential equation

$$dj_t(x) = j_t(\lambda(x))d\Lambda + j_t(\alpha(x))dA^{\dagger} + j_t(\tilde{\alpha}(x))dA + j_t(\tau(x))dt$$
$$j_0(x) = x \otimes I$$

for $x \in A$, where Λ, $A^{\dagger}$ and A are the processes of conservation, creation and annihilation defined in [6] and λ, α, $\tilde{\alpha}$ and τ are linear operators on $B(h_0)$ with common invariant domain A. A major defect of this definition is that although there is some freedom in choosing the coefficients λ, α and $\tilde{\alpha}$, once these have been selected and if the algebra A is cohomologically non-trivial, τ may not exist [5]. Indeed concrete examples in which this occurs have been studied in [7] (see also [1]).

The purpose of this paper is to investigate a new notion of quantum diffusion which generalises the one discussed above and has the advantage that, just as in the classical case, diffusions always exist. The price we have to pay for this is the introduction of a new *-algebra **B** of not necessarily bounded operators of which **A** is a subalgebra and in which at least one of the coefficients of the diffusion must have its range. **B** may be thought of as a "quantum phase space". The motivation for this analogy becomes clearer in §3 where **A** is taken to be a smooth sub-algebra of a C^*-algebra (i.e. a "non-commutative manifold"). In this case, **B** is the smallest *-algebra containing both **A** and those derivations on **A** which describe its smooth structure.

One of the main benefits of our new approach is that every quantum stochastic parallel transport process (see [1]) defined on a finitely generated projective **A**-module induces a horizontal lift.

We will employ the following notation and terminology:

All *-algebras will be taken over the complex field.

The symbol $\otimes$ will be used to denote the (algebraic) tensor product of vector spaces.

If V is a vector space, L(V) will denote the space of linear operators on V.

Let D be a dense subspace of a complex Hilbert space H. If T is a linear operator on H with domain D such that $D \subseteq$ domain of T^* we denote $T^*|_D$ by $T^\dagger$. Let $O(D)$ be the algebra of linear operators on H with common invariant domain D which have the additional property that $T \in O(D) \Rightarrow D \subseteq$ domain of T^* and $T^* D \subseteq D$ then $O(D)$ becomes a *-algebra with respect to the involution $T \rightarrow T^\dagger$.

U(H) will denote the group of all unitary operators in H.

Let A be a *-algebra. A derivation is a linear operator δ on **A** which satisfies

$$\delta(ab) = \delta(a)b + a\,\delta(b)$$

for all a, b $\in$ A. The linear space of all derivations on A will be denoted by Der(A). Aut(A) will denote the group of all *-automorphisms of A and Z(A) will be the centre of the *-algebra A.

The (covariant) functor of boson second quantisation in the category whose objects are complex Hilbert spaces and morphisms are contractions will be denoted by Γ.

Acknowledgement: My thanks to Michael Schurmann for helpful comments.

1. (A,B) Diffusions

Let D and F be dense subspaces of h_0 and H respectively such that F contains the linear span of the exponential vectors in H. We aim to make sense of the notion of diffusion on pairs **(A,B)** where **B** is a unital *-algebra and **A** a normed unital *-subalgebra of **B**. To this end we define an **(A,B) diffusion** to be a family $\{j_t, t \geq 0\}$ of injective *-homomorphisms from **B** into $\mathbf{O}(D \otimes F)$ such that

(i) $j_0 = \eta \otimes I$ where η is an injective *-homomorphism from **B** into $\mathbf{O}(D)$ with the range of $\eta|_A \subseteq B(h_0)$

(ii) For all $x \in A$,

$$dj_t(x) = j_t(\lambda(x))d\Lambda + j_t(\alpha(x))dA^\dagger + j_t(\tilde{\alpha}(x))dA + j_t(\tau(x))dt \quad \ldots(1.1)$$

where λ, α, $\tilde{\alpha}$ and τ are all linear maps from **A** into **B**.

We will postpone a proof of the existence of solutions to (1.1) to a future article and concentrate in this paper on investigating their structure. The algebraic properties of $\{j_t, t \geq 0\}$ impose the following requirements on λ, α, $\tilde{\alpha}$ and τ(c.f.[5])

(i) $\lambda(I) = \alpha(I) = \tilde{\alpha}(I) = \tau(I) = 0$

For each $x \in A$,

$$\lambda(x)^* = \lambda(x^*), \; \tau(x)^* = \tau(x^*), \; \tilde{\alpha}(x) = \alpha(x^*)^* \quad \ldots(1.2)$$

(ii) $\lambda = \sigma - \iota$ where σ is a *-homomorphism from **A** into **B** and ι is the canonical injection of **A** into **B**.

(iii) $\alpha(xy) = \alpha(x)y + \sigma(x)\alpha(y)$ for all $x, y \in A$.

(iv) $$b\tau = \eta_\alpha \quad \ldots(1.3)$$

where b is the Hothschild coboundary operator associated to the complex formed from multilinear maps on **A** taking values in **B** and $\eta_\alpha : A \times A \to B$ is defined by

$$\eta_\alpha(x,y) = -\tilde{\alpha}(x)\,\alpha(y) \quad \ldots(1.5)$$

for $x, y \in A$.

In the case $B = A$, we say that $\{j_t, t \geq 0\}$ is an **A-diffusion**. The latter is essentially the notion of "quantum diffusion" introduced by R. L. Hudson in [5].

We say that an (A,B) diffusion is **minimal** if each of λ and α have range contained in A. For such a diffusion, it is only through the map τ that the algebra B is required. For the remainder of this article, we will concern ourselves only with minimal (A,B) diffusions for which $\lambda = 0$ so that in particular α and $\tilde{\alpha} \in \mathrm{Der}(A)$.

2. Construction of Minimal (A, B) Diffusions.

From now on, we will assume that A is a norm-dense *-subalgebra of a Banach *-algebra $\mathcal{U}$. We note that α and $\tilde{\alpha}$ are both operators on $\mathcal{U}$ with common invariant domain A.

For each $\alpha \in \mathrm{Der}(A)$, let $B(\alpha)$ denote the algebra of operators on $\mathcal{U}$ with common invariant domain A generated by A itself (acting by multiplication), α and $\tilde{\alpha}$. $B(\alpha)$ is a *-algebra with respect to the involution defined by algebraic extension of the prescription:

$$\left.\begin{array}{lll} x & \rightarrow & x^* \\ \alpha & \rightarrow & -\tilde{\alpha} \end{array}\right\} \qquad \ldots(2.1)$$

c.f.[10].

We define a map $L_\alpha : A \rightarrow B(\alpha)$ by the prescription

$$L_\alpha(x) = 1/2\,(\tilde{\alpha}\,\alpha\, x - 2\,\tilde{\alpha}\, x\,\alpha + x\,\tilde{\alpha}\,\alpha) \qquad \ldots(2.2)$$

for each $x \in A$.

We will call L_α the **Lindblad map** (c.f. [8]).

Proposition 1. $bL_\alpha = \eta_\alpha$

Proof. For $x, y \in A$ we have

$$(bL_\alpha)(x,y) = L_\alpha(x)y - L_\alpha(xy) + x\, L_\alpha(y)$$

Now compute $(bL_\alpha)(x,y)z$ for $z \in A$ and use the fact that α and $\tilde{\alpha}$ are derivations.

Corollary 2 For every $\alpha \in \mathrm{Der}(A)$ there exists

(i) A *-algebra $B(\alpha)$

(ii) A minimal $(A, B(\alpha))$ diffusion $\{j_t, t \geq 0\}$ for which

$$dj_t(x) = j_t(\alpha(x))dA^\dagger + j_t(\tilde{\alpha}(x))dA + j_t(L_\alpha(x))dt \qquad \ldots(2.3)$$

for each $x \in A$.

Proof Compare the statement of proposition 1 with (1.3).

Example Let A_θ^∞ be the smooth subalgebra of the non-commutative two-torus A_θ with rotation number $\theta \in \mathbb{R}$ [4]. Let δ^1 and δ^2 denote the canonical outer derivations associated to the action of Π^2 on A_θ. From (2.3) we obtain a minimal $(A_\theta^\infty, B(\alpha))$ diffusion for all $\theta \in \mathbb{R}$ with either of the choices $\alpha = \delta^1 \pm i\delta^2$ (c.f.[1], [7]).

3. Diffusions of Lie Type and Horizontal Lifts

In this section we will take U to be a C*-algebra and take A to be the C^∞-vectors for the action $\gamma: G \to \mathrm{Aut}(U)$ of a connected Lie group on U(c.f. [4]). Denote by $\delta: L \to \mathrm{Der}(A)$, the associated Lie algebra action. We will assume that there exists a basis $\{X^j, 1 \le j \le n\}$ for L such that, writing $\delta^j = \delta(X^j)$, $1 \le j \le n$; we have

(i) $\delta^j(x)^* = \delta^j(x^*)$ for all $x \in A$.

For any faithful state ω on U we denote its restriction to A by ω_0. The action δ is said to be ω_0-**invariant** if

(ii) $\omega_0(\delta^j(x)) = 0$ for all $x \in A$, $1 \le j \le n$.

Now let (h_0, π, ψ_0) be the GNS triple associated to (U, ω). It is easily verified that $D = \{\pi(a)\psi_0, a \in A\}$ is dense in h_0.
For each $\delta \in \mathrm{Der}(A)$, we obtain a corresponding derivation $\delta_\pi \in \mathrm{Der}(\pi(A))$ by the prescription

$$\delta_\pi(\pi(x)) = \pi(\delta(x)) \qquad \text{for each } x \in A.$$

Since ω is faithful, π is injective whence δ_π is well-defined. (If U is simple we may drop the requirement that ω be faithful).

The mapping $\pi(x)\psi_0 \to \delta_\pi(\pi(x))\psi_0$

defines a linear operator on h_0 with invariant domain D which we also denote by δ_π.

The conditions (i) and (ii) of ω_0-invariance ensure that each δ_π^j $(1 \le j \le n)$ is a skew-symmetric operator and we have that

$$\tilde{\delta}_\pi^j = \delta_\pi^j \qquad \text{for } 1 \le j \le n. \qquad \ldots (3.1)$$

We now regard A as a finitely generated projective (left) A-module in the obvious way with hermitian structure given by

$$< a,b > \;=\; a^* b \quad (a,\, b \in A)$$

Remark: From a classical viewpoint this amounts to regarding the space of smooth complex-valued functions on a compact manifold M as the smooth sections of the trivial line bundle whose total space is $M \times \mathbb{C}$.

Returning to our general set-up we see that each δ^j $(1 \le j \le n)$ is a compatible connection on **A** [4]. Now let $\{M_j,\ 1 \le j \le n\}$ be a family of martingales in h with each

$$M_j = \pi(L_j^1)\, A^\dagger + \pi(L_j^2)\, A \qquad \ldots(3.2)$$

where $L_j^k \in A,\ 1 \le j \le n,\ k = 1,2$

We consider the quantum stochastic parallel transport equation [1]

$$dU = U\Big(\sum_{j=1}^{n} \delta_\pi^j\, dM_j + 1/2\big(\sum_{j=1}^{n} \delta_\pi^j\ dM_j\big)^2\Big)$$

$$U(0) = I \qquad \ldots(3.3)$$

We would like the solution of (3.3) to be a unitary process. A useful class of martingales which facilitate this requirement is obtained when we insist that

(iii) Each $L_j^k \in Z(A)$

(iv) $\delta^j(L_j^k) = 0$ for $1 \le j \le n,\ k = 1,2$.

If each δ_π^j $(1 \le j \le n)$ extends to a bounded operator on h_0, it follows from [6] that each $U(t)$ $(t \ge 0)$ is unitary if and only if

$$L_j^2 = (L_j^1)^* \ (1 \le j \le n) \qquad \ldots(3.4)$$

Under certain technical assumptions ([11]) it can be shown that this same condition guarantees unitarity in the more general (and geometrically interesting) case of unbounded derivations.

From now on, we will assume that (3.4) is satisfied and write

$$L_j^1 = L_j \ (1 \le j \le n).$$

Define

$$\alpha = \sum_{j=1}^{n} L_j\, \delta^j \qquad \ldots(3.5)$$

It follows from (iii) and (iv) above that $\alpha \in Der(A)$ with

$$\tilde{\alpha} = \sum_{j=1}^{n} L_j^*\, \delta^j$$

Furthermore, (3.3) now takes the form

$$dU = U(\alpha_\pi dA^\dagger + \tilde{\alpha}_\pi dA + 1/2\, \tilde{\alpha}_\pi \alpha_\pi dt) \qquad \ldots(3.6)$$

(To compare (3.6) with the standard form for unitary processes obtained in [6] we note that $\tilde{\alpha}_\pi = -\alpha_\pi^\dagger$).

For each $\alpha \in \mathrm{Der}(\mathbf{A})$ of the form (3.5) we see that $\mathbf{B}(\alpha)$ is a $*$-subalgebra of the "quantum phase space" $B(\delta,n)$ where $B(\delta,n)$ is the algebra of operators on U with common invariant domain $\mathbf{A}$ generated by $\mathbf{A}$ itself and $\{\delta^j,\ 1 \le j \le n\}$ with involution as described in §2.
Now let $\mathbf{B}(\delta_\pi,n)$ be the algebra of operators on h_0 with common invariant domain D generated by $\pi(\mathbf{A})$ and $\{\delta_\pi{}^j,\ 1 \le j \le n\}$. $B(\delta_\pi,n)$ is a $*$-algebra with respect to the involution $T \to T^\dagger$ for $T \in \mathbf{B}(\delta_\pi,n)$. It is not difficult to verify that the map $\varepsilon_\pi : B(\delta,n) \to B(\delta_\pi,n)$ whose action on generators is given by $\quad \varepsilon_\pi(a) = \pi(a),\ a \in \mathbf{A}\ ; \quad \varepsilon_\pi(\delta^j) = \delta_\pi{}^j,\ \ 1 \le j \le n.$
extends to a $*$-algebra isomorphism.
We define the subalgebra $\mathbf{B}(\alpha_\pi)$ of $\mathbf{B}(\delta_\pi,n)$ to be $\varepsilon_\pi(\mathbf{B}(\alpha))$.

Returning to equation (3.6), we obtain a family of $*$-homomorphisms of $B(h_0)$ into $B(h)$ $\{k_t,\ t \ge 0\}$ via the prescription

$$k_t(X) = U(t)\, X\, U(t)^* \qquad \ldots(3.7)$$

for $X \in B(h_0)$.

Generalising the definition of [1], we say that U induces **horizontal lifts** if $k_t|_{\pi(\mathbf{A})}$ yields a minimal $(\pi(\mathbf{A}),\mathbf{B}(\alpha_\pi))$ diffusion.
For $x \in \mathbf{A}$, a routine computation using the quantum Ito formula [6] yields

$$dk_t(\pi(x)) = k_t([\alpha_\pi, \pi(x)]dA^\dagger + k_t([\tilde{\alpha}_\pi, \pi(x)])dA + k_t(\varepsilon_\pi(L_\alpha(x)))dt \ldots(3.8)$$

However, since $\alpha_\pi \in \mathrm{Der}(\pi(\mathbf{A}))$ we have for $y \in \mathbf{A}$

$$[\alpha_\pi, \pi(x)]\pi(y)\ \psi_0 = \alpha_\pi(\pi(x))\pi(y)\psi_0$$

thus $\quad [\alpha_\pi, \pi(x)] \quad = \alpha_\pi(\pi(x)) \quad = \pi(\alpha(x))$

and (3.8) may be rewritten

$$dk_t(\pi(x)) = k_t(\pi(\alpha(x)))dA^\dagger + k_t(\pi(\tilde{\alpha}(x)))dA + k_t(\varepsilon_\pi(L_\alpha(x)))dt \qquad \ldots(3.9)$$

So $(k_t, t \ge 0)$ is a minimal $(\pi(\mathbf{A}),\mathbf{B}(\alpha_\pi))$ diffusion as required. Furthermore, comparing (3.9) with (2.3) we find that for $t \ge 0$

$$j_t = k_t \circ (\varepsilon_\pi \otimes I) \qquad \ldots(3.10)$$

We call $(j_t,\ t \ge 0)$ the **underlying diffusion** of the horizontal lift induced by U.
Now observe that the general solution to (1.2) is given by

$$\tau = L_\alpha + \beta \qquad \ldots(3.11)$$

where $\beta \in \mathrm{Der}(A)$ satisfies $\beta = \tilde{\beta}$. Thus for an arbitrary *-algebra A, the most general minimal $(A, B(\alpha))$ diffusion takes the form

$$dj_t(x) = j_t(\alpha(x))dA^{\dagger} + j_t(\tilde{\alpha}(x))dA + j_t(\tau(x))dt \qquad \ldots(3.12)$$

for $t \geq 0$, $x \in A$ with τ given by (3.11).

Returning to the case where A is a smooth subalgebra for the action of a Lie algebra L, we say that a minimal $(A, B(\alpha))$ diffusion is of **Lie type** if L has ω_0-invariant action and

$$\alpha = \sum_{j=1}^{n} L_j\, \delta^j \qquad \beta = \sum_{j=1}^{n} K_j\, \delta^j \qquad \ldots(3.13)$$

with $\{L_j, 1 \leq j \leq n\}$ and $\{K_j, 1 \leq j \leq n\}$ both satisfying (iii) and (iv) above.

We can now establish the following

Theorem 3. Every minimal $(A, B(\alpha))$ diffusion of Lie type $\{j_t, t \geq 0\}$ is the underlying diffusion of the horizontal lift induced by a quantum stochastic parallel transport process $Y = (Y(t), t \geq 0)$ so that for each $t \geq 0$, $x \in A$,

$$j_t(x) = Y(t)\pi(x)Y(t)^* \qquad \ldots(3.14)$$

Proof. Define the semimartingales $\{N_j, 1 \leq j \leq n\}$ by

$$N_j(t) = M_j(t) + \pi(K_j)t$$

for $t \geq 0$ where M_j is given by (3.2) and consider the quantum stochastic parallel transport equation

$$dY = Y\Big(\sum_{j=1}^{n} \delta^j\, dN_j + 1/2\, \big(\sum_{j=1}^{n} \delta^j\, dN_j\big)^2\Big) \qquad \ldots(3.15)$$

$$Y(0) = I$$

A straightforward computation shows that

$$dY = Y[\alpha_\pi\, dA^{\dagger} + \tilde{\alpha}_\pi\, dA + (\beta_\pi + 1/2\, \tilde{\alpha}_\pi \alpha_\pi)dt] \qquad \ldots(3.16)$$

The result now follows from quantum Ito's formula as above. □

Note that the process Y defined by (3.15) factorises as

$$Y = U\, V \qquad \ldots(3.17)$$

where U is given by (3.3) and $V = (V(t), t \geq 0)$ is the deterministic parallel transport [1] which satisfies

$$dV = V\, \beta_\pi\, dt \qquad \ldots(3.18)$$

Taking ω as the canonical trace on the algebra A_θ for $\theta \in \mathbb{R}$ it follows from [4] that $\mathbb{R}^2$ has ω_0-invariant action on the smooth algebra A_θ^∞ so that the diffusions constructed in the example at the end of §2 are clearly of Lie type.

4. Reduction to A-Diffusions

In this section, we'll address the problem of ascertaining those conditions under which a minimal $(A, B(\alpha))$ diffusion is in fact an A-Diffusion.
Clearly this will occur if and only if the range of the Lindblad map L_α is contained in A, in which case we may take $\tau = L_\alpha + \beta$.
We begin by obtaining a more transparent form of L_α

Lemma 4. For each $x \in A$,

$$L_\alpha(x) = 1/2(\alpha(x)\tilde{\alpha} + \tilde{\alpha}\alpha(x) - \tilde{\alpha}(x)\alpha) \qquad \ldots(4.1)$$

Proof. Apply $L_\alpha(x)$, as given by (2.2) to $y \in A$ and compute. □

From (4.1) we deduce immediately that the minimal $(A, B(\alpha))$ diffusion defined by (2.3) is an A-diffusion if and only if

$$\alpha(x)\tilde{\alpha} - \tilde{\alpha}(x)\alpha \in A \qquad \ldots(4.2)$$

for all $x \in A$.

Now suppose that $\{j_t, t \geq 0\}$ is a diffusion of Lie type so that, in particular

$$\alpha = \sum_{j=1}^{n} L_j \, \delta^j \quad \text{with each } L_j \in Z(A),\ 1 \leq j \leq n\ (n \in \mathbb{N}).$$

Let $C_{jk} = L_j L_k^* - L_k L_j^*$ for $1 \leq j, k \leq n$, then (4.2) is equivalent to the condition

$$\sum_{j,k=1}^{n} C_{jk}\, \delta_j(x)\delta_k \in A \qquad \ldots(4.3)$$

for all $x \in A$.

It is not difficult to verify that (4.3) is valid if and only if

$$\sum_{j<k} C_{jk}\, (\delta_j(x)\delta_k - \delta_k(x)\delta_j) = 0 \qquad \text{for all } x \in A \qquad \ldots(4.4)$$

Clearly a sufficient condition for (4.4) is given by

$$C_{jk} = 0, \qquad 1 \leq j < k \leq n \qquad \ldots(4.5)$$

In at least one important example, (4.5) is also a necessary condition as is shown by the following result.

Theorem 5. Let $\{j_t, t \geq 0\}$ be a minimal $(A_\theta^\infty, B(\alpha))$ diffusion of Lie type where A_θ^∞ is the smooth subalgebra of the non-commutative two torus with rotation number $\theta \in \mathbb{R}$. Then $\{j_t, t \geq 0\}$ is an A_θ^∞ - diffusion if and only if (4.5) is satisfied.

Proof. For each $\theta \in R$, A_θ^∞ contains two unitary operators U_1 and U_2 for which $\delta^p(U_q) = 2\pi i\ \delta_{pq}\ U_q$ [4].

Now (4.4) must hold in particular with $x = U_1$ which choice yields

$$2\pi i\ C_{12}\ U_1\ \delta_2 = 0$$

hence since $C_{12} \in Z(A_\theta^\infty)$ and U_1 is invertible we obtain $C_{12} = 0$ as required. □

Note that, in general, (4.5) is satisfied whenever each L_j $(1 \leq j \leq n)$ is self (skew) - adjoint but is never satisfied if k of the L_j's are self (skew)-adjoint with $k < n$ and the other n-k of them are skew (self)-adjoint. So, in particular, the minimal $(A_\theta^\infty, B(\alpha))$ diffusions discussed at the end of §2 do not reduce to A_θ^∞-diffusions, as we expect from the results of [1] and [7].

We remark that if, in general, (4.5) is satisfied, the Lindblad map reduces to the simple form

$$L_\alpha(x) = 1/2(\tilde{\alpha}\,\alpha(x)) \qquad \ldots(4.6)$$

for each $x \in A$.

A minimal $(A, B(\alpha))$ diffusion of Lie type is said to be **real Lie** if in (3.13) each $L_j = d_j I$, $K_j = e_j I$ with $d_j, e_j \in \mathbb{R}$ $(1 \leq j \leq n)$.

Let $Y_1, Y_2 \in L$ be given by

$$Y_1 = \sum_{j=1}^{n} d_j\, X^j \quad , \qquad Y_2 = \sum_{j=1}^{n} e_j\, X^j$$

then the maps $\alpha, \beta : L \to \mathrm{Der}(A)$ defined by

$$\alpha(Y_1) = \sum_{j=1}^{n} d_j\, \delta^j \quad , \qquad \beta(Y_2) = \sum_{j=1}^{n} e_j\, \delta^j$$

are Lie algebra representations.

From (4.5) and the above discussion we immediately deduce that every real Lie minimal $(A, B(\alpha))$ diffusion is an A-diffusion and by theorem 3 each such diffusion underlies the horizontal lift of the quantum stochastic parallel transport process

$$dU^g = U^g(\alpha_\pi(Y_1)dQ + (1/2\ \alpha_\pi(Y_1)^2 + \beta_\pi(Y_2))dt) \qquad \ldots(4.7)$$

$$U^g(0) = I$$

where $Q = A + A^\dagger$ is classical Brownian motion and g is the image of (Y_1, Y_2) in $G \times G$ under the exponential map. Indeed the solution to (4.7) is given by

$$U^g(t) = \exp\,[\alpha_\pi(Y_1)Q(t) + \beta_\pi(Y_2)t] \qquad \ldots(4.8)$$

for $t \geq 0$.

To justify the notation U^g in (4.7) and (4.8), suppose that G is an exponential abelian Lie group so that in particular L is abelian and the exponential map from $L \times L$ to $G \times G$ is surjective. It is now straightforward to verify that the map from $G \times G$ to $U(h)$ given by $g \to U^g(t)$ defines an adapted family of unitary representations of $G \times G$.

We immediately obtain the following:

Proposition 6. Every real Lie A-diffusion associated to a C*-dynamical system (U,G,γ) gives rise to a family of adapted representations of $G \times G$ in $\mathrm{Aut}(B(h))$ defined by

$$j_t{}^g(X) = U^g(t)\,X\,U^g(t)^*$$

for $X \in B(h)$, $t \geq 0$, $g \in G \times G$.

Note that the only noise driving $\{j_t{}^g,\ t \geq 0\}$ is classical Brownian motion.

For an example of the above construction, we again consider the non-commutative two-torus where $G = \Pi^2$ and $L = \mathbb{R}^2$.

Now let $H = \Gamma(L^2(\mathbb{R}))$ and let $Q = \{Q(t),\ t \in \mathbb{R}\}$ be two-sided Brownian motion acting on H, then all of our above results remain valid in $h = h_0 \otimes H$ (c.f. e.g. [3]) and we obtain the following analogue of infinitely divisible group representations [9].

Theorem 7. Let $U^g(t)$, for $t \geq 0$, be the representation of $G \times G$ in $U(h)$ defined by (4.8) where G is an exponential abelian Lie group. For every $n \in \mathbb{N}$, $t \geq 0$, there exists

(i) A Hilbert space $h_{1/n}(t)$

(ii) A representation $U^g{}_{1/n}(t)$, of $G \times G$ in $U(h_{1/n}(t))$

(iii) An n-tuple $(\Gamma_1(t), \ldots, \Gamma_n(t)$ of unitary maps from $h_{1/n}(t)$ to h such that for each $g \in G \times G$.

$$U^g(t) = \prod_{j=1}^{n} \Gamma_j(t)\, U^g{}_{1/n}(t)\, \Gamma_j(t)^* \qquad \ldots(4.9)$$

Proof. Define $h_{1/n}(t) = h_0 \otimes \Gamma(L^2[-\infty,t])$ and $U^g{}_{1/n}(t) = U^g(t/n)$.

Let $\Gamma(S_t)$ be the second quantised shift in H and define $\Gamma_j(t) = \Gamma(S_{(j-1)t/n})$ for $1 \le j \le n$. Recall from [3], the cocycle identity for s, t ≥ 0

$$U^g(s)^* U^g(s+t) = \Gamma(S_s) U^g(t) \Gamma(S_s)^*$$

Hence we have

$$U^g((j-1)t/n) U^g(jt/n) = \Gamma_j(t) U^g_{1/n}(t) \Gamma_j(t)^*$$

Now

$$U^g(t) = U^g_{1/n}(t) U^g_{1/n}(t)^* U^g(2t/n) U^g(2t/n)^* \ldots$$
$$\ldots U^g((n-1)t/n) U^g((n-1)t/n)^* U^g(t)$$

$$= U^g_{1/n}(t) \Gamma_1(t) U^g_{1/n}(t) \Gamma_1(t)^* \ldots \Gamma_n(t) U^g_{1/n}(t) \Gamma_n(t)^*$$

as required.

5. The case of Inner Derivations

In this section we will assume throughout that $\{j_t, t \ge 0\}$ is a minimal $(A, B(\alpha))$ diffusion of Lie type wherein each

$$\delta^j = [H_j, .] \text{ for } 1 \le j \le n \text{ where } H_j = -H_j^* \in A$$

It follows that

$$\alpha = [K, .] \text{ and } \tilde{\alpha} = [K^*, .] \text{ where } K = \sum_{j=1}^{n} L_j H_j .$$

Let us assume that (4.5) is satisfied hence $\{j_t, t \ge 0\}$ is an A-diffusion with L_α given by (4.6). An explicit computation yields

$$L_\alpha(x) = -1/2(K^*Kx - K^*xK - KxK^* + xKK^*) \qquad \ldots(5.1)$$

where $x \in A$.

Now suppose that $L_j = L_j^*$ $(1 \le j \le n)$ so that $K = -K^*$, then L_α takes the form

$$L_\alpha(x) = -KxK + 1/2(K^2x + xK^2) \qquad \ldots(5.2)$$

which we recognise as the form of the infinitesimal generator of a semigroup of completely positive maps on $B(h_0)$ [8].

To investigate this class of A-diffusions further we will work within the context of a particular example to which end we take, for $d \in \mathbb{N}$

$$U = A = M_d(\mathbb{C}),\ G = SU(d)$$
$$\gamma(x) = UxU^* \quad (U \in G,\ x \in M_d(\mathbb{C}))$$
$$\delta^j(x) = [X_j, x] \quad (1 \le j \le n,\ x \in M_d(\mathbb{C})$$

We take ω to be the trace on A so that the action δ is ω_o-invariant for any choice of basis $\{X_j, 1 \le j \le d^2 - 1\}$ in su(d). It follows from §4 that, for each $x \in A$,

$$dj_t(x) = j_t(\alpha(x))dQ + [1/2\ j_t(\alpha^2(x)) + j_t(\delta(x))]\ dt \qquad \ldots(5.3)$$

with $\delta(x) = [P, x]$ where $P = -P^* \in A$.

By (3.14) we have

$$j_t(x) = U(t)\pi(x)U(t)^* \quad \text{where}$$

$$dU(t) = U(t)(\alpha_\pi\, dQ + (1/2\ \alpha_\pi^2 + \delta_\pi)dt) \qquad \ldots(5.4)$$

However (5.3) takes the explicit form

$$dj_t(x) = j_t([K,x])dQ + [j_t(L_\alpha(x)) + j_t([P,x])]dt \qquad \ldots(5.5)$$

with L_α given by (5.2). Hence, by the results of [6] we may write

$$j_t(x) = V(t)\pi(x)V(t)^* \qquad \ldots(5.6)$$

where

$$dV = V(\pi(K)dQ + (\pi(P) + 1/2\ \pi(K^2))dt) \qquad \ldots(5.7)$$

$$V(0) = I$$

Note that $V = (V(t), t \geq 0)$ is a unitary process which is not, in general, a quantum stochastic parallel transport.

To investigate these results further we introduce the commutant representation of A

$$\pi'(x)\ y = yx^* \quad \text{for } x,y \in A$$

and the Tomita anti-isomorphism $J : A \to A'$ defined by

$$Jy = y^* \text{ for } y \in A$$

Let $W = (W(t), t \geq 0)$ be the unitary process on h defined by

$$dW(t) = W(t)(\pi'(K)dQ + (\pi'(P) + 1/2\ \pi'(K^2))dt) \qquad \ldots(5.8)$$

$$W(0) = I$$

and note that each $W(t) \in A' \otimes B(H)$.

Theorem 8. The quantum stochastic parallel transport process $U = (U(t), t \geq 0)$ defined by (5.4) factorises for each $t \geq 0$ as

$$U(t) = V(t)\ W(t) \qquad \ldots(5.9)$$

Furthermore each $W(t) = JV(t)J$ ....(5.10)

Proof. For each $t \geq 0$, $x \in A$, we have

$$V(t)\pi(x)V(t)^* = U(t)\ \pi(x)U(t)^*$$

$$\therefore V(t)^*\ U(t)\ \pi(x)\ U(t)^*\ V(t) = \pi(x)$$

so we must have $V(t)^*\ U(t) \in A' \otimes B(H)$.

Noting that

$$dV(t)^* = (-\pi(K)dQ + [1/2\ \pi(K^2) - \pi(P)]dt)\ V(t)^*$$

we obtain from Ito's formula

$$d(V(t)^*U(t)) = (-\pi(K)dQ + [1/2\,\pi(K^2) - \pi(P)]\,dt)\,V(t)^*U(t)$$
$$+ V(t)^*U(t)(\alpha_\pi\,dQ + (1/2\,\alpha_\pi^2+\delta_\pi)dt) - \pi(K)\,V(t)^*U(t)\,\alpha_\pi\,dt$$
$$= V(t)^*U(t)[(\alpha_\pi - \pi(K))dQ + (1/2\,\alpha_\pi^2 - \pi(K)\,\alpha_\pi + 1/2\,\pi(K^2)$$
$$+ \delta_\pi - \pi(P))dt]$$
$$= V(t)^*\,U(t)\,(\pi'(K)\,dQ + [1/2\,\pi'(K^2) + \pi'(P)]\,dt)$$

where we have used the fact that

$[\alpha_\pi, \pi(K)] = 0$ to write $1/2\,\alpha_\pi^2 - \pi(K)\,\alpha_\pi + 1/2\,\pi(K^2) = 1/2(\alpha_\pi - \pi(K))^2$

(5.9) now follows from the uniqueness theorem for solutions of linear quantum SDE's established in [6].
To obtain (5.10) use the fact that

$\pi(x)\,J = J\,\pi'(x)$ for all $x \in A$

to write

$$d(JV(t)J) = JdV(t)J = JV(t)[\pi(K)dQ + (\pi(P) + 1/2\,\pi(K^2))dt]\,J$$
$$= JV(t)J[\pi'(K)dQ + (\pi'(P) + 1/2\pi'(K^2))dt]$$

The required result now follows from the above quoted uniqueness theorem and the observation that

$$JV(0)\,J = J^2 = I = W(0) \qquad \square$$

From (5.9) and (5.10), we deduce that for all $t \geq 0$,

$$U(t) = (V(t)J)^2 = (J\,W(t))^2.$$

6. Projective Minimal (A,B)-Diffusions

In this section we will generalise some of the key results obtained above. To this end let Ξ be a left A-module, and let E be a Lie subalgebra of Der(A). A **connection** on Ξ is a linear map $\nabla : E \to L(\Xi)$ such that

$$\nabla_\alpha(a\,\xi) = a\,\nabla_\alpha(\xi) + \alpha(a)\,\xi \qquad \ldots(6.1)$$

for all $\alpha \in E$, $a \in A$, $\xi \in \Xi$ (c.f. [4]).

We denote by $B(\nabla_\alpha)$ the algebra of linear operators on Ξ generated by A, ∇_α and $\nabla_{\tilde\alpha}$. $B(\nabla_\alpha)$ becomes a *-algebra under the involution given by algebraic extension of the maps (for $x \in A$) $x \to x^*$, $\nabla_\alpha \to -\nabla_{\tilde\alpha}$.
We define the **generalised Lindblad map** $\mathcal{L}_\nabla^\alpha : A \to B(\nabla_\alpha)$ by

$$\mathcal{L}_\nabla^\alpha(x) = 1/2(\nabla_{\tilde\alpha}\nabla_\alpha x - 2\,\nabla_{\tilde\alpha}x\,\nabla_\alpha + x\,\nabla_{\tilde\alpha}\nabla_\alpha) \qquad \ldots(6.2)$$

A similar argument to that of proposition 1 establishes

Proposition 9. $bL_{\nabla}{}^{\alpha} = \eta_{\alpha}$

Thus we see that the construction of minimal (A,B) diffusions from a given $\alpha \in \mathrm{Der}(A)$ described in §2 is not unique.

In fact we obtain the following

Corollary 10. Given α, $\beta \in \mathrm{Der}(A)$, for every connection ∇ on an A-module Ξ, there exists a minimal $(A, B(\nabla_{\alpha}))$ diffusion $(j_t,\ t \geq 0)$ satisfying, for $x \in A$

$$dj_t(x) = j_t(\alpha(x))dA^{\dagger} + j_t(\tilde{\alpha}(x))dA + j_t(L_{\nabla}{}^{\alpha}(x) + \beta(x))dt \qquad \ldots(6.3)$$

Minimal $(A, B(\nabla_{\alpha}))$ diffusions constructed as in corollary 10 will be called **projective**.

Now suppose that all the conditions of §3 hold. We also require that Ξ is a finitely generated projective module equipped with an A-valued hermitian structure and take h_0 to be its completion as in [1]. Furthermore we will take ∇ to be compatible in the sense of [4] and impose the conditions

$$[\nabla_{\partial_j}, L_j] = 0 \qquad (1 \leq j \leq n)$$

In this case we have

$$\nabla_{\alpha} = \sum_{j=1}^{n} L_j \nabla_{\partial_j} \qquad \text{and} \qquad \nabla_{\alpha}^{\dagger} = -\nabla_{\tilde{\alpha}} .$$

Furthermore $(j_t,\ t \geq 0)$ as given by (6.3) is the underlying diffusion of the quantum stochastic parallel transport process $U = (U(t),\ t \geq 0)$ where

$$dU = U(\nabla_{\alpha}\, dA^{\dagger} + \nabla_{\tilde{\alpha}}\, dA + (1/2\ \nabla_{\tilde{\alpha}} \nabla_{\alpha} + \beta)dt) \qquad \ldots(6.4)$$

We call $(j_t,\ t \geq 0)$ a **projective minimal $(A, B(\nabla_{\alpha}))$ diffusion of Lie type.**

Applying the above considerations to the parallel transport processes driven by Yang-Mills connections on the non-commutative two tori discussed in [1] and [2] we see that we now possess a mechanism whereby these induce horizontal lifts.

References

1. D. Applebaum. **Quantum Stochastic Parallel Transport on Non-Commutative Vector Bundles** in "Quantum Probability and Applications III", Springer LNM 1303, 20-37 (1988)

2. D. Applebaum, **Stochastic Evolution of Yang-Mills Connections on the Noncommutative Two-Torus**, Lett. Math. Phys 16, 93-99 (1988)

3. D. Applebaum, A. Frigerio, **Stationary Dilations of W*-Dynamical Systems Constructed via Quantum Stochastic Differential Equations** in "From Local Times to Global Geometry, Control and Physics," Pitman Research Notes in Mathematics 150, 1-39 (1986)

4. A. Connes, M. Rieffel, **Yang-Mills for Noncommutative Two-Tori** in "Operator Algebras and Mathematical Physics," (AMS) Contemporary Mathematics 62, 237-67 (1987)

5. R. L. Hudson, **Algebraic Theory of Quantum Diffusions** in "Stochastic Mechanics and Stochastic Processes", Springer LNM 1325, 113-25 (1988)

6. R. L. Hudson, K. R. Parthasarathy, **Quantum Ito's Formula and Stochastic Evolution**, Commun. Math. Phys 93, 301-23 (1984)

7. R. L. Hudson, P. Robinson, **Quantum Diffusions and the Noncommutative Torus**, Lett. Math. Phys 15, 47-53 (1988)

8. G. Lindblad, **Dissipative Operators and Cohomology of Operator Algebras**, Lett. Math. Phys 1, 219-24 (1976)

9. R. F. Streater, **Current Commutation Relations, Continuous Tensor Products and Infinitely Divisible Group Representations**, Rendiconti di Sc. Int. di Fisica E. Fermi Vol. XI 247-63 (1969)

10. P. Jorgensen, **Approximately Inner Derivations, Decompositions and Vector Fields of Simple C*-Algebras**, to appear in Proceedings of the 1988 U.S.-Japan Seminar on Operator Algebras.

11. D. Applebaum, **Horizontal Lifts in Fock Space Stochastic Calculus**, in preparation.

This paper is in final form and no similar paper has been or is being submitted elsewhere.

SOME MARKOV SEMIGROUPS IN QUANTUM PROBABILITY

Alberto Barchielli

Dipartimento di Fisica, Università di Milano,
Via Celoria 16, 20133 Milano, Italy,
and Istituto Nazionale di Fisica Nucleare, Sezione di Milano.

1. Quantum Markov semigroups.

In order to extract probabilities and to describe state changes under measurement in a quantum theory the notion of instrument has been introduced [1], which is a certain kind of positive operator valued measure. Let us consider a quantum system described by a von Neumann subalgebra W of $\mathcal{L}(\mathcal{H})$ (the algebra of the bounded operators on a separable Hilbert space $\mathcal{H}$). Let W_* be the predual of W and (Ω,Σ) a measurable space. Then, an instrument on W with value space (Ω,Σ) is a map T from Σ into $\mathcal{L}(W)$ such that

1.1.A) $T(S)$ is completely positive [2,3] and normal ([4], p.8);

1.1.B) $T(\cdot)$ is weakly* ([4], p.7) σ-additive;

1.1.C) $T(\Omega)[\mathbb{1}] = \mathbb{1}$ (normalization).

The interpretation of instruments is that they are associated to measurement procedures: if we perform a measurement T_1 with value space (Ω_1,Σ_1) followed by a measurement T_2 and so on up to T_n, then the probability of the result $\omega_1 \in S_1$, $\omega_2 \in S_2$, ..., $\omega_n \in S_n$ $(S_j \in \Sigma_j)$ is given by

$$P(S_1,S_2,\dots,S_n|\rho) = \langle \rho, T_1(S_1)\circ T_2(S_2)\circ\dots\circ T_n(S_n)[\mathbb{1}]\rangle, \tag{1.1}$$

where ρ is the state of the system and $\langle\rho,a\rangle$ is the bilinear form giving the duality between W_* and W. The index in the instruments refers also to the instant of time or to the interval of time in which the measurement occurs; any dynamics can be included into the definition of the instruments.

Equation (1.1) refers to the case in which the jth measurement is chosen by the "experimenter" independently of what happened before, but we can think also of the case in which a measurement depends on the result of the previous one [5]. In this way one is led to introduce families of instruments that are the analogue of classical transition functions. If we include also the possibility of arbitrary times of measurement and some homogeneity in time, we are led to the notion of semigroup of quantum transition functions. We call semigroup of quantum transition functions or quantum Markov semigroup a function $T_t(S|x)$ $(t \geq 0,\ x\in\Omega,\ S\in\Sigma)$ such that

1.2.A) for fixed t and x, $T_t(\cdot|x)$ is an instrument;

1.2.B) for fixed t, S, $\rho\in W_*$ and $a\in W$, $\langle\rho,T_t(S|\cdot)[a]\rangle$ is Σ-measurable;

1.2.C) $T_0(S|x) = \delta_x(S)\mathbb{1}$, where δ_x is the Dirac measure concentrated at x;

1.2.D) $$T_{t+s}(S|x) = \int_\Omega T_t(dy|x)\circ T_s(S|y), \qquad \forall s,t \geq 0.$$

Here and in the following integrals are defined in the weak* topology.

Let now Ξ be an instrument with value space (Ω,Σ), representing a first instantaneous measurement. The probability of the result S_1 at time t_1, S_2 at time t_2, ... $(S_j \in \Sigma,\ t_j > t_{j-1} \geq 0)$, when the initial state is ρ, is given by

$$P(S_1,t_1;S_2,t_2;\dots;S_n,t_n|\rho) = \int_{y_0\in\Omega}\int_{\{y_j\in S_j\}} \times$$

$$\times \langle\rho,\Xi(dy_0)\circ T_{t_1}(dy_1|y_0)\circ T_{t_2-t_1}(dy_2|y_1)\circ\dots\circ T_{t_n-t_{n-1}}(dy_n|y_{n-1})[\mathbb{1}]\rangle. \tag{1.2}$$

Let us consider now the case in which we have some group of transformations of Ω and $T_t(S|x)$ depends only on the transformations mapping x into S and not on S and x separately. We arrive in this way to the notion of semigroup of instruments on a group.

Let G be a locally compact topological Hausdorff group and $\mathcal{B}$ its Borel σ-algebra; let us denote by e the neutral element of G. A convolution semigroup of instruments on $(G,\mathcal{B})$ is a function $\theta_t(B)$ $(t \geq 0,\ B\in\mathcal{B})$, such that

1.3.A) for fixed t, $\theta_t(\cdot)$ is an instrument on W with value space $(G,\mathcal{B})$;

1.3.B) $\forall\rho\in(W_*)_+$, $\forall a\in W_+$ (the positive cones of W_* and W), the positive and finite measure $\mu_{\rho,a}(\cdot) := \langle\rho,\theta_t(\cdot)[a]\rangle$ is regular, i.e.

$$\mu_{\rho,a}(B) = \sup\{\mu_{\rho,a}(K)\colon K\subset B,\ K \text{ compact}\};$$

1.3.C) $$\theta_0(B) = \delta_e(B)\,\mathbb{1}\,;$$

1.3.D) $$\theta_{t+s}(B) = \theta_t * \theta_s(B) := \int_G \theta_t(dg)\circ\theta_s(Bg^{-1}) \equiv \int_G \theta_t(g^{-1}B)\circ\theta_s(dg).$$

The hypoteses on G and condition 1.3.B are technical; what one really needs for the interpretation is the following. Let G be a group of transformations on Ω and let the σ-algebras Σ and $\mathcal{B}$ be such that (a) $\{g\in G\colon gx\in S\}\in\mathcal{B}$, $\forall x\in\Omega$, $\forall S\in\Sigma$, (b) $\langle\rho,\theta_t(\{g\in G\colon gx\in S\})[a]\rangle$ is a Σ-measurable function of x. Then, it is easy to verify that the equation

$$T_t(S|x) := \theta_t(\{g\in G\colon gx\in S\}), \qquad \forall x\in\Omega,\ \forall S\in\Sigma, \tag{1.3}$$

defines a quantum Markov semigroup.

The first results concerned semigroups of instruments on abelian groups [6-11]. The extension to non-abelian groups has been done in [12-14]: in [12] semigroups of instruments on "Bochner groups [15]" have been constructed via Fourier transform and quantum stochastic calculus; in [13,14] other results have been obtained via a quantum analogue of the notion of probability operator [16]. In the rest of this paper we shall present the results of [13,14] and give some examples; moreover, we shall discuss the concept of covariance under the transformations of the group and consider the case of a spin in an isotropic environment (case $G \equiv SO(3)$).

2. Semigroups of probability operators.

From now on G will be a locally compact Hausdorff group. Let ω be a left invariant Haar measure on G and consider the Banach space $L^\infty(G,\omega;W) =: N$ of all W-valued essentially bounded weakly* locally ω-measurable functions on G. Then, N is a W*-algebra under pointwise multiplication, and its predual is $N_* = L^1(G,\omega;W_*)$, the Banach space of all W_*-valued Bochner ω-integrable functions on G ([17], Theor.1.22. 13). The W*-algebra N is isomorphic to $W \otimes L^\infty(G,\omega)$ (W*-tensor product) and the algebra N_* to $W_* \otimes L^1(G,\omega)$ (tensor product with respect to the greatest cross norm) ([17], pp. 58 and 59, Def.1. 22.10, Proposition 1.22.12). The W*-algebra N can be seen as a von Neumann algebra of operators on $\mathcal{H} \otimes L^2(G,\omega)$.

Left (L_g) and right (R_g) translations are defined by setting

$$L_g[f](h) := f(gh), \quad R_g[f](h) := f(hg), \quad \omega\text{-a.e.}, \quad \forall f \in L^\infty(G,\omega). \tag{2.1}$$

Often we shall identify L_g with $\mathbb{1} \otimes L_g$ and R_g with $\mathbb{1} \otimes R_g$; therefore, L_g and R_g will denote translations in both $L^\infty(G,\omega)$ and N.

Let θ_t be a convolution semigroup of instruments as defined in the previous section. Then, one can define an operator $T_t \in \mathcal{L}(N)$ by

$$T_t := \int_G \theta_t(dg) \otimes L_g. \tag{2.2}$$

It turns out that, $\forall t,s \geq 0$,

2.1.A) $T_t \circ T_s = T_{t+s}$;

2.1.B) $T_0 = \mathbb{1}$;

2.1.C) $T_t[\mathbb{1}] = \mathbb{1}$;

2.1.D) T_t is completely positive and normal;

2.1.E) $T_t \circ R_g = R_g \circ T_t$, $\forall g \in G$.

Conversely, properties 2.1.A-2.1.E imply that there exists a unique convolution semigroup of instruments θ_t such that eq.(2.2) holds ([13], Theor.2.2; [14] Theor. 1.4). We call $\{T_t, t \geq 0\}$ a semigroup of <u>probability operators</u>.

This kind of semigroups has been studied under suitable continuity properties in t and, eventually, restrictions on W and G. Let us recall the main results.

2.2) If T_t is a norm continuous semigroup of probability operators, then its generator can be written as

$$K[A](g) = \int_{G^*} \Gamma(dg')[A(g'g)] - \frac{1}{2}\{\Gamma(G^*)[\mathbb{1}], A(g)\} + L[A(g)], \qquad \forall A \in N, \tag{2.3}$$

where L is the generator of a norm continuous completely positive quantum dynamical semigroup [2-4] on W and Γ is a quasi-instrument on W with value space $(G,\mathcal{B})$ ([13], Theor.3.1 and Coroll.3.2).

We call quasi-instrument a regular, finite, completely-positive operator valued measure, i.e. Γ satisfies conditions 1.1.A, 1.1.B, 1.3.B, but not the normalization condition 1.1.C. Moreover, we have used the notations $G^* \equiv G \setminus e$, $\{a,b\} \equiv ab + ba$.

2.3) If $\mathcal{H}$ is finite dimensional, $W = \mathcal{L}(\mathbb{C}^n)$, G is a discrete group and T_t is weakly* continuous in t, then necessarily T_t is norm continuous and its generator is given by the expression (2.3) [14].

2.4) Let $W = \mathcal{L}(\mathbb{C}^n)$, G be a Lie grup of finite dimension $d \geq 1$ and T_t be weakly* continuous in t. Then, T_t is strongly continuous in a suitable subspace of N and one can introduce an infinitesimal generator K, which determines T_t uniquely. In order to give the structure of K we need some notations.

Let $\Lambda(G)$ be the Lie algebra of G and for any $Y \in \Lambda(G)$ let us define the left derivative of a continuous function f by

$$Y[f] := \lim_{s \to 0} \frac{1}{s} (L_{\zeta(s)}[f] - f), \tag{2.4}$$

if the limit exists in the supremum norm. Here $s \in \mathbb{R}^*$ and $\zeta(s) = \exp(sY)$. Let $\{X_j, j=1,\dots,d\}$ be a basis of $\Lambda(G)$. We denote by C_2 the Banach space of all the continuous functions having a limit at infinity and for which all first and second left derivatives exist and are continuous; the norm in C_2 is given in [15], 4.1.6. Finally, let E be the Banach space of the n×n matrices of C_2-functions with norm

$$||A||_2 := \sup_{g \in G} ||A(g)|| + \sum_i \sup_{g \in G} ||X_i[A](g)|| + \sum_{ij} \sup_{g \in G} ||X_i \circ X_j[A](g)||. \tag{2.5}$$

Then, the generator K exists on E and its restriction to E is given by [14]

$$K[A](g) = K_1[A](g) + K_2[A](g) + \mathbb{1} \otimes j[A](g), \tag{2.6}$$

where K_1 is defined by the r.h.s. of eq.(2.3) and for $A \in E$, $f \in C_2$

$$K_2[A](g) := \sum_{ij} b_{ij} (C_j^* A(g) C_i - \frac{1}{2} \{C_j^* C_i, A(g)\} + X_j[A](g) C_i +$$
$$+ C_j^* X_i[A](g) + \frac{1}{2} X_j \circ X_i[A](g)), \tag{2.7}$$

$$j[f](g) = \int_{G^*} \frac{\mu(dg')}{\phi(g')} (f(g'g) - f(g) - \sum_j X_j[f](g) x_j(g')) + \sum_j b_j X_j[f](g). \tag{2.8}$$

Here $b_j \in \mathbb{R}$, $b_{ij} = b_{ji} \in \mathbb{R}$, $\{b_{ij}\}$ is a positive matrix, $C_i \in W$ and μ is a regular, finite, positive measure on G. Moreover, $\{x_j, j=1,\dots,d\}$ is a set of canonical coordinates, associated with the basis $\{X_j, j=1,\dots,d\}$ and extended to the whole G in such a way that $x_j \in C_2$, $x_j(g) \in \mathbb{R}$. Finally, ϕ is a "Hunt function": $\phi \in C_2$, $\phi(g) \in \mathbb{R}$, $\phi(g) > 0$ for $g \in G^*$, $\phi(g) = \sum_i x_i^2(g)$ in a neighborhood of e and $\phi(\infty) = c > 0$.

Conversely, an operator of the form (2.6) determines a unique semigroup of probability operators T_t.

The above results are a certain quantum version of Hunt's representation theorem for the generator of a convolution semigroup on a Lie group ([15], Theor. 4.2.8). Moreover, these results can be restated in the form of a certain operator-valued "diffusion" equation, which could be interesting for possible future developments.

Let G act transitively on a differentiable manifold Ω, let $\bar{x}$ be a fixed element of Ω and g_x be the element of G such that $g_x \bar{x} = x$. For any function A from G into W let us define a function $\tilde{A}$ from Ω into W by $\tilde{A}(x) = A(g_x)$ and for any function B from Ω into W let $\tilde{B}$: G → W be defined by $\tilde{B} = B(g\bar{x})$. Let Θ_t be the instrument associated with T_t according to eq.(2.2); from Θ_t we define $T_t(S|x)$ by eq.(1.3) and in an analogous way we construct $\Pi(S|x)$ from Γ. Finally, we introduce the derivatives on Ω by $\tilde{X}_j[B](x) := X_j[\tilde{B}](g_x)$.

If we set

$$\tilde{T}_t[B](x) := \int_\Omega T_t(dy|x)[B(y)], \tag{2.9}$$

then we obtain easily

$$\tilde{T}_t[B](x) = T_t[\tilde{B}](g_x) = \widetilde{T_t[\tilde{B}]}(x), \tag{2.10}$$

$$\frac{d}{dt}\tilde{T}_t[B](x) = KoT_t[\tilde{B}](g_x) = \widetilde{KoT_t[\tilde{B}]}(x). \tag{2.11}$$

Now we define

$$B(t,x) \equiv B_t(x) := \tilde{T}_t[B](x); \tag{2.12}$$

moreover, in order to have shorter equations, we take j≡0. Then, eq.(2.11) becomes explicitly

$$\frac{\partial}{\partial t} B(t,x) = L[B(t,x)] + \int_{\Omega\setminus x} \Pi(dy|x)[B(t,y)] - \frac{1}{2}\{\Pi(\Omega\setminus x|x)[\mathbb{1}],B(t,x)\} +$$
$$+ \sum_{ij} b_{ij}(C_j^*B(t,x)C_i - \frac{1}{2}\{C_j^*C_i,B(t,x)\} + \tilde{X}_j[B_t](x)C_i +$$
$$+ C_j^*\tilde{X}_i[B_t](x) + \frac{1}{2}\tilde{X}_j o\tilde{X}_i[B_t](x)). \tag{2.13}$$

Equation (2.13) is a kind of Fokker-Planck equation. The problem is now the following one. Let us forget the way eq.(2.13) has been obtained: Ω is now a generic differentiable manifold and no group of transformations of Ω is introduced. Moreover, assume that the various coefficients appearing in eq.(2.13) depend on the point x. Does the equation so obtained, with suitable boundary conditions, define a quantum Markov semigroup (via eqs.(2.9) and (2.12))? This would be the way in which quantum Markov semigroups on manifolds are concretely constructed.

3. Examples.

3.1) Our first example concerns counting processes in quantum mechanics [1,7,18-20].

Let us take $\Omega = \mathbb{Z}$ and let T_t be a quantum Markov semigroup on $\mathbb{Z}$. Assume that $T_t(\{m\}|n)$ depends only on m-n and set

$$N_t(n) := T_t(\{n+m\}|m). \tag{3.1}$$

Then, according to eq.(1.2) the probability of n_1 counts in the time interval

$[0,t_1)$, n_2 in $[t_1,t_2)$, ..., n_k in $[t_{k-1},t_k)$ is given by

$$P(n_1,[0,t_1);n_2,[t_1,t_2);\dots;n_k,[t_{k-1},t_k)|\rho) =$$

$$= \langle\rho, N_{t_1}(n_1) \circ N_{t_2-t_1}(n_2) \circ \dots \circ N_{t_k-t_{k-1}}(n_k)[\mathbb{1}]\rangle. \tag{3.2}$$

With respect to eq.(1.2) we have taken $\Xi(S) = \delta_0(S)\mathbb{1}$: no initial count, no initial perturbation of the system. Obviously, eq.(3.2) has the meaning of probability of counts only if

$$N_t(n) \equiv 0, \qquad \text{for } n < 0. \tag{3.3}$$

The associated semigroup of probability operators is defined by

$$T_t[A](n) = \sum_m N_t(m)[A(n+m)]; \tag{3.4}$$

let us stress that in this case $G \equiv \mathbb{Z}$. We assume that T_t is norm continuous (cases 2.2 or 2.3). Then, by setting $M(m) := \Gamma(\{m\})$, the generator (2.3) can be written as

$$K = S\otimes\mathbb{1} + R, \tag{3.5}$$

where

$$S[a] := L[a] - \frac{1}{2} \sum_{m\neq0} \{M(m)[\mathbb{1}],a\}, \tag{3.6}$$

$$R[A](n) := \sum_{m\neq0} M(m)[A(n+m)]. \tag{3.7}$$

By setting

$$R_t := R \circ \exp(S\otimes\mathbb{1}\, t), \tag{3.8}$$

we can express T_t as the Dyson expansion

$$T_t = \exp(S\otimes\mathbb{1}\, t) + \sum_{r=1}^{\infty} \int_0^t dt_r \int_0^{t_r} dt_{r-1} \dots \int_0^{t_2} dt_1 \exp(S\otimes\mathbb{1}\, t_1) \circ$$

$$\circ R_{t_2-t_1} \circ \dots \circ R_{t_r-t_{r-1}} \circ R_{t-t_r}. \tag{3.9}$$

Then, by eqs.(3.4), (3.7)-(3.9) we obtain

$$N_t(n) = \delta_{n,0}\exp(St) + \sum_{r=1}^{\infty} \int_0^t dt_r \int_0^{t_r} dt_{r-1} \dots \int_0^{t_2} dt_1 \exp(St_1) \circ$$

$$\circ \sum_{\{m_j\}}{}' M(m_1) \circ \exp(S(t_2-t_1)) \circ M(m_2) \circ \exp(S(t_3-t_2)) \circ \dots \circ M(m_r) \circ \exp(S(t-t_r)), \tag{3.10}$$

where the prime over the sign of sum means: $m_j \in \mathbb{Z}$, $m_j \neq 0$, $\sum_{j=1}^{r} m_j = n$.

In order that condition (3.3) holds, it is apparent that we must have

$$M(m) \equiv 0, \qquad \text{for } m < 0. \tag{3.11}$$

Therefore, eq.(3.10) becomes

$$N_t(0) = \exp(St), \tag{3.12}$$

$$N_t(n) = \sum_{r=1}^{n} \sum_{\{m\}}'' \int_0^t dt_r \ldots \int_0^{t_2} dt_1 \exp(St_1) \circ M(m_1) \circ \ldots \circ$$
$$\circ \exp[S(t_r - t_{r-1})] \circ M(m_r) \circ \exp(S(t-t_r)), \tag{3.13}$$

where the double prime over the sign of sum stays for $m_j = 1,\ldots,n+1-r$, $\sum_{j=1}^{r} m_j = n$.

The processes we have obtained in this way had been essentially introduced by Davies [1]. By comparing our results with that reported in [18], we see that they coincide in the case $M(r) \equiv 0$, for $r \geq 2$. The point processes considered here can be used for describing not only the counting of photons or other particles but also the emission of photons by an atom or some other system [19,20]. In this last case the generalization obtained by taking $M(r) \neq 0$ also for some $r \geq 2$ gives the possibility of treating multi-photon emission.

3.2) Now we want to show how the quantum processes introduced in [6] under the name of operation-valued stochastic processes (OVSP) can be obtained from the scheme of Sec. 1.

Let us take $\Omega \equiv \mathbb{R}^s$, Σ the Borel σ-algebra of $\mathbb{R}^s$ and let $T_t(S|x)$ be a quantum Markov semigroup on $\mathbb{R}^s$. Let us assume that $T_t(S|x)$ depends only on S-x; then, the equation

$$T_t(S|x) =: \theta_t(S-x) \tag{3.14}$$

defines a convolution semigroup of instruments.

Let us take in eq.(1.2) $\Xi(S) = \mu(S)\mathbb{1}$, where μ is a probability measure. Then, eq.(1.2) becomes

$$P(S_1,t_1;\ldots;S_n,t_n|\rho) = \int_{y_0\in\mathbb{R}^s} \mu(dy_0) \times$$
$$\times \int_{\{\sum_{r=1}^{j} y_r \in S_j\}} \langle \rho, \theta_{t_1}(dy_1)\circ\ldots\circ\theta_{t_n-t_{n-1}}(dy_n)[\mathbb{1}]\rangle. \tag{3.15}$$

Let now $\mathbb{R}_+$ be the time axis and Y the space of all functions from $\mathbb{R}_+$ into $\mathbb{R}^s$ (trajectory space). We interprete eq.(3.15) as the probability of the set of trajectories $\{y\in Y: y(t_1)\in S_1,\ldots,y(t_n)\in S_n\}$. Then, for the probability of increments we have

$$P(\{y\in Y: y(t_1)-y(t_0)\in B_1,\ y(t_2)-y(t_1)\in B_2,\ \ldots,\ y(t_n)-y(t_{n-1})\in B_n|\rho) =$$
$$= \langle \rho, \theta_{t_1}(B_1)\circ\ldots\circ\theta_{t_n-t_{n-1}}(B_n)[\mathbb{1}]\rangle. \tag{3.16}$$

Let us note that this probability does not depend on μ.

Let us consider the derivative process $x(t) = \dot{y}(t)$ and the random variables

$$x[h_j] := \int_0^{\bar{t}} h_j(t)x(t)dt, \quad h_j(t) = \alpha_j \, \chi_{[t_{j-1},t_j)}(t), \quad \alpha_j \in \mathbb{R}, \quad 0<t_1<\ldots<t_n \leq \bar{t}.$$

Then from eq.(3.16) we get for these random variables the probability

$$P(h_1,\ldots,h_n;B_1,\ldots,B_n;[0,\bar{t})|\rho) =$$

$$= \int_{\{\alpha_j z_j \in B_j\}} \langle \rho, \theta_{t_1}(dz_1)\circ\ldots\circ\theta_{t_n-t_{n-1}}(dz_n)[\mathbb{1}]\rangle. \tag{3.17}$$

It can be shown [6] that eq.(3.17) can be extended to a probability measure on a suitable trajectory space, so defining a generalized stochastic process in the sense of Gel'fand [21]. Correspondingly, there exists a family of instruments on this trajectory space enjoying particular composition properties: it is this family of instruments that has been called OVSP in [6] and that allows to describe certain continuous (in time) measurements in quantum mechanics. For a concrete application see [22].

4. Covariance.

Let W be the whole $\mathcal{L}(\mathcal{H})$, U(g) be a unitary projective representation of G on $\mathcal{H}$ and θ an instrument on (G,ß). We call the instrument θ <u>covariant</u> with respect to U(g) if, for any $a \in W$, any $B \in ß$, any $g \in G$, we have

$$\theta(B)[a] = U(g)^*\theta(gBg^{-1})[U(g)aU(g)^*]U(g). \tag{4.1}$$

Analogously, if G is a group of transformations on Ω, we call an instrument Ξ on (Ω,Σ) <u>covariant</u> if

$$U(g)^*\Xi(gS)[U(g)aU(g)^*]U(g) = \Xi(S)[a]. \tag{4.2}$$

Let now Ξ be a covariant instrument on Ω, θ_t a convolution semigroup of covariant instruments on G and let T_t be defined by eq.(1.3); then probabilities (1.2) have the following covariance property:

$$P(S_1,t_1;\ldots;S_n,t_n|\rho) = P(gS_1,t_1;\ldots;gS_n,t_n|U(g)\rho U(g)^*). \tag{4.3}$$

For related ideas (covariance of effect-valued measures) see, for instance, [23].

Let us define now the operator $\mathcal{U}(g) \in \mathcal{L}(N)$ by

$$\mathcal{U}(g)[A](h) := U(g)L_{g^{-1}}[A](h)U(g)^*, \qquad \forall g,h \in G. \tag{4.4}$$

$\mathcal{U}$ is a representation of G on N; indeed one obtains easily

$$\mathcal{U}(g_1g_2) = \mathcal{U}(g_1)\circ\mathcal{U}(g_2). \tag{4.5}$$

Then, the covariance condition (4.1) for a convolution semigroup θ_t is equivalent to

$$\mathcal{U}(g)\circ T_t = T_t\circ\mathcal{U}(g), \qquad \forall g\in G, \quad \forall t \geq 0, \tag{4.6}$$

where T_t is the semigroup of probability operators associated to θ_t according to eq. (2.2).

As an example of covariant semigroup, we consider a spin s in an isotropic environment and take $G = SO(3)$, $W = \mathcal{L}(\mathbb{C}^{2s+1})$.

First we need some notations; for the conventions see [24,25]. Let $D^J(R)$ be the standard form of the irreducible representation of SO(3) acting on $\mathbb{C}^{2J+1}$, with matrix elements $D^J_{MM'}(R) = \langle JM|D^J(R)|JM'\rangle$. Let T_{JM}, $J=0,1,\ldots,2s$, $M=-J,\ldots,J$, be the othonormal basis in $\mathcal{L}(\mathbb{C}^{2s+1})$ of the irreducible spherical tensors, which enjoy the properties

$$\mathrm{Tr}\, T^*_{JM}\, T_{J'M'} = \delta_{JJ'}\, \delta_{MM'}\,, \qquad T^*_{JM} = (-1)^M\, T_{J-M}\,, \tag{4.7}$$

$$D^s(R)\, T_{JM}\, D^s(R)^* = \sum_{M'} D^J_{M'M}(R)\, T_{JM'}\,. \tag{4.8}$$

Let T_t be a weakly* continuous semigroup of probability operators on $N = W\otimes L^\infty(SO(3),\omega)$; its generator is given by eqs.(2.3), (2.6), (2.7) and (2.8). Let X_i, i=1,2,3, be the left derivatives associated to three orthogonal directions x,y,z and set

$$Y_M := \sum_{i=1}^{3} \Psi_{Mi}\, X_i\,, \qquad Y^*_M := (-1)^M\, Y_{-M}\,, \qquad M = 1,\, 0,\, -1, \tag{4.9}$$

where Ψ is a unitary matrix with elements $\Psi_{-1,1} = -\Psi_{1,1} = 1/\sqrt{2}$, $\Psi_{1,2} = \Psi_{-1,2} = -i/\sqrt{2}$, $\Psi_{0,3} = 1$, $\Psi_{1,3} = \Psi_{-1,3} = \Psi_{0,1} = \Psi_{0,2} = 0$. Then, one has that Y_M transforms as T_{1M}, i.e.

$$L_R\circ Y_M\circ L_{R^{-1}} = \sum_{M'} D^1_{M'M}(R^{-1})\, Y_{M'}\,, \qquad L_R\circ Y^*_M\circ L_{R^{-1}} = \sum_{M'} D^1_{MM'}(R)\, Y^*_{M'}\,. \tag{4.10}$$

Moreover, we can always write ([2], Theor.2.2 and Lemma 2.2)

$$L[a] = i[H,a] + \sum_{JM}\sum_{J'M'} \lambda^{JJ'}_{MM'}\, \left(T^*_{J'M'}\, a\, T_{JM} - \tfrac{1}{2}\, \{T^*_{J'M'}\, T_{JM}\, ,a\}\right)\,, \tag{4.11}$$

$$\Gamma(dR)[a] = \sum_{JM}\sum_{J'M'} \gamma^{JJ'}_{MM'}(dR)\, T^*_{J'M'}\, a\, T_{JM}\,, \tag{4.12}$$

$$C_i = \sum_{JM} c^i_{JM}\, T_{JM}\,, \tag{4.13}$$

where $c^i_{JM}\in\mathbb{C}$, $\lambda^{JJ'}_{MM'}\in\mathbb{C}$, $H = H^*\in W$, $\mathrm{Tr}\, H = 0$, η is a positive matrix with complex entries and γ is a positive matrix of complex finite measures.

By setting

$$\begin{aligned} B_{MM'} &:= \sum_{ij} \overline{\Psi_{Mi}}\, b_{ij}\, \Psi_{M'j}\,, \qquad & B_M &:= \sum_i \overline{\Psi_{Mi}}\, b_i\,, \\ h^m_{JM} &:= \sum_i c^i_{JM}\, \Psi_{mi}\,, \qquad & y_m(R) &:= \sum_i \overline{\Psi_{Mi}}\, x_i(R)\,, \end{aligned} \tag{4.14}$$

the components of the generator K [see eq.(2.6)] can be written as

$$K_1[A](R) = L[A(R)] + \sum_{JM} \sum_{J'M'} \int_{G^*} \gamma_{MM'}^{JJ'}(dR') \times$$

$$\times \left(T_{J'M'}^* A(R'R) T_{JM} - \frac{1}{2} \{T_{J'M'}^* T_{JM}, A(R)\}\right), \tag{4.15}$$

$$K_2[A](R) = \sum_{m,m'} B_{mm'} \Big\{\sum_{JM} \sum_{J'M'} \overline{h_{J'M'}^{m'}} h_{JM}^{m} \left(T_{J'M'}^* A(R) T_{JM} - \frac{1}{2} \{T_{J'M'}^* T_{JM}, A(R)\} + \right.$$

$$\left. + \sum_{J'M'} \overline{h_{J'M'}^{m'}} T_{J'M'}^* Y_m[A](R) + \sum_{JM} h_{JM}^{m} Y_{m'}^*[A](R) T_{JM} + \frac{1}{2} Y_{m'}^* \circ Y_m[A](R)\right\}, \tag{4.16}$$

$$1\!\!1 j[A](R) = \int_{G^*} \frac{\mu(dR')}{\phi(R')} \left(A(R'R) - A(R) - \sum_M Y_M[A](R) y_M(R')\right) + \sum_M B_M Y_M[A](R). \tag{4.17}$$

Moreover, one can always choose ϕ and y_M in such a way that $\forall R, R' \in SO(3)$

$$\phi(R^{-1}R'R) = \phi(R'), \tag{4.18}$$

$$y_M(R') = \sum_{M'} D_{MM'}^{1}(R^{-1})\, y_{M'}(RR'R^{-1}). \tag{4.19}$$

<u>Theorem</u> 4.1. Let Θ_t be a convolution semigroup of instruments on $W \equiv \mathcal{L}(\mathbb{C}^{2s+1})$ with value space $(G \equiv SO(3), \mathcal{B})$ and covariant with respect to $D^s(R)$; let T_t be the associated semigroup of probability operators. Then, the generator K of T_t has the structure ($\forall A \in E$, $\forall f \in C_2$)

$$K[A](R) = K_1[A](R) + K_2[A](R) + 1\!\!1 \otimes j[A](R), \tag{4.20}$$

$$K_1[A](R) = \sum_{J=1}^{2s} \lambda_J \sum_{M=-J}^{J} \left(T_{JM}^* A(R) T_{JM} - \frac{1}{2} \{T_{JM}^* T_{JM}, A(R)\}\right) +$$

$$+ \sum_{JM} \sum_{J'M'} \int_{G^*} \gamma_{MM'}^{JJ'}(dR') \left(T_{J'M'}^* A(R'R) T_{JM} - \frac{1}{2} \{T_{J'M'}^* T_{JM}, A(R)\}\right), \tag{4.21}$$

$$K_2[A](R) = \sum_{M=-1}^{1} b\Big\{|h|^2 \left(T_{1M}^* A(R) T_{1M} - \frac{1}{2} \{T_{1M}^* T_{1M}, A(R)\}\right) +$$

$$+ \bar{h}\, T_{1M}^* Y_M[A](R) + h\, Y_M^*[A](R) T_{1M} + \frac{1}{2} Y_M^* \circ Y_M[A](R)\Big\}, \tag{4.22}$$

$$j[f](R) = \int_{G^*} \frac{\mu(dR')}{\phi(R')} \left(f(R'R) - f(R) - \sum_m Y_m[f](R) y_m(R')\right), \tag{4.23}$$

where $\lambda_J \geq 0$, $b \geq 0$, $h \in \mathbb{C}$, ϕ and y_m satisfy eqs.(4.18) and (4.19), γ is a positive $(2s+1)^2 \times (2s+1)^2$ matrix of complex finite measures such that $\forall R \in SO(3)$, $\forall B \in \mathcal{B}$

$$\gamma_{MM'}^{JJ'}(B) = \sum_{M_1 M_2} D_{MM_1}^{J}(R^{-1})\, \gamma_{M_1 M_2}^{J\,J'}(RBR^{-1})\, D_{M_2 M'}^{J'}(R); \tag{4.24}$$

finally μ is a positive finite measure such that $\forall B \in \mathcal{B}$, $\forall R \in SO(3)$,

$$\mu(RBR^{-1}) = \mu(B). \tag{4.25}$$

Conversely, if K has the structure above, then, Θ_t is covariant.

<u>Proof</u>. If K has the structure above obviously T_t is covariant and so Θ_t does. For

the necessity of this structure, we see that eq.(4.6) implies

$$K[A](e) = \mathcal{U}(R^{-1}) \circ K \circ \mathcal{U}(R)[A](e), \qquad \forall A \in E. \tag{4.26}$$

By suitable choices of A the result follows.

1) Let us choose A in such a way that $A(e) = 0$, $Y_m[A](e) = 0$, $Y_{m'} \circ Y_m[A](e) = 0$; then, eq.(4.26) becomes

$$\sum_{JM} \sum_{J'M'} \int_{G*} Y^{JJ'}_{MM'}(dR')\, T^*_{J'M'}\, A(R')\, T_{JM} + \int_{G*} \frac{\mu(dR')}{\phi(R')} A(R') =$$

$$= \sum_{JM} \sum_{J'M'} \sum_{M_1M_2} \int_{G*} Y^{J\,J'}_{M_1M_2}(RdR'R^{-1})\, D^J_{MM_1}(R^{-1})\, D^{J'}_{M_2M'}(R)\, T^*_{J'M'}\, A(R')\, T_{JM} +$$

$$+ \int_{G*} \frac{\mu(RdR'R^{-1})}{\phi(R')} A(R') \,. \tag{4.27}$$

If the integral $\int_{G*} \mu(dR)/\phi(R)$ exists, one can redefine Γ and B_m in such a way that eq.(4.25) holds (one can always take $\mu=0$). If that integral does not exist, eqs.(4.27) and (4.18) imply $\mu(RdR'R^{-1}) = \mu(dR')$ near $R' = e$. By redefining μ, Γ, B_m one can made eq.(4.25) to hold for any B.

Once eq.(4.25) is satisfied, eq.(4.27) implies eq.(4.24). Recall that the point $R' = e$ does not contribute to K and one can chose $\Gamma(\{e\})$ and $\mu(\{e\})$ arbitrarily.

2) Let us choose A in such a way that $A(e) = 0$, $Y_m[A](e) = 0$; then, eq.(4.26) becomes

$$\sum_{mm'} \Big\{B_{mm'} - \sum_{m_1m_2} D^1_{mm_1}(R^{-1})\, B_{m_1m_2}\, D^1_{m_2m'}(R)\Big\}\, Y^*_{m'} \circ Y_m[A](e) = 0.$$

This equation implies $B_{mm'} = b\, \delta_{mm'}$ and, by the positivity condition, $b \geq 0$.

3. Let us choose A in such a way that $A(e) = 0$; then, eq.(4.26) becomes

$$b \sum_{M} \sum_{J'M'} \Big\{\overline{h^{M}_{J'M'}} - \sum_{M_1M_2} D^1_{MM_1}(R^{-1})\, \overline{h^{M_1}_{J'M_2}}\, D^{J'}_{M_2M'}(R)\Big\}\, T^*_{J'M'}\, Y_M[A](e) +$$

$$+ b \sum_{JM} \sum_{M'} \Big\{h^{M'}_{JM} - \sum_{M_1M_2} D^J_{MM_1}(R^{-1})\, h^{M_2}_{JM_1}\, D^1_{M_2M'}(R)\Big\}\, Y^*_{M'}[A](e)\, T_{JM} +$$

$$+ \sum_{M} \Big\{B_M - \sum_{M'} B_{M'}\, D^1_{MM'}(R^{-1})\Big\}\, Y_M[A](e) = 0 \,.$$

This equation implies

$$B_M = 0 \,, \qquad h^m_{JM} = h\, \delta_{J1}\, \delta_{Mm} \,.$$

In this way eqs.(4.22) and (4.23) have been proved.

4. Now eq.(4.26) reduces to a condition on L, already studied in [24]. The structure of L is given in [24], Proposition 2.1; this ends the proof of eq.(4.21). Q.E.D.

A final comment on eqs.(4.24) and (4.25). Let ω be the normalized Haar measure on SO(3). Then, by eq.(4.25) we can always write

$$\mu(B) = \int_G \omega(dR)\ \nu(RBR^{-1}) , \tag{4.28}$$

for some (not unique) measure ν. The only condition on ν is that it must be a positive finite measure. Analogously, by eq.(4.24) we can write

$$\gamma^{JJ'}_{MM'}(B) = \sum_{M_1M_2} \int_G \omega(dR)\ D^J_{MM_1}(R^{-1})\ \eta^{J\,J'}_{M_1M_2}(RBR^{-1})\ D^{J'}_{M_2M'}(R) , \tag{4.29}$$

where η is a positive $(2s+1)^2\times(2s+1)^2$ matrix of complex finite measures. In particular, the quantity

$$\gamma_J(B) := \sum_M \gamma^{JJ}_{MM}(B) = \sum_M \int_G \omega(dR)\ \eta^{JJ}_{MM}(RBR^{-1}) \tag{4.30}$$

turns out to be a covariant measure in the sense that it satisfies eq.(4.25) as μ; moreover, eqs.(4.29) and (4.30) give

$$\gamma^{JJ'}_{MM'}(B) = \delta_{JJ'}\ \delta_{MM'}\ \gamma_J(B) , \qquad \forall B:\ RBR^{-1}=B\ \ \forall R\in SO(3). \tag{4.31}$$

By eqs.(4.28)-(4.31), eqs.(4.21) and (4.23) become

$$K_1[A](R) = \sum_{J=1}^{2s} \lambda_J \sum_{M=-J}^{J} \{T^*_{JM}\ A(R)\ T_{JM} - \tfrac{1}{2}\ \{T^*_{JM}\ T_{JM}\ , A(R)\} + $$

$$+ \sum_{JM} \sum_{J'M'} \sum_{M_1M_2} \int_{R_1\in G^*} \eta^{J\,J'}_{M_1M_2}(dR_1) \int_{R_2\in G} \omega(dR_2)\ D^{J'}_{M_2M'}(R_2)\ T^*_{J'M'}\ A(R_2^{-1}R_1R_2R)\ \times$$

$$\times\ T_{JM}\ D^J_{MM_1}(R_2^{-1})\ -\ \frac{1}{2}\sum_{JM} \gamma_J(G^*)\ \{T^*_{JM}\ T_{JM}\ , A(R)\} , \tag{4.32}$$

$$j[f](R) = \int_{R'\in G^*} \frac{\nu(dR')}{\phi(R')} \left\{\int_{R_1\in G} \omega(dR_1)\ f(R_1^{-1}R'R_1R)\ -\ f(R)\right\} , \tag{4.23}$$

where we have taken into account eq.(4.18) and the fact that by eq.(4.19) we have

$$\int_{R_1\in G} \omega(dR_1)\ y_m(R_1^{-1}R'R_1) = 0 . \tag{4.34}$$

References

1. E.B. Davies: Quantum Theory of Open Systems. London: Academic 1976.
2. V. Gorini, A. Kossakowski, E.C.G. Sudarshan: J. Math. Phys. 17, 821-825 (1976).
3. G. Lindblad: Commun. Math. Phys. 48, 119-130 (1976).
4. D.E. Evans, J.T. Lewis: Dilations of Irreversible Evolutions in Algebraic Quantum Theory. Commun. Dublin Inst. Adv. Studies, Series A, N.24 (1977).
5. P.A. Benioff: J. Math. Phys. 13, 908-915 (1972); 13, 1347-1355 (1972).
6. A. Barchielli, L. Lanz, G.M. Prosperi: Found. Phys. 13, 779-812 (1983).
7. A. Barchielli, G. Lupieri: J. Math. Phys. 26, 2222-2230 (1985).
8. A. Barchielli, G. Lupieri, in L. Accardi, W. von Waldenfels (eds.): Quantum Probability and Applications II. Lect. Notes Math., vol.1136, pp.57-66. Berlin:

Springer 1985.

9. K.R. Parthasarathy: Boll. Un. Mat. Ital. 5-A, 391-397 (1986).
10. A.S. Holevo: Teor. Veroyatn. Primen. 31, 560-564 (1986); 32, 142-146 (1987) [Theory Probab. Appl. 31, 493-497 (1987); 32, 131-136 (1988)].
11. A.S. Holevo, in L. Accardi, W. von Waldenfels (eds.): Quantum Probability and Applications III. Lect. Notes Math., vol. 1303, pp. 128-148. Berlin: Springer 1988.
12. A. Barchielli, G. Lupieri: Convolution semigroups in quantum probability and quantum stochastic calculus. To appear in L. Accardi, W. von Waldenfels (eds.): Quantum Probability and Applications IV.
13. A. Barchielli: Probability operators and convolution semigroups of instruments in quantum probability. To appear in Probab. Theory Relat. Fields.
14. A. Barchielli, G. Lupieri: A quantum analogue of Hunt's representation theorem for the generator of convolution semigroups on Lie groups. In preparation.
15. H. Heyer: Probability Measures on Locally Compact Groups. Berlin: Springer 1977.
16. A. Barchielli: Semesterbericht Funktionalanalysis Tubingen, Band 12, Sommersemester 1987, pp. 29-42.
17. S. Sakai: C*-Algebras and W*-Algebras. Berlin: Springer 1971.
18. M.D. Srinivas, E.B. Davies: Optica Acta 28, 981-996 (1981); 29, 235-238 (1982).
19. A. Barchielli: J. Phys. A: Math. Gen. 20, 6341-6355 (1987).
20. A. Barchielli, in Ref. [11], pp. 37-51.
21. I.M. Gel'fand, N. Ya. Vilenkin: Generalized Functions, vol. 4, Applications of Harmonic Analysis. New York and London: Academic 1964.
22. A. Barchielli: Phys. Rev. D32, 347-367 (1985).
23. A.S. Holevo: Rep. Math. Phys. 16, 385-400 (1979).
24. M. Verri, V. Gorini: J. Math. Phys. 19, 1803-1807 (1978).
25. A. Messiah: Quantum Mechanics, vol. II. Amsterdam: North-Holland 1970. Appendix C.

This paper is in final form and no similar paper has been or is being submitted elsewhere.

A QUANTUM STOCHASTIC CALCULUS IN FOCK SPACE OF INPUT AND OUTPUT NONDEMOLITION PROCESSES

Viacheslav Belavkin

Applied Mathematics Department, M.I.E.M.,

B. Vusovski 3/12 Moscow 109028 USSR

In this paper we introduce a new definition of the basic noncommutative measure for infinite-dimensional quantum noise in Fock space as the operator representation of a matrix ★-Lie algebra of a pseudo-Euclidean space with indefinite metric $(\xi|\xi) = \bar{\xi}^{-}\xi^{+} + |\xi^{\circ}|^2 + \bar{\xi}^{+}\xi^{-}$. In contrary to [1] we define the quantum stochastic integral as a uniformly continuous operator on a projective limit of Fock spaces. The described representation of the noncommutative calculus is closely connected with kernel calculus [2], but it is given directly in terms of operators instead of its kernels. The advantage of the quantum stochastic calculus gave us the possibility to prove [3] the main filtering theorem for the general output process, described by a quantum stochastic integral in Fock space.

1. A pseudo-Poissonian quantum process

Let $\mathcal{E}$ be a Hilbert integral $\mathcal{E} = \int^{\oplus} \mathcal{E}(t)dt$ of square-integrable vector-functions $t \in R^{+} \mapsto \xi(t) \in \mathcal{E}(t)$ with a scalar product

$$\langle \xi | \eta \rangle = \int_0^{\infty} \langle \xi | \eta \rangle (t) dt = \int_0^{t} \xi_0^{*}(t)\eta^{\circ}(t)dt , \qquad (1.1)$$

where $\langle\xi|\eta\rangle(t) = \xi_o^*(t)\eta^o(t)$ is the product in a vector space $\mathcal{E}(t)$, $\xi^o = (\xi^i)$ is a column, representing $\xi(t)$ in a basis of $\mathcal{E}(t)$, and $\xi_o^* = (\xi_i^*)$ is the conjugated row, defined by $\xi_i^* = \bar{\xi}^i$, $i \in J$. One can assume that all $\mathcal{E}(t)$ are equivalent to a Hilbert space $\mathcal{K}$ such that $\mathcal{E}$ can be identified with Hilbert tensor product $\mathcal{K} \otimes L^2(\mathbb{R}^+)$, in particular, $\mathcal{E} = L^2(\mathbb{R}^+)$ in the one-dimensional case $\mathcal{K} = \mathbb{C}$.

We shall denote by $\mathcal{F} = \Gamma(\mathcal{E})$ the Fock space over $\mathcal{E}$, identified with the Hilbert integral $\mathcal{F} = \int^{\oplus} \mathcal{E}(\tau)d\tau$ of square-integrable tensor-functions on $\tau \in \Omega^+ \mapsto \xi(\tau) \in \mathcal{E}(\tau)$, $\mathcal{E}(\tau) = \bigotimes_{t \in \tau} \mathcal{E}(t)$, where $\tau = (t_1, \dots, t_n)$ is a chain $0 < t_1 < \dots < t_n < \infty$ of finite length $n = |\tau| < \infty$ with natural Lebesque measure $d\tau = dt_1 \dots dt_n$. The Fock space scalar product

$$\int_0^\infty \langle\xi(\tau)|\eta(\tau)\rangle d\tau = \sum_{n=0}^{\infty} \int \underset{0<t_1<\dots<t_n<\infty}{\dots} \int \langle\xi|\eta\rangle(t_1, \dots, t_n)dt_1 \dots dt_n$$

is defined by the products

$$\langle\xi|\eta\rangle(t_1, \dots, t_n) = \xi_o^*(t_1, \dots, t_n)\,\eta^o(t_1, \dots, t_n), \quad \xi_o^*\eta^o = \xi^*_{i_1 \dots i_n}\eta^{i_1 \dots i_n}$$

of tensors $\xi_o^* = (\xi^*_{i_1 \dots i_n})$, $\xi^*_{i_1 \dots i_n} = \bar{\xi}^{i_1 \dots i_n}$, conjugated to $\xi^o = (\xi^{i_1 \dots i_n})$, which are the scalars $\xi^o(t_1, \dots, t_n) \in \mathbb{C}$ in the case of $\mathcal{F} = L^2(\Omega^+)$, corresponding to $\mathcal{K} = \mathbb{C}$.

Due to the one-to-one correspondence of the finite chain set Ω^+ and the Boolean ring of all finite subsets of $\mathbb{R}^+$, one can define on Ω^+ the Boolean operations $\cap$, $\cup$, and difference $\tau \setminus \sigma = \tau \cap \bar{\sigma}$. We shall denote by $\emptyset$ the empty chain, and by $\xi = \delta_\emptyset$ the vaccum function $\delta_\emptyset(\tau) = 0$, if $\tau \neq \emptyset$, $\delta_\emptyset(\emptyset) = 1$. According to the direct product representation $\Omega^+ = \Omega^s \times \Omega_s^+$ of any chain $\tau \in \Omega^+$ by the pair $\tau = (\tau^s, \tau_s)$, of $\tau^s = \{t \in \tau \mid t \le s\}$ and $\tau_s = \{t \in \tau \mid t > s\}$, so that $\mathcal{E}(\tau) = \mathcal{E}(\tau^s) \otimes \mathcal{E}(\tau_s)$, one can identify the Fock space $\int^{\oplus}\mathcal{E}(\tau)d\tau = \iint \mathcal{E}(\tau^s)\otimes\mathcal{E}(\tau_s)d\tau^s d\tau_s$ with the tensor product $\mathcal{F}^s\otimes\mathcal{F}_s$ of $\mathcal{F}^s = \Gamma(\mathcal{E}^s)$ and $\mathcal{F}_s = \Gamma(\mathcal{E}_s)$, where $\mathcal{E}^s = \int^{\oplus}_{t\leq s}\mathcal{E}(t)dt$ and $\mathcal{E}_s = \int^{\oplus}_{t>s}\mathcal{E}(t)dt$.

Let us consider the basic quantum creation $A^+ = (A_i^+)$, annihilation $A_- = (A_-^j)$ and number $N = (N_i^j)$ processes as operator-valued functions on $t\in\mathbb{R}_+$, defined on the tensor-product functions $\xi(\tau) = \otimes_{t\in\tau}\xi(t)$ by

$$(A_i^+(t)\xi)(\tau) = \sum_{s\in\tau}\xi(\tau^s\setminus s)\otimes\mathcal{X}_i(t,s)\otimes\xi(\tau_s)$$

$$(A_-^j(t)\xi)(\tau) = \int_0^t\xi^j(r)dr\;\xi(\tau) \tag{1.3}$$

$$(N_i^j(t)\xi)(\tau) = \sum_{s\in\tau}\xi(\tau^s\setminus s)\otimes\xi^j(s)\mathcal{X}_i(t,s)\otimes\xi(\tau_s)$$

where $\mathcal{X}_i(t,s) = 0$, if $t<s$, $\mathcal{X}_i(t,s) = \delta_i$, $t\geq s$ with $\delta_i = (\delta_i^j)$, $\delta_i^j = 0$, $i\neq j$, and $\delta_i^j = 1$, $i=j$. We shall denote these processes, acting in Hilbert tensor product $\mathcal{H} = h\otimes\mathcal{F}$ of $\mathcal{F}$ with an initial Hilbert space h as

$$\Lambda_0^+(t) = I\otimes A^+(t),\quad \Lambda_-^0(t) = I\otimes A_-(t),\quad \Lambda_0^0(t) = I\otimes N(t),$$

where I is the identity operator in h, and $\Lambda_-^+(t) = t\hat{I}$, $\hat{I} = I\otimes\hat{1}$ with $\hat{1}$ being the identity in $\mathcal{F}$. The tensor-process $\Lambda = (\Lambda_\mu^\nu)$, $\mu\in\{-,J\}$, $\nu\in\{J,+\}$ can be described by operator-valued functionals $\Lambda(c,t) = c_\nu^\mu\Lambda_\mu^\nu(t)$,

$$c_\nu^\mu\Lambda_\mu^\nu = I\otimes\left(c_+^- t\hat{1} + A^+(c_+^o,t) + A_-(c_o^-,t) + N(c_o^o,t)\right), \tag{1.4}$$

where $A^+(c_+^o) = c_+^i A_i^+$, $A_-(c_o^-) = c_j^- A_-^j$, $N(c_o^o) = c_j^i N_i^j$, satisfying the conditions of the following definition.

<u>Definition 1</u>. Let $\mathfrak{C}$ be the Lie $\bigstar$-algebra of complex matrices $c = (c^{\mu}_{\nu})$, indexed by $\mu, \nu \in \{-, J, +\}$, $c^{\mu}_{\nu} = 0$, if $\mu = +$ or $\nu = -$ with involution $c^{\bigstar} = gc^{*}g$,

$$\begin{pmatrix} 0 & c^{-}_{o} & c^{-}_{+} \\ 0 & c^{o}_{o} & c^{o}_{+} \\ 0 & 0 & 0 \end{pmatrix}^{\bigstar} = \begin{pmatrix} 0 & 0 & 1 \\ 0 & 1^{o}_{o} & 0 \\ 1 & 0 & 0 \end{pmatrix} c^{*} \begin{pmatrix} 0 & 0 & 1 \\ 0 & 1^{o}_{o} & 0 \\ 1 & 0 & 0 \end{pmatrix} = \begin{pmatrix} 0 & c^{o*}_{+} & c^{-*}_{+} \\ 0 & c^{o*}_{o} & c^{-*}_{o} \\ 0 & 0 & 0 \end{pmatrix} \tag{1.5}$$

defined by indefinite matrix $g = (g^{\mu}_{\nu})$ of pseudo-Euclidean metric in $\mathbb{C} \oplus \mathcal{K} \oplus \mathbb{C}$:

$$(\xi|\eta) = \bar{\xi}^{-}\eta^{+} + \langle\xi|\eta\rangle + \bar{\xi}^{+}\eta^{-} = \xi^{*}_{\mu} g^{\mu}_{\nu} \eta^{\nu} \tag{1.6}$$

where the row $\xi^{*} = (\xi^{*}_{-}, \xi^{*}_{o}, \xi^{*}_{+})$, $\xi^{*}_{\mu} = \overline{\xi^{\mu}}$ is conjugated to the column $\xi = (\xi^{\nu})$. The family $\Lambda = \{\Lambda(t)\}$ of representations $\Lambda(t) : c \in \mathfrak{C} \mapsto \Lambda(c, t)$, defined as linear operator-valued functionals $\Lambda(c,t) = c^{\mu}_{\nu}\Lambda^{\nu}_{\mu}(t)$, satisfying conditions $\Lambda(c,t)^{*} = \Lambda(c^{\bigstar}, t)$,

$$[\Lambda(c^{\bigstar}, t), \Lambda(c, t')] = \Lambda([c^{\bigstar}, c], t \wedge t'), \tag{1.7}$$

where $[c^{\bigstar}, c] = c^{\bigstar}c - cc^{\bigstar}$, and $t \wedge t' = \min\{t, t'\}$, is called pseudo-Poissonian quantum process in $\mathcal{H}$.

Indeed, taking into account that $A_{-}(t)^{*} = A^{+}(t)$, $N(t)^{*} = N(t)$ one obtains

$$\Lambda(c,t)^{*} = I \otimes \left(c^{-*}_{+} t\hat{1} + A_{-}(c^{-*}_{o}, t) + A^{+}(c^{o*}_{+}, t) + N(c^{o*}_{o}, t)\right),$$

that is $\Lambda(c,t)^{*} = \Lambda(c^{\bigstar}, t)$, where the conjugated matrix $c^{\bigstar}$ is defied as (1.5). The Lie-product property

$$\Lambda(b,t)\Lambda(d,t) - \Lambda(d,t)\Lambda(b,t) = \Lambda([b,d],t)$$

for arbitrary matrices $b = (b^\mu_\nu)$ and $d = (d^\mu_\nu)$ with $b^\mu_\nu = 0 = d^\mu_\nu$ for $\mu = +$ or $\nu = -$ follows directly from the canonical commutation relations

$$[A_-(b^-_o), A^+(d^o_+)] = t b^-_o d^o_+ \hat{1}, \quad [N(b^o_o), N(d^o_o)] = N([b^o_o, d^o_o])$$

$$[N(b^o_o), A^+(d^o_+)] = A^+(b^o_o d^o_+), \quad [A_-(b^-_o), N(d^o_o)] = A_-(b^-_o d^o_o)$$

for any t. Taking into account, that $\Lambda(b,t)$ commutes with increaments $\Lambda(d,t') - \Lambda(d,t)$, $t' \geq t$, one obtains

$$[\Lambda(b,t), \Lambda(d,t')] = \Lambda([b,d], t \wedge t').$$

But it is equivalent to (1.7) for all $b = c^\star$ and $d = c$ due to the polarization formula

$$bd = \sum_{n=0}^{3} (b^\star + i^n d)^\star (b^\star + i^n d)/4i^n, \quad i = \sqrt{-1},$$

giving $\Lambda(bd)$ as the sum $\sum_{n=0}^{3} \Lambda(c_n^\star c_n)/4i^n$ with $c_n = b^\star + i^n d$.

2. A quantum stochastic integral

Let us now define a QS integral with respect to Λ for a matrix process $C(t) = (C^\mu_\nu)(t)$, $\mu, \nu \in \{-, J, +\}$ in $\mathcal{H}$. Assuming that the operator-valued functions $t \mapsto C^\mu_\nu(t)$ are weakly measurable and adapted: $C(t) = C_{t]} \otimes I_t$, where $C_{t]}$ are the operators in $\mathcal{H}^t = \mathfrak{h} \otimes \mathcal{F}^t$ for all $\mu \in \{-, J\}$ and $\nu \in \{J, +\}$, one can define in the case of finite J the QS-integral

$$\int_0^t A(C(s), ds) = \int_0^t \left(C_+^- dt + C_0^- dA_- + C_+^o dA^+ + C_0^o dN\right) = \int_0^t C_\nu^\mu d\Lambda_\mu^\nu$$

as the sum of the Lebesque operator-valued integral $\int C_+^-(s)ds$ and the Ito integrals $\int C_j^- \, dA_-^j$, $\int C_+^j \, dA_j^+$, $\int C_k^i \, dN_i^k$ in Hudson-Parthasarathy sense [1]. In the general case we shall regard the QS-integral $\int C_\nu^\mu dA_\mu^\nu$ as a continuous operator $\mathcal{H}(\infty) \to \mathcal{H}(1) = \mathcal{H}$ on the projective limit $\mathcal{H}(\infty) = \bigcap_{\zeta>0} \mathcal{H}(\zeta)$, $\mathcal{H}(\zeta) = h \otimes \mathcal{F}(\zeta)$ of Fock spaces $\mathcal{F}(\zeta) = \int^\oplus \mathcal{E}(\tau)\zeta^{|\tau|} d\tau$, $\mathcal{E}(\tau) = \bigotimes_{t\in\tau} \mathcal{E}(t)$, where $\mathcal{E}(t) = \mathbb{C}^m$, if $|J| = m < \infty$, with the scalar product $\langle\varphi|\chi\rangle(\zeta) = \int \zeta^{|\tau|} \langle\varphi|\chi\rangle(\tau) d\tau$.

We shall say that a weakly measurable function $t \mapsto C(t)$ is locally QS-integrable if its components C_ν^μ, $\mu \in \{-, o\}$, $\nu \in \{o, +\}$ are locally L^p-integrable as operator-valued functions

$$C_+^-(t) : \mathcal{H}(\infty) \to \mathcal{H}, \qquad \|C_+^-(\cdot)\|_{\zeta,t}^{(1)} < \infty \quad (p = 1)$$

$$C_+^o(t) : \mathcal{H}(\infty) \to \mathcal{H} \otimes \mathcal{E}(t), \qquad \|C_+^o(\cdot)\|_{\zeta,t}^{(2)} < \infty \quad (p = 2)$$

$$C_o^-(t) : \mathcal{H}(\infty) \otimes \mathcal{E}(t) \to \mathcal{H}, \qquad \|C_o^-(\cdot)\|_{\zeta,t}^{(2)} < \infty \quad (p = 2)$$

$$C_o^o(t) : \mathcal{H}(\infty) \otimes \mathcal{E}(t) \to \mathcal{H} \otimes \mathcal{E}(t), \qquad \|C_o^o(\cdot)\|_{\zeta,t}^{(\infty)} < \infty \quad (p = \infty)$$

Here the norms are defined for any $t \geq 0$ and a sufficiently large $\zeta > 0$ by

$$\|C_+^-\|_{\zeta,t}^{(1)} = \int_0^t \|C_+^-(s)\|_\zeta ds , \qquad \|C_o^o\|_{\zeta,t}^{(\infty)} = \operatorname*{ess\,sup}_{s \leq t} \|C_o^o(s)\|_\zeta , \tag{2.1}$$

$$\|C_+^-\|_\zeta = \sup_\varphi \{\|C_+^- \varphi\| / \|\varphi\|(\zeta), \quad \|C_o^o\|_\zeta = \sup \{\|C_o^o \varphi_t^o\| / \|\varphi_t^o\|(\zeta) ,$$

where $\varphi \in \mathcal{H}(\zeta)$, $\|\varphi\|^2(\zeta) = \langle\varphi|\varphi\rangle(\zeta)$, $\varphi_t^o \in \mathcal{H}(\zeta) \otimes \mathcal{E}(t)$, $\|\varphi_t^o\|^2(\zeta) = \langle\varphi_t^o|\varphi_t^o\rangle(\zeta)$ and $\|C_+^o\|_{\zeta,t}^{(2)} = \|C_+^{ot}\|_\zeta$, $\|C_o^-\|_{\zeta,t}^{(2)} = \|C_o^{-t}\|_\zeta$ are the norms of operators

$$C_+^{ot}: \mathcal{H}(\zeta) \to \mathcal{H}\otimes\mathcal{E}^t, \qquad (C_+^{ot}\varphi)(s) = C_+^o(s)\varphi, \quad s \leqslant t,$$

$$C_o^{-t}: \mathcal{H}(\zeta)\otimes\mathcal{E}^t \to \mathcal{H}, \qquad C_o^{-t}\varphi^o = \int_0^t C_o^-(s)\varphi^o(s)ds,$$

defined by L^2-integrals

$$\|C_+^o\|_{\zeta t}^{(2)} = \left(\int_0^t \|C_+^o(s)\|_\zeta^2 ds\right)^{1/2} = \|C_+^{o*}\|_{\zeta t}^{(2)},$$
$$\|C_o^-\|_{\zeta t}^{(2)} = \left(\int_0^t \|C_{o\zeta}^{-*}(s)\|_\zeta^2 ds\right)^{1/2} = \|C_o^{-*}\|_{\zeta t}^{(2)}, \tag{2.2}$$

with $C_\zeta^*: \mathcal{H}(\infty) \to \mathcal{H}$ defined for all $\varphi \in \mathcal{H}(\infty)$, $\chi \in \mathcal{H}$ by

$$\langle C_\zeta^*\varphi|\chi\rangle = \langle\varphi\,|\,C\chi_\zeta\rangle\zeta^{1/2}), \qquad \chi_\zeta(\tau) = \chi(\tau)/\zeta^{|\tau|/2}.$$

The following theorem shows that the QS-integral for an ontegrable C can be defined on $\mathcal{H}(\infty)$ by the formula

$$\left(\int_0^t \Lambda(C(s),ds)\varphi\right)(\tau) = \int_0^t (C_+^-(s)\varphi + C_o^-(s)\varphi^o(s))(\tau)ds$$
$$+ \sum_{s\in\tau}^{s\leqslant t}(C_+^o(s)\varphi + C_o^o(s)\varphi^o(s))(\tau\setminus s), \tag{2.3}$$

where $\varphi^o(t) \in \mathcal{H}(\infty)\otimes\mathcal{E}(t)$ is defined almost everywhere as the tensor-function $\varphi^o(\tau,t) = \varphi(\tau\sqcup t)$, $\tau\sqcup t = \tau\cup t$ for $t \notin \tau$.

<u>Theorem 1</u>. Let $C(t)$ be a locally QS-integrable function, i.e. that for any $t > 0$ there exists $\zeta > 0$, such that

$$\|C_+^-\|_{\zeta t}^{(1)} < \infty, \quad \|C_+^o\|_{\zeta t}^{(2)} < \infty, \quad \|C_o^-\|_{\zeta t}^{(2)} < \infty, \quad \|C_o^o\|_{\zeta t}^{(\infty)} < \infty.$$

Then the QS-integral (2.3) is defined as the continuous operator $\mathcal{H}(\infty) \to \mathcal{H}(1-) = \bigcap_{\xi<1}\mathcal{H}(\xi)$ with

$$\|\int_0^t \Lambda(C(s),ds)\|_\eta^{(1-\varepsilon)} \leq \|C_+^-\|_{\zeta t}^{(1)} + \frac{1}{\sqrt{\varepsilon}}(\|C_o^-\|_{\zeta t}^{(2)} + \|C_+^o\|_{\zeta t}^{(2)}) + \frac{1}{\varepsilon}\|C_o^o\|_{\zeta t}^{(\infty)} \quad (2.4)$$

for $\eta \geq \zeta + \varepsilon$ and sufficiently small $\varepsilon > 0$ ($\varepsilon < 1$). Moreover, the conjugated integral, defined by

$$\langle(\int_0^t \Lambda(C(s),ds)^* \varphi|\chi\rangle = \langle\varphi|\int_0^t \Lambda(C(s),ds)\chi\rangle, \qquad \varphi,\chi \in \mathcal{H}(\infty) \quad (2.5)$$

is also densely defined on $\mathcal{H}(\infty) \subset \mathcal{H}$ as the QS-integral $\int_0^t \Lambda(C^\star(s),ds)$, if the function $C^\star(t) = gC(t)^* g$,

$$C^\star(t)_+^- = C_+^-(t)^*, \quad C^\star(t)_+^o = C_o^-(t)^*, \quad C^\star(t)_o^- = C_+^o(t)^*, \quad C^\star(t)_o^o = C_o^o(t)^*$$

is locally QS-integrable.

<u>Proof</u>. In order to show the continuity of the integral (2.3) in the projective topology of $\bigcap_{\zeta>0}\mathcal{H}(\zeta)$, one should prove that

$$\|\int_0^t \Lambda(C(s),ds)\varphi\|(1-\varepsilon) \leq c\,\|\varphi\|(\eta), \qquad \|\varphi\|(\eta) = \langle\varphi|\varphi\rangle(\eta)^{1/2}$$

for any $\varphi \in \mathcal{H}(\eta)$, $c>0$. Due to the definition (2.3)

$$\|\int_0^t \Lambda(C,ds)\varphi\| \leq \|\int_0^t C_+^- ds\varphi\| + \|\int_0^t C_o^- d\Lambda_-\varphi\| + \|\int_0^t C_+^o d\Lambda^+\varphi\| + \|\int_0^t C_o^o dN\varphi\| ,$$

where

$$\int_0^t C_o^- d\Lambda_-\varphi = \int_0^t C_o^-(s)\varphi^o(s)ds, \qquad (\int_0^t C_+^o d\Lambda^+\varphi)(\tau) = \sum_{s\in\tau}^{s\leq t}(C_+^o(s)\varphi)(\tau\backslash s),$$

and

$$(\int_0^t C_o^o(s)dN(s)\varphi)(\tau) = \sum_{s\in\tau}^{s\leq t}(C_o^o(s)\varphi^o(s))(\tau\backslash s).$$

The first two integrals in (2.3) can be easily estimated as

$$\|\int_0^t C_+^- ds\|(\xi) \leq \int_0^t \|C_+^-(s)\varphi\| ds \leq \cdot \int_0^t \|C_+^-(s)\|_\zeta\, ds\, \|\varphi\|(\zeta) = \|C_+^-\|_{\zeta t}^{(1)}\|\varphi\|(\zeta),$$

$$\|{}^t\!\!\int_0 c_o^- d\Lambda_-\varphi\|(\xi) \leq \|c_o^{-t}\varphi^o\| \leq \|c_o^{-t}\|_\zeta \|\varphi^o\|(\zeta) = \|c_o^-\|^{(2)}_{\zeta t}\left(\frac{d}{d\zeta}\|\varphi\|^2(\zeta)\right)^{1/2},$$

where we took into account that $\xi = 1-\varepsilon < 1$ and

$$\|\varphi^o\|^2(\zeta) = \iint \zeta^{|\tau|}\|\varphi(\tau \sqcup t)\|^2 d\tau dt = \int |\tau|\zeta^{|\tau|-1}\|\varphi(\tau)\|^2 d\tau = \frac{d}{d\zeta}\|\varphi\|^2(\zeta).$$

In order to estimate the integrals of c_+^o and c_o^o let us find

$$\int_0^t \|c_+^o d\Lambda^+\varphi\|^2(\xi) = \int_\Omega \xi^{|\tau|}\|\sum_{s\in\tau}^{s\leq t}(c_+^o(s)\varphi)(\tau\setminus s)\|^2 d\tau =$$

$$\int_\Omega \xi^{|\tau|}\Big(\sum_{s\in\tau}^{s\leq t}\|c_+^o(s)\varphi\ (\tau\setminus s)\|^2 + \sum_{s_1\sqcup s_2\subset\tau}^{s_1,s_2\leq t}\langle c_+^o(s_1)\varphi^o(s_2)|c_+^o(s_2)\varphi^o(s_1)\rangle(\tau\setminus$$

$$(s_1\sqcup s_2))\Big)d\tau = \xi\int_0^t\int_\Omega \|c_+^o(s)\varphi\|(\tau)^2\,\xi^{|\tau|}d\tau ds + \xi^2\int_0^t\int_0^t\int_\Omega\langle(c_+^o(s_1)\varphi)^o(\tau,s_2)|$$

$$(c_+^o(s_2)\varphi)^o(\tau,s_1)\rangle\,\xi^{|\tau|}d\tau ds_1 ds_2 \leq \xi\left(1+\xi\frac{d}{d\xi}\right)\|c_+^o\varphi\|_t^2(\xi)$$

by Schwarz inequality. In the same way we get

$$\|\int_0^t c_o^o dN\varphi\|^2(\xi) = \int_\Omega \xi^{|\tau|}\|\sum_{s\in\tau}^{s\leq t}(c_o^o(s)\varphi^o(s))(\tau\setminus s)\|^2 d\tau =$$

$$\int_\Omega \xi^{|\tau|}\Big(\sum_{s\in\tau}^{s\leq t}\|c_o^o(s)\varphi^o(s)\|(\tau\setminus s)^2 + \sum_{s_1\sqcup s_2\subset\tau}^{s_1,s_2\leq t}\langle c_o^o(s_1)\varphi^{oo}(s_1,s_2)|c_o^o(s_2)\cdot$$

$$\varphi^{oo}(s_1,s_2)\rangle(\tau\setminus(s_1\sqcup s_2))\Big)d\tau = \xi\int_0^t\int_\Omega\|c_o^o(s)\varphi^o(\tau,s)\|^2\,\xi^{|\tau|}d\tau ds +$$

$$\xi^2\int_0^t\int_0^t\int_\Omega\langle(c_o^o(s_1)\varphi^o(s_1))^o(\tau,s_2)|(c_o^o(s_2)\varphi^o(s_2))^o(\tau,s_1)\rangle\,\xi^{|\tau|}d\tau ds_1 ds_2$$

$$\leq \xi\left(1+\xi\frac{d}{d\xi}\right)\|c_o^o\varphi^o\|_t^2(\xi) \leq \left(1+\frac{d}{d\xi}\right)\|c_o^o\|^2_{\zeta t}(1-\varepsilon)\frac{d}{d\zeta}\|\varphi\|^2(\zeta).$$

Taking into account that

$$\frac{d}{d\zeta}\|\varphi\|^2(\zeta) \leq \frac{1}{\varepsilon}\left(\|\varphi\|^2(\zeta+\varepsilon) - \|\varphi\|^2(\zeta)\right) \leq \frac{1}{\varepsilon}\|\varphi\|^2(\zeta+\varepsilon)$$

$$(1+\xi\tfrac{d}{d\xi})\|\varphi^{0}\|^{2}(\xi) \leqslant \tfrac{\xi}{\varepsilon}\|\varphi^{0}\|^{2}(\xi+\varepsilon) - \tfrac{\xi-\varepsilon}{\varepsilon}\|\varphi^{0}\|^{2}(\xi) \leqslant \tfrac{\xi}{\varepsilon}\|\varphi^{0}\|^{2}(\xi+\varepsilon)$$

for $\xi > \varepsilon > 0$, one can find that

$$(1+\xi\tfrac{d}{d\xi})\|c_{+}^{0}\varphi\|_{t}^{2}(\xi) \leqslant \tfrac{\xi}{\varepsilon}\|c_{+}^{0}\|_{\zeta t}^{2}(\xi+\varepsilon)\|\varphi\|^{2}(\zeta),$$

$$(1+\xi\tfrac{d}{d\xi})\|c_{0}^{0}\varphi^{0}\|_{t}^{2}(\xi) \leqslant \tfrac{\xi}{\varepsilon}\|c_{0}^{0}\|_{\zeta}^{2}(\xi+\varepsilon)\tfrac{d}{d\zeta}\|\varphi\|^{2}(\zeta) \leqslant \tfrac{\xi}{\varepsilon^{2}}\|c_{0}^{0}\|_{\zeta t}^{2}(\xi+\varepsilon)\|\varphi\|^{2}(\zeta+\varepsilon)$$

Hence, due to $\|\varphi\|(\eta) \geqslant \|\varphi\|(\zeta+\varepsilon) \geqslant \|\varphi\|(\zeta)$ for $\eta > \zeta + \varepsilon$, we obtain for $\xi = 1-\varepsilon$

$$\Big\|\int_{0}^{t}\Lambda(c(s),ds)\varphi\Big\|(\xi) \leqslant \|c_{+}^{-}\|_{\zeta t}^{(1)} + \tfrac{1}{\sqrt{\varepsilon}}\big(\|c_{0}^{-}\|_{\zeta t}^{(2)} + \|c_{+}^{0}\|_{\zeta t}^{(2)}\big) + \tfrac{1}{\varepsilon}\|c_{0}^{0}\|_{\zeta t}^{(\infty)},$$

if $\|\varphi\|(\eta) \leqslant 1$, $0 < \varepsilon \leqslant 1$, what is equivalent to (2.4).

If the QS-matrix process $c^{\star}(t)$ is also locally QS-integrable, then there exists the conjugated integral $\int_{0}^{t}\Lambda(c(s),ds)^{*}$ defined as in (2.3) by $c^{\star}$:

$$\langle\varphi|\int_{0}^{t}\Lambda(c(s),ds)\chi\rangle = \int_{0}^{t}\langle\varphi|c_{+}^{-}(s)\chi + c_{0}^{-}(s)\chi^{0}(s)\rangle\, ds$$

$$+ \int_{0}^{t}\langle\varphi^{0}(s)|c_{+}^{0}(s)\chi + c_{0}^{0}(s)\chi^{0}(s)\rangle ds = \int_{0}^{t}\langle c_{0}^{-}(s)^{*}\varphi + c_{0}^{0}(s)^{*}\varphi^{0}(s)|\chi^{0}(s)\rangle\, ds$$

$$+ \int_{0}^{t}\langle c_{+}^{-}(s)^{*}\varphi + c_{+}^{0}(s)^{*}\varphi^{0}(s)|\chi\rangle ds = \langle\int_{0}^{t}\Lambda(c^{\star}(s),ds)|\chi\rangle .$$

Note that if $c_{\nu}^{\mu}(t)$ are bounded on $\mathcal{H} = \mathcal{H}(1)$ for all t, then the conjugated process $c(t)$ is also bounded:

$$\|c^{\star}{}_{+}^{-}\|_{t}^{(1)} = \|c_{+}^{-}\|_{t}^{(1)},\quad \|c^{\star}{}_{+}^{0}\|_{t}^{(2)} = \|c_{0}^{-}\|_{t}^{(2)},\quad \|c^{\star}{}_{0}^{-}\|_{t}^{(2)} = \|c_{0}^{+}\|_{t}^{(2)}$$

and $\|c^{\star}{}_{0}^{0}\|_{t}^{(\infty)} = \|c_{0}^{0}\|_{t}^{(\infty)}$ $(\zeta = 1)$. Hence the conjugated integral for such a locally integrable c exists as a continuous operator $\mathcal{H}(1+) \to \mathcal{H}(1-)$, $\mathcal{H}(1+) = \bigcup_{\eta>1} \mathcal{H}(\eta)$, having the same estimate for $\zeta = 1$ as (2.4) for $\zeta = 1$.

<u>Corollary</u>. If $C(t)$ is a simple integrable function, then the definition (2.3) coincides with the QS-integral, given by an integral sum in Ito sense with respect to the process (1.1). Moreover, the QS-integral (2.3) is a limit of such integral sums in the inductive operator topology, defined by the norms (2.4), if a locally QS-integrable matrix-process C can be uniformly approximated by a sequence of simple operator-valued processes with respect to the defined L^p-norms on $]0,t]$.

3. A quantum stochastic differential equation and Ito's formula

Let us consider a QS process

$$Y(t) = Y(0) + \int_0^t C^\mu_\nu(r)\, d\Lambda^\nu_\mu(r), \quad t > 0 \tag{3.1}$$

defined by a bounded initial operator $Y(0) = Y^0 \otimes \hat{1}$, $Y^0 : \mathcal{H}^0 = h \to \mathcal{H}^0$ and by the QS integral of an adapted QS matrix-process $C = (C^\mu_\nu)$, of the operators $C^\mu_\nu(t) : \mathcal{H}(\infty) \to \mathcal{H}(\infty)$. We shall prove that if the process C and $C^\star$ are locally QS-integrable with respect to the uniform topology of operators in the projective limit $\mathcal{H}(\infty) = \bigcap \mathcal{H}(\zeta)$, then $Y(t)$ and $Y^*(t)$ act on $\mathcal{H}(\infty)$ as uniformly continuous operators into $\mathcal{H}(\infty)$. Moreover, the operators $(Y^*Y)(t) = Y^*(t)Y(t)$ are defined by

$$(Y^*Y)(t) = (Y^*Y)(0) + \int_0^t d(Y^*Y)(r) \tag{3.2}$$

with $d(Y^*Y) = \left(C^{\star\mu}_\nu Y + Y^* C^\mu_\nu + (C^\star C)^\mu_\nu\right) d\Lambda^\nu_\mu$, and they are also uniformly continuous in $\mathcal{H}(\infty) \to \mathcal{H}(\infty)$. The formula (3.2) can be extended to the Ito formula [1]

$$d(XZ) = \left(B^{\mu}_{\nu}Z + XD^{\mu}_{\nu} + B^{\mu}_{\varkappa}D^{\varkappa}_{\nu}\right)d\Lambda^{\nu}_{\mu}\ , \tag{3.3}$$

which we shall write interms of $E = B + X\otimes\delta$, $G = D+Z\otimes\delta$ as $d(XZ) = \left(EG - XZ\otimes\delta\right)^{\mu}_{\nu}\, d\Lambda^{\nu}_{\mu}$. Here $B = (B^{\mu}_{\nu})$, $D = (D^{\mu}_{\nu})$ are the processes of the same kind as C, defining the QS matrix-process $F(t) = C(t) + Y(t)\otimes\delta$ in the QS differential form

$$dY(t) = \left(F(t) - Y(t)\otimes\delta\right)^{\mu}_{\nu}\, d\Lambda^{\nu}_{\mu}(t), \tag{3.4}$$

where $(Y\otimes\delta)^{\mu}_{\nu} = Y\delta^{\mu}_{\nu}$, $\delta^{\mu}_{\nu} = 1$, if $\mu = \nu$, and $\delta^{\mu}_{\nu} = 0$, if $\mu \neq \nu$ ($\delta^{0}_{0} = 1^{0}_{0}$ is the identity matrix $1^{0}_{0} = (\delta^{i}_{j})$)

In general, the form (3.4) is called the QS differential equation if the QS process $F(t) = (F^{\mu}_{\nu})(t)$ depends on $Y(t)$. We shall always assume that $F^{-}_{-}(t) = Y(t) = F^{+}_{+}(t)$, and $F^{\mu}_{\nu}(t) = 0$ if $\mu > \nu$ under the order $- < 0 < +$. The solution of the equation (3.4) is defined by a QS-integral (3.3), which is an integral QS equation, if $C^{\mu}_{\nu} = F^{\mu}_{\nu} - Y\delta^{\mu}_{\nu}$ depend on Y for $(\mu,\nu)\in\{-,0\}\times\{0,+\}$.

We shall call the QS matrix-process C projectively QS integrable, if for any $t > 0$ and $\xi > 0$ there exists $\zeta > 0$, such that

$$\|C^{-}_{+}\|^{(1)}_{\zeta t}(\xi) + \|C^{0}_{+}\|^{(2)}_{\zeta t}(\xi) + \|C^{-}_{0}\|^{(2)}_{\zeta t}(\xi) + \|C^{0}_{0}\|^{(\infty)}_{\zeta t}(\xi) < \infty \tag{3.5}$$

where $\|C\|^{(p)}_{\zeta t}(\xi)$ is defined as in (2.1) , (2.) by the norms

$$\|C(t)\|_{\zeta}(\xi) = \sup_{\varphi}\left\{\|C(t)\varphi\|(\xi) / \|\varphi\|(\zeta)\right\} \tag{3.6}$$

__Theorem 3.__ Let C and $C^{\star}$ be projectively QS integrable $\star$-conjugated matrix-processes. Then the process $(Y^{*}Y)(t) = Y(t)^{*}Y(t)$, defined by Y, satisfying the equation (3.4) with $F = C + Y\otimes\delta$, is a positive uniformly continuous process in the projective limit $\mathcal{H}(\infty)$ with

$$d(Y^{*}Y) = \left(F^{\star}F - Y^{*}Y\otimes\delta\right)^{\mu}_{\nu}\, d\Lambda^{\nu}_{\mu}\,, \tag{3.7}$$

where $F^{\star} = C^{\star} + Y^{*}\otimes\delta$. The formula (3.7) being equivalent to the QS Ito formula (3.3) and $\star$-property:

$$dY^{*} = \left(F^{\star} - Y^{*}\otimes\delta\right)^{\mu}_{\nu}\, d\Lambda^{\nu}_{\mu}\,, \tag{3.8}$$

define an algebraic isomorphism of $*$-operator associative structure of the QS processes (3.1), satisfying together with Y^{*} the conditions (3.5) for all t, ξ and $\zeta(t,\xi)$, into the $\star$-operator structure of the corresponding QS matrix-processes F. Moreover, if a QS process X is a solution of the equation

$$dX = \left(E - X\otimes\delta\right)^{\mu}_{\nu}\, d\Lambda^{\nu}_{\mu} \tag{3.9}$$

with projectively integrable $B = E - X\otimes\delta$ and f is an analytic function, for which $f(X(0))$ is defined and $f(E)$ is defined in the inductive topology of the operators on $\mathcal{H}(\infty)$ into $\mathcal{H}(\infty)$, then the process $Y = f(X)$ is defined by the equation

$$df(X) = \left(f(E) - f(X)\otimes\delta\right)^{\mu}_{\nu}\, d\Lambda^{\nu}_{\mu}\,. \tag{3.10}$$

<u>Proof</u>. If C and $C^{\star}$ satisfy the condition (3.5) for some $t>0$ and $\xi>0$, then one can find in the same way as in the proof of the theorem 2 the following estimate

$$\Big\|\int_0^t C^{\mu}_{\nu}\, d\Lambda^{\nu}_{\mu}\Big\|_{\rho}(\xi) \leq \|C^{-}_{+}\|^{(1)}_{\zeta t}(\xi) + \frac{1}{\sqrt{\varepsilon}}\left(\|C^{-}_{0}\|^{(2)}_{\zeta t}(\xi) + \xi\|C^{0}_{+}\|^{(2)}_{\zeta t}(\xi)\right)$$
$$+ \frac{\xi}{\varepsilon}\|C^{0}_{0}\|^{(\infty)}_{\zeta t}\,,$$

for all $\rho>\zeta+2\varepsilon$, $0<\varepsilon\leq\xi$, and the same for $C^{\star}$. One can write it for sufficiently large ξ $(\xi \geq 1)$ in the form of (2.4) with ε/ξ^2 instead of ε. It means, that the operator (3.1) as well as $Y(t)$ is uniformly continuous in $\mathcal{H}(\infty)$. Hence, the product $Y(t)^* Y(t)$ is defined on $\mathcal{H}(\infty)$ as uniformly continuous operator into $\mathcal{H}(\infty)$.

Using the explicit definition (2.3) of the QS integral as an operator $\mathcal{H}(\infty) \to \mathcal{H}(\infty)$, one can find the formula (3.7) for the QS differential $d(Y^*Y)$, defining

$$(Y^*Y)(t) = \left(Y^*(0) + \int_0^t \Lambda(C^\star, dr)\right)\left(Y(0) + \int_0^t \Lambda(C, dr)\right)$$

as the QS integral (3.2). It gives the Ito formula (3.3) for $X = Y^*$ and $Z = Y$, which can be uniquely extended for the arbitrary X, Z, defined by projectively QS integrable B, D and $B^\star$, $D^\star$ due to the polarization formula

$$BD = \sum_{n=0}^{3} \left(B^\star + i^n D\right)^\star \left(B^\star + i^n D\right)/4i^n .$$

Indeed, in that case $Y_n = X^* + i^n Z$, $n=0,1,2,3$, satisfies the condition of the theorem 3, hence $XZ = \sum_{n=0}^{3} Y_n^* Y_n / 4i^n$ is defined as an integral $X(0)Z(0) + \int_0^t d(XZ)$ with
$d(XZ) = \sum_{n=0}^{3} d(Y_n^* Y_n)/4i^n = \sum_{n=0}^{3} \left(F_n^\star F_n - Y_n^* Y_n \otimes \delta\right)_\nu^\mu d\Lambda_\mu^\nu / 4i^n$
$= \left(EG - XZ \otimes \delta\right)_\nu^\mu d\Lambda_\mu^\nu$, where $F_n = E^\star + i^n G$, $E = B + X \otimes \delta$, $G = D + Z \otimes \delta$. Formally, the Ito formula (3.3) can be obtained for adapted processes from the Ito differentiation rule

$$d(XZ) = dX\, Z + X\, dZ + dX\, dZ \tag{3.11}$$

by using the Hudson-Parthasarathy multiplication table [1], which can be written in terms of the $\star$-Lie representation Λ as

$$d\Lambda(c)^* d\Lambda(c) = d\Lambda(c^\star c), \quad c \in \mathcal{C} . \qquad (3.12)$$

Indeed, (3.12) can be extended to the table

$$\Lambda(B(t), dt)\, \Lambda(D(t'), dt') = \Lambda(B(t)D(t'), dt \cap dt'),$$

with $\Lambda(dt \cap dt') = 0$ for $dt \cap dt' = \emptyset$ for arbitrary adapted projectively integrable B, D and $B^\star$, $D^\star$. It can be done by linearity of the map $c \to \Lambda(c)$ and polarization formula due to commutativity of $C_n = B^\star + i^n D = C_n^t \otimes 1_t$ and $C_n^\star$ with $d\Lambda(t)$, and commutativity of $d\Lambda(t)$, $d\Lambda(t')$ with $t \neq t'$. Hence, due to (3.11)

$$d(XZ) = \Lambda(B, dt)Z + X\Lambda(D, dt) + \Lambda(BD, dt)$$

$$= \Lambda\left(B(Z \otimes \delta) + (X \otimes \delta)D + BD, dt\right) = \Lambda(EG - XZ \otimes \delta, dt).$$

The $\star$-property (3.8) can be obtained from (3.7) as the case of $X = Y^*$, $Z = \hat{I}$ by the polarization formula

$$F^\star = \sum_{n=0}^{3} (F + i^n \hat{I})^\star (F + i^n \hat{I})/4i^n .$$

The QS Ito formula, written in the form of the QS-equation (3.4), gives obviously (3.9) for any polynomial $Y = f_n(X)$. If a locally QS-integrable process C can be approximated in the norms $\|\cdot\|_{\xi t}^{(p)}$ by the polynomial $C_n = f_n(E) - f_n(X) \otimes \delta$ in $E = B + X \otimes \delta$, defined by a projectively integrable matrix-process B of a process $X(t) = X(0) + \int_0^t \Lambda(B, dr)$ in $\mathcal{H}(\infty)$, and $Y(0) = f(X(0))$ is defined by $f = \lim_{n \to \infty} f_n$ as the uniformly convergent series, then $f(X(t))$ can be defined as the uniformly continuous operator $\mathcal{F}(\infty) \to \mathcal{F}$ by

$$f(X(t)) = f(X(0)) + \int_0^t \Lambda(C(r), dr) \qquad (3.13)$$

due to the continuity of the QS-integral (3.13) in C. It proves the formula (3.10), where $f(E) = C + f(X)\otimes\delta$.

Note, that the formula (3.10) defines the exponential process

$$Y(t) = \exp\{X(t)\}, \quad X(t) = X(0) + \int_0^t \Lambda(B, dt)$$

by the equation

$$dY = \left(\exp\left(B + \ln Y\otimes\delta\right) - Y\otimes\delta\right)^{\mu}_{\nu} d\Lambda^{\nu}_{\mu}. \qquad (3.14)$$

If B and $X\otimes\delta$ commute for any t, then the equation (3.14) can be written as

$$dY = \left(\exp B - \hat{I}\otimes\delta\right)^{\mu}_{\nu} Y d\Lambda^{\nu}_{\mu}, \qquad (3.15)$$

In the noncommutative case the equation (3.14) defines the chronologically ordered stochastic exponential function $Y(t) = \overleftarrow{\exp}\{X(t)\}$, which is a product

$$Y(t) = e^{\Lambda(B_{n(t)}, t - t_{n(t)})} Y_{n(t)}, \quad Y_{n+1} = e^{\Lambda(B_n, t_{n+1} - t_n)} Y_n$$

for a simple function $B(t) = B_{n(t)}$, where $n(t) = \sum_k \chi(t, t_k)$, $\chi(t,r) = 0$, if $r \geq t$, and $= 1$, if $r < t$, $t_{k+1} < t_k$ for any $k = 1, 2, \ldots$.

<u>Corollary</u>. The QS process (3.1) with projectively QS integrable C and $C^{\star}$ is normal (Hermitian, unitary) in $\mathcal{H}(\infty)$ iff $Y(0)$ is a such one and $FF^{\star} = F^{\star}F$ $(F^{\star} = F,\ F^{\star} = F^{-1})$

for $F = C + Y\otimes\delta$, and Y is a partial isometry (isometry , orthoprojection), iff $Y(0)$ is a such one, and $FF^{\star}F = F$ ($F^{\star}F = \hat{I}\otimes\delta$, $F^{\star}F = F$).

In particular, the exponential processes (3.14), (3.15) are isometric (unitary), iff $F = \exp\{B + X\otimes\delta\}$ in (3.14) and $Z = \exp B$ in (3.15) are $\star$-isometric ($\star$-unitary) with respect to the pseudo-Euclidean metric (1.6).

Note, that in the same way one can define an analytic ordered function $Y(t) = f(X_1(t), \ldots , X_n(t))$ on a several noncommuting processes X_i , $i=1,\ldots,n$, satisfying the QS equations

$$dX_i = (E_i - X_i\otimes\delta)^{\mu}_{\nu}\, d\Lambda^{\nu}_{\mu} .$$

For example, if $F(t) = f(E_1(t), \ldots , E_n(t))$ is a QS matrix-process, defined by a convergent sequence of sequentially ordered polynomials

$$f(E_1, \ldots , E_n) = \sum_m f_{m_1\ldots m_n} E_1^{m_1} \ldots E_n^{m_n} ,$$

and $Y(0) = \sum f_{m_1\ldots m_n} X_1(0)^{m_1} \ldots X_m(0)^{m_n}$, then

$Y(t) = \sum f_{m_1\ldots m_n} X_1(t_1)^{m_1} \ldots X_n(t_n)^{m_n}$ as converging sequentially ordered polynomial in $X_i(t)$ satisfies the QS equation

$$df(X_1,\ldots,X_n) = (f(E_1,\ldots,E_n) - f(X_1,\ldots,X_n)\otimes\delta^{\mu}_{\nu})d\Lambda^{\nu}_{\mu} . \quad (3.16)$$

4. QS calculus of continuous nondemolition processes

Let us consider an isometric (unitary) QS evolution

$U(t): \mathcal{H} \to \mathcal{H}$, satisfying the equation of the form (3.15)

$$dU = (Z - \hat{I}\otimes\delta)^{\mu}_{\nu} U \, d\Lambda^{\nu}_{\mu} = U(\hat{Z} - \hat{I}\otimes\delta)^{\mu}_{\nu} d\Lambda^{\nu}_{\mu} , \qquad (4.1)$$

where $\hat{Z}(t)$ is an adapted QS matrix-process $(\hat{Z}^{\mu}_{\nu})(t)$, defined in Schrödinger picture as a pseudounitary QS process $Z(t) = (Z^{\mu}_{\nu})(t)$ with $Z^{-}_{-} = \hat{I} = Z^{+}_{+}$:

$$U(t)\,\hat{Z}^{\mu}_{\nu}(t) = Z^{\mu}_{\nu}(t)U(t). \qquad (4.2)$$

We shall suppose, that the process $Z(t)$ satisfies the sufficient conditions [4] for the existence of such a solution U of the QS equation (4.1) with $U(0) = \hat{I}$, which are weak measurability and local boundedness of the process Z. In such a case the process $\hat{Z}$ is also pseudounitary $\hat{Z}^{\star} = \hat{Z}^{-1}$ what can be expressed in terms of matrix elements $\hat{Z}^{\mu}_{\nu}$, $\mu\in\{-,o\}$, $\nu\in\{o,+\}$ as

$$\hat{Z}^{o*}_{+}\hat{Z}^{o}_{o} + \hat{Z}^{-}_{o} = 0 , \quad \hat{Z}^{o*}_{o}\hat{Z}^{o}_{+} + \hat{Z}^{-*}_{o} = 0 ,$$

$$\hat{Z}^{o*}_{o}\hat{Z}^{o}_{o} = \hat{I} , \quad \hat{Z}^{-}_{+} + \hat{Z}^{-*}_{+} + \hat{Z}^{o*}_{+}\hat{Z}^{o}_{+} = 0 , \qquad (4.3)$$

$$\hat{Z}^{o*}_{o} = \hat{Z}^{o-1}_{o} , \quad \hat{Z}^{-}_{-} = \hat{I} = \hat{Z}^{+}_{+} , \quad \hat{Z}^{\mu}_{\nu} = 0 , \text{ if } \mu>\nu .$$

We shall call a process $\hat{Y}(t)$ an output process with respect to the stochastic evolution (4.1), if it satisfies the condition

$$U(s)\hat{Y}(t) = Y(t)U(s) , \quad \forall\, s\geqslant t , \qquad (4.4)$$

where $Y(t)$ is an adapted process, defining $\hat{Y}(t)$ in Schrödinger picture by (4.3) for $s=t$.

We shall call the process $\hat{Y}(t)$ a nondemolition process with respect to a QS process $\hat{X}(t)$, if

$$[\hat{X}(s), \hat{Y}(t)] = 0 , \forall s \geqslant t . \tag{4.5}$$

The processes X and Y are called mutually nondemolition, if $[X(t) , Y(s)] = 0$ for all s and t. A nondemolition normal process Y is called a nondemolition measurement with respect to X, if it is selfnondemolition: $[Y(t) , Y(s)] = 0$ for all t and s. Let us assume that the process Y in (4.3) as well as a process X, defining the process $\hat{X}$ in the Schrödinger picture by

$$U(t)\hat{X}(t) = X(t)U(t) , \forall t \geqslant 0 \tag{4.6}$$

are the QS integrals (3.1) and

$$X(t) = X(0) + \int_0^t D^\mu_\nu(r) d\Lambda^\nu_\mu(r) \qquad t \geqslant 0 \tag{4.7}$$

with bounded $X(0) = X^o \otimes \hat{1}$ and projectively QS integrable adapted matrix-process $D = (D^\mu_\nu)$, $D^\mu_\nu = 0$ for $\mu = +$ or $\nu = -$.

Theorem 4. The QS process $\hat{X}$ is defined by (4.6), (4.7), iff it satisfies the following QS differential equation

$$d\hat{X} = (\hat{Z}^*\hat{G}\hat{Z} - \hat{X}\otimes\delta)^\mu_\nu d\Lambda^\nu_\mu \quad , \quad \hat{X}(0) = X^o \otimes \hat{1} \quad , \tag{4.8}$$

where $\hat{G}(t) = (\hat{G}^\mu_\nu)(t)$ is the QS matrix-process $\hat{G} = \hat{D} + \hat{X}\otimes\delta$, $U\hat{D}^\mu_\nu = D^\mu_\nu U$. The QS process $\hat{Y}$ is defined by (3.1) and (4.4) iff it satisfies the equation

$$d\hat{Y} = (\hat{Z}^*\hat{C}\hat{Z})^\mu_\nu d\Lambda^\nu_\mu \quad , \quad \hat{Y}(0) = Y^o \otimes \hat{1} \quad , \tag{4.9}$$

where $U\hat{C}^\mu_\nu = C^\mu_\nu U$.

The QS process $\hat{Y}$ is an output process in the sense of (4.5), iff it is nondemolition with respect to all the processes $\hat{Z}^{\mu}_{\nu}$, $\mu \in \{-,o\}$, $\nu \in \{o,+\}$, what is equivalent to the commutativity conditions

$$[Y(r), Z^{\mu}_{\nu}(t)] = 0, \quad \forall r \leq t. \tag{4.10}$$

The output process (4.9) is nondemolition with respect to an adapted QS process (4.8) iff

$$[X^o, Y^o] = 0, \; [G(t), C(t)] = 0 \quad \forall t, \; [D^{\mu}_{\nu}(s), Y(t)] = 0 \quad \forall t \geq s. \tag{4.11}$$

<u>Proof</u>. We obtain (4.8) with $\hat{G}^{\mu}_{\nu} = U^* G^{\mu}_{\nu} U$, $G^{\mu}_{\nu} = D^{\mu}_{\nu} + X \otimes \delta^{\mu}_{\nu}$ from (4.7) for U, satisfying the equation (4.1) simply by applying to $\hat{X} = U^* X U$ the QS Ito formula:

$$d(U^* X U) = (\hat{Z}^*(U^* \otimes \delta) G (U \otimes \delta)\hat{Z} - U^* X U \otimes \delta)^{\mu}_{\nu} \, d\Lambda^{\nu}_{\mu}.$$

Moreover, due to $\hat{Z}\hat{Z}^* = \hat{I} \otimes \delta$ one obtains (4.7) for $X = U\hat{X}U^*$ from the QS Ito formula

$$d(U\hat{X}U^*) = (U \otimes \delta)\, \hat{Z}\hat{Z}^* \hat{G} \hat{Z} \hat{Z}^* (U \otimes \delta) - U\hat{X}U^* \otimes \delta)^{\mu}_{\nu} \, d\Lambda^{\nu}_{\mu}.$$

If the process $\hat{Y}$ satisfies the output condition (4.4) then, by multiplying it from the left side hand by $U(t)^*$ we obtain the commutativity condition

$$[U(t,s), \hat{Y}(t)] = 0 \quad \text{for} \quad U(t,s) = U(t)^* U(s) \quad t \leq s. \tag{4.12}$$

Taking into account that the solution $U(s) = U(0,s)$ of the equation (4.1) corresponding to $t=0$ in the integral form

$$U(t,s) = \hat{I} + \int_t^s U(t,r)(Z(r) - \hat{I}\otimes\delta)^{\mu}_{\nu} d\Lambda^{\nu}_{\mu}(r) \tag{4.13}$$

satisfies the condition $U(t)U(t,s) = U(s)$ for all t and $s \geqslant t$, one can see, that the condition (4.12) is the commutativity condition of $\hat{Y}(t)$ with the QS integral in (4.13). But for any Ito integral sum

$$U_n(t,s) = \hat{I} + \sum_{i=0}^{n(s)-1} U_i(t,t_i)(\hat{Z}(t_i) - \hat{I}\otimes\delta)^{\mu}_{\nu}(\Lambda(t_{i+1}) - \Lambda(t_i))^{\nu}_{\mu} \tag{4.14}$$

where $t_o = t$, $n(s) = \sum_{i=1}^{n} \chi(s,t_i)$ with $\chi(s,t) = \begin{cases} 1 & s \geqslant t \\ 0 & s < t \end{cases}$, $t < t_1 < \ldots < t_n$, $U_o(t,s) = I$, the commutativity condition due to adapticity of $\hat{Y}$ is equivalent to the nondemolition condition with respect to $\hat{Z}^{\mu}_{\nu}$. Indeed, $[U_o(t,s), \hat{Y}(t)] = 0$, and, supposing that $[U_k(t,s), \hat{Y}(t)] = 0$ for $k < n$, one obtains

$$[U_n(t,s), \hat{Y}(t)] = \sum_{i=0}^{n(s)-1} U_i(t,t_i)[\hat{Z}^{\mu}_{\nu}(t_i), \hat{Y}(t)](\Lambda(t_{i+1}) - \Lambda(t_i)) = 0$$

due to commutativity of $\hat{Y}(t)$ with $\Lambda(t_{i+1}) - \Lambda(t_i)$. Hence, $[\hat{Z}^{\mu}_{\nu}(t_i), \hat{Y}(t)] = 0$ for any integral sum, approximating the integral (4.13) in the inductive limit of the uniform topology of operators in $\mathcal{H}(\infty)$. The commutativity of $\hat{Y}(t)$ with $\hat{Z}(t)$ gives the possibility to write the equation (4.8) for $\hat{Y}$ in the form (4.9):

$$d\hat{Y} = (\hat{Z}^{\star}\hat{F}\hat{Z} - \hat{Y}\otimes\delta)^{\mu}_{\nu} d\Lambda^{\nu}_{\mu} = (\hat{Z}^{\star}(\hat{F} - \hat{Y}\otimes\delta)\hat{Z})^{\mu}_{\nu} d\Lambda^{\nu}_{\mu} (\hat{Z}^{\star}\hat{C}\hat{Z})^{\mu}_{\nu} d\Lambda^{\nu}_{\mu}$$

Conversely, let $\hat{Y}$ be a nondemolition adapted process with respect to the operators $\hat{Z}^{\mu}_{\nu}$, $\mu \in \{-,o\}$, $\nu \in \{o,+\}$. Then we obtain by the induction with respect to $n = 0,1,2,\ldots$ that $\hat{Y}(t)$

commutes with any integral sum (4.14) . Hence it commutes with the QS integral (4.13) , which is an inductive limit of an uniformly converging sequence of such sums. So, $\hat{Y}$ is an output process for the stochastic evolution U.

If an output process $\hat{Y}$ is nondemolition with respect to a QS process $\hat{X}$, then from (4.5) one obtains

$$U(s)[\hat{X}(s), \hat{Y}(t)] = [X(s), Y(t)]U(s), \quad s \geq t , \tag{4.15}$$

where X and Y are the processes satisfying (4.4) and (4.6). It proves, in particular, the condition (4.10) as the case of $X = Z^{\mu}_{\nu}$. If the process $\hat{X}$ is defined by the equation (4.8), then it satisfies the condition (4.11) with respect to an output process $\hat{Y}$, iff

$$[\hat{X}(t), \hat{Y}(t)] = 0, \ [(\hat{Z}^{\star}\hat{G}\hat{Z})^{\mu}_{\nu}(s), \hat{Y}(t)] = 0 , \quad \forall s \geq t \tag{4.16}$$

as it follows from the QS integral representation

$$\hat{X}(s) = \hat{X}(t) + \int_{t}^{s} (\hat{Z}^{\star}\hat{G}\hat{Z} - \hat{X}\otimes\delta)^{\mu}_{\nu} \, d\Lambda^{\nu}_{\mu} .$$

The first commutativity condition in (4.16) is fulfilled for the case (4.9) iff $[\hat{X}(0), \hat{Y}(0)] = 0$ and

$$d[\hat{X} , \hat{Y}] = (\hat{Z}^{\star}[\hat{G}\hat{F}]\hat{Z} - [\hat{X} , \hat{Y}]\otimes\delta)^{\mu}_{\nu} \, d\Lambda^{\nu}_{\mu} = 0$$

what means $[X^{o}, Y^{o}] = 0$ and

$$[\hat{G} , \hat{F}] = \hat{Z}(\hat{Z}^{\star}[\hat{G}\hat{F}]\hat{Z} - [\hat{X} , \hat{Y}]\otimes\delta)\hat{Z}^{\star} = 0 \tag{4.17}$$

for all t due to adapticity of Y, and $\hat{Z}\hat{Z}^{\star} = \hat{I}$. The second condition in (4.16) can be written as

$$[\hat{G}^{\mu}_{\nu}(s), \hat{Y}(t)] = [\hat{D}^{\mu}_{\nu}(s), \hat{Y}(t)] = 0 \quad , \quad s \geqslant t \tag{4.18}$$

due to commutativity of $\hat{Y}(t)$ with $\hat{Z}(s)$ and $\hat{X}(s)$ for $s \geqslant t$. In particular, $[\hat{G}^{\mu}_{\nu}(t), \hat{Y}(t)] = 0$ and hence

$$[\hat{G}, \hat{C}] = [\hat{G}, \hat{F} - \hat{Y} \otimes \delta] = [\hat{G}, \hat{F}] = 0 \tag{4.19}$$

as it follows from (4.17). Due to the output condition (4.4) (4.18) and (4.19) can be written in Schrödinger picture as in (4.11)

$$[Y(t), D^{\mu}_{\nu}(s)]U(s) = U(s)[\hat{Y}(t), \hat{D}^{\mu}_{\nu}(s)] = 0,$$

$$[G, C](t)\, U(t) = U(t)[\hat{G}, \hat{C}](t) = 0.$$

<u>Corollary</u>. The process $\hat{X}$ is a Heisenberg transformation $U^*(X^0 \otimes \hat{1})U$ of an initial operator $X^0 \in \mathcal{B}(h)$ with respect to a QS unitary process U, described by (4.1), iff it satisfies the QS equation (4.8) with $\hat{G}^{\mu}_{\nu} = \hat{X}\,\delta^{\mu}_{\nu}$. The process $\hat{Y}$ is an output process with respect to any QS Markovian evolution, defined by

$$\hat{Z}^{\mu}_{\nu}(t) = U^*(t)(L^{\mu}_{\nu}(t) \otimes \hat{1})U(t) + \hat{I}\,\delta^{\mu}_{\nu} \tag{4.20}$$

with $L^{\mu}_{\nu}(t)$, acting in h, iff it is a Heisenberg transformation of $Y(t) = I \otimes \hat{y}(t)$, where $\hat{y}(t)$ is an adapted process in the Fock space $\mathcal{F}$. The output process $\hat{Y}(t) = U(t)^* Y(t) U(t)$, defined by a QS integral (3.1), is nondemolition with respect to the process $\hat{X}(t) = U(t)^*(X \otimes \hat{1})U(t)$, iff

$$[X^o, Y^o] = 0 \quad , \quad [X^o \otimes \hat{1}, C^\mu_\nu(t)] = 0 \quad , \forall t \tag{4.21}$$

is selfnondemolition, iff

$$[Y(t), C^\mu_\nu(s)] = 0 \quad , \forall \; t \leq s \tag{4.22}$$

and is a nondemolition measurement, iff also

$$[Y^{o*}, Y^o] = 0 \quad , \quad [C^*, C](t) = 0 \quad , \forall \; t \; . \tag{4.23}$$

Indeed, if $X(t) = X^o \otimes \hat{1}$ is a time independent process, then it satisfies the QS equation $dX = (G - X \otimes \delta)^\mu_\nu d\Lambda^\nu_\mu = 0$, corresponding to $G^\mu_\nu = X\delta^\mu_\nu$. Hence $\hat{G}^\mu_\nu$ in (4.8) is the diagonal matrix-operator $X(t)\delta^\mu_\nu$. Therefore, for such a process $D^\mu_\nu = 0$, the nondemolition conditions (4.11) can be written as (4.21).

The process $\hat{Y}$ is an output process for the generators of the Markovian form (4.20), iff

$$[Y(t), L^\mu_\nu(s) \otimes \hat{1}] = 0 \qquad \forall \; s \geq t.$$

Due to the arbitrariness of the generators L^μ_ν $[Y(t), X \otimes \hat{1}] = 0$ for all t, i.e. $Y(t) \in (\mathcal{B}(h) \otimes \hat{1})' = I \otimes \mathcal{B}(\mathcal{F})$. So, $Y(t) = I \otimes \hat{y}(t)$. For the process (3.1) it means that Y^o is a constant eI, and $C^\mu_\nu(t) = I \otimes \hat{c}^\mu_\nu(t)$, where $\hat{c} = (\hat{c}^\mu_\nu)$ is an adapted QS projectively integrable matrix process in $\mathcal{F}(\infty) = \bigcap_{\zeta > 0} \mathcal{F}(\zeta)$. For the selfnondemolition process $Y = X$ the conditions (4.11) should be written as the case $Y^o = X^o$, $G = C + Y \otimes \delta = F$, $D = C$, in which nontrivial is only the condition (4.22). Writting the conditions (4.11) for the case $X = Y^*$, one obtains the normality condition $[Y^*, Y](t) = 0$ in the form (4.23) and

$[Y^*(t), C^{\mu}_{\nu}(s)] = 0$ for $s \geqslant t$ due to (4.22) and normality of Y.

Note, that necessary and sufficient conditions (4.21) - (4.23) can be written in the same form in Heisenberg picture of the output process $\hat{Y}$, in particular, the normality condition $[\hat{C}^*, \hat{C}] = 0$ has in terms of the matrix elements $\hat{C}^{\mu}_{\nu}$, $\mu \in \{-,o\}$, $\nu \in \{o,+\}$ the form

$$\hat{C}^{o*}_{+}\hat{C}^{o}_{o} = \hat{C}^{-}_{o}\hat{C}^{o*}_{o} \quad , \quad \hat{C}^{o*}_{o}\hat{C}^{o}_{+} = \hat{C}^{o}_{o}\hat{C}^{-*}_{o}$$

$$\hat{C}^{o*}_{o}\hat{C}^{o}_{o} = \hat{C}^{o}_{o}\hat{C}^{o*}_{o} \quad , \quad \hat{C}^{o*}_{+}\hat{C}^{o}_{+} = \hat{C}^{-}_{o}\hat{C}^{-*}_{o}$$

In the case of commutativity of matrix elements $C^{\mu}_{\nu}(t)$ and $Z^{\varkappa}_{\lambda}(t)$ the output process $\hat{Y}$, defined by (3.1),(4.4) can be described as a QS integral

$$\hat{Y}(t) = \hat{Y}(0) + \int_0^t \hat{C}^{\mu}_{\nu}(r)\, d\hat{\Lambda}^{\mu}_{\nu}(r)$$

with respect to the output pseudo-Poissonian process

$$\hat{\Lambda}^{\mu}_{\nu}(t) = U^*(t)\Lambda^{\mu}_{\nu}(t)U(t) \quad :$$

$$\int_0^t \hat{C}^{\mu}_{\nu}(r)\, d\hat{\Lambda}^{\mu}_{\nu} = \int_0^t \left(\hat{C}^{-}_{+}dr + \hat{C}^{-}_{o}d\hat{A}^{o}_{-} + \hat{C}^{o}_{+}d\hat{A}^{+}_{o} + \hat{C}^{o}_{o}d\hat{N}^{o}_{o} \right)$$

The output creation $\hat{A}^+$, annihilation $\hat{A}_-$, and number process $\hat{N}$ considered by Barchielli [5] for the Markovian stationary one-dimensional case ($\mathcal{K} = \mathbb{C}$), are described by the equations (4.9) of the following form

$$A^{i}_{-}(t) = \int_0^t \left(Z^{i}_{j}dA^{j}_{-} + Z^{i}_{+}dr \right) = A^{+}_{i}(t)^*$$

$$\hat{N}^i_j(t) = \int_0^t \left\{ \left(\hat{Z}^{j*}_o \hat{Z}^i_o\right)^l_k dN^k_l + \left(\hat{Z}^{j*}_o Z^i_+\right)^l dA^+_l + \left(\hat{Z}^{j*}_+ \hat{Z}^i_o\right)_k dA^k_- + \left(\hat{Z}^{j*}_+ \hat{Z}^i_+\right) dr \right\},$$

where $Z^{j*}_o = \left(Z^{*i}_j\right)$ is a column vector of operators $Z^{*i}_j = Z^{j*}_i$, conjugated to the row vector $Z^j_o = \left(Z^j_i\right)$, $i,j \in J_o$.

The coordinate output processes $\hat{Q}_i = \hat{A}^+_i + \hat{A}^i_-$, which are Hermitian and mutually commutative, give an example of a continuous multidimensional nondemolition measurement process $\hat{Q} = \left(\hat{Q}_i\right)$.

Another example of nondemolition measurement, called the counting multidimensional measurement, give the family $\{\hat{N}_i\}$ of mutually nondemolition Hermitian processes $\hat{N}_i$, given by the diagonal elements $(j=i)$ of the matrix-valued nondemolition output number process $\hat{N} = \left(\hat{N}^i_j\right)$.

Acknowledgements

I would like to thank Professors L. Accardi, A. Barchielli, A. Frigerio and G. Lupieri for stimulating discussions on these subjects.

References

[1] Hudson, R.S. and K.R. Parthasarathy, Comm.Math.Phys. 93 (1984), 301-323.

[2] Lindsay, M. and H. Maassen, in Quantum Probability and Applications III, eds. L. Accardi and W. von Waldenfels, Springer LNM 1303, Berlin 1988, pp. 192-208.

[3] Belavkin, V.P., Quantum Stochastic Calculus and Quantum Stochastic Filtering, Preprint Centro Matematico V. Volterra, Dipartimento di Matematica, Univerità di Roma II, 1989.

[4] Accardi, L. , F. Fogna and I. Quagebeur, Representation Free Stochastic Calculus, Preprint Universitá di Roma II, 1989.

[5] Barchielli , A., in Quantum Probability and Applications III, eds. L. Accardi and W. von Waldenfels, Springer LMN 1303, Berlin 1988, pp. 37-51.

Stochastic Transitions on Preduals of von Neumann Algebras

by

Carlo Cecchini
Dipartimento di Matematica dell' Università di Genova
Via L.B. Alberti 4, 16132 Genova
Italy

and

Burkhard Kümmerer
Mathematisches Institut Universität Tübingen
Auf der Morgenstelle 10, D-7400 Tübingen
W-Germany

Abstract. The aim of this note is to prove a result on canonical state extensions from a von Neumann subalgebra to a von Neumann algebra in which it is contained. In its light the notion of a stochastic coupling for von Neumann algebras as introduced by the first named author in [5] (cf. also [1]) and the notion of transition operators introduced by the second named author in [7] appear to be particular (indeed extreme) cases of a more general theory.

In the following $\mathcal{M}$ will always denote a von Neumann algebra acting on a Hilbert space $\mathcal{H}$ with a cyclic and separating vector for $\mathcal{M}$; we shall denote by $\mathcal{S}(\mathcal{M})$ the set of normal states on $\mathcal{M}$. We will also consider a von Neumann subalgebra $\mathcal{M}_0$ of $\mathcal{M}$ which acts in a standard way on a closed subspace $\mathcal{H}_0$ of $\mathcal{H}$, i.e., $\mathcal{H}_0$ is invariant under the action of $\mathcal{M}_0$ on $\mathcal{H}$ and $\mathcal{M}_0|_{\mathcal{H}_0}$ is a standard representation of $\mathcal{M}_0$ with positive self dual cone V_0 .

For a vector $\Phi \in \mathcal{H}$ we denote by ω_Φ the corresponding vector state on $\mathcal{B}(\mathcal{H})$ or any of its subalgebras. Conversely, if $\mathcal{M}$ acts in a standard representation on $\mathcal{H}$ with positive self dual cone V and if ω is a normal state on $\mathcal{M}$ then Φ_ω denotes the unique vector in V for which $\omega = \omega_{\Phi_\omega}$.

1. Definition. A mapping $\rho : \mathcal{S}(\mathcal{M}_0) \to \mathcal{S}(\mathcal{M})$ will be called <u>canonical state extension</u> for the couple $(\mathcal{M}, \mathcal{M}_0)$ if $\mathcal{M}_0$ acts in a standard way on a closed subspace $\mathcal{H}_0 \subseteq \mathcal{H}$ with positive self dual cone V_0 such that for all $\omega_0 \in \mathcal{S}(\mathcal{M}_0)$ with representing vector Φ_{ω_0} in V_0 we have $[\rho(\omega_0)](a) = <\Phi_{\omega_0}, a\Phi_{\omega_0}>$ for all $a \in \mathcal{M}$.

For the properties of canonical state extensions and their connections with ω - conditional expectations we refer to [1], [2], [3], [4].

2. Definition. Let $\mathcal{M}_0$, $\mathcal{M}_1$ be two von Neumann algebras. A continuous linear mapping $\rho_{0,1} : (\mathcal{M}_0)_* \to (\mathcal{M}_1)_*$ will be called a ($\mathcal{M}$ implemented) stochastic transition for the couple $(\mathcal{M}_0, \mathcal{M}_1)$ if there are two faithful representations π_0 of $\mathcal{M}_0$ and π_1 of $\mathcal{M}_1$ on some Hilbert space $\mathcal{H}$ and a canonical state extension ρ for the couple $(\mathcal{M}, \pi_0(\mathcal{M}_0))$ with $\mathcal{M}$ the von Neumann algebra generated by $\pi_0(\mathcal{M}_0) \cup \pi_1(\mathcal{M}_1)$, such that $\rho_{0,1}(\omega_0) = \rho(\omega_0) \circ \pi_1$ for all $\omega_0 \in \mathcal{S}(\mathcal{M}_0)$.

It will be called even, respectively odd, if for all pairs of vectors Φ_1, Φ_2 in the cone V_0 we have $[\rho_{0,1}(\omega_{\Phi_1+i\Phi_2})](a_1) = < \Phi_1 + i\Phi_2, a_1(\Phi_1 + i\Phi_2) >$, respectively $[\rho_{0,1}(\omega_{\Phi_1-i\Phi_2})](a_1) = < \Phi_1 + i\Phi_2, a_1(\Phi_1 + i\Phi_2) >$, for all $a_1 \in \mathcal{M}_1$.

Our main result is the following

3. Theorem. Let $\mathcal{M}$ be a von Neumann algebra generated by the union of two von Neumann subalgebras $\mathcal{M}_0$ and $\mathcal{M}_1$, and let ρ a be canonical state extension for the couple $(\mathcal{M}, \mathcal{M}_0)$.

The following pairs of conditions are equivalent:

A$_1$) The mapping $\rho_{0,1} : \mathcal{S}(\mathcal{M}_0) \to \mathcal{S}(\mathcal{M}_1)$ defined by $\rho_{0,1}(\omega_0) = \rho(\omega_0)|_{\mathcal{M}_1}$ ($\omega_0 \in \mathcal{S}(\mathcal{M}_0)$) can be extended to an $\mathcal{M}$ implemented even stochastic transition for the couple $(\mathcal{M}, \mathcal{M}_0)$.

A$_2$) There is a weak operator continuous completely positive linear contraction $\epsilon_\rho : \mathcal{M}_1 \to \mathcal{M}_0$ such that for some faithful $\omega_0 \in \mathcal{S}(\mathcal{M}_0)$ we have
$$[\rho(\omega_0)](a_0 \cdot a_1 \cdot b_0) = \omega_0(a_0 \cdot \epsilon_\rho(a_1) \cdot b_0)$$
for all $a_0, b_0 \in \mathcal{M}_0$, $a_1 \in \mathcal{M}_1$.

B$_1$) The mapping $\rho_{0,1} : \mathcal{S}(\mathcal{M}_0) \to \mathcal{S}(\mathcal{M}_1)$ defined by $\rho_{0,1}(\omega_0) = \rho(\omega_0)|_{\mathcal{M}_1}$ ($\omega_0 \in \mathcal{S}(\mathcal{M}_0)$) can be extended to an $\mathcal{M}$ implemented odd stochastic transition for the couple $(\mathcal{M}, \mathcal{M}_0)$.

B$_2$) For some faithful $\omega_0 \in \mathcal{S}(\mathcal{M}_0)$ we have
$$[\rho(\omega_0)](b_0 \cdot a_0 \cdot a_1 \cdot c_0) = [\rho(\omega_0)](b_0 \cdot a_1 \cdot a_0 \cdot c_0)$$
for all $a_0, b_0, c_0 \in \mathcal{M}_0$, $a_1 \in \mathcal{M}_1$.

For the proof of our theorem we need the following

4. Lemma Let $\mathcal{M}$ be a von Neumann algebra acting on a Hilbert space $\mathcal{H}$ with cyclic and separating vector Ω for $\mathcal{M}$, and V the associated positive self dual cone with isometric involution J. Let a be a bounded positive operator on $\mathcal{H}$. The following conditions are equivalent.

a) $a \in \mathcal{M}$, respectively $a \in \mathcal{M}'$.

b) The mapping $\omega_\Phi \mapsto < \Phi, a\Phi >$ ($\Phi \in V$) admits an extension to a linear continuous functional T_a on $\mathcal{M}_*$ for which
$T_a(\omega_{\Phi_1+i\Phi_2}) = < \Phi_1 + i\Phi_2, a(\Phi_1 + i\Phi_2) >$, respectively
$T_a(\omega_{\Phi_1-i\Phi_2}) = < \Phi_1 + i\Phi_2, a(\Phi_1 + i\Phi_2) >$ for all $\Phi_1, \Phi_2 \in V$.

Proof: a) $\Rightarrow$ b) If $a \in \mathcal{M}$ (resp. $\mathcal{M}'$), set $T_a(\omega) := \omega(a)$ (resp. $T_a(\omega) := \omega(JaJ)$) for $\omega \in \mathcal{M}_*$. T_a is clearly linear and $T_a(\omega_\Phi) =< \Phi, a\Phi >$ for all $\Phi \in V$. If $a \in \mathcal{M}$ then
$$T_a(\omega_{\Phi_1+i\Phi_2}) = \omega_{\Phi_1+i\Phi_2}(a)$$
$$=< \Phi_1 + i\Phi_2, a(\Phi_1 + i\Phi_2) > .$$

If $a \in \mathcal{M}'$ then

$$T_a(\omega_{\Phi_1 - i\Phi_2}) = \omega_{\Phi_1 - i\Phi_2}(JaJ) = < \Phi_1 - i\Phi_2, JaJ(\Phi_1 - i\Phi_2) >$$
$$= < \Phi_1 + i\Phi_2, a(\Phi_1 + i\Phi_2) > .$$

b) $\Rightarrow$ a) As T_a is a linear positive continuous functional on $\mathcal{M}_*$, there is a unique operator $\overline{a} \in \mathcal{M}^+$ such that $< \Phi, a\Phi > = < \Phi, \overline{a}\Phi >$ for all $\Phi \in V$. Let now $\Phi_1, \Phi_2 \in V$. Then also $\Phi_1 + \Phi_2 \in V$ and we have

$$Re < \Phi_1, a\Phi_2 > = \frac{1}{2}(< \Phi_1 + \Phi_2, a(\Phi_1 + \Phi_2) > - < \Phi_1, a\Phi_1 > - < \Phi_2, a\Phi_2 >)$$
$$= \frac{1}{2}(< \Phi_1 + \Phi_2, \overline{a}(\Phi_1 + \Phi_2) > - < \Phi_1, \overline{a}\Phi_1 > - < \Phi_2, \overline{a}\Phi_2 >)$$
$$= Re < \Phi_1, \overline{a}\Phi_1 > .$$

As $T_a(\omega_{\Phi_1 - i\Phi_2}) = < \Phi_1 + i\Phi_2, a(\Phi_1 + i\Phi_2) > = < \Phi_1 + i\Phi_2, \overline{a}(\Phi_1 + i\Phi_2) >$, we get by a similar computation,

$$Im < \Phi_1, a\Phi_2 > = Im < \Phi_1, \overline{a}\Phi_2 >$$

for all $\Phi_1, \Phi_2 \in V$; so $< \Phi_1, a\Phi_2 > = < \Phi_1, \overline{a}\Phi_2 >$ and the density of the linear combinations of elements of V in $\mathcal{H}$ yields $a = \overline{a} \in \mathcal{M}$.

As Φ_1, Φ_2, and $\Phi_1 + \Phi_2$ are invariant under J, and $J(\Phi_1 - i\Phi_2) = \Phi_1 + i\Phi_2$, we can show similarly that if $T_a(\omega_{\Phi_1 - i\Phi_2}) = < \Phi_1 + i\Phi_2, a(\Phi_1 + i\Phi_2) >$ then $a = J\overline{a}J \in \mathcal{M}'$, and conclude the proof of our lemma.

Proof of theorem 3: We shall identify in the following the von Neumann algebras $\mathcal{M}_0$ and $\mathcal{M}_1$ with their representations $\pi_0(\mathcal{M}_0)$ and $\pi_1(\mathcal{M}_1)$ and use the notations already established at the beginning.

A_1) $\Rightarrow$ A_2) Let V_0 be the self dual positive cone for $\mathcal{M}_0$ in $\mathcal{H}_0 \subseteq \mathcal{H}$ associated with the canonical state extension ρ (i.e., for each $\Phi_0 \in V_0$ we have $< \Phi_0, a\Phi_0 > = [\rho(\omega_{\Phi_0})](a)$ for all $a \in \mathcal{M}$) and let E be the orthogonal projection from $\mathcal{H}$ to the Hilbert space $\mathcal{H}_0$, the closure of linear combinations of the elements in V_0. Put $\epsilon_\rho(a_1) := Ea_1E|_{\mathcal{H}_0}$ for $a_1 \in \mathcal{M}_1$. Then the mapping

$\omega_{\Phi_0} \mapsto < \Phi_0, \epsilon_\rho(a_1)\Phi_0 > = < \Phi_0, Ea_1E\Phi_0 > = < \Phi_0, a_1\Phi_0 > = [\rho_{0,1}(\omega_{\Phi_0})](a_1)$
($\Phi_0 \in V_0$, $\omega_{\Phi_0} \in \mathcal{S}(\mathcal{M}_0)$, $a_1 \in \mathcal{M}_1^+$) satisfies conditions b) in lemma 4.

Remark that, in particular, the eveness implies that $(\omega_{\Phi_1 + i\Phi_2})_0$ is mapped by this mapping into $< \Phi_1 + i\Phi_2, a_1(\Phi_1 + i\Phi_2) >$ for $\Phi_1, \Phi_2 \in V_0$. So $\epsilon_\rho(a_1)$ is in $\mathcal{M}_0$, which implies that for $a_0, b_0 \in \mathcal{M}_0$, $a_1 \in \mathcal{M}_1$, and $\omega_{\Phi_0} \in \mathcal{S}(\mathcal{M}_0)$:

$$[\rho(\omega_{\Phi_0})](a_0a_1b_0) = < \Phi_0, a_0a_1b_0\Phi_0 > = < \Phi_0, a_0Ea_1Eb_0\Phi_0 > = < \Phi_0, a_0\epsilon_\rho(a_1)b_0\Phi_0 >$$
$$= \omega_{\Phi_0}(a_0\epsilon_\rho(a_1)b_0) .$$

By the linearity of ϵ_ρ the same applies to all $a_1 \in \mathcal{M}_1$ (not necessarily positive), and the other properties stated for ϵ_ρ are obvious.

A_2) $\Rightarrow$ A_1) We note first that our hypothesis implies that $[\rho_{0,1}(\phi_0)](a_1) = \phi_0(\epsilon_\rho(a_1))$ for all $\phi_0 \in \mathcal{S}(\mathcal{M}_0)$. Indeed, if $\phi_0 \leq \alpha\omega_0$ for some $\alpha > 0$ and $[\phi_0/\omega_0]$ is the extension of the Connes cocycle for the couple (ϕ_0, ω_0) at the point $i/2$ (cf. [6]) we have:

$$[\rho(\phi_0)](a_1) = [\rho(\omega_0)]([\phi_0/\omega_0]^* a_1[\phi_0/\omega_0]) = \omega_0([\phi_0/\omega_0]^* \epsilon_\rho(a_1)[\phi_0/\omega_0]) = \phi_0(\epsilon_\rho(a_i)) \quad ;$$

for general ϕ_0 in $\mathcal{S}(\mathcal{M}_0)$ the claim follows by continuity.
If we now set $[\rho_{0,1}(\phi_0)](a_1) = \phi_0(\epsilon_\rho(a_1))$ for each $\phi_0 \in (\mathcal{M}_0)_*$, $a_1 \in \mathcal{M}_1$, then $\rho_{0,1}$ is clearly a linear continuous mapping from $(\mathcal{M}_0)_*$ to $(\mathcal{M}_1)_*$ and for $\phi_0 \in \mathcal{S}(\mathcal{M}_0)$ we have by construction that $\rho_{0,1}(\phi_0) = \rho(\phi_0)|_{\mathcal{M}}$. Lemma 4 guaranteees the eveness.

B$_1$) $\Rightarrow$ B$_2$) Working as in the proof of A$_1$) $\Rightarrow$ A$_2$) and using the oddness of our stochastic transition together with lemma 4 we get that $\epsilon_\rho(a_1)$ is in the commutant of the action of $\mathcal{M}_0$ on $\mathcal{H}_0$. Then for all $a_0, b_0, c_0 \in \mathcal{M}_0$, $a_1 \in \mathcal{M}_1$ we have:

$$[\rho(\omega_0)](b_0 a_0 a_1 c_0) = \omega_0(b_0 a_0 \epsilon_\rho(a_1) c_0) = \omega_0(b_0 \epsilon_\rho(a_1) a_0 c_0) = [\rho(\omega_0)](b_0 a_1 a_0 c_0) \ .$$

B$_2$) $\Rightarrow$ B$_1$) From the faithfulness of ω_0 in our hypothesis we conclude that $\epsilon_\rho(a_1) := E a_1 E|_{\mathcal{H}_0}$ is in the commutant of the action of $\mathcal{M}_0$ on $\mathcal{H}_0$; then an application of lemma 4 yields the proof.

In [5] it was shown that given two von Neumann algebras $\mathcal{M}_0$ and $\mathcal{M}_1$ generating a von Neumann algebra $\mathcal{M}$ and mutually commuting, then to each canonical state extension ρ for the couple $(\mathcal{M}, \mathcal{M}_0)$ it is possible to associate a weak operator continuous linear anti-completely positive contraction λ_ρ ("stochastic coupling") from $\mathcal{M}_1$ to $\mathcal{M}_0$ satisfying the equality $[\rho(\omega_0)](a_0 a_1) = < J\lambda_\rho(a_1)\Omega_0, a_0\Omega_0 >$ for all $a_0 \in \mathcal{M}_0$, $a_1 \in \mathcal{M}_1$ and $\omega_0 \in \mathcal{S}(\mathcal{M}_0)$.

Conversely, if λ is a weak operator continuous linear anti- completely positive contraction from a von Neumann algebra $\mathcal{M}_1$ to a von Neumann algebra $\mathcal{M}_0$ then it is always a stochastic coupling from $\mathcal{M}_1$ to $\mathcal{M}_0$ in the sense that there is a von Neumann algebra $\mathcal{M}$ generated by two mutually commuting faithful representations $\pi_0(\mathcal{M}_0)$ and $\pi_1(\mathcal{M}_1)$ on some Hilbert space and a canonical state extension ρ for the couple $(\mathcal{M}, \pi_0(\mathcal{M}_0))$ such that $\lambda = \lambda_\rho$.

In [7] on the other hand the situation of a von Neumann algebra $\mathcal{M}$ generated by two von Neumann subalgebras $\mathcal{M}_0$ and $\mathcal{M}_1$ was considered under the assumption that there exists a norm one projection ϵ from $\mathcal{M}$ to $\mathcal{M}_0$. A transition operator $T_{1,0} : \mathcal{M}_1 \to \mathcal{M}_0$ was defined by setting $T_{1,0} := \epsilon|_{\mathcal{M}_1}$.

5. Corollary (of theorem 3) Let λ be a stochastic coupling from $\mathcal{M}_1$ to $\mathcal{M}_0$. Then its predual mapping $\lambda_* : (\mathcal{M}_0)_* \to (\mathcal{M}_1)_*$ is an odd stochastic transition for the couple $(\mathcal{M}_0, \mathcal{M}_1)$. The predual mapping $(T_{1,0})_* : (\mathcal{M}_0)_* \to (\mathcal{M}_1)_*$ of a transition operator $T_{1,0}$ is an $\mathcal{M}$ implemented even stochastic transition for the couple $(\mathcal{M}_0, \mathcal{M}_1)$.

Proof. In [5] it has been proved that $(\lambda_\rho)_*(\phi_0) = \rho(\phi_0)|_{\mathcal{M}_1}$ for $\Phi_0 \in (\mathcal{M}_0)_*$. As $\mathcal{M}_0$ and $\mathcal{M}_1$ commute, condition B$_2$) in theorem 3 is satisfied and our first claim follows. In order to prove our second claim we note that as ϵ is a norm one projection we have $\epsilon(a_0 a_1 b_0) = a_0 \epsilon(a_1) b_0$ for all $a_0, b_0 \in \mathcal{M}_0$ and $a_1 \in \mathcal{M}_1$. Put $\rho(\omega_0) := \omega_0 \circ \epsilon$ for $\omega_0 \in \mathcal{S}(\mathcal{M}_0)$; then ρ is a canonical state extension for the couple $(\mathcal{M}, \mathcal{M}_0)$ for which condition A$_2$) in theorem 3 is satisfied.

Bibliography

1. L. Accardi: Cecchini's transition expectations and Markov chains. Preprint (to be published in the proceedings of the Rome year on Quantum Probability 1986/1987).
2. C. Cecchini, D. Petz: State extensions and a Radon - Nikodym theorem for conditional expectations on von Neumann algebras. Preprint (to be published in Pac. J. Math.).
3. C. Cecchini, D. Petz: Classes of conditional expectations over von Neumann algebras. Preprint (to be published in J. Funct. Anal.).
4. C. Cecchini: An abstract characterization of ω-conditional expectations. Preprint.
5. C. Cecchini: Stochastic couplings for von Neumann algebras. Preprint (to be published in the proceedings of the Rome year on Quantum Probability 1986/1987).
6. A. Connes: Sur le theoreme de Radon - Nikodym pour les poids normale fideles semifinis. Bull. Sci. Math. Sec. II, 97 (1973), 253 - 258.
7. B. Kümmerer: Survey on a theorey of non-commutative stationary Markov processes. In Quantum Probability and Applications III, Proceedings Oberwolfach 1987, (L.Accardi, W.v.Waldenfels, Eds.), Springer Lecture Notes in Mathematics 1303 (1988), 154 - 182.

Acknowledgements The first named author would like to thank warmly Prof. W. von Waldenfels for his kind hospitality at the Institut für Angewandte Mathematik at the University of Heidelberg during the preparation of this paper and the Sonderforschungsbereich 123 for the financial support.

This paper is also part of a research project of the second named author which is supported by the Deutsche Forschungsgemeinschaft.

This paper is in final form and no similar paper has been or is being submitted elsewhere.

QUANTUM STOCHASTIC CALCULUS AND A BOSON LEVY THEOREM

F. Fagnola
Dipartimento di Matematica, Università di Trento
I 38050 Povo (TN), Italy.

0. INTRODUCTION

Various theories of quantum stochastic calculus have been developed, each of them for particular quantum noises (cf.[6],[9],[10],[12]). Since in classical probability there is a general theory of stochastic calculus which does not depend on the particular features of the integrator process, it is natural to try to develop such a general theory also in quantum probability.

This programme was outlined in [1] , [5] and the first step, the definition of the stochastic integral based on the axiomatic definition of semimartingales (cf.[7],[11]), was carried on in [2]. In this note, which is part of the joint work in preparation with L.Accardi and J.Quaegebeur [3], we will briefly recall the general theory of quantum stochastic calculus (the definition of stochastic integral in [2], the existence and uniqueness theorem for quantum stochastic differential equations, the quadratic variation and the weak Itô's formula) then we will show two applications. The first one is the developement of stochastic calculus for the euclidean free-fields, namely the stochastic processes obtained in a natural way from the field operators of the Fock representation of the CCR over the Sobolev space $H^{-k}(R^n)$ (k,n>0). These are examples of quantum processes that behave like semimartingales in the sense of our definition. The other is a quantum analogue of a classical Lévy theorem: with every continous trajectory quantum martingale commuting with the past filtration it is canonically associated a Fock representation of the CCR over a real pre-hilbert space determined by the covariance of the martingale. In this representation the martingale can be identified with a linear combination of field operators.

1. NOTATIONS AND DEFINITIONS

Let $\mathcal{H}$ be a complex separable Hilbert space, $\mathcal{B}(\mathcal{H})$ the vector space of all bounded operators on $\mathcal{H}$, $\mathcal{A}$ a subalgebra of $\mathcal{B}(\mathcal{H})$ and $(\mathcal{A}_{t]})_{t\in R_+}$ an increasing family of subalgebras of $\mathcal{A}$. We write $\langle\cdot,\cdot\rangle$ and $\|\cdot\|$ to denote the scalar product and the norm in $\mathcal{H}$ and $\mathcal{A}'$ (resp.$\mathcal{A}_{t]}'$) to denote the commutant of $\mathcal{A}$ (resp.$\mathcal{A}_{t]}$) in $\mathcal{B}(\mathcal{H})$. Let $\mathcal{D}$ be a total subset of $\mathcal{H}$, Φ a unit vector in $\mathcal{D}$ and $\mathcal{H}_{t]}$ the closure in $\mathcal{H}$ of $\mathcal{A}_{t]}\Phi$. We will denote by $\mathcal{L}(\mathcal{D},\mathcal{H})$ the vector space of linear operators F with domain containing $\mathcal{D}$ such that there exists another operator F^+ in $\mathcal{L}(\mathcal{D},\mathcal{H})$ satisfying, for all $\eta,\xi\in \mathcal{D}$,

$$\langle\eta,F\xi\rangle = \langle F^+\eta,\xi\rangle .$$

If A is an element of $\mathcal{L}(\mathcal{D},\mathcal{H})$ we will say that $A^{\#}$ has some property to mean that A and A^{+} have that property.

We say that an element F of $\mathcal{L}(\mathcal{D},\mathcal{H})$ is *weakly affiliated* with $\mathcal{A}_{t]}$, and write $F \approx \mathcal{A}_{t]}$, if $\mathcal{A}_{t]}'\mathcal{D}$ is a subset of D(F) and $F'F\xi = FF'\xi$ for all $F' \in \mathcal{A}_{t]}'$ and $\xi \in \mathcal{D}$.

(1.1) DEFINITION. *A stochastic process in $\mathcal{H}$ is a family* $(X(t))_{t\in R_+}$ *of elements of* $\mathcal{L}(\mathcal{D},\mathcal{H})$. *A stochastic process is adapted if* $X^{\#}(t) \approx \mathcal{A}_{t]}$ *for all* $t \in R_+$.
An adapted stochastic process is simple if it can be written in the form:

(1.2) $$F(s) = \sum_{j=0}^{n} \chi_{(s_j,s_{j+1}]}(s)F(s_j)$$

where $0=s_0<s_1<...<s_{n+1}<\infty$

(1.3) DEFINITION. *An additive adapted process is a family* $(X(s,t))_{0\le s\le t}$ *of elements of* $\mathcal{L}(\mathcal{D},\mathcal{H})$ *with the following properties:*

(i) $X^{\#}(s,t)\Phi \in \mathcal{H}_{t]}$ *for all* $s,t \in R_+$ *with* $s\le t$,

(ii) $X^{\#}(s,t) \approx \mathcal{A}_{t]}$ *for all* $s,t \in R_+$ *with* $s \le t$.

(iii) for all $r,s,t \in R_+$ *with* $r<s<t$

$$X^{\#}(r,t) = X^{\#}(r,s) + X^{\#}(s,t)$$

on the domain $\mathcal{A}_{t]}'\mathcal{D}$.

Let $(M^{\alpha}(\cdot,\cdot))_{\alpha\in I}$ (where I is a set of indices) be a selfadjoint family of additive adapted processes that have, moreover the following properties:

(1.4) For all $\alpha \in I$ and all $s,t \in R_+$ with $s<t$, $M^{\alpha}(s,t)$ maps $\mathcal{H}_{s]}$ into $\mathcal{H}_{t]}$.

(1.5) *Commutation with the past.* For all $\alpha \in I$, $\xi \in \mathcal{D}$, $s,t \in R_+$ with $s<t$, $A_t' \in \mathcal{A}_{t]}'$ and all adapted process F, $A_t'F(s)\xi = F(s)A_t'\xi \in D(M^{\alpha}(s,t))$, $A_t'M^{\alpha}(s,t)\xi = M^{\alpha}(s,t)A_t'\xi \in D(F(s))$ and:

$$M^{\alpha}(s,t)F(s)A_t'\xi = F(s)A_t'M^{\alpha}(s,t)\xi = A_t'F(s)M^{\alpha}(s,t)\xi = A_t'M^{\alpha}(s,t)F(s)\xi$$

(1.6) *Semimartingale property.* For all $\alpha \in I$, $\xi \in \mathcal{D}$, and all $t \in R_+$ there exists a positive constant $c_{t,\alpha,\xi}$ and a positive measure $\mu_{\alpha,\xi}$ on the Borel σ-field on R_+, finite on bounded Borel sets, without atomic part, such that, for all simple process F we have:

(1.7) $$\left\| \int_0^t dM^{\alpha}(s)F(s)\xi \right\|^2 \le c_{t,\alpha,\xi} \int_0^t \| F(s)\xi \|^2 \mu_{\alpha,\xi}(ds) .$$

(1.8) REMARK. As a consequence of the semimartingale property we can deduce that, for all $\alpha \in I$, $\xi \in \mathcal{D}$, all $s,t \in R_+$ with $s<t$ and all adapted process F:

$$\| M^{\alpha}(s,t)F(s)\xi \|^2 \le c_{t,\alpha,\xi}\, \mu_{\alpha,\xi}(s,t) \| F(s)\xi \|^2$$

(This is just (1.7) applied to the simple adapted process $G(r) = F(s)\chi_{(s,t)}(r)$).Then, for all $\alpha \in I$, $\xi \in \mathcal{D}$ and all $s,t \in R_+$ with $s<t$ the map

$$\{(r,r') \mid s\le r\le r'\le t\} \ni (r,r') \to \| M^{\alpha}(r,r')F(s)\xi \|^2 \in R_+$$

is uniformly continous.

2. STOCHASTIC INTEGRALS

In this section we will denote by M and M^+ two semimartingales of the family $(M^{\alpha}(\cdot,\cdot))_{\alpha\in I}$ satisfying the condition $\langle \eta, M(s,t)\xi\rangle = \langle M^+(s,t)\eta,\xi\rangle$ for all $\eta,\xi \in \mathcal{D}$ and all $s,t\in R_+$ with $s\leq t$. We will write also $c_{t,\xi}$ for $\max\{c_{t,\alpha,\xi}, c_{t,\alpha+,\xi}\}$.

Let F be a simple adapted process written in the form (1.2). For all $\alpha\in I$ and all $t>0$ we define the operator in $\mathcal{L}(\mathcal{D},\mathcal{H})$:

$$\int_0^t dM(s)F(s) = \sum_{j=1}^{n} \chi_{(t_j,t_{j+1}]}(t)M(t_j,t_{j+1})F(t_j).$$

The process $\left(\int_0^t dM(s)F(s)\right)_{t\in R_+}$ is the simple stochastic integral of F with respect to M and is denoted by $\int dM(s)F(s)$.

Using the semimartingale property we can extend the simple stochastic integral with respect to M to larger classes of adapted processes. Let $\mathcal{F}$ be the vector space of all adapted processes $(F(t))_{t\in R_+}$ such that the functions

$$s\rightarrow |F(s)\xi|^2 \qquad s\rightarrow |F^+(s)\xi|^2$$

are continous on R_+ for all $\xi\in\mathcal{D}$. For $F\in\mathcal{F}$, all $t>0$, all integer n and all partition $\mathcal{P}$ of $[0,t]$ by points $0=t_0<t_1<\ldots<t_{n+1}<\infty$ let $|\mathcal{P}| = \sup_{0\leq j\leq n} |t_{j+1}-t_j|$ and consider the simple adapted process

$$F^{(\mathcal{P})}(s) = \sum_{j=0}^{n} \chi_{(t_j,t_{j+1})}(s)F(t_j).$$

From the continuity property of F, for all $t\in R_+$ and all $\xi\in\mathcal{D}$, we have:

$$\lim_{|\mathcal{P}|\rightarrow 0} \sup_{0\leq s\leq t} |F^{(\mathcal{P})}(s)\xi - F(s)\xi| = 0 .$$

It follows then from the semimartingale inequality (1.7) and Lebesgue's theorem that, for all $t\in R_+$, the net $\left(\int_0^t dM(s)F^{(\mathcal{P})}(s)\right)_{\mathcal{P}}$ is Cauchy in $\mathcal{L}(\mathcal{D},\mathcal{H})$ for the topology of strong convergence on $\mathcal{D}$. We can therefore define:

$$\int_0^t dM(s)F(s) = \lim_{|\mathcal{P}|\rightarrow 0} \int_0^t dM(s)F^{(\mathcal{P})}(s)$$

In a similar way we define $\int_0^t dM^+(s)F^+(s)$. The process $\left(\int_0^t dM(s)F(s)\right)_{t\in R_+}$ is the *stochastic integral of* F *with respect to* M and will be denoted by $\int dM(s)F(s)$. It is to be noted that, for all $t>0$, the operator in $\mathcal{L}(\mathcal{D},\mathcal{H})$ $\int_0^t dM(s)F(s)$ can be defined on the larger domain $\mathcal{A}_{t]}{}'\mathcal{D}$ as the strong limit of the net we considered because of property (1.5) of M.

The following proposition is easily checked verifying the conditions of definition (1.3) on simple stochastic integrals and taking limits as $|\mathcal{P}|\rightarrow 0$.

(2.1) PROPOSITION. *For all element* F *of* $\mathcal{F}$, *the stochastic integral* $\int dM(s)F(s)$ *is an additive adapted process strongly continous on* $\mathcal{D}$. *For all* $\xi,\eta \in \mathcal{D}$ *and all* $t \in R_+$ *we have* :

$$\langle \eta, \int_0^t dM(s)F(s)\,\xi \rangle = \langle \int_0^t dM^+(s)F^+(s)\eta, \xi \rangle$$

(2.2)
$$\left\| \int_0^t dM(s)F(s)\xi \right\|^2 \leq c_{t,\xi} \int_0^t \| F(s)\xi \|^2 \mu_\xi(ds)$$

Moreover, if G *is another element of* $\mathcal{F}$, *then*

$$\int dM(s)(F(s)+G(s)) = \int dM(s)F(s) + \int dM(s)G(s).$$

3. STOCHASTIC DIFFERENTIAL EQUATIONS

From now on we will suppose that the set I is finite and we will denote by $|I|$ the number of elements of I. We will also indicate by $c_{t,\xi}$ the constant $\max_{\alpha\in I} c_{t,\alpha,\xi}$ and, for all $\xi \in \mathcal{D}$, by μ_ξ the sum of the measures $\mu_{\alpha,\xi}$. Summation over repeated greek indices will be understood.

Let $(F_\alpha)_{\alpha\in I}$ be a family of elements of $\mathcal{A}_{0]}$ leaving the domain $\mathcal{D}$ invariant and $(f_\alpha)_{\alpha\in I}$ be a family of complex-valued continous functions on R_+. The coefficients of the stochastic differential equations we consider are processes $(F_\alpha(t))_{t\in R_+}$, $(F^+_\alpha(t))_{t\in R_+}$ where $F_\alpha(t)=f_\alpha(t)F_\alpha$ for all $\alpha\in I$. Denote $K=\sup_{\alpha\in I}\|F_\alpha\|$ and $g(t)=\sup_{\alpha\in I}|f_\alpha(t)|^2$. We can prove the following:

(3.1) THEOREM. *Suppose that, for all* $\xi\in\mathcal{D}$, *all* $t\in R_+$ *and all* $\alpha\in I$ *we have* :

(3.2)
$$\sup_{\beta\in I} c_{t,\alpha,F^+_\beta\xi} \leq c_{t,\alpha,\xi} \qquad \sup_{\beta\in I} \mu_{\alpha,F^+_\beta\xi} \leq \mu_{\alpha,\xi}.$$

Then, for all $Y_0\in\mathcal{A}_{0]}$, *there exists unique additive adapted processes* $(Y^\#(t))_{t\in R_+}$, *strongly continous on* $\mathcal{D}$, *satisfying the stochastic differential equations* :

(3.3)
$$Y(t) = Y_0 + \int_0^t dM^\alpha(s)F_\alpha(s)Y(s) \quad , \quad Y^+(t) = Y^+_0 + \int_0^t dM^{\alpha+}(s)Y^+(s)F^+_\alpha(s).$$

Moreover, for all $\xi\in\mathcal{D}$, *the following inequality hold:*

(3.4)
$$\| Y(t)\xi \|^2 \leq 2\| Y_0\xi \|^2 \exp\left\{ 2c_{t,\xi}|I|K^2 \int_0^t g(s)\mu_\xi(ds) \right\}$$

Proof. Define by induction $Y^{(0)}(t) = Y_0$, $Y^{(0)+}(t) = Y^+_0$ and:

(3.5)
$$Y^{(n+1)}(t) = \int_0^t dM^\alpha(s)F_\alpha(s)Y^{(n)}(s) \qquad Y^{(n+1)+}(t) = \int_0^t dM^{\alpha+}(s)Y^{(n)+}(s)F_\alpha^+(s).$$

It is easy to show by induction that the sequences $(Y^{(n)}(t))_{t\in R_+}$, $(Y^{(n)+}(t))_{t\in R_+}$ are well defined. As a matter of fact the processes $(F_\alpha(t)Y_0)_{t\in R_+}$, $(Y^+_0F^+_\alpha(t))_{t\in R_+}$ are elements of $\mathcal{F}$. Suppose that

$(F_\alpha(t)Y^{(n)}(t))_{t\in R_+}$ and $(Y^{(n)+}(t)F^+_\alpha(t))_{t\in R_+}$ are in the class $\mathcal{F}$; by proposition (2.1), their stochastic integrals with respect to M^α are well defined and are in the class $\mathcal{F}$. It follows then, from the properties of processes $(F_\alpha(t))_{t\in R_+}$, $(F^+_\alpha(t))_{t\in R_+}$, that $(F_\alpha(t)Y^{(n+1)}(t))_{t\in R_+}$, $(Y^{(n+1)}(t)F^+_\alpha(t))_{t\in R_+}$ are elements of $\mathcal{F}$.

For all $\xi\in \mathcal{D}$ and all $n\in N$, the following inequalities hold:

$$\| Y^{(n)\#}(t)\xi \|^2 \leq (\| I \| c_{t,\xi} K^2)^n \frac{1}{n!}\Big(\int_0^t g(s)\mu_\xi(ds)\Big)^n \| Y_0{}^\# \|^2 \|\xi\|^2. \tag{3.6}$$

The inequality without superscripts # can be easily proven by induction using (2.2) and the inequality for Hilbert space vectors $\| x_1+x_2+...+x_k \|^2 \leq k(\| x_1 \|^2+\| x_2 \|^2+...+\| x_k \|^2)$. We will prove the inequality with superscripts # equal to + . Applying (2.2) to the process $(Y^{(n)}(t))_{t\in R_+}$, in view of (3.2), we have:

$$\| Y^{(n)+}(t)\xi \|^2 \leq \| I \| c_{t,\xi} \int_0^t \| Y^{(n-1)+}(t_1)F^+_{\alpha_1}\xi \|^2 g(t_1)\mu_{\alpha_1,\xi}(dt_1) ;$$

using again (2.2) and assumption (3.2) for the measures $\mu_{\alpha,\xi}$, we majorize the right-hand side by:

$$(\| I \| c_{t,\xi})^2 \int_0^t g(t_1)\mu_{\alpha_1,\xi}(dt_1) \int_0^{t_1} \| Y^{(n-1)+}(t_2)F^+_{\alpha_2}F^+_{\alpha_1}\xi \|^2 g(t_2)\mu_{\alpha_2,\xi}(dt_2)$$

Iterating, we majorize $\| Y^{(n)+}(t)\xi \|^2$ by:

$$(\| I \| c_{t,\xi})^n \int_0^t g(t_1)\mu_{\alpha_1,\xi}(dt_1) \int_0^{t_1} g(t_2)\mu_{\alpha_2,\xi}(dt_2) ... \int_0^{t_{n-1}} g(t_n)\mu_{\alpha_n,\xi}(dt_n) \| Y^+_0F^+_{\alpha_n}...F^+_{\alpha_2}F^+_{\alpha_1}\xi \|^2.$$

Majorizing the factor $\| Y^+_0F^+_{\alpha_n}...F^+_{\alpha_2}F^+_{\alpha_1}\xi \|^2$ by $K^{2n}\| Y^+_0 \|^2 \|\xi\|^2$ and summing over all indices $\alpha_1,\alpha_2,...\alpha_n$, we have:

$$\| Y^{(n)+}(t)\xi \|^2 \leq (\| I \| c_{t,\xi}K^2)^n \int_0^t g(t_1)\mu_\xi(dt_1) \int_0^{t_1} g(t_2)\mu_\xi(dt_2) \int_0^{t_{n-1}} g(t_n)\mu_\xi(dt_n) \| Y^+_0 \|^2 \|\xi\|^2$$

and (3.6) follow.

Then the series $\sum_{n=0}^{\infty} Y^{(n)\#}(t)$ converge for the strong topology on $\mathcal{D}$ uniformly on bounded intervals of R_+ to an element $(Y^\#(t)))_{t\in R_+}$ of $\mathcal{F}$. We show now that these processes are solutions of (3.3). From inequality (2.2) we have, for all n:

$$\Big\| \int_0^t dM^{\alpha+}(s)\Big(Y^+(s)F^+_\alpha(s) - \sum_{k=0}^{n} Y^{(k)+}(t)F^+_\alpha(s)\Big)\xi\Big) \Big\|^2 \leq$$

$$\leq c_{t,\xi} \int_0^t \Big\| \sum_{k=n+1}^{\infty} Y^{(k)+}(t)F^+_\alpha\xi \Big\|^2 | f_\alpha(s) |^2 \mu_{\alpha,\xi}(ds)$$

and then:

$$\lim_{n\to\infty} \int_0^t dM^{\alpha+}(s)\Big(\sum_{k=0}^{n} Y^{(k)+}(t)F^{+}_{\alpha}(s)\Big)\xi\Big) = \int_0^t dM^{\alpha+}(s)Y^{+}(s)F^{+}_{\alpha}(s))\xi.$$

From the identity

$$\sum_{k=0}^{n+1} Y^{(k)+}(t)\xi = Y_0{}^{+}\xi + \int_0^t dM^{\alpha+}(s)\Big(\sum_{k=0}^{n} Y^{(k)+}(t)F^{+}_{\alpha}(s)\Big)\xi\Big)$$

it follows that $(Y^{+}(t))_{t\in R_+}$ satisfies the second of the equations (3.3). In a similar way we can prove that $(Y(t))_{t\in R_+}$ is a solution of the first of the equations (3.3). Moreover , for all $\xi,\eta\in\mathcal{D}$ and all $t\in R_+$, we have $\langle\eta, Y(t)\xi\rangle = \langle\eta, Y^{+}(t)\xi\rangle$.

We prove now the estimate (3.4); as a consequence we obtain uniqueness of solutions of (3.3). Applying (2.2) we have the inequalities:

$$\| Y(t)\xi - Y(0)\xi \|^2 \leq |I|\, c_{t,\xi} \int_0^t \| F_\alpha(s)Y(s)\xi \|^2 \mu_{\alpha,\xi}(ds)$$

$$\| Y(t)\xi \|^2 \leq 2\| Y(0)\xi \|^2 + 2|I|\, c_{t,\xi} \int_0^t \| Y(s)\xi \|^2\, g(s)\mu_\xi(ds)$$

and, from Gronwall lemma, we obtain (3.4).

4. QUADRATIC VARIATION AND WEAK ITO FORMULA

The goal of this section is to show how a weak Itô formula can be obtained in the abstract scheme we developed for quantum stochastic calculus. This formula will be applied to solutions of quantum stochastic differential equations to find unitarity conditions.

(4.1) PROPOSITION. *Let us suppose that, for all* $\alpha,\beta\in I$, *all* $\xi\in\mathcal{D}$, *all* $s,t\in R_+$ *with* $s<t$ *and all adapted process* F, $M^{\alpha}(s,t)F(s)\xi$ *is in the domain of* $M^{\beta}(s,t)$. *Then there exists an additive adapted process* $[|M^{\beta},M^{\alpha}|]$ *satisfying the following equality* :

$$M^{\beta}(s,t)M^{\alpha}(s,t)F(s)\xi = \Big\{ \int_s^t dM^{\beta}(r)M^{\alpha}(s,r) + \int_s^t dM^{\alpha}(r)M^{\beta}(s,r) + [|M^{\beta},M^{\alpha}|](s,t) \Big\} F(s)\xi \tag{4.2}$$

Proof. For all partition $\mathcal{P}$ of $[s,t]$ by points $s=t_0<t_1<\ldots<t_{n+1}=t$ we have:

$$M^{\beta}(s,t)M^{\alpha}(s,t)F(s)\xi = \sum_{k=0}^{n} M^{\beta}(s,t_k)M^{\alpha}(t_k,t_{k+1})F(s)\xi + \sum_{k=0}^{n} M^{\beta}(t_k,t_{k+1})M^{\alpha}(s,t_k)F(s)\xi +$$

$$+ \sum_{k=0}^{n} M^{\beta}(t_k,t_{k+1})M^{\alpha}(t_k,t_{k+1})F(s)\xi.$$

Due to remark (1.8), as $\sup_{0\leq k\leq n} | t_{k+1}-t_k |$ converges to zero the first and the second sum of the right-hand side converge to the stochastic integrals in (4.2). As a consequence the third sum converge; denoting $[|M^{\beta},M^{\alpha}|](s,t)F(s)\xi$ its limit we obtain (4.2).

The process $[|M^\beta,M^\alpha|]$ is called the *quadratic variation* of M^β and M^α.

We assume now that, for all $\alpha,\beta\in I$:

(4.3) $$[|M^\beta,M^\alpha|](s,t) = \int_s^t dM^\gamma(r)c_\gamma^{\beta\alpha}(r) .$$

Moreover we introduce the following technical assumptions:

(4.4) For all $\alpha\in I$ and all $\xi,\eta\in \mathcal{D}$ there exixts a measure $\upsilon_{\alpha,\xi,\eta}$ on the Borel σ-field on R_+, finite on bounded Borel sets such that, for all adapted processes A, B, C, D and all $s,t\in R_+$ with $s<t$:

$$|\langle A(s)\xi,M^\alpha(s,t)B(s)\eta\rangle - \langle C(s)\xi,M^\alpha(s,t)D(s)\eta\rangle| \leq \upsilon_{\alpha,\xi,\eta}(s,t)\, |\langle A(s)\xi,B(s)\eta\rangle - \langle C(s)\xi,D(s)\eta\rangle|$$

(4.5) For all $\alpha,\beta\in I$, $\xi,\eta\in \mathcal{D}$, $t>0$, and all adapted process A, B:

$$\lim_{|\mathcal{P}|\to 0} \sum_{k=0}^{n} \{\langle A(t_k)\xi,M^\beta(t_k,t_{k+1})M^\alpha(t_k,t_{k+1})B(t_k)\eta\rangle - \langle A(t_k)\xi,c_\gamma^{\beta\alpha}(t_k)M^\gamma(t_k,t_{k+1})B(t_k)\eta\rangle\} = 0$$

where $\mathcal{P}$ is a partition of $[0,t]$ by points $0=t_0<t_1<...<t_{n+1}=t$.

(4.6) PROPOSITION *Let* H_α , K_α $(\alpha\in I)$ *be elements of the class* $\mathcal{F}$ *and let* :

$$X(t) = X(0) + \int_0^t dM^\alpha(s)H_\alpha(s) \quad , \quad Y(t) = Y(0) + \int_0^t dM^\alpha(s)K_\alpha(s).$$

We have then :

$$\langle Y^+(t)\xi,X(t)\eta\rangle = \lim_{|\mathcal{P}|\to 0} \sum_{k=0}^{n} \{\langle Y^+(t_k)\xi,M^\alpha(t_k,t_{k+1})H_\alpha(t_k)\eta\rangle + \langle K_\beta^+(t_k)\xi,M^\beta(t_k,t_{k+1})X(t_k)\eta\rangle + $$
$$+ \langle K_\beta^+(t_k)\xi,c_\gamma^{\beta\alpha}(t_k)M^\gamma(t_k,t_{k+1})H_\alpha(t_k)\eta\rangle\}.$$

or, in differential notation :

$$d(Y(t)X(t)) = dM^\beta(t)(K_\beta(t)X(t)) + dM^\alpha(t)(Y(t)H_\alpha(t)) + dM^\gamma(t)c_\gamma^{\beta\alpha}(t)K_\beta(t)H_\alpha(t).$$

The proof is rather technical and will be omitted.

5. STOCHASTIC CALCULUS ON THE EUCLIDEAN FREE FIELD

In this section we introduce the Euclidean free field as a new model of a Fock space in which a quantum stochastic calculus can be developed following the general scheme we outlined.

Let $\hbar$ be the complex separable Hilbert space $H^{-k}(R^n)$ $(k,n>0)$. $\hbar$ can be described as the vector space of all tempered distributions f such that:

$$\int_{R^n} \frac{|\hat{f}(\xi)|^2}{(1+|\xi|^2)^k} d\xi < +\infty$$

where $\hat{f}$ denotes the Fourier transform of f. The scalar product in $\mathcal{h}$ is

$$\langle g,f\rangle = \int_{R^n} \frac{\overline{\hat{g}(\xi)}\hat{f}(\xi)}{(1+|\xi|^2)^k} d\xi$$

and the norm induced by this scalar product will be denoted by $|\cdot|$.

If A is a closed subset of R, let e_A denote the orthogonal projection in $\mathcal{h}$ onto the family of elements with support in A. Projections corresponding to the sets $\{(x_1,\dots,x_n)\,|\,x_n\leq t\}$, $\{(x_1,\dots,x_n)\,|\,x_n=t\}$ and $\{(x_1,\dots x_n)\,|\,t\leq x_n\}$ will be denoted by $e_{t]}$, $e_{\{t\}}$ and $e_{[t}$. We shall need the following properties of these projections:

(5.1) PROPOSITION (a) *Let* A *and* B *be closed subsets of* R^n *with* $B\subseteq A$ *then we have* :

$$e_A e_B = e_B e_A = e_B .$$

(b) *If* A *and* C *are closed subsets of* R^n, *then* :

$$e_A e_C = e_{\partial A\cup(A\cap C)} e_C .$$

In particular :

$$e_{[t}e_{t]} = e_{t]}e_{[t} = e_{\{t\}}.$$

(c) *The following orthogonal decomposition holds for all* $f\in\mathcal{h}$ *and* $t\in R_+$:

$$f = (1-e_{[t})f + e_{\{t\}}f + (1-e_{t]})f .$$

Proof. (a) Clearly $e_A e_B = e_B$. Taking the adjoints operators we have $e_B e_A = e_B$.(b) We refer to [8](proposition II.3 (i)) for the proof. (c) Since $e_{[t}$ and $e_{t]}$are commuting projections, $(1-e_{t]})$ (resp.$(1-e_{[t})$) maps the range of $e_{[t}$ (resp.$e_{t]}$) in itself. We can therefore write down the following orthogonal decompositions:

$$f = (1-e_{[t})f + e_{[t}f = (1-e_{[t})f + e_{\{t\}}f + (1-e_{t]})e_{[t}f$$
$$f = e_{t]}f + (1-e_{t]})f = (1-e_{[t})e_{t]}f + e_{\{t\}}f + (1-e_{t]})f .$$

The uniqueness of the orthogonal decomposition implies that:

$$(1-e_{t]})e_{[t} = (1-e_{t]}) \qquad (1-e_{[t})e_{t]} = (1-e_{[t})$$

and the conclusion follows.

(5.2) PROPOSITION. *For all* $f\in\mathcal{h}$ *and all* $t\in R_+$, *we have* :

$$\lim_{s\to t} e_{s]}f = e_{t]}f \quad , \quad \lim_{s\to t} e_{[s}f = e_{[t}f \quad , \quad \lim_{s\to+\infty} e_{s]}f = f.$$

Proof. Let us first consider the left translation by h units in the last coordinate τ_h : $(\tau_h g)(x_1,\dots,x_n) = g(x_1,\dots,x_n+h)$ and $\widehat{\tau_h g}(\xi) = \exp(ih\xi_n)\hat{g}(\xi)$. A simple application of Schwarz inequality and Lebesgue's theorem shows that, for all $f\in\mathcal{h}$, the function $h\to\langle f,\tau_h g\rangle$ is continous uniformly for g in a bounded subset of $\mathcal{h}$. Let ε and δ be positive numbers such that, for all h with $|2h|<\delta$ and all $s\in R$

$$|\langle f,e_{s]}f-\tau_{2h}e_{s]}f\rangle|<\varepsilon^2 ,$$

Taking into account the fact that $\mathrm{supp}(\tau_{2h}e_{s]}f)\subseteq\{(x_1,\dots,x_n)\,|x_n\leq s-2h\}$ and that τ_{2h} is an isometry of $\mathcal{h}$, we have:

$$|e_{t+\delta]}f - \tau_{2\delta}e_{t+\delta]}f|^2 = 2\mathrm{Re}\langle f, e_{t+\delta]}f - \tau_{2\delta}e_{t+\delta]}f\rangle \leq 2\varepsilon^2.$$

Therefore, for all $s\in(t-\delta,t+\delta)$:

$$|e_{t]}f-e_{s]}f| = |e_{t]}e_{t+\delta]}f-e_{s]}e_{t+\delta]}f| \leq$$

$$\leq \| e_{t]}e_{t+\delta]}f - e_{t]}(\tau_{2\delta}e_{t+\delta]}f) \| + \| e_{s]}e_{t+\delta]}f - e_{s]}(\tau_{2\delta}e_{t+\delta]}f) \| + \| e_{t]}(\tau_{2\delta}e_{t+\delta]}f) - e_{s]}(\tau_{2\delta}e_{t+\delta]}f) \| \leq$$
$$\leq 2 \| e_{t+\delta]}f - \tau_{2\delta}e_{t+\delta]}f \| < 2\sqrt{2}\varepsilon$$

and the first equality is proven. The second can be proven in the same way and the third using the density of compact support elements of h in h.

As a consequence of proposition (5.2), for all $g,f \in h$, there exists a unique complex-valued measure $\mu_{g,f}$ on the Borel σ-field on R, without atoms such that, for all $a,b \in R$ with $a<b$

$$\mu_{g,f}(a,b) = \langle g, e_{b]}f - e_{a]}f \rangle.$$

The measure $\mu_{f,f}$ is also positive and will be denoted by μ_f. The total variation $|\mu_{g,f}|$ of $\mu_{g,f}$ satisfies the inequality : $2\,|\mu_{g,f}| \leq \mu_f + \mu_g$.

We will now briefly describe the Fock space $\mathcal{H} = \Gamma(h)$ over the one particle space h. Let us denote:

- $\mathcal{D} = \{\psi(f) \mid f \in h\}$ the set of exponential vectors in $\mathcal{H}$.
- $W(f)$ ($f \in h$) the Weyl unitary operator characterized by

$$W(f)\psi(g) = \exp(-\tfrac{1}{2}\|f\|^2 - \langle f,g \rangle)\, \psi(f+g).$$

- For all contraction C in h, $\Gamma(C)$ the contraction in $\mathcal{H}$ defined by $\Gamma(C)\psi(f) = \psi(Cf)$.
- For all $t \in R$, $\mathcal{H}_{t]} = \Gamma(e_{t]})\mathcal{H}$, $\mathcal{H}_{\{t\}} = \Gamma(e_{\{t\}})\mathcal{H}$, $\mathcal{H}_{[t} = \Gamma(e_{[t})\mathcal{H}$, $\mathcal{H}_{t)} = \Gamma(1-e_{[t})\mathcal{H}$, $\mathcal{H}_{(t} = \Gamma(1-e_{t]})\mathcal{H}$.

We can think to these spaces respectively as the past, present, future, strict past and strict future space. It is to be noted that the present space $\mathcal{H}_{\{t\}}$ is non-trivial since $e_{\{t\}}h$ contains the δ distribution supported on the hyperplane $\{(x_1,\ldots,x_n) \mid x_n = t\}$ and its first k-1 derivatives. We will consider also the space $\mathcal{H}_{(s,t]} = \Gamma(e_{t]} - e_{s]})\mathcal{H}$ ($s,t \in R$, $s<t$).

Due to proposition (5.2),the following tensor product factorizations hold:

$$\mathcal{H}_{t]} = \mathcal{H}_{t)} \otimes \mathcal{H}_{\{t\}}, \qquad \mathcal{H}_{[t} = \mathcal{H}_{\{t\}} \otimes \mathcal{H}_{(t}$$
$$(5.3) \qquad \mathcal{H} = \mathcal{H}_{t)} \otimes \mathcal{H}_{\{t\}} \otimes \mathcal{H}_{(t}, \qquad \mathcal{H} = \mathcal{H}_{s)} \otimes \mathcal{H}_{(s,t]} \otimes \mathcal{H}_{(t}.$$

Let $I_{\mathcal{K}}$ denote the identity operator on the suspace $\mathcal{K}$ of $\mathcal{H}$ and let

$$\mathcal{A}_{t]} = \mathcal{B}(\mathcal{H}_{t]}) \otimes I_{\mathcal{H}_{(t}}, \qquad \mathcal{A}_{[t} = I_{\mathcal{H}_{t)}} \otimes \mathcal{B}(\mathcal{H}_{[t}), \qquad \mathcal{A}_{\{t\}} = I_{\mathcal{H}_{t)}} \otimes \mathcal{B}(\mathcal{H}_{\{t\}}) \otimes I_{\mathcal{H}_{(t}}.$$

be the past, future and present algebra. The maps $E_{t]}: \mathcal{A} \to \mathcal{A}_{t]}$, $E_{[t}: \mathcal{A} \to \mathcal{A}_{[t}$, $E_{\{t\}}: \mathcal{A} \to \mathcal{A}_{\{t\}}$ defined respectively by:

$$E_{t]}(A) = \Gamma(e_{t]})A\Gamma(e_{t]}) \otimes I_{\mathcal{H}_{(t}}, \quad E_{[t}(A) = I_{\mathcal{H}_{t)}} \otimes \Gamma(e_{[t})A\Gamma(e_{[t}), \quad E_{\{t\}}(A) = I_{\mathcal{H}_{t)}} \otimes \Gamma(e_{\{t\}})A\Gamma(e_{\{t\}}) \otimes I_{\mathcal{H}_{(t}}$$

are conditional expectations. The following fact is easily checked:

(5.4) PROPOSITION. *For all* $B \in \mathcal{A}_{[t}$ *we have* : $E_{t]}(B) = E_{\{t\}}(B)$.

This property translates into the *Markov property* : for questions about the future, knowledge of the present is as good as knowledge of the past.

Let $A(f)$, $A^+(f)$, N be the annihilation, creation and number fields defined on $\mathcal{D}$ by the relations:

$$A(f)\psi(g) = \langle f,g\rangle\psi(g)\ ,\ A^+(f)\psi(g) = \frac{d}{d\varepsilon}\psi(g+\varepsilon f)\Big|_{\varepsilon=0}\ ,\ N(t)\psi(g) = -i\frac{d}{d\varepsilon}\psi(\exp(i\varepsilon e_{t]})g)\Big|_{\varepsilon=0}$$

We will write also $A^\#_f(t)$ for $A^\#(e_{t]}f)$, $A^\#_f(s,t)$ for $A^\#(e_{t]}f-e_{s]}f)$ and $N(s,t)$ for $N(t)-N(s)$.

$(A_f(s,t))_{0\le s\le t}$, $(A^+_f(s,t))_{0\le s\le t}$ and $(N(s,t))_{0\le s\le t}$ are additive adapted processes in the sense of definition (1.3) and have the property (1.4). Moreover, since $\mathcal{A}_{t]}'=1_{\mathcal{H}_{t]}}\otimes\mathcal{B}(\mathcal{H}_{(t})$, they have the commutation with the past property (1.5). We can prove also the semimartingale estimates:

(5.5) PROPOSITION. *For all* $f\in h$ *and all simple adapted process* F *we have the following inequalities* :

$$\Big|\int_0^t dA_f(s)F(s)\psi(g)\Big|^2 \le |\mu_{g,f}|(0,t)\int_0^t |F(s)\psi(g)|^2\,|\mu_{g,f}|(ds)$$

$$\Big|\int_0^t dA^+_f(s)F(s)\psi(g)\Big|^2 \le \left(\sqrt{|\mu_{g,f}|(0,t)}+\sqrt{1+|\mu_{g,f}|(0,t)}\right)\int_0^t |F(s)\psi(g)|^2(\mu_f+|\mu_{g,f}|)(ds)$$

$$\Big|\int_0^t dN(s)F(s)\psi(g)\Big\|^2 \le \left(\sqrt{\mu_g(0,t)}+\sqrt{1+\mu_g(0,t)}\right)\int_0^t |F(s)\psi(g)|^2\mu_g(ds).$$

Proof. We will prove, for example, the second estimate. Let us suppose, for simplicity, that F is written in the form (1.2) and $s_{n+1}=t$. We can write the left-hand side in the form:

$$\sum_{1\le j\le n}|A^+_f(s_j,s_{j+1})F(s_j)\psi(g)|^2 + 2\mathrm{Re}\sum_{1\le k<j\le n}\langle A^+_f(s_k,s_{k+1})F(s_k)\psi(g),A^+_f(s_j,s_{j+1})F(s_j)\psi(g)\rangle$$

Due to the tensor product factorization of $\mathcal{H}$ for all j we have

$$A^+_f(s_j,s_{j+1})F(s_j)\psi(g) = F(s_j)\psi(e_{s_j]}g)\otimes A^+_f(s_j,s_{j+1})\psi((1-e_{s_j]})g).$$

and the sum can be rewritten, with an easy computation, as

$$\sum_{j=1}^n |F(s_j)\psi(g)|^2\mu_f(s_j,s_{j+1}) + 2\mathrm{Re}\sum_{j=1}^n\Big\langle\int_0^{s_j} dA^+_f(s)F(s)\psi(g),F(s_j)\psi(g)\Big\rangle\mu_{g,f}(s_j,s_{j+1}).$$

and the modulus can be majorized, using the Schwarz inequality, by:

$$\sum_{j=1}^n |F(s_j)\psi(g)|^2\mu_f(s_j,s_{j+1}) + 2\sum_{j=1}^n |F(s_j)\psi(g)|\,\Big\|\int_0^{s_j} dA^+_f(s)F(s)\psi(g)\Big\|\,|\mu_{g,f}|(s_j,s_{j+1}).$$

Let

$$R(t)=\sup_{s\le t}\Big|\int_0^s dA^+_f(s)F(s)\psi(g)\Big| \qquad a(t)=\Big(\int_0^t |F(s)\psi(g)|^2(\mu_f+|\mu_{g,f}|)(ds)\Big)^{1/2};$$

we have then:

$$R^2(t) \leq 2 \int_0^t |F(s)\psi(g)| R(s)\, |\mu_{g,f}|\,(ds) + a^2(t).$$

Using again the Schwarz inequality and majorizing R(s) by R(t):

$$R^2(t) \leq 2\sqrt{|\mu_{g,f}|\,(0,t)}\,a(t)R(t) + a^2(t);$$

then $R(t) \leq (\sqrt{|\mu_{g,f}|\,(0,t)} + \sqrt{1+|\mu_{g,f}|\,(0,t)})a(t)$ and the second inequality follows.

The results of [4] (corollary (5.6), theorem (7.2)) on quadratic variation give the following:

(5.7) PROPOSITION *For all* $f,g \in h$ *and all* $s,t \in R_+$ *with* $s<t$ *we have* :

$$[|A_g,A_f^+|](s,t) = \langle g,(e_{t]}-e_{s]})f\rangle, \qquad [|A_f,N|](s,t) = A_f(s,t), \qquad [|A_g,A_f^+|](s,t) = A_f^+(s,t).$$

The quadratic variation of any other pair of processes A_g, A_f^+, N *is zero.*

It can be shown that the processes A_g, A_f^+, N satisfy assumptions (4.4), (4.5) and then we can apply the weak Itô formula (4.6) to prove unitarity conditions for the solutions of stochastic differential equations driven by them. Details will be published in [3].

6. A QUANTUM LEVY THEOREM.

In this section we will prove a quantum extension of a characterisation of classical brownian motion due to P.Lévy (cf.[7],[11]). Let $(\Omega,\mathcal{F},P)$ be a probability space and $(\mathcal{F}_t)_{t\in R_+}$ be a filtration. A real valued $(\mathcal{F}_t)_{t\in R_+}$-martingale $(X(t))_{t\in R_+}$ with continous sample paths and with quadratic variation $[|X,X|](s,t) = (t-s)$ is a standard brownian motion.

We keep the notations of sections 1,3,4 . Let us suppose that:

(6.1) $|I| = 3$ and $M^0(t)=tI$, $M^1(t)=M(t)$, $M^2(t)=M^+(t)$ where I denotes the identity operator on $\mathcal{H}$ and M is an additive adapted process satisfying conditions (1.4), (1.5), (1.6).

(6.2) There exists complex-valued continous functions $\sigma^{i,j}$ $(1\leq i,j\leq 2)$ on R_+ such that:

$$[|M^+,M|](s,t) = \int_s^t \sigma^{1,1}(r)dr, \qquad [|M,M|](s,t) = \int_s^t \sigma^{1,2}(r)dr,$$

$$[|M,M^+|](s,t) = \int_s^t \sigma^{2,2}(r)dr, \qquad [|M^+,M^+|](s,t) = \int_s^t \sigma^{2,1}(r)dr;$$

so that the matrix $(\sigma^{i,j}(t))_{1\leq i,j\leq 2}$ is positive definite.

(6.3) M and M^+ are martingales with respect to the vector Φ, i.e. for all $s,t\in R_+$ with $s<t$ and all adapted process H, K we have:

$$\langle K(s)\Phi, M^{\#}(s,t)H(s)\Phi\rangle = 0\,.$$

(6.4) REMARK. As a corollary of the semimartingale definition, we have $[|M^0,M^\#|] = [|M^\#,M^0|] = 0$. As a matter of fact, for all $\xi \in \mathcal{D}$, all $s,t \in R_+$ with $s<t$ and all adapted process F:

$$\left| \sum_{k=0}^{n} (t_{k+1}-t_k) M^\#(t_k,t_{k+1}) F(s)\xi \right| \le \sum_{k=0}^{n} (t_{k+1}-t_k) \left| M^\#(t_k,t_{k+1}) F(s)\xi \right|$$

where $s=t_0<t_1<...<t_n<t_{n+1}=t$. Then, by remark (1.8),

$$\lim_{|\mathcal{P}|\to 0} \sum_{k=0}^{n} (t_{k+1}-t_k) M^\#(t_k,t_{k+1}) F(s)\xi = 0 .$$

We will denote $(f,g)_\sigma(t)=\sigma^{1,1}(t)f(t)g^+(t)+\sigma^{1,2}(t)f(t)g(t)+\sigma^{2,1}(t)f^+(t)g^+(t)+\sigma^{2,2}(t)f^+(t)g(t)$ and:

$$I(f,g)_\sigma(t) = \frac{1}{2i}\{(f,g)_\sigma(t)-(g,f)_\sigma(t)\}$$

for all $f,g \in C_c(R_+;C)$ the vector space of complex-valued compact support continous functions on R_+. We will prove the following:

(6.5) THEOREM. *The solutions of the stochastic differential equations*

$$U_f(t) = I + \int_0^t (if(s)dM^+(s)+if^+(s)dM(s) - \tfrac{1}{2}(f,f)_\sigma(s))U_f(s) \qquad (f \in C_c(R_+;C))$$

are a unitary representation $(\mathcal{H}, \{U_f \mid f \in C_c(R_+;C)\})$ *of the* CCR *algebra on* $C_c(R_+;C)$ *associated with the real bilinear symplectic form* $(f,g) \to \int_0^\infty I(f,g)_\sigma(t)dt$.

The unitary group $(U_{tf})_{t\in R_+}$ *is strongly continous on* $\mathcal{H}$ *and its infinitesimal generator* $iB(f)$ *coincides, on the domain* $\mathcal{D}$, *with the operator*

$$\int_0^\infty (f(s)dM^+(s)+f^+(s)dM(s)) . \tag{6.6}$$

The state Φ *is gaussian and we have*

$$\langle \Phi, U_f(t)\Phi\rangle = \exp\left(-\tfrac{1}{2}(f,g)_\sigma(t)\right). \tag{6.7}$$

(6.8) REMARK. In this theorem there is no sample path continuity assumption since sample paths do not make sense in the context of quantum probability theory. However from assumptions (6.2), (6.4) it follows that, for all $s,t \in R_+$ with $s<t$

$$\left| M^\#(s,t)M^\#(s,t)\Phi \right|^2 \le \left(\int_s^t |\sigma|(r)dr\right)^2$$

where $|\sigma|(r) = \sup_{1\le i,j\le 2} |\sigma^{i,j}(r)|$.

This is a fourth order moment condition that can be easily translated in the language of classical probability and is sufficient for the existence of a version, of a classical stochastic process with continous sample paths by the well known theorem of Kolmogorov.

Proof. The process U_f^+ is the solution of the stochastic differential equation

$$U_f^+(t) = I + \int_0^t (-if^+(s)dM(s) - if(s)dM^+(s) - \tfrac{1}{2}(f,f)_\sigma(s))U_f^+(s).$$

Applying the weak Itô formula (4.6) we can easily see that, for all $\xi,\eta \in \mathcal{D}$:

$$\langle U_f(t)\xi, U_f(t)\eta\rangle = \langle \xi,\eta\rangle = \langle U_f^+(t)\,\xi, U_f^+(t)\,\eta\rangle$$

and then $U_f(t)$ can be extended to a unitary operator on $\mathcal{H}$; we will denote still by $U_f(t)$ this extension. Using again the weak Itô formula we can show that, for all $f,g \in C_c(R_+;C)$, processes $(U_f(t)U_g(t))_{t\in R_+}$ and $(U_{f+g}(t)\exp(-i\int_0^t I(f,g)_\sigma(t)dt))_{t\in R_+}$ are solution of the same stochastic differential equation with the same initial condition; they are the same process and we have the CCR

$$U_f(t)U_g(t) = \exp(-i\int_0^t I(f,g)_\sigma(t)dt)U_{f+g}(t).$$

For all $f \in C_c(R_+;C)$ we denote by U_f the unitary operator $\lim_{t\to\infty} U_f(t)$ (for the strong topology). It follows from (6.1) that $(U_{sf})_{s\in R_+}$ is a group. Moreover, for all $\varepsilon \in R$, $U_{\varepsilon f}$ satisfies the stochastic differential equation:

$$U_{\varepsilon f}(t) = I + \int_0^t (i\varepsilon f(s)dM^+(s) + i\varepsilon f^+(s)dM(s) - \frac{\varepsilon^2}{2}(f,f)_\sigma(s))U_f(s)$$

and then, using the semimartingale estimate (2.2), for all $\xi \in \mathcal{D}$

$$| U_{\varepsilon f}\xi - \xi |^2 \le 2\varepsilon c_f \{ \int_0^\infty |f(s)|^2 \mu_\xi(ds) + \frac{\varepsilon}{2}| \int_0^\infty (f,f)_\sigma(s)ds |^2 | \xi |^2$$

where c_f is a constant depending on f. It follows that $(U_{sf})_{s\in R_+}$ is strongly continous on $\mathcal{D}$ and then strongly continous on $\mathcal{H}$. For all $\xi \in \mathcal{D}$ we have also

$$\| \{\frac{U_{\varepsilon f} - I}{\varepsilon} - \int_0^t (if(s)dM^+(s) + if^+(s)dM(s))\}\xi \|^2 \le$$

$$\le 2\int_0^\infty | U_{\varepsilon f}\xi - \xi |^2 |f(s)|^2 \mu_\xi(ds) + \varepsilon | \int_0^\infty (f,f)_\sigma(s)ds |^2 | \xi |^2.$$

This clearly implies that the infinitesimal generator coincides with the operator (6.6) on the domain $\mathcal{D}$. Finally $\langle \Phi, U_f(t)\Phi\rangle$ is the only solutuion of the ordinary differential equation

$$\langle \Phi, U_f(t)\Phi\rangle = 1 - \frac{1}{2}\int_0^t (f,f)_\sigma(s)\langle \Phi, U_f(s)\Phi\rangle ds$$

because of the martingale property (6.3) and then (6.7) holds.

REFERENCES

[1] Accardi L. Quantum stochastic calculus. Proceedings IV Vilnius Conference on Probability and Mathematical Statistics. VNU Science Press (1987).

[2] Accardi L., Fagnola F. Stochastic integration. Quantum Probability and Applications III. Proceedings, Oberwolfach. LN 1303 Springer (1988).

[3] Accardi L.,Fagnola F.,Quaegebeur J. Quantum stochastic calculus. To appear

[4] Accardi L., Quaegebeur J. The Itô algebra of quantum gaussian fields. To appear.

[5] Accardi L., Parthasarathy K.R. Stochastic calculus on local algebras. Quantum Probability and Applications II Proceedings, Heidelberg. LN 1136 (1985).

[6] Barnett C., Streater R., Wilde I.F. The Itô-Clifford integral. J.Funct.Anal.48 (1982).

[7] Dellacherie C., Meyer P.-A. Probabilités et potentiel. Chap.V à VIII. Hermann (1980).

[8] Guerra F.,Rosen L., Simon B. $P(\phi)_2$ euclidean quantum field theory. Ann. of Math. 101 (1975).

[9] Hudson R.L., Lindsay J.M. Stochastic integration and a martingale representation theorem for non-Fock quantum brownian motion. J. Funct. Anal. 61 (1985).

[10] Hudson R.L., Parthasarathy K.R. Quantum Itô's formula and stochastic evolutions. Commun Math. Phys. 93 (1984).

[11] Letta G. Martingales et intégration stochastique. Quaderni Scuola Normale Superiore. Pisa. (1984).

[12] Lindsay J.M., Maassen H. An integral kernel approach to noise. Quantum Probability and Applications III. Proceedings, Oberwolfach. LN 1303 Springer (1988).

This paper is in final form and no similar paper has been or is being submitted elsewhere.

LOCALLY INDEPENDENT BOSON SYSTEMS

by

Karl-Heinz Fichtner and Uwe Schreiter

Friedrich-Schiller-Universität Jena

Sektion Mathematik, GDR - 6900 Jena

1. Introduction

In their paper [10] GLAUBER and TITULAER deal with coherent states of photon systems. They discuss a characterization of these states by the property of local independence.

Basing on the algebraic approach to quantum statistical mechanics FICHTNER and FREUDENBERG have proposed a new description of locally normal states of boson systems. They proved that one may assign to each locally normal state a uniquely determined point process which contains all informations about position measurements and therefore is called the position distribution of the system (cf. [5]). Under certain assumptions there exists a special function - the so called conditional reduced density matrix - which together with the position distribution determines the state completely (cf. [3], [4]). We consider boson systems in an arbitrary Polish space G equipped with a locally finite measure ν on G such that $\nu(\{x\}) = 0$ for all $x \in G$. Using FICHTNER/ FREUDENBERG [4] we introduce a class of locally normal states of boson systems which can be considered as a generalization of normal coherent states. We prove that this type of states is characterized by local independence too. In the lattice case (i.e. $\nu(\{x\}) > 0$ for some $x \in G$) the notions "coherent state" and "locally independent state" don't coincide any more.

In 2. we collect some basic notions and notations. In 3. and 4. we introduce the position distribution and the conditional reduced density matrix of locally normal states of boson systems. In 5. one can find the main result proved in chapter 6.

2. Locally normal states of boson systems

Let $[G,\mathcal{G}]$ be a complete separable metric space equipped with the σ-algebra of BOREL subsets, $\mathcal{B}$ the ring of bounded sets in $\mathcal{G}$, ν a locally finite diffuse measure on $[G,\mathcal{G}]$, i.e. $\nu(\Lambda) < \infty$ for all $\Lambda \in \mathcal{B}$ and $\nu(\{x\}) = 0$ for all $x \in G$. We denote by δ_x the DIRAC measure corresponding to $x \in G$ and by $\mathbb{N}$ the set of non-negative integers. Furthermore, let M be the set of all locally finite integer-valued measures on $[G,\mathcal{G}]$, i.e.

$$M := \left\{\varphi\colon \varphi \text{ measure on } [G,\mathcal{G}],\ \varphi(\Lambda) \in \mathbb{N} \text{ for all } \Lambda \in \mathcal{B}\right\}$$

The elements of M which are called counting measures can be interpreted as locally finite point configurations in G.

For arbitrary $\Lambda \in \mathcal{G}$ we denote by ${}_\Lambda\mathfrak{M}$ the smallest σ-algebra of subsets of M such that for each $\Lambda' \in \mathcal{B}$, $\Lambda' \subseteq \Lambda$ the mapping $\varphi \to \varphi(\Lambda')$ is measurable. Especially we set $\mathfrak{M} := {}_G\mathfrak{M}$.

We still introduce some important subsets of M. By M^s we denote the set of simple counting measures, i.e.

$$M^s := \left\{\varphi \in M\colon \varphi(\{x\}) \leq 1 \text{ for all } x \in G\right\}$$

M^f is the set of finite counting measures, i.e.

$$M^f := \left\{\varphi \in M\colon \varphi(G) < \infty\right\}$$

For arbitrary $\Lambda \in \mathcal{G}$ we put

$$M_\Lambda := \left\{\varphi \in M\colon \varphi(\Lambda^c) = 0\right\}$$

$$\mathfrak{M}_\Lambda := \mathfrak{M} \cap M_\Lambda$$

$$\varphi_\Lambda := \varphi(\cdot \cap \Lambda) \qquad ; \varphi \in M$$

$$F_\Lambda(Y) := \chi_Y(\emptyset) + \sum_{n\geq 1} \frac{1}{n!} \int_{\Lambda^n} \nu^n(d\bar{x}^n)\, \chi_Y(\delta_{\bar{x}^n}) \qquad ; Y \in \mathfrak{M}$$

($\emptyset$ denotes the void point configuration, i.e. $\emptyset(G) = 0$, χ_Y is the indicator function of the set Y, $\delta_{\bar{x}^n} = \delta_{(x_1,\ldots,x_n)}$ is an abbreviation of $\sum_{j=1}^{n} \delta_{x_j}$.)

F_Λ is a σ-finite measure on $[M,\mathfrak{M}]$. It is concentrated on $M_\Lambda \cap M^s \cap M^f$ and finite if Λ is bounded. We set $F := F_G$.

By $\mathbb{C}$ we denote the set of complex numbers. For $\Lambda \in \mathcal{G}$ we set

$$\mathcal{M}_\Lambda := \left\{\psi: M \to \mathbb{C},\ \text{measurable},\ \operatorname{supp}\psi \subseteq M_\Lambda \cap M^f,\ \int F_\Lambda(d\varphi)|\psi(\varphi)|^2 < \infty\right\}$$

The HILBERT space $\mathcal{M}_\Lambda$ equipped with the scalar product

$$< \psi_1,\psi_2 > := \int F_\Lambda(d\varphi)\ \overline{\psi_1(\varphi)}\ \psi_2(\varphi)$$

($\overline{\psi}$ denotes the complex conjugate of ψ) is called the symmetric FOCK space over Λ (cf. [4]). Obviously, we can identify $\mathcal{M}_\Lambda$ and $L_2(M,F_\Lambda)$. For brevity we put $\mathcal{M} := \mathcal{M}_G$.

Let $\Lambda,\Lambda' \in \mathcal{G}$, $\Lambda \cap \Lambda' = \emptyset$. There exists a unique isomorphism $I_{\Lambda,\Lambda'}$ between $\mathcal{M}_\Lambda \otimes \mathcal{M}_{\Lambda'}$ and $\mathcal{M}_{\Lambda\cup\Lambda'}$ such that

$$I_{\Lambda,\Lambda'}(\psi \otimes \psi')(\varphi) = \psi(\varphi_\Lambda)\cdot\psi'(\varphi_{\Lambda'}) \qquad ; \psi \in \mathcal{M}_\Lambda\ ,\ \psi' \in \mathcal{M}_{\Lambda'}\ ,\ \varphi \in M_{\Lambda\cup\Lambda'}$$

(cf.[4], proposition 2.6.).

Let $\mathcal{A}_\Lambda := \mathcal{L}(\mathcal{M}_\Lambda)$ be the algebra of bounded linear operators on $\mathcal{M}_\Lambda$ ($\Lambda \in \mathcal{G}$). For $\Lambda \in \mathcal{B}$ we denote by ${}_\Lambda\mathcal{A}$ the natural embedding of $\mathcal{A}_\Lambda$ into $\mathcal{L}(\mathcal{M})$. More precisely, we define a mapping ${}_\Lambda J: \mathcal{A}_\Lambda \to \mathcal{L}(\mathcal{M})$ by

$${}_\Lambda JA := I_{\Lambda,\Lambda^c}(A \otimes \mathbb{1}_{\Lambda^c})\, I^{-1}_{\Lambda,\Lambda^c} \qquad ; A \in \mathcal{A}_\Lambda$$

($\mathbb{1}_{\Lambda^c}$ is the identity in $\mathcal{A}_{\Lambda^c}$) and put ${}_\Lambda\mathcal{A} := \{{}_\Lambda JA: A \in \mathcal{A}_\Lambda\}$.

${}_\Lambda\mathcal{A}$ is called the local algebra on Λ. Now we put

$$\mathcal{A} := \overline{\bigcup_{\Lambda\in\mathcal{B}} {}_\Lambda\mathcal{A}}$$

($\overline{}$ denotes the uniform closure)

The pair $[\mathcal{A},({}_\Lambda\mathcal{A})_{\Lambda\in\mathcal{B}}]$ (or for brevity $\mathcal{A}$) is called the quasilocal algebra. A positive normalized linear functional on an operator algebra is called a state. A state ω on $\mathcal{A}$ is said to be normal (more precisely, ω has a normal extension to $\mathcal{L}(\mathcal{M})$) if there exists a positive trace-class operator ρ on $\mathcal{M}$ with tr ρ = 1 such that

$$\omega(A) = \operatorname{tr}\rho A \qquad ; A \in \mathcal{A}$$

A state ω on $\mathcal{A}$ is called locally normal if for each $\Lambda \in \mathcal{B}$ there exists a positive trace-class operator ${}_\Lambda\rho$ such that

$$\omega(A) = \mathrm{tr}\ {}_\Lambda\rho A \qquad ; A \in {}_\Lambda\mathcal{A}$$

For a general survey about theory of operator algebras and states of quantum systems we refer to BRATTELI / ROBINSON [1].

3. The position distribution of locally normal boson systems

A probability measure on $[M,\mathcal{M}]$ is called a point process. According to the interpretation of $\varphi \in M$ a point process can be considered as the distribution law of a locally finite random point sýstem.
A point process Q is said to be simple if $Q(M^s) = 1$, it is called finite if $Q(M^f) = 1$. Q is called continuous if for all $x \in G$ there holds $Q(\{\varphi \in M: \varphi(\{x\} > 0\}) = 0$.
The intensity measure I_Q of a point process Q is defined by

$$I_Q(\Lambda) := \int Q(d\varphi)\ \varphi(\Lambda) \qquad ; \Lambda \in \mathfrak{G}$$

We denote by O_g the operator of multiplication corresponding to a bounded measurable function $g: M \to \mathbb{C}$, i.e.

$$O_g\psi(\varphi) := g(\varphi)\cdot\psi(\varphi) \qquad ; \psi \in \mathcal{M},\ \varphi \in M$$

The following statement holds (cf. FICHTNER / FREUDENBERG [5])

<u>Proposition 1:</u> Let ω be a locally normal state on $\mathcal{A}$. There exists an uniquely determined point process Q_ω such that for all $\Lambda \in \mathcal{B}$ and real functions g such that $O_g \in {}_\Lambda\mathcal{A}$ there holds

$$\omega(O_g) = \int Q_\omega(d\varphi)\ g(\varphi)$$

The selfadjoint operators O_g correspond to position measurements. Therefore we call Q_ω the position distribution of ω (cf. [3], [4]). Proposition 1 is also true in the lattice case (cf. [2]), but for the next lemma we need the assumption that ν is a diffuse measure on $[G,\mathfrak{G}]$.

Lemma 1: The position distribution Q_ω of a locally normal state ω on $\mathcal{A}$ is a simple and continuous point process, there holds $I_{Q_\omega} \ll \nu$.

We still need some more notions from point process theory.

A point process Q is called a POISSON process if for all finite sequences $\Lambda_1,..,\Lambda_m$ of pairwise disjoint sets from $\mathcal{B}$ and all $n_1,..,n_m \in \mathbb{N}$ the following equality holds:

$$Q(\{\varphi \in M: \varphi(\Lambda_1) = n_1,..,\varphi(\Lambda_m) = n_m\}) = \exp(-I_Q(\bigcup_{s=1}^{m} \Lambda_s)) \prod_{s=1}^{m} \frac{(I_Q(\Lambda_s))^{n_s}}{n_s!}$$

Let μ: G $\to$ [0,+∞) be a locally integrable function. We denote by P^μ the POISSON process with intensity measure

$$I_{P^\mu}(\Lambda) = \int_\Lambda \nu(dx)\, \mu(x) \qquad ; \Lambda \in \mathcal{B}$$

The reduced CAMPBELL measure C_Q of a point process Q is a measure on $\mathcal{G} \otimes \mathcal{M}$ characterized by

$$C_Q(\Lambda \times Y) := \int Q(d\varphi) \int_\Lambda \varphi(dx)\, \chi_Y(\varphi-\delta_x) \qquad ; \Lambda \in \mathcal{G},\ Y \in \mathcal{M}$$

A point process Q is called a Σ'-process if $C_Q \ll \nu \times Q$. P^μ is a Σ'-process because we have $C_Q = I_Q \times Q$ if Q is a POISSON process.

For further and more detailed informations on point process theory we refer to MATTHES, KERSTAN, MECKE [8] and KALLENBERG [6]. Properties and characterizations of Σ'-processes one can find in KERSTAN, MATTHES, MECKE [7] and in RAUCHENSCHWANDTNER [9].

4. The conditional reduced density matrix

The following theorems one can find in [4]:

Proposition 2: Let Q be a Σ'-point process and let k: $M^f \times M^f \times M \to \mathbb{C}$ be a measurable function such that

(a) For each $\Lambda \in \mathcal{B}$, $\varphi \in M_{\Lambda^c}$ there exists a positive trace class operator $K_{\Lambda,\varphi}$ with kernel $k(\cdot,\cdot,\varphi)\chi_{M_\Lambda \times M_\Lambda}(\cdot,\cdot)$.

(b) $\operatorname{tr} K_{\Lambda,\varphi} O_{\chi_Y} = \int_Y F(d\hat{\varphi})\, k(\hat{\varphi},\hat{\varphi},\varphi)$; $Y \in \mathfrak{B}_\Lambda$, $\varphi \in M_{\Lambda^c}$

(c) $k(\delta_x,\delta_x,\varphi) = \dfrac{dC_Q}{d(\nu\times Q)}(x,\varphi)$; $x \in G$, $\varphi \in M$

(d) $k(\varphi_1+\hat{\varphi},\varphi_2+\hat{\varphi},\varphi) = k(\hat{\varphi},\hat{\varphi},\varphi)\cdot k(\varphi_1,\varphi_2,\varphi+\hat{\varphi})$; $\varphi_1,\varphi_2,\hat{\varphi} \in M^f$, $\varphi \in M$

Then there exists exactly one locally normal state ω on $\mathcal{A}$ such that

(e) $\omega({}_\Lambda JA) = \int_{M_{\Lambda^c}} Q_\omega(d\varphi) \int F_\Lambda(d\varphi_1) \int F_\Lambda(d\varphi_2)\, a(\varphi_1,\varphi_2)\, k(\varphi_2,\varphi_1,\varphi)$

for all $\Lambda \in \mathfrak{B}$ and all integral operators $A \in \mathcal{A}_\Lambda$ with kernel a. The point process Q is the position distribution of ω.

Under certain assumptions locally normal states may be characterized by their position distribution and a function k_ω which fulfils (c) -(e). Because the position distribution of these states is always a Σ'-point process we call them Σ'-states. We have

Proposition 3: Let ω be a Σ'-state. If the reduced density matrix (cf. [1]) μ_m of order m exists then it holds

$$\mu_m(\bar{x}^m,\bar{y}^m) = \int Q(d\varphi)\, k_\omega(\delta_{\bar{y}^m},\delta_{\bar{x}^m},\varphi) \quad ; \bar{x}^m,\bar{y}^m \in G^m$$

For that reason k_ω is called the conditional reduced density matrix (cf. FICHTNER / FREUDENBERG [4], chapter 5).

Now we will introduce the notion of a coherent or GLAUBER state. Let $g: G \to \mathbb{C}$ be a locally square integrable function, i.e.

$$\int_\Lambda \nu(dx)|g(x)|^2 < \infty \quad \text{for all } \Lambda \in \mathfrak{B}.$$

We set

$$k_g(\varphi_1,\varphi_2,\varphi) := \prod_{x\in\varphi_1} g(x) \prod_{y\in\varphi_2} \overline{g(y)} \quad ; \varphi_1,\varphi_2 \in M^f,\ \varphi \in M$$

($x \in \varphi$ means $\delta_x \ll \varphi$, $\prod_{x\in\emptyset} := 1$)

In case of $Q = P^{|g|^2}$ we can apply proposition 2 (cf. [4], ch. 7.2). Hence there exists exactly one locally normal state ω_g having the position distribution $P^{|g|^2}$ and the conditional reduced density matrix k_g. ω_g is called the coherent or GLAUBER state corresponding to g. This notation is justified by the fact that in case $g \in L_2(G)$ the state ω_g is a pure normal state corresponding to the normalized wave function ψ_g given by

$$\psi_g(\varphi) = \begin{cases} \exp(-\frac{\|g\|^2}{2}) \prod_{x\in\varphi} g(x) & ; \varphi \in M^f \\ 0 & \text{otherwise} \end{cases}$$

Coherent states are well known in quantum optics. They are states of particle systems which are nearest to the classical wave picture, i.e. states of minimal uncertainty with respect to the HEISENBERG relation of uncertainty. In the case $g \in L_2(G)$ the normal state ω_g can be considered as a state of free bosons being all in the same one-particle state. If $g \notin L_2(G)$ the state ω_g may be interpreted as a wave packet of independent particles. These particles do not exist as single particles but as a part of the infinite system with particle density $|g|^2$.

5. Local independence

Let ω be a locally normal state on $\mathcal{A}$. ω is said to be locally independent if for all $\Lambda_1, \Lambda_2 \in \mathcal{B}$, $\Lambda_1 \cap \Lambda_2 = \emptyset$ there holds

$$\omega(A_1 A_2) = \omega(A_1)\cdot\omega(A_2) \qquad ; A_1 \in {}_{\Lambda_1}\mathcal{A},\ A_2 \in {}_{\Lambda_2}\mathcal{A}.$$

Theorem: Let ω be a locally normal state on $\mathcal{A}$. ω is a coherent state iff ω is locally independent.

Corollary: Let $g \in L_2^{loc}(G)$. Then there exists a uniquely determined locally normal state ω_g - the coherent state corresponding to g - such that

(a) The reduced density matrix of first order of ω_g is given by

$$\mu_1(x,y) = g(x)\cdot\overline{g(y)} \qquad ; x,y \in G$$

(b) ω_g is locally independent.

Remark 1: The theorem was proven by TITULAER and GLAUBER in the case of normal photon states on algebras of creation and annihilation operators (cf. [10]).

Remark 2: The theorem can be considered in a certain sense as a generalization of a well known theorem in point process theory characterizing POISSON processes by local independence (cf. [8], theorem 1.11.8.).

Remark 3: If ω is locally independent then for each sequence $\Lambda_1,\ldots,\Lambda_m$ of pairwise disjoint sets from $\mathcal{B}$ and $A_j \in {}_{\Lambda_j}\mathcal{A}$ $(j = 1,\ldots,m)$ it holds $\omega(A_1\cdots A_m) = \omega(A_1)\cdots\omega(A_m)$.

6. Proofs

6.1. Proof of lemma 1: Let $\Lambda \in \mathcal{B}$. We define a point process $(Q_\omega)_\Lambda$ by

$$(Q_\omega)_\Lambda(Y) := Q_\omega(\{\varphi: \varphi_\Lambda \in Y\}) \qquad ; Y \in \mathfrak{M}$$

Then from proposition 3.1. in [4] we can conclude that $(Q_\omega)_\Lambda$ is a finite Σ^c-point process, i.e. there exists a σ-finite measure S on $[M,\mathfrak{M}]$ such that $C_{(Q_\omega)_\Lambda} \ll \nu \times S$. Hence we have

$$(Q_\omega)_\Lambda \ll F$$

(cf [3], remark 2.7). Because of $(Q_\omega)_\Lambda(M_\Lambda) = 1$ we can conclude that

$$(Q_\omega)_\Lambda \ll F_\Lambda$$

Since F_Λ is concentrated on M^s (i.e. $F_\Lambda(M \setminus M^s) = 0$) that implies

$$(Q_\omega)_\Lambda(M^s) = 1$$

Thus we obtain

$$Q_\omega(M^s) = 1$$

i.e. Q_ω is simple.

On the other hand it holds

$$\nu(B) = \exp(-\nu(B)) \int F_\Lambda(d\varphi)\, \varphi(B) \qquad ; B,\Lambda \in \mathcal{B},\ B \subseteq \Lambda$$

For that reason we obtain

$$F_\Lambda(\{\varphi \in M: \varphi(B) \neq 0\}) = 0 \qquad ; B \in \mathcal{G},\ \nu(B) = 0$$

Because of $(Q_\omega)_\Lambda \ll F_\Lambda$ this implies

$$(Q_\omega)_\Lambda(\{\varphi \in M: \varphi(B) \neq 0\}) = 1 \qquad ; B \in \mathcal{G},\ \nu(B) = 0$$

Therefore it is $I_Q(B) = 0$ if $\nu(B) = 0$, i.e. $I_Q \ll \nu$. Finally we have to prove that Q_ω is continuous. But this is an immediate consequence of $I_Q \ll \nu$. ■

Before we are going to prove the theorem we still need some preparations.

6.2. Let $\Lambda_1, \Lambda_2 \in \mathcal{G}$, $\Lambda_1 \cap \Lambda_2 = \emptyset$, and $g: M \to \mathbb{C}$ a measurable, $F_{\Lambda_1 \cup \Lambda_2}$-integrable function. Then the following equality holds

$$\int F_{\Lambda_1}(d\varphi^1) \int F_{\Lambda_2}(d\varphi^2)\, g(\varphi^1+\varphi^2) = \int F_{\Lambda_1 \cup \Lambda_2}(d\varphi) g(\varphi)$$

Proof: This is an immediate consequence of 2.7. in [4]. ■

6.3. Let $\Lambda_1, \Lambda_2 \in \mathcal{B}$, $\Lambda_1 \cap \Lambda_2 = \emptyset$, $A_1 \in {}_{\Lambda_1}\mathcal{A}$, $A_2 \in {}_{\Lambda_2}\mathcal{A}$ such that ${}_{\Lambda_1}J^{-1}A_1$ and ${}_{\Lambda_2}J^{-1}A_2$ are integral operators with kernels a_1 and a_2 respectively. Then ${}_{\Lambda_1 \cup \Lambda_2}J^{-1}(A_1 A_2) \in \mathcal{A}_{\Lambda_1 \cup \Lambda_2}$ is an integral operator too, a kernel of it is given by

$$a(\varphi^1, \varphi^2) = a_1(\varphi^1_{\Lambda_1}, \varphi^2_{\Lambda_1})\, a_2(\varphi^1_{\Lambda_2}, \varphi^2_{\Lambda_2}) \qquad ; \varphi^1, \varphi^2 \in M_{\Lambda_1 \cup \Lambda_2}$$

Proof: We have already stated in 2. that there exists a unique isomorphism I_{Λ_1, Λ_2} between $\mathcal{M}_{\Lambda_1} \otimes \mathcal{M}_{\Lambda_2}$ and $\mathcal{M}_{\Lambda_1 \cup \Lambda_2}$ such that

$$I_{\Lambda_1, \Lambda_2}(\psi_1 \otimes \psi_2)(\varphi) = \psi_1(\varphi_{\Lambda_1})\psi_2(\varphi_{\Lambda_2}) \quad ; \psi_1 \in \mathcal{M}_{\Lambda_1},\ \psi_2 \in \mathcal{M}_{\Lambda_2},\ \varphi \in M_{\Lambda_1 \cup \Lambda_2}$$

For that reason we have to prove

$$({}_{\Lambda_1\cup\Lambda_2}J^{-1}(A_1A_2))\psi(\varphi) = \int F_{\Lambda_1\cup\Lambda_2}(d\hat{\varphi})a_1(\varphi_{\Lambda_1},\hat{\varphi}_{\Lambda_1})a_2(\varphi_{\Lambda_2},\hat{\varphi}_{\Lambda_2})\psi(\hat{\varphi}) \ ,$$

for all $\psi_1 \in \mathcal{M}_{\Lambda_1}$, $\psi_2 \in \mathcal{M}_{\Lambda_2}$, $\psi \in \mathcal{M}_{\Lambda_1\cup\Lambda_2}$ such that

$$\psi(\varphi) = \psi_1(\varphi_{\Lambda_1})\cdot\psi_2(\varphi_{\Lambda_2}) \qquad ; \ \varphi \in M_{\Lambda_1\cup\Lambda_2}$$

Because of

$$({}_{\Lambda_1\cup\Lambda_2}J^{-1}A_1A_2)\psi(\varphi) = ({}_{\Lambda_1}J^{-1}A_1)\psi_1(\varphi_{\Lambda_1})\cdot({}_{\Lambda_2}J^{-1}A_2)\psi_2(\varphi_{\Lambda_2})$$

$$= \int F_{\Lambda_1}(d\hat{\varphi})a_1(\varphi_{\Lambda_1},\hat{\varphi})\psi_1(\hat{\varphi}) \int F_{\Lambda_2}(d\tilde{\varphi})a_2(\varphi_{\Lambda_2},\tilde{\varphi})\psi_2(\tilde{\varphi})$$

and

$$\int F_{\Lambda_1\cup\Lambda_2}(d\hat{\varphi})a_1(\varphi_{\Lambda_1},\hat{\varphi}_{\Lambda_1})a_2(\varphi_{\Lambda_2},\hat{\varphi}_{\Lambda_2})\psi_1(\hat{\varphi}_{\Lambda_1})\psi_2(\hat{\varphi}_{\Lambda_2}) =$$

$$= \int F_{\Lambda_1\cup\Lambda_2}(d\hat{\varphi})a_1(\varphi_{\Lambda_1},\hat{\varphi}_{\Lambda_1})a_2(\varphi_{\Lambda_2},\hat{\varphi}_{\Lambda_2})\psi(\hat{\varphi})$$

it is sufficient to prove

$$\int F_{\Lambda_1}(d\hat{\varphi})a_1(\varphi_{\Lambda_1},\hat{\varphi})\psi_1(\hat{\varphi}) \int F_{\Lambda_2}(d\tilde{\varphi})a_2(\varphi_{\Lambda_2},\tilde{\varphi})\psi_2(\tilde{\varphi}) =$$

$$= \int F_{\Lambda_1\cup\Lambda_2}(d\hat{\varphi})a_1(\varphi_{\Lambda_1},\hat{\varphi}_{\Lambda_1})a_2(\varphi_{\Lambda_2},\hat{\varphi}_{\Lambda_2})\psi_1(\hat{\varphi}_{\Lambda_1})\psi_2(\hat{\varphi}_{\Lambda_2})$$

But this is an immediate consequence of 6.2. ■

Now we are going to prove the first part of the theorem.

<u>6.4.</u> Let $\Lambda_1, \Lambda_2 \in \mathcal{B}$, $\Lambda_1 \cap \Lambda_2 = \emptyset$ and $g \in L_2^{loc}(G)$. Then for all $A_1 \in {}_{\Lambda_1}\mathcal{A}$, $A_2 \in {}_{\Lambda_2}\mathcal{A}$ the following equality holds

$$\omega_g(A_1A_2) = \omega_g(A_1)\cdot\omega_g(A_2)$$

<u>Proof:</u> For each $\Lambda \in \mathcal{B}$ the set of all integral operators from $\mathcal{A}_\Lambda$ is dense in $\mathcal{A}_\Lambda$ with respect to the σ-strong topology (cf. [1], example 2.4.9.). Thus we can approximate each operator $A \in {}_\Lambda\mathcal{A}$ in the σ-strong topology by operators of the type ${}_\Lambda JA'$, A' integral operator from $\mathcal{A}_\Lambda$. The restriction ${}_{\Lambda_1\cup\Lambda_2}\omega_g$ of ω_g to ${}_{\Lambda_1\cup\Lambda_2}\mathcal{A}$ is a normal state, hence ${}_{\Lambda_1\cup\Lambda_2}\omega_g$ is σ-weakly continuous (cf. [1],

theorem 2.4.21.). Then ${}_{\Lambda_1\cup\Lambda_2}\omega_g$ is σ-strongly continuous too. For that reason it is sufficient to prove that

$$\omega_g({}_{\Lambda_1}JA_1\cdot{}_{\Lambda_2}JA_2) = \omega_g({}_{\Lambda_1}JA_1)\cdot\omega_g({}_{\Lambda_2}JA_2)$$

holds for all integral operators $A_1 \in \mathscr{A}_{\Lambda_1}$, $A_2 \in \mathscr{A}_{\Lambda_2}$ with kernels a_1 and a_2 respectively. Using 6.3. we can conclude that there exists an integral operator $A \in \mathscr{A}_{\Lambda_1\cup\Lambda_2}$ with kernel a such that

$${}_{\Lambda_1}JA_1\cdot{}_{\Lambda_2}JA_2 = {}_{\Lambda_1\cup\Lambda_2}JA$$

$$a(\varphi^1,\varphi^2) = a_1(\varphi^1_{\Lambda_1},\varphi^2_{\Lambda_1})\ a_2(\varphi^1_{\Lambda_2},\varphi^2_{\Lambda_2}) \qquad ;\ \varphi^1,\varphi^2 \in M_{\Lambda_1\cup\Lambda_2}$$

Now from the definition of ω_g it follows

$$\omega_g({}_{\Lambda_1\cup\Lambda_2}JA) =$$

$$= P^{|g|^2}(M_{(\Lambda_1\cup\Lambda_2)^c}) \int F_{\Lambda_1\cup\Lambda_2}(d\varphi^1) \int F_{\Lambda_1\cup\Lambda_2}(d\varphi^2)\ a(\varphi^1,\varphi^2)\ k_g(\varphi^2,\varphi^1,\varphi)$$

$$\omega_g({}_{\Lambda_j}JA_j) =$$

$$= P^{|g|^2}(M_{\Lambda_j^c}) \int F_{\Lambda_j}(d\varphi^1) \int F_{\Lambda_j}(d\varphi^2)\ a_j(\varphi^1,\varphi^2)\ k_g(\varphi^2,\varphi^1,\varphi) \qquad ;\ j = 1,2$$

For that reason it is sufficient to prove that

$$P^{|g|^2}(M_{(\Lambda_1\cup\Lambda_2)^c}) = P^{|g|^2}(M_{\Lambda_1^c})\ P^{|g|^2}(M_{\Lambda_2^c}) \quad \text{and}$$

$$\int F_{\Lambda_1\cup\Lambda_2}(d\varphi^1) \int F_{\Lambda_1\cup\Lambda_2}(d\varphi^2)\ a(\varphi^1,\varphi^2)\ k_g(\varphi^2,\varphi^1,\varphi) =$$

$$= \prod_{j=1}^{2} \int F_{\Lambda_j}(d\varphi^1) \int F_{\Lambda_j}(d\varphi^2)\ a_j(\varphi^1,\varphi^2)\ k_g(\varphi^2,\varphi^1,\varphi)$$

The first equality follows from

$$P^{|g|^2}(M_{\Lambda^c}) = \exp(-\int_\Lambda |g(x)|^2\nu(dx)) \qquad ;\ \Lambda \in \mathscr{B}$$

Using 6.2. we obtain the second equality from

$$k_g(\varphi^1+\varphi^3,\varphi^2+\varphi^4,\varphi) = k_g(\varphi^1,\varphi^2,\varphi)\cdot k_g(\varphi^3,\varphi^4,\varphi) \qquad ;\ \varphi^1,\varphi^2,\varphi^3,\varphi^4 \in M^f,\ \varphi \in M$$

and

$$\varphi_{\Lambda_1 \cup \Lambda_2} = \varphi_{\Lambda_1} + \varphi_{\Lambda_2} \qquad ; \varphi \in M \qquad \blacksquare$$

In order to complete the proof of the theorem we have to prove that each locally independent state is a coherent one. We will do this in several steps.

In the following, let ω be a locally independent state.

<u>6.5.</u> There exists a locally integrable function $\mu: G \to [0,+\infty)$ such that $Q_\omega = P^\mu$.

<u>Proof:</u> Let $\Lambda_1,\ldots,\Lambda_m \in \mathcal{B}$, be a sequence of pairwise disjoint sets, $n_1,\ldots,n_m \in \mathbb{N}$. We set

$$Y_j := \{\varphi \in M: \varphi(\Lambda_j) = n_j\} \qquad ; j = 1,\ldots,m$$

Since $Y_j \in {}_{\Lambda_j}\mathfrak{M}$ we get $O_{\chi_{Y_j}} \in {}_{\Lambda_j}\mathcal{A}$ (cf. [4], 2.11.). For that reason we get

$$\omega(O_{\chi\left(\bigcap_{j=1}^m Y_j\right)}) = \omega(\prod_{j=1}^m O_{\chi_{Y_j}})$$
$$= \prod_{j=1}^m \omega(O_{\chi_{Y_j}})$$

Using the definition of the position distribution Q_ω we can conclude from this equality that

$$Q_\omega(\{\varphi \in M: \varphi(\Lambda_j) = n_j,\ 1 \le j \le m\}) = \prod_{j=1}^m Q_\omega(\{\varphi: \varphi(\Lambda_j) = n_j\})$$

i.e. the point process Q_ω is free from after-effects (cf. [8], 1.3.). We additionally know from lemma 1 that Q_ω is continuous and simple. For that reason we can conclude that Q_ω is a POISSON process (cf. [8], theorem 1.11.8.). Furthermore, lemma 1 implies $I_{Q_\omega} \ll \nu$. This proves 6.5. $\blacksquare$

Let $\Lambda \in \mathcal{B}$. The local state ω_Λ of ω defined by

$$\omega_\Lambda(A) := \omega({}_\Lambda JA) \qquad ; A \in \mathcal{A}_\Lambda$$

is a normal state on $\mathcal{A}_\Lambda$ (cf. [4], chapter 2). We denote by ρ_Λ the density matrix according to ω_Λ, let r_Λ be a kernel of ρ_Λ (such a kernel always exists).

6.6. Let $\Lambda, \Lambda' \in \mathcal{B}$, $\Lambda' \subsetneq \Lambda$. Then the following equality holds for $F_\Lambda \times F_\Lambda$-a.a. (φ^1, φ^2)

$$r_\Lambda(\varphi^1,\varphi^2)\cdot \exp\Big(-\int_\Lambda \nu(dx)\mu(x)\Big) = r_\Lambda(\varphi^1_{\Lambda'},\varphi^2_{\Lambda'})\cdot r_\Lambda(\varphi^1_{\Lambda\setminus\Lambda'},\varphi^2_{\Lambda\setminus\Lambda'})$$

Proof: For integral operators $A_1 \in \mathcal{A}_{\Lambda'}$, $A_2 \in \mathcal{A}_{\Lambda\setminus\Lambda'}$ with kernels a_1 and a_2 respectively we have

$$(6.1) \qquad \omega_{\Lambda'}(A_1) = \int F_{\Lambda'}(d\varphi^1) \int F_{\Lambda'}(d\varphi^2) a_1(\varphi^1,\varphi^2) r_{\Lambda'}(\varphi^1,\varphi^2)$$

$$(6.2) \qquad \omega_{\Lambda\setminus\Lambda'}(A_2) = \int F_{\Lambda\setminus\Lambda'}(d\varphi^1) \int F_{\Lambda\setminus\Lambda'}(d\varphi^2) a_2(\varphi^1,\varphi^2) r_{\Lambda\setminus\Lambda'}(\varphi^1,\varphi^2)$$

Local independence of ω implies

$$(6.3) \qquad \omega_\Lambda(({}_\Lambda J^{-1}{}_{\Lambda'} JA_1)\cdot({}_\Lambda J^{-1}{}_{\Lambda\setminus\Lambda'} JA_2)) = \omega_{\Lambda'}(A_1)\cdot\omega_{\Lambda\setminus\Lambda'}(A_2)$$

Using 6.2. and 6.3. we obtain from (6.1) - (6.3)

$$\int F_\Lambda(d\varphi^1) \int F_\Lambda(d\varphi^2) a_1(\varphi^1_{\Lambda'},\varphi^2_{\Lambda'}) a_2(\varphi^1_{\Lambda\setminus\Lambda'},\varphi^2_{\Lambda\setminus\Lambda'}) r_\Lambda(\varphi^1,\varphi^2) =$$

$$= \int F_\Lambda(d\varphi^1) \int F_\Lambda(d\varphi^2) a_1(\varphi^1_{\Lambda'},\varphi^2_{\Lambda'}) a_2(\varphi^1_{\Lambda\setminus\Lambda'},\varphi^2_{\Lambda\setminus\Lambda'})\cdot$$

$$r_{\Lambda'}(\varphi^1_{\Lambda'},\varphi^2_{\Lambda'}) r_{\Lambda\setminus\Lambda'}(\varphi^1_{\Lambda\setminus\Lambda'},\varphi^2_{\Lambda\setminus\Lambda'})$$

For that reason we can conclude that

$$(6.4) \qquad r_\Lambda(\varphi^1,\varphi^2) = r_{\Lambda'}(\varphi^1_{\Lambda'},\varphi^2_{\Lambda'}) r_{\Lambda\setminus\Lambda'}(\varphi^1_{\Lambda\setminus\Lambda'},\varphi^2_{\Lambda\setminus\Lambda'})$$

holds for $F_\Lambda \times F_\Lambda$-a.a. (φ^1,φ^2). This implies

$$(6.5) \qquad r_\Lambda(\varphi^1,\varphi^2) = r_{\Lambda'}(\varphi^1,\varphi^2) r_{\Lambda\setminus\Lambda'}(\emptyset,\emptyset) \qquad ; F_{\Lambda'} \times F_{\Lambda'}\text{-a.e.}$$

$$(6.6) \qquad r_\Lambda(\varphi^1,\varphi^2) = r_{\Lambda'}(\emptyset,\emptyset) r_{\Lambda\setminus\Lambda'}(\varphi^1,\varphi^2) \qquad ; F_{\Lambda\setminus\Lambda'} \times F_{\Lambda\setminus\Lambda'}\text{-a.e.}$$

On the other hand we obtain using 6.5.

$$(6.7)\quad r_{\Lambda\setminus\Lambda'}(\emptyset,\emptyset) = tr(\rho_{\Lambda\setminus\Lambda'} O_{\chi_{\{\emptyset\}}})$$
$$= \omega(_{\Lambda\setminus\Lambda'} JO_{\chi_{\{\emptyset\}}})$$
$$= Q_\omega(\{\varphi:\ \varphi(\Lambda\setminus\Lambda') = 0\})$$
$$= \exp(-\int_{\Lambda\setminus\Lambda'} \nu(dx)\mu(x))$$

$$(6.8)\quad r_{\Lambda'}(\emptyset,\emptyset) = \exp(-\int_{\Lambda'} \nu(dx)\mu(x))$$

From (6.4) - (6.8) follows that 6.6. is true. ∎

6.7. Let $\Lambda \in \mathcal{B}$. There holds for $F_\Lambda x F_\Lambda$-a.a. $[\varphi^1,\varphi^2]$

$$(6.9)\quad r_\Lambda(\varphi^1,\varphi^2)\cdot \exp(\int_\Lambda \nu(dx)\mu(x)) =$$

$$= \prod_{x\in\varphi^1} r_\Lambda(\delta_x,\emptyset)\cdot \prod_{y\in\varphi^2} \overline{r_\Lambda(\delta_y,\emptyset)}\cdot \exp((\varphi^1+\varphi^2)(G)\cdot \int_\Lambda \nu(dx)\mu(x))$$

whereby we assume $\prod_{x\in\emptyset} r_\Lambda(\delta_x,\emptyset) = 1$.

Proof: There exists a countable set D of finite partitions of Λ into measurable subsets such that

$$F_\Lambda \times F_\Lambda(M_\Lambda \times M_\Lambda) =$$
$$= F_\Lambda \times F_\Lambda(\{[\emptyset,\emptyset]\} \cup \bigcup_{(\Lambda_j)_1^n\in D} \{[\varphi^1,\varphi^2]:\ \varphi^1+\varphi^2(\Lambda_j) = 1,\ j=1,..,n\})$$

In the case $\varphi^1 = \varphi^2 = \emptyset$ (6.9) follows immediately from (6.8). For that reason we have to prove that (6.9) holds for each partition $(\Lambda_j)_1^n$ of Λ and $F_\Lambda \times F_\Lambda$-a.a. $[\varphi^1,\varphi^2]$ with

$$(6.10)\quad \varphi^1+\varphi^2(\Lambda_j) = 1 \qquad ;\ j = 1,..,n$$

Now ρ_Λ is selfadjoint. Thus we have

$$(6.11)\quad r_\Lambda(\varphi,\emptyset) = \overline{r_\Lambda(\emptyset,\varphi)} \qquad ;\ F_\Lambda\text{-a.a. } \varphi \in M_\Lambda$$

Because of (6.10) and (6.11) the equality (6.9) is true iff the following equality holds

$$(6.12)\quad r_\Lambda(\varphi^1,\varphi^2)\cdot \exp(-(n-1)\int_\Lambda \nu(dx)\mu(x)) = \prod_{j=1}^n r_\Lambda(\varphi^1_{\Lambda_j},\varphi^2_{\Lambda_j})$$

From 6.3. we get that this equation holds for each finite partition $(\Lambda_j)_{j=1}^n$ of Λ and $F_\Lambda \times F_\Lambda$-a.a. $[\varphi^1,\varphi^2]$. This proves 6.7. ∎

<u>6.8.</u> Let $\Lambda \in \mathcal{B}$. We set

$$(6.13) \quad g(x) := \begin{cases} r_\Lambda(\delta_x,\mathcal{O})\cdot\exp(\int_\Lambda \nu(dx)\mu(x)) & ; x \in \Lambda \\ 0 & ; x \in \Lambda^c \end{cases}$$

Then we have

$$(6.14) \quad \omega_g(A) = \omega(A) \qquad ; A \in {}_\Lambda\mathcal{A}$$

<u>Proof:</u> From the definition of ω_g we obtain

$$(6.15) \quad \omega_g(A) = P^{|g|^2}(M_{\Lambda^c}) \int F_\Lambda(d\varphi^1) \int F_\Lambda(d\varphi^2)\, a(\varphi^1,\varphi^2)\, k_g(\varphi^2,\varphi^1,\varphi)$$

for each $A \in {}_\Lambda\mathcal{A}$ such that ${}_\Lambda J^{-1}A$ is an integral operator with kernel a. From (6.9) and (6.13) follows that

$$k_g(\varphi^1,\varphi^2,\varphi) = r_\Lambda(\varphi^1,\varphi^2)\cdot\exp(\int_\Lambda \nu(dx)\mu(x)) \qquad ; \varphi^1,\varphi^2 \in M_\Lambda,\ \varphi \in M$$

Further we know that

$$\omega(A) = tr(\rho_\Lambda A) = \int F_\Lambda(d\varphi^1) \int F_\Lambda(d\varphi^2)\, r_\Lambda(\varphi^2,\varphi^1,\varphi)\, a(\varphi^1,\varphi^2)$$

Thus we get

$$\omega_g(A) = P^{|g|^2}(M_{\Lambda^c})\cdot\exp(\int_\Lambda \nu(dx)\mu(x))\cdot\omega(A)$$

From this follows

$$(6.16) \quad \omega_g(A) = P^{|g|^2}(M_{\Lambda^c})\cdot(P^\mu(M_{\Lambda^c}))^{-1}\cdot\omega(A) \qquad ; A \in {}_\Lambda\mathcal{A}.$$

If we put $A = \mathbf{1}$, we obtain from (6.16)

$$P^{|g|^2}(M_{\Lambda^c}) = P^\mu(M_{\Lambda^c})$$

Hence from (6.16) follows 6.8. ∎

In order to complete the proof of the theorem we consider a sequence $(\Lambda_n)_{n=1}^\infty$ from $\mathcal{B}$, such that

$\Lambda_n \subseteq \Lambda_{n+1}$; $n = 1,2,\ldots$,

$\bigcup_{n=1}^{\infty} \Lambda_n = G$

Using 6.8. we obtain for each $n \in \mathbb{N}$ a function $g_n \in L_2^{loc}(G,\nu)$ such that

$$(6.17) \quad \omega_{g_n}(A) = \omega(A) \qquad ; A \in {}_{\Lambda_n}\mathcal{A},\ n \in \mathbb{N}$$

From this follows

$$\omega_{g_{n+m}}(A) = \omega_{g_n}(A) \qquad ; A \in {}_{\Lambda_n}\mathcal{A},\ n,m \in \mathbb{N}$$

Hence we have ν-a.e.

$$g_{n+m}\cdot\chi_{\Lambda_n} = g_n\cdot\chi_{\Lambda_n}$$

We set

$$g := g_1\cdot\chi_{\Lambda_1} + \sum_{n=2}^{\infty} g_n\cdot\chi_{\Lambda_n\setminus\Lambda_{n-1}}$$

Then there holds ν-a.e.

$$g\cdot\chi_{\Lambda_n} = g_n$$

Because of (6.17) we get from this

$$\omega_g(A) = \omega_{g_n}(A) = \omega(A) \qquad ; A \in {}_{\Lambda_n}\mathcal{A},\ n \in \mathbb{N}$$

ω_g und ω are states and therefore they are continuous. Thus we obtain

$$\omega_g(A) = \omega(A) \qquad ; A \in \mathcal{A}$$

This completes the proof of the theorem. ■

References

[1] O. Bratteli, D.W. Robinson, Operator algebras and quantum statistical mechanics I, II. Springer-Verlag, New York - Heidelberg - Berlin 1979 / 1981

[2] T. Damaschke, W. Freudenberg, On the position distribution of infinite boson systems on a lattice, Forschungsergebnisse der FSU Jena Nr. N/88/27, 1988

[3] K.-H. Fichtner, W. Freudenberg, Point processes and normal states of boson systems, preprint NTZ K.-Marx-Univ. Leipzig 1986

[4] K.-H. Fichtner, W. Freudenberg, Point processes and states of infinite boson systems, preprint NTZ K.-Marx-Univ. Leipzig 1986

[5] K.-H. Fichtner, W. Freudenberg, Point processes and the position distribution of infinite boson systems, Journal of statistical physics 47 No. 5/6 (1987), 959 - 978

[6] O. Kallenberg, Random measures, Akademie-Verlag, Berlin, and Academic Press, London - New York 1983

[7] И. Керстан, К. Маттес, И. Мекке, Безгранично делимые точечные процессы, изд. наука, Москва 1982

[8] K. Matthes, J. Kerstan, J. Mecke, Infinitely divisible point processes , J.Wiley, London - New York - Sydney - Toronto 1978

[9] B. Rauchensohwandtner, Gibbsprozesse und Papangeloukerne, VWGÖ Wien 1980

[10] U.M. Titulaer, R.J.Glauber, Density operators for coherent fields, Phys. Rev. 145 (4) 1966.

This paper is in final form and no similar paper has been or is being submitted elsewhere.

TIME-INHOMOGENEOUS AND NONLINEAR QUANTUM EVOLUTIONS

Alberto Frigerio
Dipartimento di Matematica e Informatica
Universita' di Udine, Via Zanon 6, 33100 Udine, Italy

1. Introduction.

The theory of unitary dilations of irreversible quantum evolutions [9,22] has received great impulse from quantum stochastic calculus [20, 19,5,27] and from the construction of a quantum Poisson process [23,24].

In the present paper we consider the dilation problem for time-inhomogeneous quantum evolutions (such as may arise in a quantum version [13] of simulated annealing [17,18]) and for nonlinear quantum evolutions [3] (Boltzmann-like, for instance). Our key tool is the quantum Poisson process [23,15,10,12,24,27,1], which we briefly describe in Section 2; Theorem 2 in that Section describes a class of time-inhomogeneous quantum evolutions which possess, in an obvious sense, a unitary dilation. As an interesting example, in Section 3 we give a sketchy account of some results [13] on simulated annealing for quantum evolutions. Note that, in spite of the similarity of the name, this has nothing to do with "quantum stochastic optimization" as in [4], but is more similar to "traditional" simulated annealing. In Section 4 we construct a unitary dilation for the corresponding evolution. Finally, in Section 5 we consider the dilation problem for nonlinear evolutions, by means of a slight generalization of Theorem 2 of Section 2. Proofs are given in a sketchy form.

The present paper is still preliminary. Therefore we shall not try to give a general, formal definition of what should be considered to be a dilation for a time-inhomogeneous or for a nonlinear evolution. We shall give only examples, from which we hope that something may be learned.

Acknowledgements. The results of the present paper originate ultimately from our joint work with H. Maassen [15]; some results mentioned here have been obtained in collaboration with C. Aratari [14]. We wish to thank them both. We are also grateful to P. Serafini and M. Colombetti, who aroused our interest in simulated annealing.

2. Preliminaries.

We shall consider quantum stochastic differential equations for operator processes living in the Hilbert space $H \otimes F(h)$, where the separable Hilbert space H is the so-called initial space, and where $F(h)$ is the symmetric Fock space over $h = L^2(R;K)$, K being a separable Hilbert space to be interpreted as the instantaneous noise space or the circumambient space [10]. The symmetric Fock space $F(h)$ over a Hilbert space h is generated by the coherent vectors (normalized exponential vectors) $\{C(f) : f \in h\}$ satisfying

$$\langle C(f),C(g)\rangle_{F(h)} = \exp[\langle f,g\rangle_h - \|f\|_h^2/2 - \|g\|_h^2/2] . \qquad (2.1)$$

The Weyl operators $\{W(g) : g \in h\}$ are defined by

$$W(g)\, C(f) = \exp[- i \operatorname{Im} \langle g,f\rangle]\, C(f + g) : \quad f \in h \quad ; \qquad (2.2)$$

they extend to unitary operators on $F(h)$.

We shall work with the Weyl algebra $W(L_c^2(R;K))$ generated by the Weyl operators $W(f)$ with test function f in the space $L_c^2(R;K)$ of square-integrable K-valued functions of compact support in R. Then, for every g in $L_{loc}^2(R;K)$, we can define the locally normal coherent state φ_g [11] via

$$\varphi_g(W(f)) = \langle C(\chi_{supp(f)}g),W(f)C(\chi_{supp(f)}g)\rangle : f \in L_c^2(R;K). \qquad (2.3)$$

If H, K are Hilbert spaces, and $f \in K$, we denote by E_f the completely positive linear map of $B(H \otimes K)$ into $B(H)$ such that

$$\langle u,E_f(X)v\rangle_H = \langle u \otimes f,X(v \otimes f)\rangle_{H \otimes K} \quad : u , v \in H , \quad X \in B(H \otimes K); \qquad (2.4)$$

its norm is given by $\|f\|_K^2$. If $K = F(h)$ and $f = C(0)$ is the vacuum vector, we shall write just E for $E_{C(0)}$. Finally, if $h = L^2(R;K)$, we can define a map of $B(H) \otimes W(L_c^2(R;K))$ into $B(H)$ as $E_{C(g)}$ also if g is only in $L_{loc}^2(R;K)$, although, strictly speaking, $C(g)$ does not exist unless $g \in L^2(R;K)$. We have

$$W(f) = \exp[a^+(f) - a(f)] \ , \qquad f \in h \ ; \tag{2.5}$$

where $a(f)$ and $a^+(f)$ are the annihilation and the creation operator corresponding to f . For any bounded operator X on h , there exists a densely defined operator $F(X)$ in $F(h)$ such that

$$F(X)\,C(f) = C(Xf)\exp[-\|f\|^2/2 + \|Xf\|^2/2] : \quad f \in h \ . \tag{2.6}$$

Let also $\lambda(X)$ be the densely defined operator given by

$$\lambda(X)\,C(f) = \frac{d}{d\varepsilon} F(\exp[\varepsilon X])\,C(f)\Big|_{\varepsilon=0} : \qquad f \in h. \tag{2.7}$$

Denote by $A_j(t) = M_{0j}(t)$ and by $A_j^+(t) = M_{j0}(t)$ the annihilation and the creation operator in $F(h)$ corresponding to the test function $e_j \otimes \chi_{[0,t]}$, $(e_1,\dots,e_n,\dots)$ being a complete orthonormal set in K. Define also the gauge process $\{M_{ij}(t) : i,j=1,2,\dots;\ t \in R^+\}$ by

$$M_{ij}(t) = \lambda(e_{ij} \otimes \chi_{[0,t]}) \ , \tag{2.8}$$

where $\{e_{ij} : i,j=1,2,\dots\}$ are the matrix units $|e_i\rangle\langle e_j|$ in $B(K)$. We have the following quantum Ito table [20,27,1]:

$$dM_{\alpha\beta}(t)\,dM_{\gamma\delta}(t) = \delta_{\beta\gamma}(1 - \delta_{\beta 0})\,dM_{\alpha\delta}(t) \tag{2.9}$$

where $dM_{00}(t) = dt$; the remaining products of differentials vanish. Another convenient compact notation for quantum stochastic differentials has been introduced in [8].

Now we shall describe the quantum Poisson process in the sense of [15,12]. Unfortunately, [12] contains some misprints and mistakes, which obscure its content. Therefore we feel it appropriate to conclude this paper with an Erratum to [12].

Let $t \mapsto X(t) = \sum_{j,k=1}^{\infty} X_{jk}(t) \otimes e_{jk}$, $X_{jk}(t) \in B(H)$, be a bounded measurable function mapping R^+ into the bounded operators on $H \otimes K$,

and let $g = \sum_{j=1}^{\infty} e_j \otimes g_j$, $g_j \in L^2_{loc}(R)$, be an element of $L^2_{loc}(R,K)$.
We define [18]

$$dN_t(X,g)$$

$$= \sum_{j,k=1}^{\infty} X_{jk}(t)[g_k(t)dA_j^+(t) + dM_{jk}(t) + \overline{g_j}(t)dA_k(t) + \overline{g_j}(t)g_k(t)dt] \tag{2.10}$$

and (for $r < t \in R$)

$$N_{r,t}(X,g) = \int_r^t dN_s(X,g). \tag{2.11}$$

$N_{r,t}(X,g)$ will be called the <u>quantum Poisson process</u>.

We have, for all u_1, u_2 in H and f_1 , f_2 in h ,

$$\langle u_1 \otimes C(f_1), N_{r,t}(X,g)\, u_2 \otimes C(f_2)\rangle_{H \otimes F(h)} \tag{2.12}$$

$$= \int_r^t \langle u_1 \otimes [f_1(s)+g(s)], X(s)\, u_2 \otimes [f_2(s)+g(s)]\rangle_{H \otimes K} ds \langle C(f_1), C(f_2)\rangle_{F(h)}$$

and, for all admissible X and g , and for $S \leq s \leq t \leq T$,

$$N_{s,t}(X,g) = W(\chi_{[S,T]}g)^{-1} N_{s,t}(X,0)\, W(\chi_{[S,T]}g) . \tag{2.13}$$

Then, up to a unitary transformation, we can reduce the study of the quantum Poisson process $N_{s,t}(X,g)$ to the special case $g = 0$, for which we have simply

$$dN_t(X,0) = \sum_{j,k=1}^{\infty} X_{jk}(t)\, dM_{jk}(t) , \tag{2.14}$$

provided we consider as a reference state the locally normal coherent state φ_g . Quantum Poisson processes as number processes in coherent states have been considered in the physical literature, see e.g. [6].
From [20] and [27] we obtain

$$dN_t(X,0)\, dN_t(Y,0) = dN_t(XY,0) ; \tag{2.15}$$

Equation (2.15) extends immediately to arbitrary g .

Also, if $t \mapsto X(t)$ is a function as above and $t \mapsto A(t)$ is a bounded measurable function from R to $B(H)$, we have

$$A(t) \, dN_t(X,g) \;=\; dN_t(AX,g) \;,\; dN_t(X,g) \, A(t) \;=\; dN_t(XA,g) \;. \quad (2.16)$$

Theorem 1. The quantum stochastic differential equation

$$d \, U_g(t,s) \;=\; dN_t(S - \mathbb{1},g) \, U_g(t,s) \;,\; U_g(s,s) \;=\; \mathbb{1} \;,(t \geq s) \quad (2.17)$$

has a unique adapted solution, which is unitary if and only if $S(t)$ is unitary on $H \otimes K$ for all t. Moreover, for $S \leq s \leq t \leq T$, we have

$$U_g(t,s) \;=\; W(\chi_{[S,T]}g)^{-1} \, U_0(t,s) \, W(\chi_{[S,T]}g) \;, \quad (2.18)$$

where $U_0(t,s)$, the solution of Equation (2.19) with $g = 0$, is a unitary martingale as described in [27].

Proof. See for example [10,15,27].

Theorem 2. For all A in $B(H)$, let

$$T_{s,t}(A) \;=\; E[U_g^+(t,s)(A \otimes \mathbb{1})U_g(t,s)]$$

$$\;=\; E_{C(g)}[U_0^+(t,s)(A \otimes \mathbb{1})U_0(t,s)] \;. \quad (2.19)$$

Then we have $(d/dt) \, T_{s,t}(A) \;=\; T_{s,t} \, L_t(A)$, where

$$L_t(A) \;=\; E_{g(t)}[S^+(t)(A \otimes \mathbb{1})S(t)] \;-\; \| g(t) \|_K^2 \, A \;. \quad (2.20)$$

Proof. From the general theory of quantum stochastic differential equations, we have

$$\begin{aligned}
d \, T_{s,t}(A) = E_{C(g)}[& dU_0^+(t,s)(A \otimes \mathbb{1})U_0(t,s) \\
& + U_0(t,s)(A \otimes \mathbb{1})dU_0(t,s) \\
& + dU_0(t,s)(A \otimes \mathbb{1})dU_0(t,s)] \;.
\end{aligned}$$

From (2.17) and (2.16) after some simple algebraic manipulation we obtain

$$d \, T_{s,t}(A) = E_{C(g)}[U_0^+(t,s)dN_t(S^+(A \otimes \mathbb{1})S - (A \otimes \mathbb{1}))U_0(t,s)].$$

Then, using adaptedness of $U_0(s,t)$ and the tensor product structure of the coherent vectors $C(g)$, we obtain

$$d\,T_{s,t}(A) = E_{C(g)}[U_0^+(t,s)\,E_{C(g)}[dN_t(S^+(A \otimes 1)S - (A \otimes 1))]U_0(t,s)]$$

that is, recalling the definitions (2.19) and (2.20) and taking into account (2.12),

$$d\,T_{s,t}(A) = T_{s,t}(L_t(A))\,dt\ ,$$

which proves the Theorem.

A different, but related mathematical object, the <u>non-commutative Poisson process</u>, was introduced by Kümmerer [23,24]. His construction is more general, but in this special case both constructions are essentially equivalent. In collaboration with C. Aratari, we have found a proof of this equivalence; however, we shall omit it since the result appears to be known, see [24].

3. Time-inhomogeneous evolutions: simulated annealing.

By a <u>time-inhomogeneous evolution</u> on $B(H)$ we mean a two-parameter family $\{T_{s,t} : s \leq t \in R^+\}$ of completely positive identity preserving normal linear maps of $B(H)$ into itself, satisfying the evolution property $T_{r,s}\,T_{s,t} = T_{r,t}$: $r \leq s \leq t \in R^+$ and such that

$$(d/dt)\,T_{s,t}(A) = T_{s,t}\,L_t(A) \tag{3.1}$$

for all A in $B(H)$ and for all $s \leq t \in R^+$, where, for each fixed t, L_t is the generator of a quantum dynamical semigroup $\{T_s^{(t)} : s \in R^+\}$. As an example, we may consider $L_t = L_{\beta(t)}$, where $t \mapsto \beta(t)$ is a function mapping R^+ into R^+ and where, for each positive β,

$$L_\beta(A) = \sum_{j=1}^{n} c_j(\beta)\,\Big\{ (V_j^+ A V_j - (1/2)[V_j^+ V_j, A]_+) + \exp(-\beta w_j)(V_j A V_j^+ - (1/2)[V_j V_j^+, A]_+)\Big\}\ , \tag{3.2}$$

where $c_j(\beta) > 0$, $c_j(\beta) \to 1$ as $\beta \to \infty$, and $V_j \in B(H)$, $w_j \in R^+$ satisfy

$$[V_0, V_j] = - w_j V_j \quad : j = 1, \ldots, n \quad , \tag{3.3}$$

V_0 being a self-adjoint operator in H such that $\exp[-\beta V_0]$ is trace class for all positive . Then $T_s^{(t)}$ satisfies the quantum detailed balance condition [2,21,30] with respect to the canonical state

$$\rho_{\beta(t)} = \exp[-\beta(t)V_0]/\mathrm{Tr}[\exp[-\beta(t)V_0]] \tag{3.4}$$

at inverse temperature $\beta(t)$. Moreover, if the von Neumann algebra generated by the V_j is the whole of $B(H)$, then $\rho_{\beta(t)}$ is the unique normal stationary state for $T_s^{(t)}$ and all normal states approach $\rho_{\beta(t)}$ in the limit as $s \to \infty$ [16]. Under suitable conditions on the cooling schedule $t \mapsto \beta(t)$, $\beta(t) \to +\infty$ as $t \to +\infty$, it can be shown [13] that $T_{s,t}$ induces approach to the uniform distribution on the states which minimize V_0 (at least when H is finite-dimensional). This is a quantum version [13] of simulated annealing [17,18].

We sketch the argument for finite-dimensional H. We have, for all A in $B(H)$,

$$- \mathrm{Tr}\,[\rho_\beta A^+ L_\beta(A)] = - \mathrm{Tr}\,[\rho_\beta L_\beta(A)^+ A]$$

$$= \frac{1}{2} \sum_{j=1}^{n} c_j(\beta)\left\{\mathrm{Tr}(\rho_\beta [V_j, A]^+ [V_j, A]) + \exp(-\beta w_j)\,\mathrm{Tr}(\rho_\beta [V_j^+, A]^+ [V_j^+, A])\right\} \geq 0. \tag{3.5}$$

(the first line of (3.5) expresses the detailed balance condition). If $\{V_j, V_j : j = 1, \ldots, n\}'' = B(H)$, then (3.5) is strictly positive unless A is a multiple of the identity. Since H is finite-dimensional, there exists a strictly positive number $\Gamma(\beta)$ (the spectral gap) such that

$$- \mathrm{Tr}\,[\rho_\beta A^+ L_\beta(A)] \geq \Gamma(\beta)[\mathrm{Tr}[\rho_\beta A^+ A] - \mathrm{Tr}[\rho_\beta A^+]\,\mathrm{Tr}[\rho_\beta A]] \tag{3.6}$$

for all A in $B(H)$. Estimates on the spectral gap can be given in several circumstances [18,30]; typically $\Gamma(\beta)$ decreases to 0 as

$\beta \to \infty$, and for infinitely extended systems one might have $\Gamma(\beta) \to 0$ as β tends to some critical $\beta_c > 0$. (critical slowing down).

Theorem 3. Let $\beta(t)$ be piecewise constant, equal to β_k on each interval of the form $(s_{k-1}, s_k]$: $0 = s_0 < s_1 < \ldots < s_k \to +\infty$, with β_k monotonically increasing to $+\infty$; assume also that the differences $\beta_k - \beta_{k-1}$ are uniformly bounded in k. Let $t_k = s_k - s_{k-1}$ be the time spent at inverse temperature β_k. Assume that

$$[\Gamma(\beta_k)t_k - \frac{1}{2}(\beta_k - \beta_{k-1})\|V_0\|] \geq a > 0 \quad \text{for all } k . \tag{3.7}$$

Then for all A in B(H) and for all density operators ρ on H we have

$$\lim_{t\to\infty} \mathrm{Tr}[\rho\, T_{0,t}(A)] = \lim_{\beta\to\infty} \mathrm{Tr}[\rho_\beta A] . \tag{3.8}$$

Proof. Let K be the Hilbert space of all linear operators on the finite-dimensional Hilbert space H , equipped with the scalar product $\langle A,B\rangle_K = (\dim H)^{-1} \mathrm{Tr}_H[A^+ B]$. Let B(H) be identified with its left regular representation acting on K . For each k = 1, 2,... , let

$$S_k\, g = \exp[L_{\beta_k} t_k](g \exp[\beta_k V_0/2]) \exp[-\beta_k V_0/2] \quad : \quad g \in K , \tag{3.9}$$

$$R_{k,k+1} g = Z(\beta_k)^{-1/2} Z(\beta_{k+1})^{1/2} g \exp[(\beta_{k+1} - \beta_k)V_0/2] \quad : g \in K , \tag{3.10}$$

where $Z(\beta) = \mathrm{Tr}\exp[-\beta V_0]$ is the partition function. Then we have

$$\mathrm{Tr}\,[\rho\, T_{0,s_k}(A)] = \langle \rho\, \rho_{\beta_1}^{-1/2}, T_{0,s_k}(A)\, \rho_{\beta_1}^{1/2}\rangle_K$$

$$= \langle \rho\, \rho_{\beta_1}^{-1/2}, S_1 R_{1,2} S_2 \ldots R_{k-1,k} S_k (A\, \rho_{\beta_k}^{1/2})\rangle_K \tag{3.11}$$

and

$$\mathrm{Tr}\,[\rho_{\beta_k}(A)] = \langle \rho_{\beta_k}^{1/2}, A\, \rho_{\beta_k}^{1/2}\rangle_K . \tag{3.12}$$

Since S_k and $R_{k-1,k}$ are self-adjoint operators in K for all k, (for S_k this follows from detailed balance) we have the bound

$$\left|\operatorname{Tr}[\rho\, T_{0,s_k}(A)] - \operatorname{Tr}[\rho_{\beta_k}(A)]\right| \leq d_k \operatorname{Tr}[\rho_{\beta_k} A^+ A]^{1/2} , \qquad (3.13)$$

where

$$d_k = \| S_k R_{k,k-1} \cdots S_2 R_{1,2} S_1 \rho \rho_{\beta_1}^{-1/2} - \rho_{\beta_k}^{1/2} \|_K . \qquad (3.14)$$

We may assume without loss of generality that the spectrum of V_0 is contained in $[0, \|V_0\|]$. Then, using the spectral gap condition and (3.7), we can easily prove that

$$d_k \leq e^{-a}(d_{k-1} + b_{k-1}) : k = 1, 2, \ldots , \qquad (3.15)$$

where $d_0 = \| \rho \rho_{\beta_1}^{-1/2} - \rho_{\beta_1}^{1/2} \|_K$, $b_0 = 0$, and where, for $k > 1$,

$$b_{k-1} = \| R_{k-1,k}\, \rho_{\beta_{k-1}}^{1/2} - \rho_{\beta_k}^{1/2} \|_K$$

$$= [Z(2\beta_{k-1} - \beta_k)\, Z(\beta_k)\, Z(\beta_{k-1})^{-2} - 1]^{1/2} , \qquad (3.16)$$

which tends to 0 as $k \to \infty$. Since $e^{-a} < 1$, it follows from (3.15) that $d_k \to 0$ as $k \to \infty$, which concludes the proof.

4. Dilations of time-inhomogeneous evolutions.

Theorem 2 describes an example of a time-inhomogeneous evolution which admits a unitary dilation. We shall now describe a construction of a dilation of the time-inhomogeneous evolution determined by the family of generators (3.2).

Let $K = C^{2n}$, and introduce the short-hand notation

$$a_j = |e_{2j-1}\rangle\langle e_{2j}| \qquad a_j^+ = |e_{2j}\rangle\langle e_{2j-1}| . \qquad (4.1)$$

For each t let $S(t) = S_\alpha$, where S_α is the unitary operator given by

$$S_\alpha = \exp[i\alpha V] \tag{4.2}$$

where V is the self-adjoint operator on $H \times K$ defined by

$$V = -i \sum_{j=1}^{n} (V_j \otimes a_j^+ - V_j^+ \otimes a_j) \quad . \tag{4.3}$$

Let also $g(t) = g_\alpha(t)$ be of the form

$$g_\alpha(t) = \alpha^{-1} \sum_{j=1}^{n} c_j(\beta(t))\left\{e_{2j-1} + \exp[-\beta(t)w_j]\, e_{2j}\right\}. \tag{4.4}$$

Let $U_g^{(\alpha)}(t,s)$ be the solution of the quantum stochastic differential equation as in Theorem 1, with $S = S_\alpha$ and $g = g_\alpha$, and let $T_{s,t}^{(\alpha)}$ be the corresponding time-inhomogeneous evolution as in Theorem 2.

<u>Theorem 4.</u> In the limit as $\alpha \to 0$, $U_g^{(\alpha)}(t,s)$ and $T_{s,t}^{(\alpha)}$ converge to a family $U(t,s)$ of unitary operators and to a time-inhomogeneous evolution $T_{s,t}$ respectively. $T_{s,t}$ is the evolution determined by (3.2), and $U(t,s)$ is the solution of the following quantum stochastic differential equation

$$d\,U(t,s)$$

$$= \sum_{j=1}^{n} [V_j\, dA_j^{(\beta)+} - V_j^+\, dA_j^{(\beta)} - \tfrac{1}{2} c_j(\beta(t))(V_j^+V_j + e^{-\beta(t)w_j} V_j V_j^+)dt]U(t,s) \tag{4.5}$$

with $U(s,s) = \mathbb{1}$, where $A_j^{(\beta)}$, $A_j^{(\beta)+}$ are quasi-free Brownian motions [5] satisfying the following Ito table, which is a time-dependent version of the one of [19]:

$$\begin{aligned} dA_j^{(\beta)}(t)\, dA_k^{(\beta)+}(t) &= \delta_{jk}\, c_j(\beta(t))\, dt, \\ dA_j^{(\beta)+}(t)\, dA_k^{(\beta)}(t) &= \delta_{jk}\, c_j(\beta(t)) \exp[-\beta(t)w_j]\, dt. \end{aligned} \tag{4.6}$$

<u>Proof.</u> For $\alpha > 0$ and for all j we have

$$(e^{i\alpha V} - \mathbb{1})_{2j-1,2j-1} = \cos(\alpha|V_j|) - \mathbb{1} = \alpha^2 K_j(\alpha), \qquad (4.7)$$

$$(e^{i\alpha V} - \mathbb{1})_{2j,2j-1} = J_j \sin(\alpha|V_j|) = \alpha L_j(\alpha). \qquad (4.8)$$

$$(e^{i\alpha V} - \mathbb{1})_{2j,2j} = \cos(\alpha|V_j^+|) - \mathbb{1} = \alpha^2 \tilde{K}_j(\alpha), \qquad (4.9)$$

$$(e^{i\alpha V} - \mathbb{1})_{2j-1,2j} = - J_j^+ \sin(\alpha|V_j^+|) = \alpha \tilde{L}_j(\alpha), \qquad (4.10)$$

where we have used the polar decompositions $V_j = J_j |V_j|$ of V_j and $V_j^+ = J_j^+ |V_j^+|$ of V_j^+. $K_j(\alpha)$ and $L_j(\alpha)$ converge to $(1/2)V_j^+V_j$ and to V_j; similarly, $\tilde{K}_j(\alpha)$ and $\tilde{L}_j(\alpha)$ converge to $(1/2)V_jV_j^+$ and to V_j^+ respectively.

Let $U_g^{(\alpha)}(t,s)$ be the solution of the quantum stochastic differential equation (2.17) (Theorem 1) with $S = S_\alpha$ and $g = g_\alpha$. Using the explicit form (2.10) of the stochastic differential of the quantum Poisson process and taking into account (4.4) - (4.10), we see that

$$d\,U_g^{(\alpha)}(t,s)$$

$$= \left\{ \sum_{j=1}^{n} [V_j\, dA_j^{(\beta)+} - V_j^+\, dA_j^{(\beta)} - \tfrac{1}{2} c_j(\beta)(V_j^+V_j + e^{-\beta(t)w_j} V_jV_j^+)dt] + \sigma(\alpha) \right\} U_g^{(\alpha)}(t,s), \qquad (4.11)$$

where $\sigma(\alpha)$ is a stochastic differential whose coefficients are bounded operators with norm of the order of α, and where

$$dA_j^{(\beta)}(t) = c_j(\beta(t))\, dA_{2j-1}(t) + \exp[-\beta w_j]\, dA_{2j}^+(t)\,, \qquad (4.12)$$

with the adjoint expression for $dA_j^{(\beta)+}(t)$. These stochastic differentials do satisfy the Ito table (4.6). Since all the operators on H involved are bounded, it is easy to prove that (4.11) implies that $U_g^{(\alpha)}(t,s)$ converges strongly to the solution of (4.5).

Remark 1. If we work with the quantum Poisson process (non-zero α), then we may regard $T^{(\alpha)}_{s,t}$ as

$$T^{(\alpha)}_{s,t}(A) = E_{C(g_\alpha)}[U^{(\alpha)+}_0(t,s)F(S_{t-s})(A \otimes 1)F^+(S_{t-s})U^{(\alpha)}_0(t,s)]$$

where $\{S_t : t \in R\}$ is the right shift on $L^2(R;K)$, and where $\{F^+(S_t)U(t,0) : t \in R^+\}$ extends to a group of unitary operators on $H \otimes F(h)$; then we have a group dilation, where the time evolution of the global system is time-homogeneous and time dependence enters only through the locally normal coherent states which serves to define the conditional expectation $E_{C(g)}$. On the other hand, we may also write

$$T^{(\alpha)}_{s,t}(A) = E[U^{(\alpha)+}_g(t,s)F(S_{t-s})(A \otimes 1)F^+(S_{t-s})U^{(\alpha)}_g(t,s)]$$

so that the state which defines the conditional expectation E is time-homogeneous, but the global time evolution cannot be extended to a group.

Remark 2. In (3.10) the commutation relations of $dA^{(\beta)}_j$ and $dA^{(\beta)+}_k$ are time-independent if and only if $c_j(\beta) = (1 - \exp[-\beta w_j])^{-1}$. We could obtain time-independent anticommutation relations for finite-temperature fermionic Brownian motion with $c_j(\beta) = (1 + \exp[-\beta w_j])^{-1}$.

5. Unitary dilations of a quantum Boltzmann equation.

We consider the following quantum Boltzmann equation:

$$\frac{d}{dt}\rho(t) = -i[H,\rho(t)] + tr_1[S\,\rho(t) \otimes \rho(t)\,S^+] - Tr[\rho(t)]\,\rho(t) . \tag{5.1}$$

The unknown $\rho(t)$ is a positive trace class operator on a Hilbert space H_0, H is a self-adjoint operator on H_0, S is a unitary operator on $H_0 \otimes H_1$, H_0 and H_1 being isomorphic, and tr_1 is the partial trace over H_1. This is essentially the same equation as considered in Snider

[28], and is also a continuous time version of the iterated Boltzmann map of Streater [29]. No effect of quantum statistics has been included in (5.1); this should be permissible for very dilute gases. Existence and uniqueness of the solution follows from work of Alicki and Messer [3].

Following some considerations in [12] we can show that the quantum Poisson process can provide a unitary dilation of the above equation. We use as circumambient space K the space of Hilbert-Schmidt operators on H_0, so that every positive normal functional $\mathrm{Tr}_{H_0}[\rho\ .\ ]$ on $B(H_0)$ can be represented by a vector $g = \rho^{1/2}$ in K as

$$\mathrm{Tr}_{H_0}[\rho A] = \langle \rho^{1/2}, \pi(A)\, \rho^{1/2} \rangle_K \qquad : A \in B(H)\ , \tag{5.2}$$

where π is the left regular representation of $B(H)$. In the following we shall omit all explicit indication of π and all subscripts H_0 and K, identifying $B(H_0 \otimes H_1)$ with $B(H_0) \otimes \pi(B(H_1))$.

<u>Theorem 5</u>. Let $g(t) = \rho(t)^{1/2}$, where $\rho(t)$ is the solution of (5.1) with initial condition ρ, and let $S(t) = S$ for all t. Then we have

$$\mathrm{Tr}\,[\rho(t)\, A] = \mathrm{Tr}\,[\rho\ E_{C(g)}[U_0^+(t)(A \otimes \mathbb{1})U_0(t)]] \ : A \in B(H) \tag{5.3}$$

where $U_0(t)$ is the solution of

$$d\, U_0(t) = [dN_t(S - \mathbb{1}, 0) - i\, H\, dt]\, U_0(t)\ , \quad U_0(0) = \mathbb{1}. \tag{5.4}$$

<u>Proof</u>. It is a straightforward variant of Theorem 2.

<u>Remark 1</u>. In the above dilation we have coupled the system of interest to a heat bath whose initial state contains the history of the system itself through the solution of the quantum Boltzmann equation; this bears some analogy with the construction of Lewis and Thomas [25]. Another unitary dilation, which does not require the knowledge of the solution of the quantum Boltzmann equation, has been constructed in [14], elaborating on old results of Wild [32], McKean [26].

<u>Remark 2</u>. A general theory of nonlinear quantum evolutions has been given recently by Belavkin [7]. It would be interesting to investigate its possible connections with the present framework.

References:

1. Accardi, L., Journé, J.-L., and Lindsay, J.M.: On multi-dimensional Markovian cocycles. In: Accardi, L., and von Waldenfels, W. (Eds.): Quantum Probability and Applications IV (Proceedings, Roma 1987/88). To appear in Lecture Notes in Mathematics, Springer-Verlag.

2. Alicki, R.: On the quantum detailed balance condition for non-Hamiltonian systems. Rept. Math. Phys. 10 (1976) 249-258.

3. Alicki, R., and Messer, J.: Nonlinear quantum dynamical semigroups for many-body open systems. J.Stat. Phys. 32 (1983) 299-312.

4. Apolloni, B., Carvalho, C., and de Falco, D.: Quantum stochastic optimization. Preprint BiBoS nr. 327/88 (1988).

5. Applebaum, D.: Quasi-free stochastic evolutions. In: Accardi, L., and von Waldenfels, W. (Eds.): Quantum Probability and Applications II (Proceedings, Heidelberg 1984). Lecture Notes in Mathematics vol. 1136, Springer-Verlag (1985) pp. 46-56.

6. Barchielli, A.: Quantum stochastic differential equations: an application to the electron shelving effect. J. Phys. A: Math. Gen. 20 (1987) 6341-6355.

7. Belavkin, V.P.: Quantum branching processes and nonlinear dynamics of multiquantum systems. (in Russian). Doklady Akademia Nauk 301, 1348-1351 (1988).

8. Belavkin, V.P.: Nondemolition measurements, nonlinear filtering and dynamic programming of quantum stochastic processes. Preprint (1988).

9. Evans, D.E., and Lewis, J.T.: Dilations of irreversible evolutions in algebraic quantum theory. Commun. DIAS, series A, No. 24 (1977).

10. Evans, M., and Hudson, R.L.: Multidimensional quantum diffusions. In: Accardi, L., and von Waldenfels, W. (Eds.): Quantum Probability and Applications III (Proceedings, Oberwolfach 1987). Lecture Notes in Mathematics vol. 1303, Springer-Verlag (1988) pp. 69-88.

11. Fichtner, K.H., and Freudenberg, W.: Point processes and the position distribution of infinite boson systems. J.Stat.Phys. 47(1987)959-978.

12. Frigerio, A.: Quantum Poisson processes: physical motivations and applications. In: Accardi, L., and von Waldenfels, W. (Eds.): Quantum Probability and Applications III (Proceedings, Oberwolfach 1987). Lecture Notes in Mathematics vol. 1303, Springer-Verlag (1988) pp. 107-127.

13. Frigerio, A.: Simulated annealing and quantum detailed balance. In preparation.

14. Frigerio, A., and Aratari, C.: Unitary dilations of a nonlinear quantum Boltzmann equation. In:Accardi,L.,and von Waldenfels,W.(Eds): Quantum Probability and Applications IV (Proceedings, Roma 1987/88). To appear in Lecture Notes in Mathematics, Springer-Verlag.

15. Frigerio, A., and Maassen, H.: Quantum Poisson processes and dilations of dynamical semigroups. Preprint Nijmegen (1987).

16. Frigerio, A., and Verri, M.: Long-time asymptotic properties of dynamical semigroups on W*-algebras. Math. Z. 180 (1982) 275-286.

17. Geman, S., and Geman, D.: Stochastic relaxation, Gibbs distribution, and the Bayesian restoration of images. IEEE Trans. Pattern Anal. Mach. Intell. 6 (1984) 721-741.

18. Holley, R., and Stroock, D.: Simulated annealing via Sobolev inequalities. Commun. Math. Phys. 115 (1988) 553-569.

19. Hudson, R. L., and Lindsay, J. M.: Stochastic integration and a martingale representation theorem for non-Fock quantum Brownian motion. J. Funct. Anal. 61 (1985) 202-221.

20. Hudson, R. L., and Parthasarathy, K. R.: Quantum Ito's formula and stochastic evolutions. Commun. Math. Phys. 93 (1984) 301-323.

21. Kossakowski, A., Frigerio, A., Gorini, V., and Verri, M.: Quantum detailed balance and KMS condition. Commun. Math. Phys. 57 (1977) 97-110.

22. Kümmerer, B.: Markov dilations on W*-algebras. J. Funct. Anal. 63 (1985) 139-177.

23. Kümmerer, B.: Markov dilations and non-commutative Poisson processes. Preprint Tubingen (1986).

24. Kümmerer, B.: Survey on a theory of non-commutative stationary Markov processes. In: Accardi, L., and von Waldenfels, W. (Eds.): Quantum Probability and Applications III (Proceedings, Oberwolfach 1987). Lecture Notes in Mathematics vol. 1303, Springer-Verlag (1988) pp. 154-182.

25. Lewis, J. T., and Thomas, L. C.: How to make a heat bath. In: Functional Integration. Proceedings of International Conference, Cumberland Lodge, London, April 1974. Oxford, Clarendon Press (1975).

26. McKean, H. P. (Jr.): An exponential formula for solving Boltzmann equation for a Maxwellian gas. J. Comb. Theory 2 (1967) 358-382.

27. Parthasarathy, K. R., and Sinha, K. B.: Representation of a class of quantum martingales II. In: Accardi, L., and von Waldenfels, W. (Eds.): Quantum Probability and Applications III (Proceedings, Oberwolfach 1987). Lecture Notes in Mathematics vol. 1303, Springer-Verlag (1988) pp. 232- 250.

28. Snider, R.F.: Quantum-mechanical modified Boltzmann equation for degenerate internal states. J. Chem. Phys. 32 (1960) 1051-1060.

29. Streater, R.F.: Convergence of the iterated quantum Boltzmann map Commun. Math. Phys. 98 (1985) 177-185.

30. Verbeure, A.: Detailed balance and critical slowing down. In: Accardi, L., and von Waldenfels, W. (Eds.): Quantum Probability and Applications III (Proceedings, Oberwolfach 1987). Lecture Notes in Mathematics vol. 1303, Springer-Verlag (1988) pp. 69-88.

31. Wild, E.: On Boltzmann equation in the kinetic theory of gases. Proc. Cambridge Phil. Soc. 47 (1951) 602-609.

Erratum: QUANTUM POISSON PROCESSES : PHYSICAL MOTIVATIONS AND APPLICATIONS .

Alberto Frigerio

(in: QUANTUM PROBABILITY AND APPLICATIONS III
Springer Lecture Notes in Mathematics vol 1303)

1) Line 11 of page 117 should read:

"parameter $|z|^2$:"

2) Equation (5.11) on page 119 should read:

$$" \quad dN_t(X) = dA_t^+(X\xi) + d\Lambda_t(X) + dA_t(X^+\xi) + \langle \xi | X\xi \rangle \, dt \quad (5.11)"$$

Similarly, two lines below one should have:

"annihilation martingales A_t^+ , Λ_t and A_t of Hudson and"

3) The line before Equation (5.14) on page 120 should read:

"estimate (for X of the form $\sum_{j=1}^{\infty} B_j \otimes C_j$):"

Equation (5.14) itself should read:

$$" \quad \| \hat{N}_t(X)(\phi \otimes \Psi(0)) \|^2$$

$$= \sum_{i,j} \langle B_i\phi \otimes [A_t^+(C_i\xi) + \langle \xi | C_i\xi \rangle t]\Psi(0) | B_j\phi \otimes [A_t^+(C_j\xi) + \langle \xi | C_j\xi \rangle t]\Psi(0) \rangle$$

$$= \sum_{i,j} \langle B_i\phi | B_j\phi \rangle [\langle A_t^+(C_i\xi)\Psi(0) | A_t^+(C_j\xi)\Psi(0) \rangle + \overline{\langle \xi | C_i\xi \rangle} \langle \xi | C_j\xi \rangle t^2]$$

$$\leq \quad t(1 + \|\xi\|^2 t) \sum_{i,j} \langle B_i\phi \otimes C_i\xi | B_j\phi \otimes C_j\xi \rangle$$

$$= \quad t(1 + \|\xi\|^2 t) \, \| X(\phi \otimes \xi) \|^2 \quad (5.14)"$$

4) Equation (6.6) on page 122 should read:

$$" \quad \| \hat{N}_t(X)(\phi \otimes \Psi(0)) \|^2$$

$$\leq \quad t(1 + \|\xi\|^2 t) \, \| X(\phi \otimes \xi) \|^2$$

$$(\quad \leq \quad 2\, t \|X\|^2 \, \|\phi\|^2 \, \|\xi\|^2 \qquad \text{if} \quad t \leq \|\xi\|^{-2} \;) \qquad (6.6)"$$

Quantum Central Limit and Coarse Graining

D. Goderis*, A. Verbeure , P. Vets†
Instituut voor Theoretische Fysica
Universiteit Leuven
B-3030 Leuven, Belgium

February 20, 1990

Abstract

For quantum systems we prove a central limit theorem for products of fluctuations and give a mathematically rigorous description of coarse graining.

1 Introduction

The problem of central limit theorems for non-commuting observables or random variables has already a long story. It came first to our attention by the work of Hepp and Lieb [1,2,3], studying fluctuations in the Dicke Maser model. Different results have been obtained in different settings [see e.g. 4-10]. Recently we were able to complete these results by giving a full mathematical characterisation of the objects obtained by the central limit [11]. In particular we proved that the central limits yield a representation of a C*-algebra of canonical commutation relations induced by a quasi-free state. Moreover we proved the stability of the relative entropy under the central limit theorem and gave the physical interpretation of all this for mean field models [12]. The minimal mixing conditions for the existence of such central limits can be found in [13].

We take the occasion of these lecture notes to extend our results by proving a central limit theorem for products of fluctuations under general mixing conditions of the system. This result is proved in section 4. In section 3 we are able to give a precise mathematical description of the physical notion of coarse graining due to the central limit.

---

*Onderzoeker IIKW,Belgium
†Onderzoeker IIKW, Belgium

2 Preliminaries about the CCR and quasi-free states

The essential ingredient in the description of normal fluctuations of a quantum system is the notion of CCR-C*-algebra and its quasi-free states. Here we repeat the definitions and results, necessary for later use. Details about the CCR-algebra can be found in [14] and about quasi-free states in [15].

Suppose that a real symplectic space (H,σ) is given, i.e. H is a real vector space with a (possibly degenerated) symplectic form σ:

$$\sigma : H \times H \to \mathbb{R} \; : \; (x,y) \to \sigma(x,y)$$

such that σ is bilinear and antisymmetric.

Denote by $W(H,\sigma)$ the complex vector space generated by the functions $W(x), x \in H$ defined by

$$W(x) \; : \; H \to \mathbb{C}$$
$$y \to W(x)y = \begin{cases} 0 & \text{if } x \neq y \\ 1 & \text{if } x = y \end{cases}$$

$W(H,\sigma)$ becomes an algebra with unit $W(0)$ for the product:

$$W(x)W(y) = W(x+y)\exp - \frac{i}{2}\sigma(x,y) \; ; \; x,y \in H \tag{2.1}$$

(H,σ) is a C*-algebra for the involution:

$$W(x) \to W(x)^* = W(-x)$$

We equip $W(H,\sigma)$ with the minimal regular norm [14] and assume that $W(H,\sigma)$ is complete for this norm topology. All this means that $W(H,\sigma)$ is a C*-algebra,called the CCR-C*-algebra.

A symplectic operator S on a symplectic space (H,σ) is a linear operator on H such that for all $x,y \in H$:

$$\sigma(Sx,Sy) = \sigma(x,y) \tag{2.2}$$

Any symplectic operator S defines a *-automorphism α_S of $W(H,\sigma)$ by the relation:

$$\alpha_S W(x) = W(Sx) \tag{2.3}$$

This type of automorphisms are called quasi-free automorphisms

A state of $W(H,\sigma)$ is a positive linear normalized functional of it. A state ω of $W(H,\sigma)$ is called regular if for all $x,y \in H$ the map

$$\lambda \in \mathbb{R} \to \omega(W(\lambda x + y))$$

is continuous. Let ω be a regular state and let $(\mathcal{H},\pi,\Omega)$ be the GNS-triplet induced by ω, then there exists a linear map

$$B : H \to \mathcal{L}(\mathcal{H}); x \to B(x)$$

such that $B(x) = B(x)^*$ and $\pi(W(x)) = \exp iB(x)$. The map B is called the Bose field, satisfying the Bose field commutation relations :

$$[B(x), B(y)] = i\sigma(x,y)\mathbb{1} \;\; ; \; x, y \in H \tag{2.4}$$

for a dense domain in $\mathcal{H}$. Let s be a real symmetric positive bilinear form on H such that

$$\frac{1}{4}|\sigma(x,y)|^2 \leq s(x,x)s(y,y) \tag{2.5}$$

and let χ be any character of the additive group H, then the state $\omega_{s,\chi}$ defined by

$$\omega_{s,\chi}(W(x)) = \chi(x)\exp - \frac{1}{2}s(x,x) \tag{2.6}$$

is called a quasi-free state. If the character χ is trivial, i.e. $\chi(x) = 1$ for all $x \in H$, then we denote $\omega_{s,1}$ simply by ω_s.

Remark that all quasi-free states are regular states. If $(\mathcal{H}_s, \pi_s, \Omega_s)$ is the GNS-triplet of ω_s and B_s the associated Bose field,then one easily calculates:

$$(\Omega_s, B_s(x)\Omega_s) \;=\; 0 \; ; \; x \in H \tag{2.7}$$

$$(\Omega_s, B_s(x)B_s(y)\Omega_s) \;=\; s(x,y) + \frac{i}{2}\sigma(x,y) \, ; \; x, y \in H \tag{2.8}$$

3 Normal fluctuations and coarse graining

We develop the theory for systems which are defined on a ν-dimensional lattice $\mathbb{Z}^\nu$ and which have a quasi-local structure [16]. Let $\mathcal{D}(\mathbb{Z}^\nu)$ be the directed set of finite subsets of $\mathbb{Z}^\nu$,where the direction is the inclusion. With each $x \in \mathbb{Z}^\nu$ we associate the algebra $\mathcal{A}_x$, a copy of a C*-algebra $\mathcal{A}$. For all $\Lambda \in \mathcal{D}(\mathbb{Z}^\nu)$, the tensor product $\otimes_{x\in\Lambda}\mathcal{A}_x$ defines a C*-algebra $\mathcal{A}_\Lambda$. The family $\mathcal{A}_\Lambda$, $\Lambda \in \mathcal{D}(\mathbb{Z}^\nu)$ has the usual properties of locality and isotony:

$$[\mathcal{A}_{\Lambda_1}, \mathcal{A}_{\Lambda_2}] = 0 \quad \text{if } \Lambda_1 \cap \Lambda_2 = \emptyset$$

$$\mathcal{A}_{\Lambda_1} \subseteq \mathcal{A}_{\Lambda_2} \qquad \text{if } \Lambda_1 \subseteq \Lambda_2$$

Denote by $\mathcal{A}_L$ all local observables :

$$\mathcal{A}_L = \bigcup_{\Lambda\in\mathcal{D}(\mathbb{Z}^\nu)} \mathcal{A}_\Lambda \tag{3.1}$$

The norm closure $\mathcal{B}$ of $\mathcal{A}_L$ is again a C*-algebra :

$$\mathcal{B} = \overline{\mathcal{A}_L} = \overline{\bigcup_{\Lambda\in\mathcal{D}(\mathbb{Z}^\nu)} \mathcal{A}_\Lambda} \tag{3.2}$$

and taken to be the algebra of quasi local observables of our system.

The group $\mathbb{Z}^\nu$ of space-translations of the lattice acts as a group of *-automorphisms on the algebra of observables $\mathcal{B}$ denoted by :

$$\tau_x : A \in \mathcal{A}_\Lambda \to \tau_x(A) \in \mathcal{A}_{\Lambda+x} \quad , x \in \mathbb{Z}^\nu \tag{3.3}$$

In the following we consider the physical system $(\mathcal{B}, \omega)$ where ω is a translation invariant state of $\mathcal{B}$, i.e. $\omega \circ \tau_x = \omega$ for all $x \in \mathbb{Z}^\nu$.

At this point we are able to introduce the local fluctuations of the system $(\mathcal{B}, \omega)$. Denote by Λ_n the cube centered around the origin with edges of length $2n+1$. For any $A \in \mathcal{B}$, the local fluctuation $\tilde{A}^n$ of A is given by:

$$\tilde{A}^n = \frac{1}{|\Lambda_n|^{1/2}} \sum_{x \in \Lambda_n} \tau_x(A - \omega(A)) \tag{3.4}$$

where $|\Lambda_n|$ denotes the number of points in Λ_n. The main problem is to give a rigorous meaning to the limits

$$\lim_{n \to \infty} \tilde{A}^n = A_\omega$$

For A an element of $\mathcal{B}$, the limits A_ω are called the macroscopic fluctuations of the system $(\mathcal{B}, \omega)$. Furthermore we want to give a complete mathematical description of the set of macroscopic fluctuations of the system $(\mathcal{B}, \omega)$. Already Hepp and Lieb [1,2] suggested that the fluctuations of one-point observables ($A \in \mathcal{A}_0$) form some abstract Lie-algebra and that their commutators are c-numbers. Later work [8,9] suggested that the fluctuations behave like bosons, i.e. they satisfy the canonical commutation relations. We complete this characterization by proving that one gets a well-defined representation of a CCR-C*-algebra, uniquely associated with the original system $(\mathcal{B}, \omega)$.

Definition 3.1 *Let $\mathcal{B}_0$ be any *-subspace of $\mathcal{B}$; the system $(\mathcal{B}, \mathcal{B}_0, \omega)$ is said to have normal fluctuations for $\mathcal{B}_0$ if*

(i) $\forall A, B \in \mathcal{B}_0$:

$$\sum_{x \in \mathbb{Z}^\nu} |\omega(A \tau_x B) - \omega(A)\omega(B)| < \infty \tag{3.5}$$

(ii) $\forall A = A^* \in \mathcal{B}_0; t \in \mathbb{R}$:

$$\lim_{n \to \infty} \omega(e^{it\tilde{A}^n}) = e^{-\frac{t^2}{2} s_\omega(A,A)} \tag{3.6}$$

where

$$s_\omega(A, B) = \lim_{n \to \infty} Re\, \omega(\tilde{A}^n, \tilde{B}^n)$$

for all A, B self-adjoint elements of $\mathcal{B}_0$. Remark that this limit exists because of (i). The limit (ii) is called the central limit.

Now we define the algebra of normal fluctuations of the system $(\mathcal{B}, \mathcal{B}_0, \omega)$ in a canonical way as the following CCR-C*-algebra. In view of section 2, we have to fix first a symplectic space (H, σ). Let $\mathcal{B}_{0,sa}$ be the real vectorspace of self-adjoint elements of $\mathcal{B}_0$, take

$$H = \mathcal{B}_{0,sa} \tag{3.7}$$

and for all $A, B \in \mathcal{B}_{0,sa}$:

$$\sigma_\omega(A,B) = -i \lim_{n\to\infty} \omega([\tilde{A}^n, \tilde{B}^n]) \tag{3.8}$$

this limit exists because of (i) of 3.1. Clearly σ_ω is a symplectic form on $H = \mathcal{B}_{0,sa}$. Hence $(\mathcal{B}_{0,sa}, \sigma_\omega)$ is a symplectic space. The CCR-C*-algebra

$$W(\mathcal{B}_{0,sa}, \sigma_\omega) \tag{3.9}$$

is called the *algebra of normal fluctuations* of the system $(\mathcal{B}, \mathcal{B}_{0,sa}, \omega)$.

Furthermore, the central limit fixes also a representation of this CCR-algebra (3.9). This is proved in the following theorem.

Theorem 3.2 *If the system $(\mathcal{B}, \mathcal{B}_0, \omega)$ has normal fluctuations for $\mathcal{B}_0$, then ω_s is a quasi-free state on the algebra of normal fluctuations (3.9), and*

$$\omega_s(W(A)) = e^{-\frac{1}{2}s_\omega(A,A)} \tag{3.10}$$

*Moreover if α is a *-automorphism of $\mathcal{B}$ leaving invariant $\mathcal{B}_0$, commuting with space-translations ($[\alpha, \tau_x] = 0$ for all $x \in \mathbb{Z}^\nu$) and leaving the state ω invariant ($\omega \circ \alpha = \omega$) then $\tilde{\alpha}$ defined by:*

$$\tilde{\alpha}(W(A)) = W(\alpha(A)) \,,\; A \in \mathcal{B}_{0,sa}$$

*defines a quasi-free *-automorphism of $W(\mathcal{B}_{0,sa}, \sigma_\omega)$.*

Proof: By definition (3.6), s_ω is a real, symmetric bilinear form on $\mathcal{B}_{0,sa}$. Hence the formula (3.10) defines a quasi-free state ω_s if the inequality (see 2.5) holds:

$$\frac{1}{4}|\sigma_\omega(A,B)|^2 \le s_\omega(A,A)s_\omega(B,B) \;\; ; A, B \in \mathcal{B}_{0,sa}$$

But for each $n \in \mathbb{N}$, Schwartz inequality yields:

$$\begin{aligned} \frac{1}{4}|-i\omega([\tilde{A}^n, \tilde{B}^n])|^2 &= \frac{1}{4}|\omega([\tilde{A}^n, \tilde{B}^n])|^2 \\ &= |\, Im\, \omega(\tilde{A}^n \tilde{B}^n)|^2 \\ &\le \omega((\tilde{A}^n)^2)\omega((\tilde{B}^n)^2). \end{aligned}$$

and by taking the limit $n \to \infty$, one gets the above inequality, proving the first statement.

To prove the second statement we have to show that α defines a symplectic operator (see 2.2) on $(\mathcal{B}_{0,sa}, \sigma_\omega)$. For all $n \in \mathbb{N}_0$:

$$\begin{aligned}\widetilde{(\alpha A)}^n &= \frac{1}{|\Lambda_n|^{1/2}} \sum_{x\in\Lambda_n} \tau_x(\alpha A - \omega(\alpha A)) \\ &= \frac{1}{|\Lambda_n|^{1/2}} \sum_{x\in\Lambda_n} \alpha(\tau_x A - \omega(A)) \\ &= \alpha(\tilde{A}^n)\end{aligned}$$

Hence

$$\begin{aligned}-i\omega([\widetilde{(\alpha A)}^n, \widetilde{(\alpha B)}^n]) &= -i\omega(\alpha[\tilde{A}^n, \tilde{B}^n]) \\ &= -i\omega([\tilde{A}^n, \tilde{B}^n])\end{aligned}$$

By taking the limit $n \to \infty$, one gets

$$\sigma_\omega(\alpha A, \alpha B) = \sigma_\omega(A, B)$$

The central limit theorem (3.1) guarantees the limit of the characteristic function defined on $\mathcal{B}_0$:

$$\lim_{n\to\infty} \omega(e^{i\tilde{A}^n})$$

By Theorem 3.2 we show that this limit is given by:

$$\lim_{n\to\infty} \omega(e^{i\tilde{A}^n}) = \omega_s(e^{iB_\omega(A)})$$

This enables us to make the identification:

$$\lim_{n\to\infty} \tilde{A}^n \equiv B_\omega(A)$$

where B_ω is a representation of the Bose field induced by the quasi-free state ω_s. In particular we have the commutation relations

$$[B_\omega(A), B_\omega(B)] = i\sigma_\omega(A, B)\mathbb{1}$$

The other part of the theorem shows in which way transformations of the micro-system $(\mathcal{B},\omega)$ are transported to the algebra of fluctuations. For example if a translation invariant time evolution α_t is given on the micro-system such that $\omega \circ \alpha_t = \omega$ and $\alpha_t\mathcal{B}_0 \subseteq \mathcal{B}_0$, then it induces a quasi-free evolution on the normal fluctuations. In practical applications this macro-evolution yields information on the micro-evolution.

Another application of this theorem is the description of the coarse graining procedure. Therefore assume that $\tau_x(\mathcal{B}_0) \subseteq \mathcal{B}_0$ for all $x \in \mathbb{Z}^\nu$. Mathematically coarse graining is the property that the map:

$$A \in \mathcal{B}_0 \to B_\omega(A)$$

is not injective, i.e. many micro-observables yield the same fluctuation as a result of the central limit theorem. As a matter of example, it is clear that each observable $A \in \mathcal{B}_0$ and its space-translates $\tau_x A$ yield the same fluctuation field: $B_\omega(A) = B_\omega(\tau_x A)$, $x \in \mathbb{Z}^\nu$.

Here we call A and B in $\mathcal{B}_0$ equivalent, denoted by $A \sim B$ if

$$s_\omega(A - B, A - B) = 0 \tag{3.11}$$

It is easily checked that (3.11) defines an equivalence relation on $\mathcal{B}_0$.

Remark also that e.g. $A \sim \tau_x A$ for all $x \in \mathbb{Z}^\nu$.

Consider now the fluctuation algebra $W(\mathcal{B}_0, \sigma_\omega)$, the associated state ω_s and the induced GNS-triplet $(\mathcal{H}, \pi, \Omega)$:

$$\omega_s(W(X)) = (\Omega, \pi(W(X))\Omega) \quad ; X \in \mathcal{B}_0$$

Denote by $\mathcal{M}$ the von Neumann algebra generated by the representation $\pi(W(\mathcal{B}_0, \sigma_\omega))$ and by $\mathcal{M}'$ its commutant. The property of coarse graining is characterized in the following theorem.

Theorem 3.3 *Let $A, B \in \mathcal{B}_0$, then the following is equivalent*

(i) $A \sim B$

(ii) $\pi(W(A)) = \pi(W(B))$

Proof: Suppose first (ii) is satisfied, then

$[\pi(W(A)), \pi(W(B))] = 0$ and hence $\sigma_\omega(A, B) = 0$

Further

$$\begin{aligned} 1 &= \omega_s(W(A)W(B)^*) &= \omega_s(W(A)W(-B)) \\ &= \omega_s(W(A - B)) &= e^{-\frac{1}{2}s_\omega(A-B,A-B)} \end{aligned}$$

or $s_\omega(A - B, A - B) = 0$, proving (i). Conversely, suppose $A \sim B$ then

$$\frac{1}{4}|\sigma_\omega(A - B, X)|^2 \leq s_\omega(A - B, A - B)s_\omega(X, X)$$

implies that $\sigma_\omega(A - B, X) = 0$ for all $X \in \mathcal{B}_0$, i.e. $\pi(W(A - B)) \in \mathcal{M}'$.

Also

$$\begin{aligned} \|(\pi(W(A - B)) - \mathbb{1})\Omega\|^2 &= \omega_s((W(A - B)) - \mathbb{1})(W(A - B) - \mathbb{1})^*) \\ &= 2 - \omega_s(W(A - B)) - \omega_s(W(B - A)) \\ &= 0 \end{aligned}$$

As Ω is cyclic for $\mathcal{M}$, it is separating for $\mathcal{M}'$. Hence

$$\pi(W(A-B)) = \mathbb{1}$$

or

$$\pi(W(A)) = \pi(W(B))$$

■

Denote by $[\mathcal{B}_0]$ the equivalence classes of $\mathcal{B}_0$ with respect to the equivalence relation (3.11). This theorem proves that the central limit can be considered as a one-to-one map of the elements ξ of $[\mathcal{B}_0]$ onto the elements $\pi(W(\xi))$. This is the precise mathematical description of the notion of coarse graining. Furthermore remark also that under the central limit all correlations for the fluctuations vanish. This fact is mathematically expressed by π being a quasi-free representation of the fluctuation algebra.

By taking $\alpha = \tau_x$ $(x \in \mathbb{Z}^\nu)$, it follows from Theorem (3.2) that there exist quasi-free automorphisms $\tilde{\tau}_x$ of the fluctuation algebra such that

$$\tilde{\tau}_x(W(A)) = W(\tau_x A) \ ; A \in \mathcal{B}_0$$

Moreover as $A \sim \tau_x(A)$ it follows from theorem (3.3) that

$$\pi(\tilde{\tau}_x W(A)) = \pi(W(\tau_x A)) = \pi(W(A)) \tag{3.12}$$

and $\omega_s \circ \tilde{\tau}_x = \omega_s$ for all $x \in \mathbb{Z}^\nu$.

By the GNS-theorem, there exists unitaries $\tilde{U}_x$ such that

$$\pi(\tilde{\tau}_x W(A)) = \tilde{U}_x \pi(W(A)) \tilde{U}_x^* \quad , \tilde{U}_x \Omega = \Omega.$$

But by (3.12) : $\tilde{U}_x = \mathbb{1}$, i.e. the fluctuation algebra belongs to the translation invariant subspace. In particular

$$\tilde{\tau}_x B_\omega(A) = B_\omega(A)$$

In physical language one says that the fluctuations belong to the $k = 0$ mode.

4 Central limit theorem for mixing quantum systems

Above we introduced the quasi-local system $(\mathcal{B}, \omega)$ and gave the definition of normal fluctuations of this system. The conditions (3.5) and (3.6) state that the characteristic function $\omega(\exp it\tilde{A}^n)$ must converge pointwise to a gaussian functional on a $*$-subspace $\mathcal{B}_{0,sa}$. In probability theory one says that the central limit theorem holds for all random variables $A \in \mathcal{B}_{0,sa}$.

In this section, first we set up a reasonable set of clusterconditions and then we actually prove the existence of the limit (3.6).

Difficulties due to the quantum mechanical nature of the system arise because already the set of observables $\{\tau_x(A) \mid x \in \mathbb{Z}^\nu\}$ consists in general of non-commuting observables. Directly related to the non-commutativity of the system is the convergence of products of exponentials :

$$\lim_{n\to\infty} \omega(e^{i\tilde{A}^n}\, e^{i\tilde{B}^n} \cdots)\ ;\ A, B \in \mathcal{B}_{0,sa} \tag{4.1}$$

Obviously in the commutative case, the limit in (4.1) converges if the limit in (3.6)converges, because

$$\lim_{n\to\infty} \omega(e^{i\tilde{A}^n}\, e^{i\tilde{B}^n} \cdots) = \lim_{n\to\infty} \omega(e^{i(\widetilde{A+B+\cdots})^n})$$

The main theorem in this section states that a similar result holds in the non-commutative case. We prove that the L^1-clustering conditions (see
[13; Theorem 4.1]) are also sufficient to get the limit (4.1). It is clear that we can limit ourselves to the case of products of two. The general case is then obtained by induction.

First we introduce these clusterconditions. Let ω be a translation invariant state on the quasi-local algebra $\mathcal{B}$. Denote :

$$\alpha^\omega(\Lambda, \tilde{\Lambda}) = \sup_{\substack{A\in\mathcal{A}_\Lambda;\|A\|=1\\ B\in\mathcal{A}_{\tilde{\Lambda}};\|B\|=1}} |\,\omega(AB) - \omega(A)\omega(B)\,| \tag{4.2}$$

$$\alpha^\omega_N(d) = \sup_{\Lambda,\tilde{\Lambda}}\{\alpha^\omega(\Lambda,\tilde{\Lambda})\ ;\ \mathrm{dist}(\Lambda,\tilde{\Lambda}) \geq d \ \text{ and } \ \max\,(|\Lambda|, |\tilde{\Lambda}|) \leq N\}$$

Obviously $\alpha^\omega_N(d)$ has the properties :

$$\begin{aligned}\alpha^\omega_N(d) &\leq \alpha^\omega_N(d') \ \text{ for } N \in \mathbb{N}_0 \text{ and } d \geq d'\\ \alpha^\omega_N(d) &\leq \alpha^\omega_{N'}(d) \ \text{ for } d \in \mathbb{R}_0 \text{ and } N \leq N'\end{aligned}$$

In order to avoid to much technicalities we will prove the theorem for
$\mathcal{B}_0 = \mathcal{A}_L$, the local observables. The clusterconditions are :

CL 1 : ω is a L^1-clustering state of $\mathcal{A}_L$:

$$\sum_{x\in\mathbb{Z}^\nu} |\omega(A\tau_x B) - \omega(A)\omega(B)| < \infty \quad A, B \in \mathcal{A}_L$$

CL 2 : for all $A = A^* \in \mathcal{A}_L$ there exists an $\epsilon > 0$ such that

$$\lim_{n\to\infty} \frac{1}{|\Lambda_n|^{1-\epsilon}} \omega((\tilde{A}^n)^4) < \infty$$

CL 3 : there exists a $\delta > 0$ such that:

$$\lim_{N\to\infty} N^{1/2} \alpha_N^\omega(N^{\frac{1}{2\nu}-\delta}) = 0$$

Now we formulate our main result :

Theorem 4.1 *If the state ω satisfies the clusterconditions (CL 1,2,3), then for all $A, B \in \mathcal{A}_{L,sa}$*

$$\begin{aligned}\lim_{n\to\infty} \omega(e^{i\tilde{A}^n} e^{i\tilde{B}^n}) &= \lim_{n\to\infty} \omega(e^{i(\widetilde{A+B})^n}) e^{-\frac{1}{2}\omega([\tilde{A}^n, \tilde{B}^n])} \\ &= \exp\{-\frac{1}{2} s_\omega(A+B, A+B) - \frac{i}{2}\sigma_\omega(A, B)\}\end{aligned}$$

where the notation of section 3 is used.

In particular the state ω has normal fluctuations.

Before proceeding to the proof of the theorem we make some comments on the conditions. Condition CL1 is clearly necessary, it guarantees the existence of the second moment. It is certainly not typical for the quantum mechanical nature of the system. Condition CL2 indicates that the existence of the fourth moment (i.e. $\epsilon = 1$) is not required. In fact CL2 is a very weak condition because by CL1

$$\begin{aligned}\lim_{n\to\infty} \frac{\omega((\tilde{A}^n)^4)}{|\Lambda_n|} &\leq \lim_{n\to\infty} \frac{\|\tilde{A}^n\|^2}{|\Lambda_n|} \omega((\tilde{A}^n)^2) \\ &= \lim_{n\to\infty} \omega((\tilde{A}^n)^2) \|A\| < \infty\end{aligned}$$

About CL3 remark that in commutative probability theory, one uses a uniform clusterfunction. Translated in our notation this means that the function $\alpha_N^\omega(.)$ is independent of N, or that the clustering depends only on the separation between the two volumes and not on their shape. This situation is also encountered in [17], where $\alpha^\omega(.)$ is called the modulus of decoupling. It is equivalent to

$$\alpha^\omega(d) = O\left(\frac{1}{d^{\nu+\delta}}\right), \delta > 0 \tag{4.3}$$

implying that $\alpha^\omega(d) \in L^1(\mathbb{R}^+, dx)$; $\nu = 1, 2, 3, \cdots$

This corresponds to the uniform mixing condition in the commutative case [18; Theorem 18.5.4]. In general however one can not expect uniform clustering and therefore condition (4.3) is replaced by CL3.

Proof of Theorem 4.1: We have to prove two equalities. The first consists in showing that the limit of products of exponentials reduces essentially to the limit of one exponential. The limit of the latter one is proved to exist in [13;Theorem 4.1].Hence essentially we have to prove here the first equality. As far as this is concerned, it will be clear from the proof that it can be done mostly by norm estimations.

As A and B are strictly local observables, we may assume that $A, B \in \mathcal{A}_{\Lambda_N}$ where Λ_N is a cube centered around the origin with edges of length $2N+1$. Also without loss of generality we assume that $\omega(A) = \omega(B) = 0$.In particular we prove the following two formulae:

a) $\lim_{n\to\infty} \| e^{i\tilde{A}^n} e^{i\tilde{B}^n} - e^{i(\tilde{A}^n+\tilde{B}^n)} e^{-\frac{1}{2}[\tilde{A}^n,\tilde{B}^n]} \| = 0$

b)
$$\begin{aligned}
\lim_{n\to\infty} \omega(e^{i(\tilde{A}^n+\tilde{B}^n)}) e^{-\frac{1}{2}[\tilde{A}^n,\tilde{B}^n]}) &= \lim_{n\to\infty} \omega(e^{i(\tilde{A}^n+B^n)}) e^{-\frac{1}{2}\omega([\tilde{A}^n,\tilde{B}^n])} \\
&= \lim_{n\to\infty} \omega(e^{i(\widetilde{A+B})^n}) e^{-\frac{1}{2}\sigma_\omega(A,B)}
\end{aligned}$$

Proof of a):

For $n \in \mathbb{N}_0$, let

$$\begin{aligned}
p(n) &= [\log(2n+1)] \\
k(n) &= [\frac{2n+1}{2p(n)}]
\end{aligned}$$

where $[x]$ denotes the integer part of $x \in \mathbb{R}^+$. For A and B, consider:

$$\xi_0^n(A) = \frac{1}{|\Lambda_n|^{1/2}} \sum_{x\in\Gamma_p} \tau_x A$$

where Γ_p is the cube: $\Gamma_p = [-n, -n+p-1]^\nu$, and

$$\xi_i^n(A) = \tau_{2pi}\xi_0^n(A)$$

with $i = (i_1, \cdots, i_\nu) \in \{0, 1, \cdots, k-1\}^\nu$. Furthermore introduce

$$\begin{aligned}
S_0^n(A) &= \sum_i \xi_i^n(A) \\
S_\alpha^n(A) &= \tau_{p\alpha} S_0^n(A)\, ; \; \alpha \in \{0, 1\}^\nu
\end{aligned}$$

and finally

$$\begin{aligned}
C^n(A) &= \sum_\alpha S_\alpha^n(A) \\
R^n(A) &= \tilde{A}^n - C^n(A)
\end{aligned}$$

Clearly $R^n(A)$ is a surface term with a number of translation terms $|R^n|$ bounded by:

$$|R^n| \leq 2\nu p(n)(2n+1)^{\nu-1}$$

The first step in the proof of a) consists in the separation of this surface term from the bulk contribution $C^n(A)$. We prove

$$\begin{aligned} I(\tilde{A}^n, \tilde{B}^n) &\equiv \|e^{i\tilde{A}^n} e^{i\tilde{B}^n} - e^{i(\tilde{A}^n+\tilde{B})^n} e^{-\frac{1}{2}[\tilde{A}^n, \tilde{B}^n]}\| \\ &\leq I(C^n(A), C^n(B)) + I(R^n(A), R^n(B)) + O\left(\frac{\log n}{n}\right) \end{aligned} \quad \text{(a.1)}$$

Using the locality of A and B one computes the bounds

$$\begin{aligned} \|[C^n(A), R^n(B)]\| &\leq \frac{1}{|\Lambda_n|} \sum_{x \in C^n} \sum_{y \in R^n} \|[\tau_x A, \tau_y B]\| \\ &\leq \frac{|R^n|}{|\Lambda_n|} \sum_z \|[A, \tau_z B]\| = O\left(\frac{\log n}{n}\right) \end{aligned}$$

and analogously

$$\|[R^n(A), R^n(B)]\| \leq \frac{|R^n|}{|\Lambda_n|} \sum_z \|[A, \tau_z B]\| = O\left(\frac{\log n}{n}\right)$$

$$\|[C^n(A), C^n(B)]\| \leq \sum_z \|[A, \tau_z B]\| = O(1)$$

$$\|\,[[C^n(A), C^n(B)], R^n(A)]\,\| \leq O\left(\frac{\log n}{n^{\nu/2+1}}\right)$$

Using the bounds

$$\|e^{i(x+y)} - e^{ix}\| \leq \|y\|$$

$$\|[e^{ix}, e^{iy}]\| \leq \|[x, y]\|$$

$$\|e^{i(x+y)} - e^{ix} e^{iy}\| \leq \frac{1}{2}\|[x, y]\|$$

for $x = x^*$, $y = y^*$ elements of a C*-algebra, one gets:

$$\begin{aligned} \|e^{i\tilde{A}^n} e^{i\tilde{B}^n} &- e^{iC^n(A)} e^{iC^n(B)} e^{iR^n(A)} e^{iR^n(B)}\| \\ &\leq \frac{1}{2}\|[C^n(A), R^n(A)]\| + \frac{1}{2}\|[C^n(B), R^n(B)]\| + \|[R^n(A), C^n(B)]\| \\ &\leq O\left(\frac{\log n}{n}\right) \end{aligned}$$

and

$$\begin{aligned} \|e^{i(\tilde{A}^n+\tilde{B}^n)} e^{-\frac{1}{2}[\tilde{A}^n, \tilde{B}^n]} &- e^{iC^n(A+B)} e^{-\frac{1}{2}[C^n(A), C^n(B)]} e^{iR^n(A+B)} e^{-\frac{1}{2}[R^n(A), R^n(B)]}\| \\ &\leq \frac{1}{2}\|[C^n(A+B), R^n(A+B)]\| \\ &+ \frac{1}{2}\|[C^n(A), R^n(B)] + [R^n(A), C^n(B)]\| \\ &+ \frac{1}{8}\|\,[[C^n(A), C^n(B)], [R^n(A), R^n(B)]]\,\| \end{aligned}$$

$$+ \frac{1}{2}\| [R^n(A+B), [C^n(A), C^n(B)]] \| \leq O\left(\frac{\log n}{n}\right)$$

and (a.1) follows immediately.

Now we concentrate on the term $I(C^n(A), C^n(B))$ in (a.1).

We prove

$$I(C^n(A), C^n(B)) \leq O\left(\frac{1}{\log n}\right) + 2^\nu I(S_0^n(A), S_0^n(B)) \tag{a.2}$$

To prove (a.2) one uses the bounds : for $\alpha \neq \beta$

$$\|[S_\alpha^n(A), S_\beta^n(B)]\| \leq O\left(\frac{1}{\log n}\right)$$

$$\| \left[[S_\alpha^n(A), S_\alpha^n(B)], S_\beta^n(A)\right] \| \leq O\left(\frac{1}{(\log n)|\Lambda_n|^{1/2}}\right)$$

$$\| \left[[S_\alpha^n(A), S_\alpha^n(B)], [S_\beta^n(A), S_\beta^n(B)]\right] \| \leq O\left(\frac{1}{|\Lambda_n| \log n}\right)$$

We indicate the proof of this bounds.

Denote by $e = (1, 0, \ldots, 0) \in \{0,1\}^\nu$. Then remark that

$$\begin{aligned} \|[\tau_{p\alpha}\xi_i^n(A), \tau_{p\beta}\xi_j^n(B)]\| &= 0 \quad \text{if } \max\{|i_r - j_r| \,;\, 1 \leq r \leq \nu\} > 1 \\ &\leq \|[\xi_0^n(A), \tau_{pe}\xi_0^n(B)]\| \quad \text{otherwise} \end{aligned}$$

and

$$\|[\xi_i^n(A), \xi_j^n(B)]\| = 0 \qquad \text{if} \qquad i \neq j$$

It follows that, indeed :

$$\begin{aligned} \|[S_\alpha^n(A), S_\beta^n(B)]\| &\leq \sum_{i,j} \|[\tau_{p\alpha}\xi_i^n(A), \tau_{p\beta}\xi_j^n(B)]\| \\ &\leq 3^\nu k^\nu \|[\xi_0^n(A), \tau_{pe}\xi_0^n(B)]\| \\ &\leq \frac{3^\nu k^\nu}{|\Lambda_n|} \sum_{x,y\in\Gamma_p} \|[\tau_x A, \tau_{y+pe} B]\| \\ &\leq \frac{3^\nu k^\nu}{|\Lambda_n|}(2N+1)p^{\nu-1} \sum_x \|[A, \tau_x B]\| \\ &\leq O\left(\frac{1}{p}\right) = O\left(\frac{1}{\log n}\right) \end{aligned}$$

The other bounds are found analogously.

Now by a straightforward but tedious computation

$$\begin{aligned} & I(C^n(A), C^n(B)) \\ \leq\ & \|e^{iC^n(A)}e^{iC^n(B)} - \prod_\alpha e^{iS_\alpha^n(A)}e^{iS_\alpha^n(B)}\| \\ +\ & \|\prod_\alpha e^{iS_\alpha^n(A)}e^{iS_\alpha^n(B)} - \prod_\alpha e^{iS_\alpha^n(A+B)}e^{-\frac{1}{2}[S_\alpha^n(A), S_\alpha^n(B)]}\| \end{aligned}$$

$$
\begin{aligned}
&+ \; \|\prod_{\alpha} e^{iS_\alpha^n(A+B)} e^{-\frac{1}{2}[S_\alpha^n(A),S_\alpha^n(B)]} - e^{iC^n(A+B)} e^{-\frac{1}{2}[C^n(A),C^n(B)]}\| \\
&\leq \; O\left(\frac{1}{\log n}\right) + 2^\nu I(S_0^n(A), S_0^n(B))
\end{aligned}
$$

proving (a.2).

Finally

$$\lim_{n\to\infty} I(S_0^n(A), S_0^n(B)) = 0 \tag{a.3}$$

Indeed, for n large enough and $i \neq j$: $[\xi_i^n(A), \xi_j^n(B)] = 0$, hence

$$
\begin{aligned}
I(S_0^n(A), S_0^n(B)) &= \|\prod_j e^{i\xi_j^n(A)} e^{i\xi_j^n(B)} - \prod_j e^{i\xi_j^n(A+B)} e^{-\frac{1}{2}[\xi_j^n(A),\xi_j^n(B)]}\| \\
&\leq k^\nu I(\xi_0^n(A), \xi_0^n(B))
\end{aligned}
$$

Remark that

$$\|\xi_0^n(A)\| \leq \frac{|\Gamma_p|}{|\Lambda_n|^{1/2}}\|A\| \leq O\left(\frac{p^\nu}{n^{\nu/2}}\right)$$

Using the power series of the exponential function one gets

$$I(S_0^n(A), S_0^n(B)) \leq O\left(\frac{(p^\nu)^2}{n^{\nu/2}}\right)$$

and (a.3) follows.

Finally we concentrate on the restterm contribution in (a.1) and show

$$\lim_{n\to\infty} I(R^n(A), R^n(B)) = 0 \tag{a.4}$$

In order to prove this one proceeds along the same lines of above, substituting $R^n(A)$ and $R^n(B)$ for $\tilde{A}^n$ and $\tilde{B}^n$. Repeating the procedure once, one gets new restterms denoted by $R_2^n(A)$ and $R_2^n(B)$. By iteration after m steps one gets R_m^n bounded by :

$$\|R_m^n(A)\| \leq O\left(\frac{n^{\nu-m}(2p)^m}{|\Lambda_n|^{1/2}}\right) \leq O\left(\frac{(\log n)^m}{n^{m-\nu/2}}\right)$$

For $m = \left[\frac{\nu}{2}\right] + 1$:

$$\lim_{n\to\infty} \|R_m^n(A)\| = 0$$

yielding (a.4).This completes the proof of a).

Finally we prove formula b).

Because of the cluster property CL1, the state ω is ergodic implying that for all $C \in \mathcal{B}$, the following weak operator limit in the GNS-representation induced by ω, yields :

$$\text{weak} - \lim_{\Lambda} C_\Lambda = \omega(C)$$

where $C_\Lambda = \frac{1}{|\Lambda|}\sum_{x\in\Lambda}\tau_x C$.
A straightforward computation gives

$$\lim_\Lambda \omega\left(e^{C_\Lambda}\right) = e^{\omega(C)}$$

For $A, B \in \mathcal{A}_{L,sa}$ one easily computes

$$[\tilde{A}^n, \tilde{B}^n] = \frac{1}{|\Lambda_n|}\sum_{x\in\Lambda_n}\tau_x(\sum_{y\in\mathbb{Z}^\nu}[A, \tau_y B])$$

with $\sum_{x\in\mathbb{Z}^\nu}[A, \tau_y B] \in \mathcal{B}$. Hence

$$\begin{aligned}\lim_{n\to\infty}\omega\left(e^{-\frac{1}{2}[\tilde{A}^n,\tilde{B}^n]}\right) &= e^{-\frac{1}{2}\sum_x \omega([A,\tau_x B])}\\ &= e^{-\frac{1}{2}\sigma_\omega(A,B)}\end{aligned}$$

Then by Schwartz inequality :

$$\begin{aligned}&\left|\omega\left(e^{i(\tilde{A}^n+\tilde{B}^n)}e^{-\frac{1}{2}[\tilde{A}^n,\tilde{B}^n]}\right) - \omega\left(e^{i(\tilde{A}^n+\tilde{B}^n)}\right)e^{-\frac{1}{2}\sigma_\omega(A,B)}\right|^2\\ \leq\ &\omega\left(\left(e^{+\frac{1}{2}[\tilde{A}^n,\tilde{B}^n]} - e^{\frac{1}{2}\sigma_\omega(A,B)}\right)\left(e^{-\frac{1}{2}[\tilde{A}^n,\tilde{B}^n]} - e^{-\frac{1}{2}\sigma_\omega(A,B)}\right)\right)\end{aligned}$$

which completes the proof of b).

■

References

1. Hepp K., Lieb E.H.; Ann. Phys. **76**, 360 (1973)

2. Hepp K., Lieb E.H.; Helv. Phys. Acta **46**, 573 (1974)

3. Wreszinski W.F.; Helv. Phys. Acta **46**, 844 (1974)

4. Cushon C.D., Hudson R.L.; J. Appl. Probability **8**, 454 (1971)

5. Fannes M., Quaegebeur J.; Public. RIMS Kyoto **19**, 469 (1983)

6. Quaegebeur J.; J. Funct. Anal. **56**, 1 (1984)

7. Parthasarathy K.R.; Central Limit Theorems for Positive Definite Functions on Lie groups; Symposia Math. XXI, p. 245, ed. Zanichelli (1977)

8. Giri N., von Waldenfels W.; Z. Wahrscheinl. Verw. Gebiete **42**, 129 (1978)

9. Accardi L., Bach A.; Z. Wahrscheinl. Verw. Gebiete **68**, 393 (1985)

10. Hegerfeldt G.C.; J. Funct. Anal. 64, 436 (1985)

11. Goderis D., Verbeure A., Vets P.; Non-commutative Central Limits; Prob. Theory Rel. Fields, to appear

12. Goderis D., Verbeure A., Vets P.; J. Math. Phys. 29, 2581 (1988)

13. Goderis D., Vets P.;Central Limit Theorem for Mixing Quantum Systems and the CCR-Algebra of Fluctuations; Comm. Math. Phys., to appear

14. Manuceau J., Sirugue M., Testard D., Verbeure A.; Comm. Math. Phys. 32, 231 (1973)

15. Manuceau J., Sirugue M., Rocca F., Verbeure A.; Quasi-free States; Cargèse Lecture Notes in Physics; vol. 4, Ed. D. Kastler, New-York; Gordon-Breach 1970

16. Bratteli O., Robinson D.W.; Operator Algebras and Quantum Statistical Mechanics, vol. I,II; Springer Verlag 1979

17. Haag R., Kadison R.V., Kastler D.; Comm. Math. Phys. 32, 231 (1973)

18. Ibragimov I.A., Linnick Yu.V.; Independent and stationary sequences of Random Variables, Wolters-Noordhoff, 1971.

The paper is in final form and no similar paper has been or is being submitted elsewhere.

AN OPEN PROBLEM IN QUANTUM SHOT NOISE

Gerhard C. Hegerfeldt
Institut für Theoretische Physik, Universität Göttingen
Bunsenstr. 9, D 3400 Göttingen

1. Introduction

Quantum shot noise[1] is a noncommutative generalization of classical shot noise considered in physics and electrical engineering. In the physical situation underlying quantum shot noise "particles" arrive randomly with average frequency ν ("pulse density") at some location, and a particle arriving at time τ contributes a single "pulse" $\varphi(t;\tau)$ which is a time-dependent matrix or operator ("potential"). The arrival times of the particles are governed by a Poisson process. Quantum shot noise is the sum, $V(t)$, of the individual pulses,

$$V(t) = \sum_k \varphi(t;\tau_k)$$

In ordinay shot noise, φ is just a real-valued function. In the following it is assumed, as in Ref. 1, that $\varphi(t;t_1)$ decreases to 0 sufficiently rapidly for $|t_1| \to \infty$.

In applications[2] one encounters the differential equation

$$\frac{d}{dt} U(t,t_o) = V(t)\, U(t) \tag{1.1}$$

with $U(t_o,t_o) = 1$, and one is interested in the expectation $\langle U(t,t_o)\rangle$. For classical shot noise this presents no problem, but in the quantum case the possible noncommutativity of the pulses - and also of a single pulse at different times t - causes difficulties which prevent in general a solution in closed form.

Therefore in Ref. 1 two types of expansions for $\langle U\rangle$ in terms of the pulse density ν were presented. In the first expansion $\langle U\rangle$ is determined from a differential equation,

$$\frac{d}{dt} U(t,t_o) = \sum_{n=1}^{\infty} \nu^n K_n(t,t_o)\, \langle U(t,t_o)\rangle \tag{1.2}$$

where the K_n's are obtained as follows. For fixed $t_1,\ldots,t_n \in \mathbb{R}$, $U(t,t_o;t_1,\ldots,t_n)$ denotes the solution of

$$\frac{d}{dt} U(t,t_o;t_1,\ldots,t_n) = \sum_{k=1}^{n} \varphi(t;t_k)\, U(t,t_o;t_1,\ldots,t_n) \tag{1.3}$$

and $U^K(t,t_o;t_1,\ldots,t_n)$ denotes its K-cumulant with respect to $t_1,\ldots,t_n$ (cf. Appendix). Then one has

$$(1.4)\qquad K_n(t,t_o) = \frac{1}{(n-1)!}\int_{\mathbb{R}^n} d^nt\ \varphi(t;t_1)U^K(t,t_o;t_1,\ldots,t_n)$$

In the second expansion $\langle U\rangle$ is determined from an integral equation,

$$(1.5)\qquad \langle U(t,t_o)\rangle = 1 + \int_{t_o}^{t} ds \sum_{n=1}^{\infty} \nu^n\, G_n(t,s)\ \langle U(s,t_o)\rangle$$

where the G_n's are given as follows. For an index set $\Lambda = \{\lambda_1,\ldots,\lambda_r\}$ one puts $t_\Lambda = \{t_{\lambda_1},\ldots,t_{\lambda_r}\}$. One then defines $U(t,t_o;t_\emptyset) := 1$ and, for $n\geq 1$,

$$(1.6)\qquad U_n^\Delta(t,t_o) = \frac{1}{n!}\int_{\mathbb{R}^n} d^nt \sum_{\Lambda\subset\{1,\cdots,n\}} (-1)^{n-|\Lambda|}\, U(t,t_o;t_\Lambda)$$

where $\Lambda = \emptyset$ is included in the sum. Now one introduces $H_1 := U_1^\Delta$ and $H_n(t,t_o)$ by the recursion relation

$$(1.7)\qquad H_n(t,t_o) = U_n^\Delta(t,t_o) - \sum_{i=1}^{n-1}\int_{t_o}^{t} ds\ H_i(t,s)\,\frac{\partial}{\partial s}\,U_{n-i}^\Delta(s,t_o)$$

Then one finally has

$$(1.8)\qquad G_n(t,s) = -\frac{\partial}{\partial s}H_n(t,s)$$

If the individual pulse is of the form

$$(1.9)\qquad \varphi(t;\tau) = \varphi(t-\tau)$$

then the process is stationary, and the quantities $\langle U\rangle$, U_n^Δ, and G_n depend on difference variables only. In the stationary case Eq.(1.5) is easily solved by Laplace transformation, and the Laplace transform $\hat{G}_n$ of $G(t,o)$ is given by[1]

$$(1.10)\qquad \hat{G}_n = \sum_{n_1+\ldots+n_k=n} (-1)^{k-1}\,\hat{\dot{U}}_{n_1}^\Delta\cdots\hat{\dot{U}}_{n_k}^\Delta$$

where $n_i\geq 1$ and where $\dot{}$ denotes time derivative.

One may now obtain N-th-order approximations to $\langle U\rangle$ by terminating at N the sum over n in Eq.(1.2) and Eq.(1.5), respectively.

In this contribution I would like to pose the following problem: *When is the first-order approximation exact,* either for the differential equation or the integral equation for $\langle U\rangle$? In other words, what are necessary and sufficient conditions on the single pulse $\varphi(t;\tau)$ for this to happen? In the following I will derive sufficient conditions, and I will end with a conjecture about necessary and sufficient ones.

2. Commuting pulses and classical shot noise

In this section the case of commuting smooth pulses will be considered, i.e. it will be assumed that $\varphi(t;t_1)$ and $\varphi(t';t_1')$ commute for all t,t_1 and $t';t_1'$. Clearly classical shot noise is a special case of this.

Proposition 1 In case of commuting (smooth) pulses the first-order approximation to the differential equation for $\langle U\rangle$ is exact, and one has

$$\text{(2.1)} \qquad \langle U(t,t_o)\rangle = \exp\left\{\nu \int dt_1 [U(t,t_o;t_1)-1]\right\}$$

where

$$\text{(2.2.)} \qquad U(t,t_o;t_1) = \exp\int_{t_o}^{t} ds\ \varphi(s;t_1)$$

Proof By commutativity the solution of Eq. (1.3) is given by

$$\text{(2.3)} \qquad U(t,t_o;t_1,\ldots,t_n) = \prod_{k=1}^{n} U(t,t_o;t_k).$$

Then one has

$$\text{(2.4)} \qquad U^K(t,t_o;t_1,t_2) = U(t,t_o;t_1,t_2) - U(t,t_o;t_1)U(t,t_o;t_2) = 0$$

and more generally, by the cluster property of K-cumulants (cf. Appendix) for $n \geq 2$,

(2.5) $U^K(t,t_o;t_1,\ldots,t_n) = 0.$

Hence by Eq. (1.4) one has for $n \geq 2$

(2.6) $K_n(t,t_o) = 0$

so that the first order is exact. From Eq. (1.4), K_1 can be written as

(2.7) $K_1(t,t_o) = \frac{d}{dt}\int dt_1[U(t,t_o;t_1)-1]$

where the subtraction of 1 ensures the convergence of the integral. From Eqs.(2.6) and (2.8) one immediately obtains Eq.(2.1), by commutativity.

It is noteworthy that the first-order approximation to the *integral* equation Eq.(1.5) is in general not exact for commuting pulses.

3. Pulses of δ-like type

In this section pulses are considered which are δ-like in time, i.e. of the form

(3.1) $\varphi(t;\tau) = \phi(\tau)\delta(t-\tau)$

In this case the equation $\dot{U} = VU$ has to be *reinterpreted.* For smooth pulses the solution U can be written, with the time-ordering symbol $\mathcal{T}$, as

(3.2) $$U(t,t_o) = \mathcal{T}\exp\int_{t_o}^{t} ds\, V(s)$$

If the smooth pulses were nonoverlapping in time t this would factor into contributions from individual pulses,

(3.3) $$U(t,t_o) = \prod_k \mathcal{T}\exp\int_{t_o}^{t} ds\, \varphi(s;\tau_k)$$

where the factors of the product are ordered from *right to left* with respect to increasing arrival times τ_k of the pulses. In case of δ-like pulses as in Eq.(3.1) the pulses are indeed nonoverlapping in time, with probability 1, and Eq.(3.3) becomes

$$(3.4) \qquad U(t,t_o) = \prod_k \exp\left\{\phi(\tau_k)\,\chi_{[t_o,t]}(\tau_k)\right\}$$

where the ordering is as in Eq.(3.3) and where χ denotes the characteristic function of a set. One has $\tau_k \neq t_o,t$ with probability 1. Eq.(3.4) means that only pulses arriving in the time interval $[t_o,t]$ contribute to $U(t,t_o)$, and the individual contribution is $\exp\phi(\tau_k)$. In a similar way, $U(t,t_o;\, t_1,\ldots,t_n)$ of Eq. (1.3) has to be reinterpreted. By symmetry one can assume $t_1 > \ldots > t_n$. Then, for $t_i \neq s,t$,

$$(3.5) \qquad \begin{aligned} U(t,s;\, t_1,\ldots,t_n) &= \prod_{i=1}^{n} U(t,s;t_i) \\ &= \prod_{i=1}^{n} \exp\left\{\phi(t_i)\,\chi_{[s,t]}(t_i)\right\} \end{aligned}$$

<u>**Proposition 2**</u> Let the pulses be δ-like in time, as in Eq.(3.1), and let U be as in Eq. (3.4). Then the first-order approximation to the integral equation, Eq. (1.5), is *exact*. One has

$$(3.6) \qquad G_1(t,s) = e^{\phi(s)} - 1$$

and

$$(3.7) \qquad \langle U(t,t_o)\rangle = \mathcal{T} \exp\left\{\nu \int_{t_o}^{t} ds\, [e^{\phi(s)} - 1]\right\} .$$

For the proof the following lemma is needed.

<u>**Lemma 1**</u> If the pulses are δ-like in time then, for $n \geq 2$ and $t_o \leq t$, one has

$$(3.8) \qquad U_n^{\Delta}(t,t_o) = \int_{t_o}^{t} ds\, G_1(t,s)\, U_{n-1}^{\Delta}(s,t_o)$$

<u>**Proof**</u> From Eq. (3.5) one has, for $t_1 > \ldots > t_n$ and $n \geq 1$,

$$(3.9)\qquad \prod_{i=1}^{n} [U(t,t_o;t_i)-1] = \sum_{\Lambda \subset \{1,\dots,n\}} (-1)^{n-|\Lambda|}\, U(t,t_o;t_\Lambda)$$

Now, $U(t,t_o;t_1,\dots,t_n)$ is symmetric in $t_1,\dots,t_n$ for any n, and therefore the integrand of Eq. (1.6), which is the right-hand side of Eq. (3.9), is also symmetric. Hence the integration in Eq. (1.6) can be restricted to the time-ordered domain $t_1>\dots>t_n$ which gives a factor of n! and yields, by Eq. (3.9),

$$(3.10)\qquad U_n^{\Delta}(t,t_o) = \int_{t>t_1>\dots>t_n>t_o} dt_1\dots dt_n \prod_{i=1}^{n} [U(t,t_o;t_i) - 1]$$

From Eq. (3.8) one has, for $t_o < t_i < t$,

$$(3.11)\qquad U(t,t_o;\, t_i) - 1 = e^{\phi(t_i)} - 1$$

Therefore Eq. (3.10) becomes

$$(3.12)\qquad U_n^{\Delta}(t,t_o) = \int_{t_o}^{t} dt_1 \int_{t_o}^{t_1} dt_2 \cdots \int_{t_o}^{t_{n-1}} dt_n \prod_{i=1}^{n} [e^{\phi(t_i)} - 1]$$

Recursion gives, for $n \geq 2$,

$$(3.13)\qquad U_n^{\Delta}(t,t_o) = \int_{t_o}^{t} dt_1\, [e^{\phi(t_1)} - 1]\, U_{n-1}^{\Delta}(t_1,t_o)$$

while for n = 1 one obtains

$$(3.14)\qquad U_1^{\Delta}(t,t_o) = \int_{t_o}^{t} dt_1\, [e^{\phi(t_1)} - 1].$$

From $H_1 = U_1^{\Delta}$ and from Eq. (1.8) one now has

(3.15) $$G_1(t,s) = -\frac{\partial}{\partial s} U_1^{\Delta}(t,s) = e^{\phi(s)} - 1.$$

Inserting this into Eq. (3.13) proves the lemma.

Proof of Proposition 2 Eq.(3.6) for G_1 coincides with Eq. (3.15). It will now be shown that $G_n = 0$ for $n \geq 2$. It suffices to show that the choice $H_1 = U_1^{\Delta}$ and $H_n = 0$ for $n \geq 2$ satisfies Eq. (1.7) since that recursion relation has a unique solution. Inserting this choice into Eq. (1.7) one thus has to verify that for $n \geq 2$

(3.16) $$0 = U_n^{\Delta}(t,t_o) - \int_{t_o}^{t} ds\, H_1(t,s) \frac{\partial}{\partial s} U_{n-1}^{\Delta}(s,t_o)$$

holds. But after a partial integration this becomes Eq. (3.8) since the boundary terms vanish, due to

$$H_1(t,t) = 0$$

and to

$$U_n^{\Delta}(t_o,t_o) = 0$$

where the latter is true generally but in the present case most easily seen from Eq. (3.8).

Inserting for G_n in the integral equation for $\langle U \rangle$, Eq. (1.5), one obtains

(3.17) $$\langle U(t,t_o)\rangle = 1 + \nu \int_{t_o}^{t} ds\, [e^{\phi(s)} - 1]\, \langle U(s,t_o)\rangle.$$

Taking the derivative yields

(3.18) $$\frac{d}{dt} \langle U(t,t_o)\rangle = \nu\, [e^{\phi(t)} - 1]\, \langle U(t),t_o)\rangle.$$

The solution of this differential equation, with initial condition $\langle U(t_o,t_o)\rangle = 1$, is just given by Eq. (3.7). This completes the proof.

Corollary If the pulses are δ-like in time then also the first-order approximation to the differential equation for $\langle U \rangle$ is exact, with K_1 given by

(3.19) $$K_1(t,t_o) = e^{\phi(t)} - 1.$$

<u>Proof</u> This follows immediately from Eq. (3.18).

<u>Note</u> Intuitively, δ-like pulses commute for all times since at any given moment at most a single pulse is present. However, one cannot apply the argument of Section 2 to this case since Eq.(1.4) for K_n does not hold for δ-like pulses, due to the ill-defined nature of the integrand. And indeed, the resulting expression for ⟨U⟩ in Eq. (3.7) differs from that in Eq. (2.1) by a time-ordering symbol.

4. When is the first-order approximation exact: A conjecture

What are necessary and sufficient conditions on $\varphi(t;\tau)$ such that for $n \geq 2$ either all K_n or all G_n vanish? At least for smooth pulses a look at Eq. (1.4) suggests that for $K_n = 0$ one possibly needs, for all $t, t_o, t_1, ... t_n$, $n \geq 2$,

(4.1) $$U^K(t,t_o;t_1,...,t_n) = 0$$

By the recursion relation Eq. (A.1) this entails

$$U(t,t_o;t_1,...,t_n) = U(t,t_o;t_1)\ U(t,t_o;t_2,...,t_n)$$

Thus

$$U(t,t_o;t_1...,t_n) = \prod_{i=1}^{n} U(t,t_o;t_i)$$

and by symmetry in $t_1,...,t_n$ the individual factors commute, suggesting commutativity of the pulses.

If one tries to find models for which $G_n = 0$, $n \geq 2$, one always seems to run into difficulties unless one assumes δ-like pulses. I therefore arrive at the following.

<u>Conjecture</u> The first-order approximation to the differential equation or ⟨U⟩ is exact if and only if the pulses commute or are δ-like in time. The first-order approximation to the integral equation

for $\langle U\rangle$ becomes exact if and only if the pulses are δ-like in time.

A solution of this problem would be useful for practical applications because for a particular model of quantum shot noise a low-order approximation is expected to be the better the closer the model is to one for which the approximation is exact.

Appendix

K-cumulants[4] for symbols $\langle 1\cdots n\rangle$ can be recursively defined by[5] $\langle 1\rangle^K = \langle 1\rangle$ and

$$\text{(A.1)} \qquad \langle 1\cdots n\rangle = \langle 1\cdots n\rangle^K + \sum_{\substack{\Lambda_1 \dot{\cup} \Lambda_2 = \{2,\cdots,n\} \\ \Lambda_2 \neq \emptyset}} \langle 1\Lambda_1\rangle^K \langle \Lambda_2\rangle$$

where Λ_1 may be empty. Thus, e.g.,

$$U^K(t,t_o;t_1,t_2) = U(t,t_o;t_1,t_2) - U(t,t_o;t_1)\,U(t,t_o;t_2)$$

which agrees with the ordinary cumulant. For higher K-cumulants this is no longer true in the noncommutative case.

The *cluster property*[5] is related to statistical independence. Let $\langle 1\cdots n\rangle$ be fixed and let there be an m, $1 \le m < n$, such that whenever

$$\text{(A.2)} \qquad \Lambda_1 \subset \{1,\cdots,m\}\ ,\ \Lambda_2 \subset \{m+1,\cdots,n\}$$

one has

$$\text{(A.3)} \qquad \langle \Lambda_1 \cup \Lambda_2\rangle = \langle \Lambda_1\rangle \langle \Lambda_2\rangle\ .$$

Then it follows that

$$\text{(A.4)} \qquad \langle 1\cdots n\rangle^K = 0$$

References

1) G.C. Hegerfeldt and H. Schulze, Quantum shot noise: Expansions in powers of the pulse density, J. Statist. Physics 51 (1988), 711-728

2) G.C. Hegerfeldt and H. Schulze, Density expansions for the autocorrelation function of spectral-line profiles. Phys. Rev. A (1989, in press)

3) W. von Waldenfels, in *Séminaire des Probabilités IX, Université de Strasbourg,* P.A. Meyer, ed., Lecture Notes in Mathematics 465, Berlin 1975

4) N.G. van Kampen, A cumulant expansion for stochastic linear differential equations, I and II; Physica 74 (1974), 215-238 and 239-247

5) G.C. Hegerfeldt and H. Schulze, Cumulants for stochastic differential equations and for generalized Dyson series. J. Statist. Physics 51 (1988), 691-710.

This paper is in final form and no similar paper has been or is being submitted elsewhere.

A METHOD OF OPERATOR ESTIMATION AND A STRONG LAW OF LARGE NUMBERS IN VON NEUMANN ALGEBRAS

Ewa Hensz

Institute of Mathematics, University of Łódź, ul. S. Banacha 22, 90-238 Łódź, Poland

1. Introduction. In the classical probability theory, a great number of strong limit theorems especially concerning the pairwise orthogonal random variables, can be proved by the scheme as follows:

Let (Ω,F,μ) be a probability space and $(Y_n)_{n\in N}$ be a sequence of random variables from $L_2(\Omega,F,\mu)$. In order to prove that $Y_n\to 0$ a.s. $(n\to\infty)$ it is enough to show that

(i) $\quad Y_{2^n} \to 0$ a.s. $\quad n\to\infty$

and

(ii) $\quad \max_{2^n<m\leq 2^{n+1}} |Y_m - Y_{2^n}| \to 0$ a.s. as $n\to\infty$.

In practice, instead of (i), we often prove that

(i´) $\quad \sum_{n=0}^{\infty} \|Y_{2^n}\|^2 < \infty$

and apply the Beppo Lévy theorem.

On the other hand, a great deal of work has been done lately to generalize various strong limit theorems to the von Neumann algebra context (cf. [5]).

Roughly speaking, in this case, the algebra L_∞ over a probability space (Ω,F,μ) is replaced by an arbitrary von Neumann algebra M of operators acting in a Hilbert space. The role of the measure μ is played by a normal faithful state ϕ on M. The classical Egorov theorem makes it possible to introduce the analogue of the almost sure convergence which is known as the almost uniform convergence of operators.

Many classical results have been extended for traces, i.e. for finite von Neumann algebras. However, for states, the proofs are more complicated and need some new approach (because, in general, the states are not subadditive on the lattice of projections in a von Neumann algebra). One of the possibilities, very useful, in particular, in proving strong laws of large numbers, is a powerful method of operator estimation which follows the classical procedure described above.

The aim of this paper is to give a strong law of large numbers in a von Neumann algebra which, in classical probability, is a simple consequence (via the Kronecker lemma) of the celebrated Rademacher-Menshov theorem. It is worth-while noticing that the analogue of the Rademacher-Menshov theorem does not hold in this context (because even in the classical case the limit of bounded random variables can be unbounded).

The main result of this paper extends strong laws of large numbers given [4,2] and corresponds with applying the general Kronecker lemma to the Rademacher-Menshov theorem in classical case.

2. Preliminaries. Let us begin with some notations and definitions. Throughout the paper M will denote a σ-finite von Neumann algebra of operators with a faithful normal state ϕ. For a projection $p \in M$, always $p^{\perp} = 1 - p$ and $|x| = (x^*x)^{1/2}$ for $x \in M$. We shall often use the inequality

$$(1) \qquad \left|\sum_{k=1}^{n} x_k\right|^2 \leq n \sum_{k=1}^{n} |x_k|^2$$

where $x_1,\dots,x_n \in M$ which is implied by the Schwartz inequality $xy + y^*x \leq x^*x + y^*y$ for $x,y \in M$.

We recall that a sequence (x_n) in M converges almost uniformly to $x \in M$ ($x_n \to x$ a.u.) if, for each $\varepsilon > 0$, there exists a projection $p \in M$ with $\phi(p^{\perp}) < \varepsilon$ and such that $\|(x_n - x)p\| \to 0$ as $n \to \infty$.

We say that a sequence (x_n) in M is orthogonal (relative to the state ϕ) if $\phi(x_k^* x_l) = 0$ for $k \neq l$.

3. The method of operator estimation. In the sequel, we shall need the following theorem which is a special case of the non-commutative maximal ergodic theorem of Goldstein ([1]).

THEOREM 0. Let (a_n) be a sequence of positive operators in M and (ε_n) be a sequence of positive numbers. Then there exists a projection $p \in M$ with

$$\phi(p^{\perp}) \leq 2 \sum_{k=1}^{\infty} \varepsilon_k^{-1} \phi(a_k)$$

and such that $\|pa_n p\| < 2\varepsilon_n$ for all $n \geq 1$.

The method of operator estimation can be expressed as the following

THEOREM 1. Let $(y_n)_{n\in N}$ be a sequence of operators in M such that the following conditions are fulfilled:

(i) $\sum_{n=0}^{\infty} \phi(|y_{2^n}|^2) < \infty$

(ii) there exists a sequence of positive operators $(D_n)_{n\in N}$ in M with

(2) $$\sum_{n=1}^{\infty} \phi(D_n) < \infty$$

such that

(3) $$|y_m - y_{2^n}|^2 \leq D_n \quad \text{for} \quad 2^n < m \leq 2^{n+1}.$$

Then

$$y_n \to 0 \quad \text{a.u.}$$

P r o o f. For each projection $p \in M$, by (1) and (3) we have, for $2^n < m \leq 2^{n+1}$

(4) $$\begin{aligned} \|y_m p\|^2 &\leq 2\|(y_m - y_{2^n})p\|^2 + 2\|y_{2^n}p\|^2 \\ &\leq 2\|p|y_{2^n}|^2 p\| + 2\|p|y_m - y_{2^n}|^2 p\| \\ &\leq 2\|p|y_{2^n}|^2 p\| + 2\|pD_n p\|. \end{aligned}$$

Let $\varepsilon > 0$ be given. By (i) and (2), we can find a sequence (ε_n) of positive numbers such that $\varepsilon_n \to 0$ and

$$\sum_{n=1}^{\infty} \varepsilon_n^{-1}\phi(|y_{2^n}|^2 + D_n) < \frac{\varepsilon}{2}.$$

Then, by Theorem 0, there is a projection $p \in M$ with $\phi(p^{\perp}) < \varepsilon$ and such that

$$\|p|y_{2^n}|^2 p\| < 2\varepsilon_n$$

$$\|pD_n p\| < 2\varepsilon_n, \qquad n \geq 1.$$

Now, by (4) we have

$$\|y_m p\|^2 \le 2\|p|y_{2^n}|^2 p\| + 2\|pD_n p\| < 8\varepsilon_n \to 0 \quad \text{as} \quad m \to \infty.$$

This means that $y_n \to 0$ a.u.

Applying the above method to proving of the almost uniform convergence, we have to, in each case individually obviously, find suitable estimation operators D_n (cf. [4], [2], [3]). In our paper, to use the method just described we exercise the following

PROPOSITION ([4]). Let $z_1, \ldots, z_{2^n}$ be a system of pairwise orthogonal operators in M. Put

$$s_m = \sum_{k=1}^{m} z_k, \qquad 1 \le m \le 2^n.$$

Then there exists in M a positive operator B_n such that

$$|s_m|^2 \le B_n, \qquad 1 \le m \le 2^n$$

and

$$\phi(B_n) \le (n+1)^2 \sum_{k=1}^{2^n} \phi(|z_k|^2).$$

4. Main result. Now we are in position to prove the following strong law of large numbers.

THEOREM 2. Let $(x_n)_{n\in N}$ be a sequence of pairwise orthogonal operators in a von Neumann algebra M with a faithful normal state ϕ. Let $(b_n)_{n\in N}$ be an increasing to infinity sequence of positive numbers such that the sequence $(\log n/b_n)_{n\ge 2}$ is non-increasing. If

$$(5) \qquad \sum_{k=1}^{\infty} \frac{\log^2 k}{b_k^2} \phi(|x_k|^2) < \infty$$

then the averages

$$u_n = \frac{1}{b_n} \sum_{k=1}^{n} x_k$$

converge to 0 almost uniformly as $n \to \infty$. (All the logarithms are of base 2.)

In particular, putting either $b_n = n + 1$ or $b_n = \log n$ or $b_n = n \log n$, we obtain the non-commutative analogue of the Rademacher-Menshov and Moricz-Tandori laws of large numbers, respectively ([4], [2]), cf. also [7], [6]).

P r o o f of theorem. It is enough to show the conditions (i) and (ii) of Theorem 1. In fact, by the orthogonality of (x_n) and properties of the sequence of (b_n) we have

$$\sum_{n=1}^{\infty} \phi(|u_{2^n}|^2) = \sum_{n=1}^{\infty} \frac{1}{b_{2^n}^2} \sum_{k=1}^{2^n} \phi(|x_k|^2)$$

$$\leq \sum_{n=1}^{\infty} \frac{1}{b_{2^n}^2} \cdot \frac{b_{2^n}^2}{\log^2 2^n} \sum_{k=1}^{2^n} \frac{\log^2 k}{b_k^2} \phi(|x_k|^2)$$

$$= \sum_{n=1}^{\infty} \frac{1}{n^2} \left(\sum_{k=1}^{\infty} \frac{\log^2 k}{b_k^2} \phi(|x_k|^2)\right) < \infty$$

by (5).

To show the condition (ii) we shall estimate $|u_m - u_{2^n}|^2$ for $2^n < m \leq 2^{n+1}$. We have, by (1)

$$(6) \qquad |u_m - u_{2^n}|^2 = \left|\left(\frac{1}{b_m} \sum_{k=1}^{m} x_k - \frac{1}{b_{2^n}} \sum_{k=1}^{2^n} x_k\right|^2\right.$$

$$= \left|\left(\frac{1}{b_m} - \frac{1}{b_{2^n}}\right) \sum_{k=1}^{2^n} x_k + \frac{1}{b_m} \sum_{k=2^n+1}^{m} x_k\right|^2$$

$$\leq 2\left(\frac{1}{b_m} - \frac{1}{b_{2^n}}\right)^2 \left| \sum_{k=1}^{2^n} x_k\right|^2 + \frac{2}{b_{2^n}^2} \left| \sum_{k=2^n+1}^{m} x_k\right|^2 .$$

Applying Proposition to the system $x_{2^n+1}, \ldots, x_{2^{n+1}}$ of mutually orthogonal operators we can find a positive operator B_n in M such that

$$(7) \qquad \left| \sum_{k=2^n+1}^{m} x_k\right|^2 \leq B_n \qquad (2^n < m \leq 2^{n+1})$$

and

$$(8) \qquad \phi(B_n) \leq (n+1)^2 \sum_{k=2^n+1}^{2^{n+1}} \phi(|x_k|^2).$$

Now, putting for $n \in N$

(9) $$D_n = \frac{2(b_{2^{n+1}} - b_{2^n})^2}{b_{2^n}^2 b_{2^{n+1}}^2} \left| \sum_{k=1}^{2^n} x_k \right|^2 + \frac{2}{b_{2^n}^2} B_n,$$

we obtain $D_n \in M$, $D_n \geq 0$ and

$$|u_m - u_{2^n}|^2 \leq D_n \quad \text{for} \quad 2^n < m \leq 2^{n+1}$$

by (6) and (7).

It remains to prove

$$\sum_{n=1}^{\infty} \phi(D_n) < \infty.$$

In fact, we have from (8) and (9)

$$\sum_{n=1}^{\infty} \phi(D_n) = \sum_{n=1}^{\infty} \frac{2(b_{2^{n+1}} - b_{2^n})^2}{b_{2^n}^2 b_{2^{n+1}}^2} \sum_{k=1}^{2^n} \phi(|x_k|^2)$$

$$+ \sum_{n=1}^{\infty} \frac{2(n+1)^2}{b_{2^n}^2} \sum_{k=2^n+1}^{2^{n+1}} \phi(|x_k|^2)$$

$$\leq 2 \sum_{n=1}^{\infty} \left(1 - \frac{b_{2^n}}{b_{2^{n+1}}}\right)^2 \cdot \frac{1}{n^2} \sum_{k=1}^{2^n} \frac{\log^2 k}{b_k^2} \phi(|x_k|^2)$$

$$\leq 2 \sum_{n=1}^{\infty} \frac{1}{n^2} \sum_{k=1}^{\infty} \frac{\log^2 k}{b_k^2} \phi(|x_k|^2)$$

$$+ 8 \sum_{n=1}^{\infty} \frac{\log^2 k}{b_k^2} \phi(|x_k|^2) < \infty$$

by (5) which ends the proof.

REFERENCES

[1] M.S. Goldstein, Theorems in almost everywhere convergence in von Neumann algebras (in Russian), J. Oper. Theory 6 (1981), 233--311.

[2] E. Hensz, Orthogonal series and strong laws of large numbers in von Neumann algebras, to appear in Proc. Rome II Quantum Probability Symposium, Lecture Notes in Math., Springer-Verlag.

[3] E. Hensz, The Cesaro means and strong laws of large numbers for orthogonal sequences in von Neumann algebras, to appear.

[4] R. Jajte, Strong limit theorems for orthogonal sequences in von Neumann algebras, Proc. AMS, vol. 94, No. 2, (1985), 229-236.

[5] R. Jajte, Strong limit theorems in non-commutative probability, Lecture Notes in Math., vol. 1110, Springer-Verlag, Berlin-Heidelberg-New York-Tokyo, 1985.

[6] F. Moricz, K. Tandori, Counterexamples in the theory of orthogonal series, Acta Math. Hung. 49, 1-2, 1987, 283-290.

[7] P. Révész, The laws of large numbers, Akadémiai Kiado, Budapest, 1967.

This paper is in final form and no similar paper has been or is being submitted elsewhere.

An analog of the Ito decomposition for multiplicative processes with values in a Lie group.

A.S.Holevo

Steklov Mathematical Institute
Academy of Sciences of the USSR

The structure of a process with independent increments can be described in two different ways: one is the Levi-Khinchin formula for the associated Markov semigroup and the other is the Ito decomposition for the process itself. For a multiplicative process with values in a (nonabelian) Lie group the extension of the Levi-Khinchin formula was found long ago by Hunt (see e.g. [1]). In this note we suggest a formula which can be regarded as an analog of the Ito decomposition.

Let G be a matrix Lie group of n dimensions, $\mathcal{G}$ - its Lie algebra. Fix a basis X_j; $j=1,\dots,n$ in $\mathcal{G}$ and let $x_j(X)$; $j=1,\dots,n$ be the corresponding canonical coordinates in a neighbourhood of the unit matrix I. We choose $h>0$ such that $x_j(X)=0$ outside the neighbourhood $\{X: \|X-I\|<h\}$. For $A\in\mathcal{G}$ the corresponding right derivative of a function f on G is defined by

$$\tilde{A}f(X)=\frac{d}{d\varepsilon}\Big|_{\varepsilon=0} f(X\exp\varepsilon A).$$

Let $\{X_t\}$ be stochastically continuous process with independent stationary multiplicative increments (SCISMI-process) in G and let

$$T_t f(X)=M\{f(X_t)\,|\,X_0=X\},\ t\in\mathbb{R}_+;\ f\in C(G) \quad (1)$$

be the corresponding Markov semigroup on $C(G)$. The Hunt theorem says that the infinitesimal generator $\mathcal{L}$ of $T_t(X)$ is defined on $\tilde{C}_2^0(G)$ and has the form

$$\mathcal{L}f(X) = \sum_{j=1}^{n} a_j \tilde{X}_j f(X) + \sum_{j,k=1}^{n} a_{jk} \tilde{X}_j \tilde{X}_k f(X) +$$
$$+ \int_{G\setminus\{I\}} [f(XY) - f(X) - \sum_{j=1}^{n} \tilde{X}_j f(X) x_j(Y)] \mu(dY), \tag{2}$$

where a_j, $a_{jk} \in R$, $[a_{jk}] \geqslant 0$, and μ is a measure on $G\setminus\{I\}$ with properties specified in [1]. Conversely any generator of this form defines the Markov semigroup, to which corresponds a SCISMI - process $\{X_t\}$ satisfying (1).

Theorem. Let $\{X_t\}$ be a SCISMI - process in G. Then $\{X_t\}$ satisfies the stochastic differential equation

$$dX_t = X_t \Big\{ [A_0 + \frac{1}{2}\sum_{j=1}^{n} A_j^2 + \int_{0<\|Y-I\|<h} (Y - I - \sum_{j=1}^{n} X_j x_j(Y))\mu(dY)]dt +$$
$$+ \sum_{j=1}^{n} A_j dW_{jt} + \int_{0<\|Y-I\|<h} (Y-I)[\Pi(dY dt) - \mu(dY)dt] + \tag{3}$$
$$+ \int_{h\leqslant\|Y-I\|} (Y-I)\Pi(dY dt) \Big\},$$

where $A_j \in \mathcal{G}$; $j = 0,1,\dots,n$, are such that

$$\tilde{A}_0 f = \sum_{j=1}^{n} a_j \tilde{X}_j f, \quad \sum_{j=1}^{n} \tilde{A}_j^2 f = \sum_{j,k=1}^{n} a_{jk} \tilde{X}_j \tilde{X}_k f;$$

W_{jt}; $j = 1,\dots,n$ are independent standard Wiener processes; μ is the measure on $G\setminus\{I\}$ from the Hunt representation (2); $\Pi(dY dt)$ is a Poisson random measure on $G\setminus\{I\}$ such that $M\{\Pi(dYdt)\} = \mu(dY)dt$. Conversely, any solution of (3) is a SCISMI-process with the Markov semigroup defined by (1), (2).

Sketch of proof. To prove the converse statement first note that according to [2] solution of (3) with the nonrandom initial condition $X_0 = X$ exists and is a matrix semimartingale. Then take $f \in \tilde{C}_2^0(G)$ and apply a generalization of the Ito formula as given in [3] to $f(X_t)$. After some calculations one gets

$$f(X_t) = f(X) + \int_0^t \{ \mathcal{L} f(X_s)\, ds + \sum_{j=1}^{n} \tilde{A}_j f(X_s)\, dW_{js} +$$

$$\int_{G \setminus \{I\}} [f(XY) - f(X)][\Pi(dY ds) - \mu(dY) ds]\},$$

whence by (1)

$$T_t f(X) = f(X) + \int_0^t T_s(\mathcal{L} f)(X)\, ds,$$

from which (2) follows.

To prove direct statement one may use the results on "stochastic semigroups" [4] from which it follows that a SCISMI-process $\{X_t\}$ satisfies $dX_t = X_t\, dZ_t$ where $\{Z_t\}$ is a process with stationary independent additive increments in the matrix algebra containing G. Moreover

$$Z_t = \lim_{N \to \infty} \sum_{i=0}^{N-1} \left(X_{t\frac{i}{N}}^{-1} X_{t\frac{(i+1)}{N}} - I \right)$$

in probability. Let $\mathcal{L}$ be the infinitesimal generator of the Markov semigroup corresponding to $\{X_t\}$ and consider the solution $\{\hat{X}_t\}$ of the equation (3) corresponding to $\mathcal{L}$. Write it as $d\hat{X}_t = \hat{X}_t\, d\hat{Z}_t$. By the converse statement $\{\hat{X}_t\}$ and $\{X_t\}$ are stochastically equivalent so are $\{\hat{Z}_t\}$ and $\{Z_t\}$. Therefore $\{Z_t\}$ is the additive process of the required form and $\{X_t\}$ satisfies the equation of the type (3).

It is noteworthy that although the formula (3) is entirely in the scope of the classical (Kolmogorov) probability theory, the author came to it by considering a problem in quantum probability.

As shown in [5] any noncommutative unitary process $\{U_t\}$ with stationary independent increments satisfies the quantum stochastic differential equation

$$dU_t = U_t\{dA_t(\underline{\ell}) - dA_t^{+}(\underline{W}\underline{\ell}) + d\Lambda_t(\underline{W}-I) + + (iH - \langle \underline{\ell}, \underline{\ell} \rangle)\, dt\}, \tag{4}$$

where $(\underline{\ell} \in M_d \otimes \mathcal{H})$ $\underline{W} \in M_d \otimes \mathcal{L}(\mathcal{H})$, $H \in M_d$, $\underline{W}$ is unitary, H is hermitean, M_d is the algebra of all complex $d \times d$ -matrices, $\mathcal{L}(\mathcal{H})$ is the algebra of bounded operators in a Hilbert space $\mathcal{H}$. It was noticed in [5] that any SCISMI-process with values in the Lie group $\mathcal{U}_d$ of unitary $d \times d$ -matrices is a particular case, and thus it satisfies (4). However the concrete form of $\mathcal{H}$, $\underline{W}$, $\underline{\ell}$ for this case was not specified. It can be shown that one can take

$$\mathcal{H} = \mathbb{C}_n \oplus L^2(\mathcal{U}_d \setminus \{I\}, \mu);\ \underline{W} = [I, \dots, I] \oplus \{Y\};$$
$$\underline{\ell} = [L_1, \dots, L_n] \oplus \{Y^* - I\},$$

where $L_j \in M_d$ are hermitean and $\{\psi(Y)\}$ denotes the function on $\mathcal{U}_d \setminus \{I\}$ taking value $\psi(Y) \in M_d$ at the point Y. A calculation then shows that (4) can be transformed to

$$dU_t = U_t\{[iH - \tfrac{1}{2}\sum_{j=1}^{n} L_j^2 - \tfrac{1}{2}\int_{\mathcal{U}_d \setminus \{I\}} (Y-I)^*(Y-I)\mu(dY)]\,dt + + i\sum_{j=1}^{n} L_j\, dQ_{jt} + \int_{\mathcal{U}_d \setminus \{I\}} (Y-I)[N(dY dt) - \mu(dY)dt], \tag{5}$$

where $dQ_j = i^{-1}(dA_j - dA_j^{+})$ in $\Gamma(L^2(\mathbb{R}_+) \otimes \mathbb{C}_n)$ and $N(E \times dt) = d\Lambda_t(1_E) + dA_t(1_E) + dA_t^{+}(1_E) + \mu(E)dt$ for $E \in \mathcal{B}(\mathcal{U}_d \setminus \{I\})$ in $\Gamma(L^2(\mathbb{R}_+) \otimes L^2(\mathcal{U}_d \setminus \{I\}, \mu))$. Since Q_j; $j = 1, \dots, n$ are stochastically equivalent to independent standard Wiener processes W_j and N is equivalent to the Poisson measure Π, the relation (5) is a particular case of (3) for $G = \mathcal{U}_d$

ovided that the last integral in (3) disappears for h large ough due to the compactness of $\mathcal{U}_d$ and

$$\frac{1}{2}\int_{\mathcal{U}_d\setminus\{I\}}(Y-I)^*(Y-I)\mu(dY)=\int_{\mathcal{U}_d\setminus\{I\}}(Y-I-\sum_{j=1}^{n}X_j x_j(Y))\mu(dY)+iH',$$

here $H'=i^{-1}\int_{\mathcal{U}_d\setminus\{I\}}(\sum X_j x_j(Y)+\frac{Y^*-Y}{2})\mu(dY)$ is hermitean by nitarity of Y.

The author acknowledges fruitful discussions with Prof. W.von aldenfelds, Dr.A.Barchielli, and Dr. M.Schurmann.

References

1 Heyer H. Probability measures on locally compact groups. Berlin etc.: Springer 1977.

2 Emery M. Stabilite der solutions de equations differentielles stochastiques: applications aux integrales multiplicatives stochastiques. Z. Wahr. (1978), b 41, 241-262.

3 Ikeda N.,Watanabe S.,Stochastic differential equations and diffusion processes.Amsterdam etc.North Holland,1931

4 Skorohod A.V. Operator stochastic differential equations and stochastic semigroups. UMN, (1982), v. 37, № 6, 157-183 (In Russian).

5 Schurmann M. Noncommutative stochastic processes with independent and stationary increment satisfy quantum stochastic differential equations. Preprint № 478, Heidelberg University, 1988.

STOCHASTIC DILATIONS OF QUANTUM DYNAMICAL SEMIGROUPS USING ONE-DIMENSIONAL QUANTUM STOCHASTIC CALCULUS

R L Hudson and P Shepperson*
Department of Mathematics, University of Nottingham
Nottingham NG7 2RD, England

Abstract

We show that every norm continuous one-parameter semigroup $(\mathcal{T}_t: t \in \mathbb{R}_+)$ of unital ultraweakly continuous completely positive maps on the algebra $\mathcal{B}(\mathcal{H}_0)$ of bounded operators on the separable infinite dimensional Hilbert space $\mathcal{H}_0$ can be expressed as

$$\mathcal{T}_t(x) = \mathbb{E}_0[j_t(x)] \qquad (x \in \mathcal{B}(\mathcal{H}_0),\ t \geq 0)$$

where $(j_t: t \in \mathbb{R}_+)$ is a quantum diffusion over $\mathcal{B}(\mathcal{H}_0)$ constructed using only one dimensional quantum stochastic calculus. In general the diffusion is not inner.

§1. Introduction

One of the successful applications of quantum stochastic calculus has been to the construction of dilations of quantum dynamical semigroups[1,6]. In [5] it was observed that, if U is the unique unitary process which solves the quantum stochastic differential equation

$$dU = U((w-1)\,d\Lambda + l\,dA^\dagger - l^* w\,dA + (ih - \tfrac{1}{2}l^* l)\,dt), \qquad U(0) = 1, \tag{1.1}$$

where w, l, h are bounded operators in the initial space $\mathcal{H}_0$ with w unitary and h self-adjoint, then, if $\mathcal{T}_t(x)$ denotes the vacuum conditional expectation

$$\mathcal{T}_t(x) = \mathbb{E}_0[U(t)x \otimes 1 U^{-1}(t)] \qquad (x \in \mathcal{B}(\mathcal{H}_0),\ t \geq 0),$$

$(\mathcal{T}_t: t \geq 0)$ is a norm continuous one-parameter semigroup of unital, ultraweakly continuous, completely positive maps (that is a quantum dynamical semigroup) over $\mathcal{B}(\mathcal{H}_0)$ of which the infinitesimal generator is

$$\tau(x) = i[h,x] - \tfrac{1}{2}(l^* l x - 2l^* x l + x l^* l). \tag{1.2}$$

Note that choice of the gauge term in (1.1) has no bearing on the generator (1.2).

More generally, in [6] a unitary process was constructed which provides a similar stochastic dilation of the general norm continuous quantum dynamical semigroup over $\mathcal{B}(\mathcal{H}_0)$. When the infinitesimal generator is expressed in the Lindblad form

$$\tau(x) = i[h,x] - \tfrac{1}{2}\sum_j (l_j^* l_j x - 2l_j^* x l_j + x l_j^* l_j) \tag{1.3}$$

the corresponding unitary process may be regarded heuristicly as the solution of the stochastic differential equation

$$dU = U\left(\sum_j (l_j\,dA_j^\dagger - l_j^*\,dA_j) + \left\{ih - \tfrac{1}{2}\sum_j l_j^* l_j\right\} dt\right), \qquad U(0) = 1 \tag{1.4}$$

where the $A_j^\dagger$ and A_j are components of multidimensional creation and annihilation process whose dimension is the (possibly infinite) number of dissipative terms occurring in the generator (1.3). Again

*PS is supported by an SERC Research Studentship.

the construction of the process (1.4) is not canonical, both because the form (1.3) of the generator is far from unique and because multidimensional gauge terms generalising that of (1.1) could also have been included.

The purpose of this note is to show that, provided $\mathcal{H}_0$ is infinite dimensional, there is a simple construction of a stochastic dilation of the general norm continuous quantum dynamical semigroup over $\mathcal{B}(\mathcal{H}_0)$ using only one-dimensional quantum stochastic calculus—in other words only the gauge, creation and annihilation processes occurring in (1.1). The price paid for the simplicity of the construction is that it is at the algebraic level rather than at the level of unitary processes. In other words we construct a quantum diffusion* in the sense of [4] which (unless the generator can be expressed in the form (1.2)) is not inner.

§2. A canonical form for the infinitesimal generator of a quantum dynamical semigroup.

The standard form for the infinitesimal generator τ of a norm continuous quantum dynamical semigroup over $\mathcal{B}(\mathcal{H}_0)$ is

$$\tau(x) = \Psi(x) + K^*x + xK \qquad (x \in \mathcal{B}(\mathcal{H}_0)) \tag{2.1}$$

where Ψ: $\mathcal{B}(\mathcal{H}_0) \to \mathcal{B}(\mathcal{H}_0)$ is completely positive and K is an element of $\mathcal{B}(\mathcal{H}_0)$ satisfying

$$\Psi(1) + K^* + K = 0. \tag{2.2}$$

According to Stinespring's theorem Ψ can be expressed as

$$\Psi(x) = v^*\pi(x)v$$

where π is a representation of $\mathcal{B}(\mathcal{H}_0)$ in a Hilbert space $\mathcal{K}$ and v is a bounded operator from $\mathcal{H}_0$ to $\mathcal{K}$. Then

$$\Psi(1) = v^*v$$

and (2.1) can be rewritten using (2.2) as

$$\tau(x) = \mathrm{i}[h,x] - \tfrac{1}{2}(v^*vx - 2v^*\pi(x)v + xv^*v) \tag{2.3}$$

where h is a self-adjoint element of $\mathcal{B}(\mathcal{H}_0)$.

The general representation π of $\mathcal{B}(\mathcal{H})$ is unitarily equivalent to a direct sum $\pi_1 + \pi_2$, where π_1 is the ampliation

$$\pi_1(x) = x \otimes 1_{\mathfrak{K}} \qquad (x \in \mathcal{B}(\mathcal{H}_0))$$

from $\mathcal{H}_0$ to a Hilbert space tensor product $\mathcal{H} \otimes \mathfrak{K}$ in which (by restricting ourselves to the closure of the range of v if need be) we can assume that $\mathfrak{K}$ is separable, and the representation π_2 vanishes on compact operators in $\mathcal{B}(\mathcal{H}_0)$. An argument of Lindblad [8] shows that the generator τ of a norm-continuous semigroup of ultraweakly continuous maps is itself ultraweakly continuous. Since each of the maps $x \mapsto \tau(x), -\mathrm{i}[h,x], \frac{1}{2}v^*vx, \frac{1}{2}xv^*v, -v^*\pi_1(x)v$, so too is their sum $x \mapsto v^*\pi_2(x)v$. Since every element of $\mathcal{B}(\mathcal{H}_0)$ can be approximated ultraweakly by compact operators, on which π_2 vanishes, the term $l^*\pi_2(x)l$ vanishes, so that (2.3) can be rewritten

$$\tau(x) = \mathrm{i}[h,x] - \tfrac{1}{2}(v^*vx - 2v^*(x \otimes 1)v + xv^*v). \tag{2.4}$$

Since $\mathcal{H}_0$ and $\mathcal{H}_0 \otimes \mathfrak{K}$ are both infinite dimensional separable Hilbert spaces, we can find a unitary operator w from $\mathcal{H}_0 \otimes \mathfrak{K}$ to $\mathcal{H}_0$. Then, denoting by σ the endomorphism $x \mapsto wx \otimes 1w^*$ of $\mathcal{B}(\mathcal{H}_0)$, and by l the element wv of $\mathcal{B}(\mathcal{H}_0)$, (2.4) assumes the form we require, namely

$$\tau(x) = \mathrm{i}[h,x] - \tfrac{1}{2}(l^*lx - 2l^*\sigma(x)l + xl^*l) \tag{2.5}$$

where h, l are elements of $\mathcal{B}(\mathcal{H}_0)$ with h self-adjoint, and σ is an endomorphism of $\mathcal{B}(\mathcal{H}_0)$.

---

*'Quantum flow' would be a better terminology, however we maintain the established use of 'diffusion' in this paper.

§3 Construction of the stochastic dilation

In [4] the general form of quantum diffusion over $\mathscr{B}(\mathscr{H}_0)$ was found to be governed by the system of stochastic integral equations

$$j_t(x) = x\otimes 1 + \int_0^t [j_s(\sigma(x)-x)\,dA + j_s(lx-\sigma(x)l)\,dA^\dagger$$
$$+ j_s(xl^* - l^*\sigma(x))\,dA + j_s(\tau(x))\,ds] \tag{3.1}$$

where σ is a unital endomorphism of $\mathscr{B}(\mathscr{H}_0)$, l an element of $\mathscr{B}(\mathscr{H}_0)$ and τ is given by (2.5). The existence of a unique quantum diffusion satisfying (3.1) follows from [2]; an explicit construction may be found in [4].

Taking the vacuum conditional expectation in (3.1), we obtain

$$\mathbb{E}_0[j_t(x)] = x\otimes 1 + \int_0^t \mathbb{E}_0[j_s(\tau(x))]\,ds$$

showing that

$$\mathbb{E}_0[j_t(x)] = e^{t\tau}(x) = \mathcal{T}_t(x).$$

Thus the diffusion given by (3.1) gives a stochastic dilation, at the algebraic level, of the general norm-continuous quantum dynamical semigroup of which the infinitesimal generator is given by (2.5).

References

[1] Applebaum D B and Frigerio A, Stochastic dilations of w^*-dynamical systems constructed via quantum stochastic differential equations, in *From local times to global geometry, control and physics*, ed. Elworthy K D, Longman (1986).

[2] Evans M P, Existence of quantum diffusions, Probability Theory and Related Fields, **81**, 473–483 (1989).

[3] Evans M P and Hudson R L, Perturbations of quantum diffusions, to appear in Proc. London Math. Soc.

[4] Hudson R L, Quantum diffusions on the algebra of all bounded operators on a Hilbert space, in *Quantum Probability* IV, ed. Accardi L and von Waldenfels W, Springer L N M (1989).

[5] Hudson R L and Parthasarathy K R, Quantum Itô's formula and stochastic evolutions, Commun. Math. Phys. **93**, 301–321 (1984).

[6] Hudson R L and Parthasarathy K R, Stochastic dilations of uniformly continuous completely positive semigroups, Acta Applicandae Math. **2** 353–379 (1984).

[7] Kadison R V and Ringrose J R, *Fundamentals of the theory of operator algebras*, volume II, Academic Press (1986).

[8] Lindblad G, On the generators of quantum dynamical semigroups, Commun. Math. Phys. **48**, 119–130 (1976).

This paper is in final form and no similar paper has been or is being submitted elsewhere.

SLUGGISH DECAY OF PREPARATION EFFECTS IN LOW TEMPERATURE QUANTUM SYSTEMS

Gert-Ludwig Ingold and Hermann Grabert
Fachbereich Physik, Universität-GHS Essen
Universitätsstraße 5, D-4300 Essen 1, Germany

1. Introduction

Dissipation is a widespread phenomenon in physics as well as in neighboring sciences. It arises from the coupling of the system under investigation to an environment with a large number of degrees of freedom. The familiar approach to discuss dissipative quantum systems, as e.g. in quantum optics or magnetic resonance, is based on master equations and quantum Langevin equations. However, in the last few years the interest has focussed on systems where the coupling to the heat bath may be strong or where the temperature may be very low as e.g. in superconducting devices. The familiar methods fail in these cases. It turned out that the functional integral representation of quantum mechanics provides an appropriate tool to describe such systems.

In the following we will discuss the effects of the preparation of the initial state on the dynamics of the system. Using a functional integral technique which allows for the description of a very large class of initial states we will discuss the time evolution of an exactly solvable model.

As an example we consider squeezed states which are of great interest in connection with quantum measurements near the quantum limit of resolution. It will turn out that there can be qualitative differences between the evolution of a factorizing initial state which neglects correlations between system and heat bath and an initial state where such correlations are present as it is usually the case. Then the relaxation may show algebraic long time tails which are not present for factorizing initial states and which cannot be obtained within a weak coupling theory.

2. The Model

A widely used model for dissipative systems represents the heat bath by harmonic oscillators which are linearly coupled to the system coordinate [1, 2]. The corresponding Hamiltonian

$$H = H_S + H_R + H_{SR} \tag{1}$$

consists of three parts describing the system

$$H_S = \frac{p^2}{2M} + V(q,t), \tag{2}$$

the heat bath or reservoir

$$H_R = \sum_{n=1}^{\infty}\left(\frac{p_n^2}{2m_n} + \frac{1}{2}m_n\omega_n^2 x_n^2\right), \tag{3}$$

and the coupling

$$H_{SR} = -q\sum_{n=1}^{\infty} c_n x_n + q^2 \sum_{n=1}^{\infty} \frac{c_n^2}{2m_n\omega_n^2}. \tag{4}$$

Here we introduced the coordinate q and momentum p of the system, the coordinate x_n, momentum p_n and frequency ω_n of the n-th bath oscillator, and the coupling constant c_n between system and reservoir. The system is subject to an external potential $V(q,t)$ which may be time dependent for $t > 0$. This model is simple enough to be mathematically tractable. Furthermore it can often be used to describe even situations where the coupling between system and heat bath is nonlinear. In this case the model becomes exact if the perturbation of a single bath degree of freedom by the system is only weak and can be treated within a linear response theory. It should be emphasized that due to the large number of degrees of freedom, linear response of the bath does not imply a restriction to weak damping.

In general, one is not interested in the details of the behavior of the heat bath. If one integrates out these degrees of freedom one finds that the relevant quantity describing the heat bath is the spectral density of bath oscillators

$$I(\omega) = \pi \sum_{n=1}^{\infty} \frac{c_n^2}{2m_n\omega_n}\delta(\omega - \omega_n) \tag{5}$$

which is determined by the density of bath modes and the coupling strength between the system and the bath modes. For the reduced classical dynamics of the system one obtains the equation of motion

$$M\ddot{q} + M\int_0^t ds\,\gamma(t-s)\dot{q}(s) + \frac{\partial V(q,t)}{\partial q} = \xi(t) \tag{6}$$

where $\xi(t)$ is a noise term and the damping kernel

$$\gamma(t) = \frac{2}{M}\int_0^{\infty} \frac{d\omega}{\pi}\frac{I(\omega)}{\omega}\cos(\omega s) \tag{7}$$

is uniquely determined by the spectral density $I(\omega)$. This connection between the Hamiltonian (1) and the classical equation of motion (6) allows us to model a dissipative quantum system provided we know the phenomenological classical equation of motion of the system.

While the damping kernel $\gamma(t)$ in general describes damping with memory, it also contains the special case of memoryless or so-called Ohmic damping which is obtained for

$$I(\omega) = M\gamma\omega \tag{8}$$

which corresponds to

$$\gamma(t) = 2\gamma\delta(t). \tag{9}$$

In a real system the spectral density will not diverge as $\omega \to \infty$ but there will be a high frequency cutoff. We call a heat bath Ohmic if for low frequencies the spectral density is proportional to ω because it is the low frequency behavior which dominates the dynamics of the system.

3. Initial States

The dynamics of the system is governed by the Hamiltonian (1). Before calculating the time evolution explicitly, we have to discuss the preparation of an initial state. For $t < 0$ we assume that the total system consisting of the system degree of freedom and the heat bath be in thermal equilibrium at temperature $k_BT = 1/\beta$. It is then described by the density matrix

$$W_\beta = Z_\beta^{-1} \exp(-\beta H) \tag{10}$$

where Z_β is the partition function. At $t = 0$ we prepare the initial state by applying operators O_j and O'_j to W_β which act only in the Hilbert space of the system [2] according to

$$W_0 = \sum_j O_j W_\beta O'_j. \tag{11}$$

In coordinate representation the initial state is then given by

$$W_0(q, x_n, q', x'_n) = \int d\bar{q} \int d\bar{q}'\, \lambda(q, \bar{q}, q', \bar{q}')\, W_\beta(\bar{q}, x_n, \bar{q}', x'_n) \tag{12}$$

with the preparation function

$$\lambda(q, \bar{q}, q', \bar{q}') = \sum_j \langle q|O_j|\bar{q}\rangle\langle\bar{q}'|O'_j|q'\rangle. \tag{13}$$

For $t > 0$ the state propagates according to

$$W(t) = \exp(-\frac{i}{\hbar}Ht) W_0 \exp(\frac{i}{\hbar}Ht). \tag{14}$$

From the definition (11) of the initial state it is clear that in general it contains correlations between the system and the heat bath in contrast to the factorizing initial state [3]

$$W_F(q, x_n, q', x'_n) = \rho_0(q, q') \cdot W_R(x_n, x'_n) \tag{15}$$

where the system is in an arbitrary state ρ_0 and the heat bath is in thermal equilibrium. Such states can also be described within the framework of our more general approach and we can thus compare the dynamics of factorizing and nonfactorizing initial states. We note that although factorizing states are often used because the theory becomes simpler from a mathematical point of view [3,4], this assumption is in most cases somewhat unrealistic since the heat bath usually cannot be decoupled from the system.

Let us now discuss a few examples which are contained in the class of initial states (11). First we can replace the operators $\mathcal{O}_j$ and $\mathcal{O}'_j$ by the identity. Obviously, we then get the equilibrium density matrix as initial state for which we may calculate the time evolution under a time dependent force for $t > 0$. An extension would be to interpret W_β as a constrained equilibrium state under a constant external force F and to discuss the relaxation into the unconstrained equilibrium with $F = 0$. Another possibility is to use projection operators for $\mathcal{O}_j$ and $\mathcal{O}'_j$. We then could make a position measurement and by choosing an appropriate weight function we could construct a wave packet. Below we will discuss the dynamics of a state where the density matrix was projected on a squeezed state. As a last example we mention the state BW_β which in general is not a proper density matrix. If B operates in the Hilbert space of the system only we still can calculate the dynamics and from the expectation value of another system operator A we are able to determine the equilibrium correlation function $\langle A(t)B\rangle_\beta$ which is connected to physically measurable quantities.

4. The Dynamics

We do not want to present the elimination of the heat bath within the functional integral formalism. Rather we only mention the results for the dynamics since the calculation is quite tedious and has been expounded elsewhere [2]. Since the heat bath is harmonic it can be eliminated exactly and after a lengthy calculation one is left with the so-called influence functional which describes the influence of the reservoir on the system. An exact solution for the time evolution of the system is only possible for a few special cases like the harmonic oscillator and the free damped particle. In the rest of the paper we will use for $V(q,t)$ in (2) the time independent harmonic potential

$$V(q) = \frac{1}{2}M\omega_0^2 q^2. \tag{16}$$

Then the dynamics for a Gaussian initial state is fully determined by the first and second moments and we will restrict the following discussion to these quantities. It is convenient to introduce sum and difference coordinates

$$r = \frac{q+q'}{2}, \quad x = q - q'. \tag{17}$$

The initial density matrix of the system then reads

$$\rho_i(x_i, r_i) = \int d\bar{r}d\bar{x}\, \lambda(x_i, r_i, \bar{x}, \bar{r})\rho_\beta(\bar{x}, \bar{r}) \tag{18}$$

where

$$\rho_\beta = Tr_R W_\beta \tag{19}$$

is the equilibrium density matrix traced over the reservoir. The first and second moments can be written in a compact form by introducing

$$q_1(t) = \dot{G}_+(t)q, \tag{20}$$

$$q_2(t) = G_+(t)\frac{p}{M}, \tag{21}$$

$$q_3(t) = [\frac{S(t)}{\langle q^2\rangle} - \dot{G}_+(t)]\bar{r}, \tag{22}$$

$$q_4(t) = -\frac{i}{\hbar}[M\dot{S}(t) + \frac{\langle p^2\rangle}{M}G_+(t)]\bar{x}. \tag{23}$$

We then have for the first moments at time t

$$\langle q\rangle_t = \sum_{\alpha=1}^{4}\langle q_\alpha(t)\rangle_0, \tag{24}$$

$$\langle p\rangle_t = M\sum_{\alpha=1}^{4}\langle \dot{q}_\alpha(t)\rangle_0 \tag{25}$$

where we need to calculate the following first moments at $t = 0$

$$\langle q\rangle_0 = \int dr_i d\bar{x} d\bar{r}\, \lambda(0, r_i, \bar{x}, \bar{r})\, r_i\, \rho_\beta(\bar{x}, \bar{r}), \tag{26}$$

$$\langle p\rangle_0 = \int dr_i d\bar{x} d\bar{r}\, [\frac{\hbar}{i}\frac{\partial}{\partial x_i}\lambda(x_i, r_i, \bar{x}, \bar{r})\rho_\beta(\bar{x}, \bar{r})]\Big|_{x_i=0}, \tag{27}$$

$$\langle \bar{r}\rangle_0 = \int dr_i d\bar{x} d\bar{r}\, \lambda(0, r_i, \bar{x}, \bar{r})\, \bar{r}\, \rho_\beta(\bar{x}, \bar{r}), \tag{28}$$

$$\langle \bar{x}\rangle_0 = \int dr_i d\bar{x} d\bar{r}\, \lambda(0, r_i, \bar{x}, \bar{r})\, \bar{x}\, \rho_\beta(\bar{x}, \bar{r}). \tag{29}$$

For the second moments one obtains

$$\langle q^2\rangle_t = \langle q^2\rangle[1 - \frac{S^2(t)}{\langle q^2\rangle^2}] + \frac{\langle p^2\rangle}{M^2}G_+^2(t) + 2G_+(t)\dot{S}(t)$$

$$+ \sum_{\nu,\mu=1}^{4}\langle q_\nu(t)q_\mu(t)\rangle_0, \tag{30}$$

$$\tfrac{1}{2}\langle pq + qp\rangle_t = \frac{1}{2}M\frac{d}{dt}\langle q^2\rangle_t, \tag{31}$$

$$\langle p^2\rangle_t = \langle p^2\rangle[\dot{G}_+^2(t) + 1] - M^2\frac{\dot{S}^2(t)}{\langle q^2\rangle} + 2M^2\dot{G}_+(t)\ddot{S}(t)$$

$$+ M^2\sum_{\nu,\mu=1}^{4}\langle \dot{q}_\nu(t)\dot{q}_\mu(t)\rangle_0. \tag{32}$$

The initial second moments are defined as obvious generalization of the expressions (26–29) for the first moments. The dynamics of the moments for factorizing initial conditions is obtained from these results by neglecting all moments containing $\bar{r}$ or $\bar{x}$ which is equivalent to vanishing initial correlations between system and reservoir.

In eqs. (20–23) and (30–32) we have made use of the functions $G_+(t)$ and $S(t)$ which represent the Green's function and the symmetrized position autocorrelation function, respectively. In the following section we will discuss their properties. Furthermore we introduced the equilibrium correlations $\langle q^2 \rangle$ and $\langle p^2 \rangle$ which can be obtained from the correlation function $S(t)$ by

$$\langle q^2 \rangle = S(0) \tag{33}$$

$$\langle p^2 \rangle = -M^2 \ddot{S}(0). \tag{34}$$

5. The Position Autocorrelation Function of the Harmonic Oscillator

The explicit evaluation of the correlation function $S(t)$ and the Green's function $G_+(t)$ may proceed along various lines [2,5,6]. The most straightforward way to obtain these quantities is the original method [5] making use of Ehrenfest's theorem and the fluctuation dissipation theorem. The Laplace transform of the Green's function can easily be found from the equation of motion (6) to be

$$\hat{G}_+(z) = \frac{1}{z^2 + z\hat{\gamma}(z) + \omega_0^2}. \tag{35}$$

Since $\hat{G}_+(z)$ does not depend on temperature it always shows classical behavior even in the quantum regime. Therefore the linear response to an external force is classical. While in general it is difficult to explicitly calculate $G_+(t)$ from (35) it can be done for the Ohmic case where the Laplace transform of the damping kernel $\hat{\gamma}(z)$ takes the frequency independent value γ. One gets

$$G_+(t) = \Theta(t)\frac{1}{\zeta}\exp(-\frac{\gamma}{2}t)\sinh(\zeta t) \tag{36}$$

where $\Theta(t)$ is the unit step function which makes $G_+(t)$ a causal Green's function and where

$$\zeta = \sqrt{\frac{1}{4}\gamma^2 - \omega_0^2}. \tag{37}$$

For long times the behavior of $G_+(t)$ is therefore determined by an exponential decay with time constant $\gamma/2$ in the underdamped case ($\gamma < 2\omega_0$) and $\gamma/2 - \zeta$ in the overdamped case.

As a consequence of the fluctuation dissipation theorem, the Laplace transform $\hat{S}(z)$ of the symmetrized position autocorrelation function can be expressed through the Green's function $\hat{G}_+(z)$. Introducing the thermal frequencies

$$\nu_n = \frac{2\pi}{\hbar\beta}n \tag{38}$$

one gets

$$\hat{S}(z) = \frac{1}{\beta M}\sum_{n=-\infty}^{\infty}\frac{z}{\nu_n^2 - z^2}[\hat{G}_+(z) - \hat{G}_+(|\nu_n|)]. \tag{39}$$

This quantity is temperature dependent and its classical and quantum properties are quite different. In the Ohmic case one can again obtain an explicit expression for $S(t)$ which is given by [5]

$$S(t) = \frac{\hbar}{4M\zeta}[\cot(\frac{\beta\hbar\lambda_2}{2})e^{-\lambda_2 t} - \cot(\frac{\beta\hbar\lambda_1}{2})e^{-\lambda_1 t}]$$
$$- \frac{2\gamma}{M\beta}\sum_{n=1}^{\infty}\frac{\nu_n e^{-\nu_n t}}{(\nu_n^2-\lambda_1^2)(\nu_n^2-\lambda_2^2)} \tag{40}$$

where

$$\lambda_{1,2} = \frac{\gamma}{2} \pm \zeta. \tag{41}$$

For high temperatures, i.e. in the classical regime, one again finds exponential decay of the correlation function with the time scale determined by the damping strength γ. As temperature is decreased the thermal frequencies ν_n become smaller and for low temperatures the long time behavior is dominated by the first thermal frequency ν_1. As temperature approaches absolute zero more and more thermal frequencies contribute in the sum in (40) and for $T = 0$ one finds the algebraic long time decay [5, 6]

$$S(t) \sim -\frac{\hbar\gamma}{\pi M\omega_0^4}\frac{1}{t^2}. \tag{42}$$

At finite temperatures this specific behavior can be found on an intermediate time scale before the exponential decay takes over [7]. The observation of these quantum effects may be difficult for very weak damping since the sum in (40) vanishes for $\gamma \to 0$.

6. Coherent and Squeezed States

In the rest of this paper we apply the theory of dissipative quantum systems to four different initial states where we project the equilibrium density matrix onto a squeezed state or where we let a displacement and squeezing operator act on it. In order to introduce the terminology we discuss a few properties of coherent and squeezed states which will be needed in the following [8].

We define the annihilation operator of the undamped quantum harmonic oscillator characterized by the potential (16) as

$$a = \sqrt{\frac{M\omega_0}{2\hbar}}q + i\sqrt{\frac{1}{2\hbar M\omega_0}}p \tag{43}$$

and the creation operator a^+ to be its hermitian conjugate. Then a coherent state $|\alpha\rangle$ is defined to be an eigenstate of a

$$a|\alpha\rangle = \alpha|\alpha\rangle \tag{44}$$

with complex eigenvalue α. This state can be generated from the ground state using the displacement operator

$$\mathcal{D}(\alpha) = \exp(\alpha a^+ - \alpha^* a), \tag{45}$$

i.e.

$$|\alpha\rangle = \mathcal{D}(\alpha)|0\rangle. \tag{46}$$

The coordinate representation of the displacement operator which will be needed later reads

$$\langle q|\mathcal{D}(p_0,q_0)|q'\rangle = \exp(\frac{i}{\hbar}p_0q)\exp(\frac{i}{2\hbar}p_0q_0)\delta(q-q_0-q'). \tag{47}$$

The effect of the displacement operator on the ground state is to displace the state in p–q space to the mean value of the coordinate

$$q_0 = \sqrt{\frac{2\hbar}{M\omega_0}}\,\mathrm{Re}(\alpha) \tag{48}$$

and the mean value of the momentum

$$p_0 = \sqrt{2\hbar M\omega_0}\,\mathrm{Im}(\alpha). \tag{49}$$

This can readily be verified by inspection of the coordinate representation of the coherent state

$$\langle q|\alpha\rangle = (\frac{M\omega_0}{\pi\hbar})^{1/4}\exp[-\frac{M\omega_0}{2\hbar}(q-q_0)^2 + i\frac{p_0}{\hbar}(q+\frac{q_0}{2})]. \tag{50}$$

This state represents a minimum uncertainty state with equal variances in p and q. In order to produce a minimum uncertainty state with different variances one needs the squeezing operator

$$\mathcal{S}(z) = \exp[\frac{z}{2}a^2 - \frac{z^*}{2}a^{+2}] \tag{51}$$

which in coordinate representation is of the form

$$\langle q|\mathcal{S}(z)|q'\rangle = \zeta^{1/2}\delta(\zeta q - q') \tag{52}$$

where we introduced

$$\zeta = \exp(z). \tag{53}$$

In (52) we restricted ourselves to real z which means that the variance σ_{pq} vanishes initially. The squeezed state defined through

$$|\alpha, z\rangle = \mathcal{D}(\alpha)\mathcal{S}(z)|0\rangle \tag{54}$$

has the coordinate representation

$$\langle q|\alpha, z\rangle = (\frac{M\omega_0\zeta^2}{\pi\hbar})^{1/4}\exp[-\frac{M\omega_0}{2\hbar}\zeta^2(q-q_0)^2 + \frac{i}{\hbar}p_0(q+\frac{q_0}{2})]. \tag{55}$$

For $\zeta > 1$ the position uncertainty is reduced at the expense of a larger uncertainty in the momentum. We note that the squeezed state (55) for fixed q_0 and p_0 can be obtained from the coherent state (50) by formally scaling the frequency ω_0 of the harmonic oscillator by a factor of ζ^2. This is due to the fact that the ground state of a given harmonic oscillator represents a squeezed state with respect to a harmonic oscillator with different frequency. On the other hand, one can easily derive the results for coherent states by setting $\zeta = 1$ in the results for squeezed states.

7. Four Special Initial States

The preceding section dealt with coherent and squeezed states for an undamped harmonic oscillator. In real systems the oscillator may be coupled to a heat bath which introduces dissipation. Therefore one has to find an initial condition which replaces the squeezed state of the undamped case. One may think of two different preparations. The first one is a generalization of (54) to the form (11) with $\mathcal{O} = \mathcal{D}(\alpha)\mathcal{S}(z), \mathcal{O}' = \mathcal{O}^+$ where the vacuum is replaced by the equilibrium density matrix. This state reduces to a squeezed state in the limit of vanishing damping and zero temperature. A second possibility is to project the equilibrium density matrix onto a squeezed state. We may now define four different initial states

$$W_i^{(1)} \simeq \mathcal{D}(\alpha)\mathcal{S}(z)W_\beta \mathcal{S}^+(z)\mathcal{D}^+(\alpha), \tag{56}$$

$$W_i^{(2)} \simeq |\alpha, z\rangle\langle\alpha, z|W_\beta|\alpha, z\rangle\langle\alpha, z|, \tag{57}$$

$$W_i^{(3)} \simeq \mathcal{D}(\alpha)\mathcal{S}(z)\rho_\beta \mathcal{S}^+(z)\mathcal{D}^+(\alpha)\cdot W_R, \tag{58}$$

and

$$W_i^{(4)} \simeq |\alpha, z\rangle\langle\alpha, z|\rho_\beta|\alpha, z\rangle\langle\alpha, z|\cdot W_R, \tag{59}$$

where $\simeq$ means equal apart from an appropriate normalization factor. Here, ρ_β denotes the equilibrium density matrix of the damped harmonic oscillator. In coordinate representation this density matrix is given by

$$\rho_\beta(\bar{x}, \bar{r}) = (2\pi\langle q^2\rangle)^{-1/2} \exp\left[-\left(\frac{1}{2\langle q^2\rangle}\bar{r}^2 + \frac{\langle p^2\rangle}{2\hbar^2}\bar{x}^2\right)\right]. \tag{60}$$

$W_i^{(3)}$ ($W_i^{(4)}$) represents the factorizing initial state corresponding to the nonfactorizing initial state $W_i^{(1)}$ ($W_i^{(2)}$). While the initial reduced density matrices are equal, i.e.

$$\rho_i^{(1)} = \rho_i^{(3)} \tag{61}$$

and

$$\rho_i^{(2)} = \rho_i^{(4)}, \tag{62}$$

the dynamics of these states is different as will be shown below. The preparations 1 and 3 were discussed already earlier [2, 9] while preparations 2 and 4 have not been considered as yet.

In the framework of the theory presented in section 3 we need the preparation function λ in order to describe the initial states and their dynamics. For the first initial state (56) we use the coordinate representations of the displacement operator (47) and the squeezing operator (52) and obtain with (13)

$$\lambda^{(1)}(x_i, r_i, \bar{x}, \bar{r}) = \zeta \exp\left(\frac{i}{\hbar}p_0 x_i\right)\delta(\zeta x_i - \bar{x})\delta(\zeta(r_i - q_0) - \bar{r}) \tag{63}$$

where we used sum and difference coordinates as introduced in (17). From (55) we derive the coordinate representation of the projector $|\alpha, z\rangle\langle\alpha, z|$ and obtain

$$\lambda^{(2)}(x_i, r_i, \bar{x}, \bar{r}) = N \exp[-\frac{M\omega_0\zeta^2}{\hbar}[(r_i - q_0)^2 + \frac{x_i^2}{4} + (\bar{r} - q_0)^2 + \frac{\bar{x}^2}{4}]]$$
$$\times \exp[i\frac{p_0}{\hbar}(x_i - \bar{x})] \tag{64}$$

where the normalization factor has to be chosen such that the initial density matrix is normalized to unity, i.e.

$$\int dr_i d\bar{x} d\bar{r}\, \lambda(0, r_i, \bar{x}, \bar{r}) \rho_\beta(\bar{x}, \bar{r}) = 1. \tag{65}$$

The corresponding factorizing initial states are obtained by setting $\bar{x}$ and $\bar{r}$ equals to zero. Thus we have

$$\lambda^{(3)}(x_i, r_i, \bar{x}, \bar{r}) \simeq \lambda^{(1)}(x_i, r_i, \bar{x}, \bar{r}) \delta(\bar{x}) \delta(\bar{r}) \tag{66}$$

and

$$\lambda^{(4)}(x_i, r_i, \bar{x}, \bar{r}) \simeq \lambda^{(2)}(x_i, r_i, \bar{x}, \bar{r}) \delta(\bar{x}) \delta(\bar{r}). \tag{67}$$

Together with the definitions of the moments (24), (25), (30–32) we have all what is needed in order to calculate the dynamics of the four initial states defined in this section.

8. Decay of Coherent and Squeezed States in the Presence of Dissipation

According to (24) and (25) the dynamics of the first moments $\langle q \rangle$ and $\langle p \rangle$ is determined by the initial first moments $\langle q \rangle_0$, $\langle p \rangle_0$, $\langle \bar{r} \rangle_0$, and $\langle \bar{x} \rangle_0$ which can easily be calculated using (26–29). For the first initial state (56) we obtain [2, 9]

$$\langle q \rangle_0^{(1)} = q_0 \tag{68}$$
$$\langle p \rangle_0^{(1)} = p_0 \tag{69}$$
$$\langle \bar{r} \rangle_0^{(1)} = 0 \tag{70}$$
$$\langle \bar{x} \rangle_0^{(1)} = 0. \tag{71}$$

On the other hand, for the second initial state (57) we get

$$\langle q \rangle_0^{(2)} = q_0 \tag{72}$$
$$\langle p \rangle_0^{(2)} = p_0 \tag{73}$$
$$\langle \bar{r} \rangle_0^{(2)} = \frac{q_0}{1 + \dfrac{\hbar}{2M\omega_0\zeta^2\langle q^2 \rangle}} \tag{74}$$
$$\langle \bar{x} \rangle_0^{(2)} = -i\frac{2}{M\omega_0\zeta^2} \frac{p_0}{1 + \dfrac{2\langle p^2 \rangle}{M\hbar\omega_0\zeta^2}}. \tag{75}$$

For the factorizing inital states we again find the result (68–71) since the moments $\langle \bar{r} \rangle_0^{(3,4)}$ and $\langle \bar{x} \rangle_0^{(3,4)}$ vanish per definition.

The result for $\langle q \rangle_0$ and $\langle p \rangle_0$ for all four states is given by the displacements (48) and (49). As far as the first moments are concerned there is no difference between the nonfactorizing preparation 1 and its factorizing counterpart 3. In both cases one finds according to (20), (21), (24), and (25) that the first moments decay classically. This does however not mean that there are no initial correlations between system and heat bath in the initial state 1 as we will see below.

The situation is quite different in the case of the preparations 2 and 4. Here we have nonvanishing moments $\langle \bar{r} \rangle_0$ and $\langle \bar{x} \rangle_0$ for the nonfactorizing case. As a consequence the dynamics of $\langle q \rangle_t^{(2)}$ and $\langle p \rangle_t^{(2)}$ is not only determined by the Green's function $G_+(t)$ but also by the position autocorrelation function $S(t)$. This means that in contrast to the factorizing case we may observe in the evolution of the expectation values of q and p the specific quantum effects discussed in section 5. For instance we have

$$\langle q \rangle_t^{(2)} - \langle q \rangle_t^{(4)} = [\frac{S(t)}{\langle q^2 \rangle} - \dot{G}_+(t)] \frac{q_0}{1 + \frac{\hbar}{2M\omega_0 \zeta^2 \langle q^2 \rangle}} - \frac{1}{\hbar}[M\dot{S}(t) + \frac{\langle p^2 \rangle}{M} G_+(t)] \frac{2}{M\omega_0 \zeta^2} \frac{p_0}{1 + \frac{2\langle p^2 \rangle}{M\hbar\omega_0 \zeta^2}} \tag{76}$$

This difference between nonfactorizing and factorizing initial states is not only quantitative but both states show qualitatively different time behavior at low temperatures. While the factorizing state decays exponentially with a time constant determined by the damping constant γ, one has exponential decay with ν_1 or even algebraic decay $\propto t^{-2}$ for the nonfactorizing state.

The calculation of the second moments according to (30–32) is straightforward. We restrict the discussion to the second moments for preparations 1 and 3 in order to show that there are differences between the two preparations although they do not show up in the first moments.

The nonvanishing second moments for preparation 1 are obtained as [2,9]

$$\langle q^2 \rangle_0 = q_0^2 + \zeta^{-2} \langle q^2 \rangle \tag{77}$$

$$\tfrac{1}{2}\langle pq + qp \rangle_0 = p_0 q_0 \tag{78}$$

$$\langle p^2 \rangle_0 = p_0^2 + \zeta^2 \langle p^2 \rangle \tag{79}$$

$$\langle \bar{r}^2 \rangle_0 = \langle q^2 \rangle \tag{80}$$

$$\langle q\bar{r} \rangle_0 = \zeta^{-1} \langle q^2 \rangle \tag{81}$$

$$\langle p\bar{x} \rangle_0 = \frac{\hbar}{i} \zeta . \tag{82}$$

For preparation 3 the last three moments vanish. From (80–82) one concludes that there are indeed correlations between system and heat bath in the case of the nonfactorizing preparation 1. Therefore the dynamics of this initial state differs from the corresponding factorizing state.

We want to mention another effect concerning squeezed states which is due to damping and cannot be obtained within a weak coupling theory. While the squeezing due to the preparation decays on a certain time scale there exists a static contribution to squeezing which survives even in the limit $t \to \infty$. The reason is that for the harmonic potential (16) one always finds the equilibrium density matrix ρ_β (60) for long enough times. The variances in q and p differ from the corresponding values in the undamped case. Therefore there will be squeezing with respect to the undamped harmonic oscillator with a significant contribution for strong damping.

9. Conclusions

We have discussed the time evolution of squeezed states coupled to an environment. By way of example we showed that the influence of the initial preparation may decay on a time scale much longer than the time scale for the classical relaxation to equilibrium. We have proposed two preparations (2 and 4) with the same reduced density matrix at $t = 0$ where the initial correlations between system and bath influence the average relaxation for extremely long times at low temperatures. In another example (1 and 3) the correlation effects do not show up in the average relaxation of $\langle p \rangle$ and $\langle q \rangle$ but they are apparent in the time evolution of the second moments. These effects cannot be obtained within a weak coupling theory and become more pronounced as temperature is decreased. The commonly used approximation of factorizing initial states is then not necessarily well justified. It is expected that this behavior is of general relevance for dissipative quantum systems and not restricted to the special case considered here.

Acknowledgements

This work was supported by the Deutsche Forschungsgemeinschaft through the Sonderforschungsbereich 237.

References

[1] R. Zwanzig, J.Stat.Phys.**9**,215(1973)
A. O. Caldeira, A. J. Leggett, Ann.Phys.(N.Y.)**149**,374(1983)
[2] H. Grabert, P. Schramm, G.-L. Ingold, Phys.Rep.**168**,115(1988)
[3] R. P. Feynman, F. L. Vernon, Ann.Phys.(N.Y.)**24**,118(1963)
[4] A. O. Caldeira, A. J. Leggett, Physica **A121**,587(1983)
[5] H. Grabert, U. Weiss, P. Talkner, Z.Phys.**B55**,87(1984);
[6] P. S. Riseborough, P. Hanggi, U. Weiss, Phys.Rev.**A31**,471(1985);
F. Haake, R. Reibold, Phys.Rev.**A32**,2462(1985)
C. Aslangul, N. Pottier, D. Saint-James, J.Stat.Phys.**40**,167(1985)
[7] R. Jung, G.-L. Ingold, H. Grabert, Phys.Rev.**A32**,2510(1985)
[8] see e.g. G. J. Milburn, D. F. Walls, Am.J.Phys.**51**,1134(1983);
D. F. Walls, Nature **306**,141(1983)
[9] P. Schramm, H. Grabert, Phys.Rev.**A34**,4515(1986);
P. Schramm, thesis, Universität Stuttgart (1987)

For a more complete list of references related to this subject see [2].

This paper is in final form and no similar paper has been or is being submitted elsewhere.

Almost sure convergence of iterates of contractions in noncommutative L_2-spaces

Ryszard Jajte

Institute of Mathematics of Łódź University

Banacha 22, 90-238 Łódź, Poland

1. Preliminaries

Recently a remarkable progress has been made in the individual ergodic theory of positive contractions in von Neumann algebras [6, 7, 8, 10, 12, 13, 16, 17, 18, 21, 23]. Many pointwise ergodic theorems have been extended to the operator algebra context. The study of such problematics is motivated by the theory of open (irreversible) quantum dynamical systems [2]. From the physical point of view the most important are (semigroups of) completely positive maps on C*- or W*-algebras [14] but in the context of this paper it seems to be more natural to consider a larger class of positive contractions. We shall discuss the asymptotic behaviour of Schwarz maps in von Neumann algebras. More exactly, we are going to prove some results concerning the iterates of contractions in L_2 over a von Neumann algebra M (with respect to a faithful normal state Φ on M). These contractions in $L_2(M,\Phi)$ are generated by some Schwarz maps on M. The result presented here are closely related to [3, 19, 7]. Our main goal is to prove a noncommutative analogue of the well-known classical theorem of Burkholder and Chow on the almost sure convergence of the iterates of two conditional expectations. Our proof is entirely different in comparison with that of Burkholder and Chow in the classical commutative case.

We rather follow some ideas of Gaposhkin [4, 5] and use the techniques developed in [7]. In fact, we prove a more general result concerning two special positive contractions but we do not know a simpler more direct proof for the special case of conditional expectations. It would be interesting to simplify our rather long argument.

Let us begin with some notation and definitions. Throughout the paper, M is a σ-finite von Neumann algebra with a faithful normal state ϕ. We assume that M acts, in a standard way, on the Hilbert space $H = L_2(M,\phi)$ the completion of M under the norm $x \to \phi(x^*x)^{\frac{1}{2}}$ with a cyclic and separating vector Ω such that $\phi(x) = (x\Omega,\Omega)$, for $x \in M$. We put, for $x \in M$, $|x|^2 = x^*x$. We call $\alpha \in L(M)$ a Schwarz map if α satisfies the inequality $|\alpha(x)|^2 \le \alpha(|x|^2)$, for $x \in M$. Note that α is necessarily a contraction in M. A map $\alpha \in L(M)$ is said to be ϕ-contractive if $\phi(\alpha x) \le \phi(x)$, for all $x \in M_+$. A ϕ-contractive Schwarz map in M will be called a kernel. In the sequel the norm in H will be denoted by $\|.\|$ and the norm in M by $\|.\|_\infty$.

For a projection $p \in M$ and $\xi \in H$, we put

$$S_{\xi,p} = \{(x_k) \subset M : \sum_{k=1}^{\infty} x_k\Omega = \xi \text{ in } H \text{ and } \sum_{k=1}^{\infty} x_k p \text{ converges in norm in } M\}$$

and

$$\|\xi\|_p = \inf \{\|\sum_{k=1}^{\infty} x_k p\|_\infty : (x_k) \in S_{\xi,p}\}.$$

We adopt the following definition.

1.1. Definition [7]. A sequence (ξ_n) in $H = L_2(M,\phi)$ is said to be almost surely (a.s.) convergent to $\xi \in H$ if for every $\varepsilon > 0$ there exists a projection $p \in M$ such that $\phi(1 - p) < \varepsilon$

and $\|\xi_n - \xi\|_p \to 0$ as $n \to \infty$. It is easily seen that in the classical commutative case of $M = L_\infty$ (over a probability space) the convergence just defined coincides with the usual almost everywhere convergence (via Egorov s theorem).

2. Maximal Ergodic Lemma

Let α_o be a kernel in M. Then one can extend α_o (in a unique way) to be a contraction α in H (via the formula $\alpha(x\Omega) = \alpha_o(x)\Omega$, for $x \in M$). In this case we shall say that the contraction α in H is generated by a kernel in M. We shall need the following result.

2.1. Theorem (Goldstein s maximal theorem for several kernels). Let $\alpha_1, \alpha_2, \ldots, \alpha_r$ be kernels in M, (ε_k) sequence of positive numbers, $(y_k) \subset M_+$. Set

$$s_n^{(i)} = n^{-1} \sum_{k=0}^{n-1} \alpha_i^k, \qquad \text{for} \quad i = 1,2,\ldots,r, \quad n = 1,2,\ldots$$

Then there exists a projection $p \in M$ such that

$$\phi(1 - p) \le 2 \sum_{n=1}^{\infty} \varepsilon_n^{-1} \phi(y_n)$$

and

$$\|ps_n^{(i)}(y_k)p\|_\infty \le 2\varepsilon_k, \qquad \text{for} \quad n,k = 1,2,\ldots; \quad i = 1,\ldots,r.$$

Proof. This is the trivial extension of Goldstein s result [6]. The proof can be easily obtained (mutatis mutandis) from the proof for one kernel [6, 8]. We shall refer to [8, p. 19-21]. Instead of L defined on page 19, we consider

$$\mathcal{L} = \{(y_{n,k,i}) : 1 \le n,k \le N; \ 1 \le i \le r, \ y_{n,k,i} \in M' \text{ and } \sum_{n,k,i} y_{n,k,i} \le 1\}.$$

Then we put

$$g_n^{(i)} = \sum_{k=1}^{N} k[(s_k^{(i)}(\tilde{x}_n)y_{n,k,i}\Omega,\Omega) - (y_{n,k,i}\Omega,\Omega)]$$

with $\tilde{x}_n = \varepsilon_n^{-1}x_n$, for $i = 1,2,\dots,r$ and set

$$g_n = \sum_{i=1}^{r} g_n^{(i)}, \qquad g(y) = \sum_{n=1}^{N} g_n(y).$$

Then it is enough to repeat the reasoning in [8, p. 19-20]. ▪

The above theorem implies the following

2.2. Maximal Ergodic Lemma. For $i = 1,2,\dots,N$, let $\alpha_{i,o}$ be a kernel in M. Denote by α_i the contraction in H generated by $\alpha_{i,o}$. Put

$$s_n^{(i)} = \frac{1}{n}\sum_{k=0}^{n-1}\alpha_i^k, \qquad \sigma_n^{(i)} = \frac{1}{n}\sum_{k=0}^{n-1}\alpha_{i,o}^k \quad (i=1,2,\dots,N).$$

Then, for every sequence (δ_k) of positive numbers, every $(\xi_k) \subset H$ and every $(a_k) \subset M_+$, there exists a projection $p \in M$ such that

$$\phi(1-p) \leq 4\sum \delta_k^{-1}(\phi(a_k) + \|\xi_k\|^2)$$

and

$$\|p\sigma_n^{(i)}(a_k)p\|_\infty < 2\delta_k$$

and

$$\|s_n^{(i)}(\xi_k)\|_p < 5\delta_k^{1/2}, \qquad \text{for } n,k = 1,2,\dots \text{ and } i = 1,2,\dots,N.$$

Proof. The proof is by reducing to Theorem 2.1. This can be done by a rather easy modification of reasoning in [7, p. 416-417] and [9, p. 137-138]. We omit the details. ▪

3. Main Results

We are in the position to formulate and prove our main results concerning the convergence of the iterates of contractions.

3.1. Theorem. Let α_o and β_o be kernels in M and α and β be contractions in H generated by α_o and β_o, respectively. Let us assume that the contraction α is positive in H (i.e. $(\alpha\xi,\xi) \geq 0$ for $\xi \in H$). Then, for every $\xi \in H$, the sequence $\{\beta\alpha^n\xi\}$ converges almost surely to $\bar{\xi}$ (given by the mean ergodic theorem) as $n \to \infty$.

The following theorem is a noncommutative version of the classical result of Burkholder and Chow ([3]).

3.2. Theorem. Let E and F be two conditional expectations in M preserving ϕ and let P and Q be orthogonal projections in H generated by E and F, respectively. Then, for every $\xi \in H$, the sequence $\{(PQ)^n\xi\}$ converges almost surely to $(P \wedge Q)\xi$ as $n \to \infty$.

Proof. This theorem follows immediately from Theorem 3.1. Indeed, let us put $\alpha = QPQ$ (with $\alpha_o = FEF$) and $\beta = P$. Then $\alpha^n = Q(PQ)^n$, so $\beta\alpha^{n-1} = (PQ)^n$ and it is enough to apply Theorem 3.1. ∎

Let us remark that the limit $\lim_{n\to\infty} (PQ)^n\xi = (P \wedge Q)\xi$ exists in L_2-norm (i.e. in H), for every $\xi \in H$. This is well-known and is also an immediate consequence of a theorem of von Neumann [15, p. 55].

4. Proof of Theorem 3.1.

Let

$$\alpha = \int_o^1 xE(dx) \tag{1}$$

be the spectral decomposition of α, with the spectral measure $E(\cdot)$.

Let us put

$$s_n = \frac{1}{n} \sum_{k=0}^{n-1} \alpha^k, \tag{2}$$

$$\sigma_n = \frac{1}{n} \sum_{k=0}^{n-1} \alpha_o^k$$

and write the formula

$$\alpha^n\xi - \bar{\xi} = (\alpha^n\xi - s_n\xi) + (s_n\xi - \bar{\xi}), \tag{3}$$

for some fixed $\xi \in H$, where $\bar{\xi} = E(\{1\})\xi$.

Putting

$$g_n = \alpha^n\xi - s_n\xi, \tag{4}$$

we have

$$g_n = \int_o^1 k_n(x)Z(dx), \tag{5}$$

where

$$k_n(x) = x^n - \frac{1}{n} \sum_{k=0}^{n-1} x^k \tag{6}$$

and

$$Z(\cdot) = E(\cdot)\xi.$$

Moreover,

$$\|g_{2^n}\|^2 = \int_o^1 |k_{2^n}(x)|^2 F(x) = \sum_{k=1}^{\infty} \int_A |k_{2^n}(x)|^2 F(dx) \tag{7}$$

where

$$A_k = \{x : 1 - 2^{-k+1} \leq x < 1 - 2^{-k}\}$$

and

$$F(\cdot) = \|E(\cdot)\xi\|^2.$$

Putting $a_k = F(A_k)$ and using the estimation

$$|K_n(x)| \leq 2 \min (n(1 - x), n^{-1}(1 - x)^{-1}) \tag{8}$$

for $0 < x < 1$, we obtain

$$\|g_{2^n}\|^2 \leq C \sum_{k=1}^{n} 2^{2(k-n)} a_k + \sum_{k=n+1}^{\infty} 2^{2(n-k)} a_k,$$

so we easily get

$$\sum_{n=1}^{\infty} \|g_{2^n}\|^2 < \infty. \tag{9}$$

In the sequel, we use the techniques well-known in the theory of orthogonal series ([1]). We are going to keep the notation of [7] because we want to refer to some similar calculations made in [7], Put

$$\delta_{n,m} = g_{2^n+m} - g_{2^n} \qquad (m = 1,2,\ldots,2^n-1). \tag{10}$$

Writing m in the form

$$m = \sum_{q=1}^{n} \varepsilon_q 2^{n-q} \qquad \text{with} \quad \varepsilon_q = 0 \quad \text{or} \quad 1, \tag{11}$$

we get

$$\delta_{n,m} = \sum_{q=1}^{n} \varepsilon_q \Delta_q^{j_q}, \tag{12}$$

where

$$\Delta_q^j = \int_0^1 R_{n,q,j}(x) Z(dx) \tag{13}$$

and

$$R_{n,q,j}(x) = k_{2^n+j2^{n-q}}(x) - k_{2^n+(j-1)2^{n-q}}(x) \tag{14}$$

$j = 1,2,\ldots,2^q$; $q = = 1,2,\ldots,n$.

Taking a suitable partition of $[0,1]$, we can write

$$\Delta_q^j = \eta_q^j + \sum_{t=1}^{t_n} R_{n,q,j}(z_t^{(n)}) \zeta_t^{(n)}, \tag{15}$$

with mutually orthogonal vectors $\zeta_t^{(n)} \in H$, such that

$$\sum_{t=1}^{t_n} \|\zeta_t^{(n)}\|^2 = F([0,1]) \tag{16}$$

and

$$\|\eta_q^{(j)}\| < 2^{-2n}, \qquad j = 1,2,\dots,2^q; \tag{17}$$

$j = 1,2,\dots,2^q;$ $q = 1,2,\dots,n.$

Now, we fix $x_{n,t} \in M$ and $\xi_t^{(n)} \in H$ $(t = 1,2,\dots,t_n)$ such that the following formulae hold.

$$\zeta_t^{(n)} = \xi_t^{(n)} + x_{n,t}\Omega$$

$$\|\xi_t^{(n)}\|^2 < 2^{-2n} t_n^{-3} \tag{18}$$

$$|\phi(x_{n,t}^* x_{n,v})| < 2^{-2n} t_n^{-3},$$

$t,v = 1,2,\dots,t_n,$ $t \neq v.$

Finally, we obtain

$$\delta_{n,m} = \eta_{n,m} + \xi_{n,m} + y_{n,m}\Omega, \tag{19}$$

where

$$\eta_{n,m} = \sum_{q=1}^{n} \varepsilon_q \eta_q^{j_q}$$

$$\xi_{n,m} = \sum_{q=1}^{n} \varepsilon_q \sum_{t=1}^{t_n} R_{n,q,j}(z_t^{(n)})\xi_t^{(n)} \tag{20}$$

$$y_{n,m} = \sum_{q=1}^{n} \varepsilon_q \sum_{t=1}^{t_n} R_{n,q,j}(z_t^{(n)})x_{n,t},$$

$m = 1,\dots,2^{n-1};$ $n = 1,2,\dots).$

Putting $\gamma_{n,m} = \eta_{n,m} + \xi_{n,m},$ exactly in the same way as in [7, p. 423] we show that

$$\sum_{n=1}^{\infty} \sum_{m=1}^{2^n-1} \|\gamma_{n,m}\|^2 < \infty. \tag{21}$$

Setting, for $j = 1,2,\dots,2^q;$ $q = 1,2,\dots,n$

$$d_{n,q,j} = \sum_{t=1}^{t_n} R_{n,q,j}(z_t^{(n)})x_{n,t}, \tag{22}$$

we get (exactly as in [7, p. 421]) that

$$|y_{n,m}|^2 \leq D_n, \tag{23}$$

where

$$D_n = 2 \sum_{q=1}^{n} q^2 \sum_{j=1}^{2^q} |d_{n,q,j}|^2 \tag{24}$$

(using the inequality

$$|\sum_{i=1}^{n} \alpha_i x_i|^2 \leq \sum_{i=1}^{n} |\alpha_i|^2 \sum_{i=1}^{n} |x_i|^2,$$

for complex numbers α_i and $x_j \in M$).

Consequently,

$$\beta_o(|y_{n,m}|^2) \leq \beta_o(D_n). \tag{25}$$

Now, we are in a position to show the almost sure convergence to zero of the sequence

$$\beta\alpha^n\xi - \tilde{\xi} = \beta(\alpha^n - s_n)\xi + (\beta s_n\xi - \beta\tilde{\xi}), \tag{26}$$

for some fixed $\xi \in H$.

Let $\varepsilon > 0$ be given. Let us assume for a moment that

$$\sum_{s=1}^{\infty} \phi(D_s) < \infty \tag{27}$$

holds.

Fix $(h_k) \subset H$ and $(y_k) \subset M$ such that

$$\xi - \tilde{\xi} = h_k + (y_k - \alpha_o^{\ell} y_k)\Omega, \tag{28}$$

with

$$\sum_k \|h_k\|^2 < \infty. \tag{29}$$

This is always possible because $H = H_1 \oplus H_2$, where

$$H_1 = [\xi \in H : \alpha\xi = \xi]$$

and

$$H_2 = [(y - \alpha_o y)\Omega : y \in M]^-.$$

By (9), (22), (27), (28) and (29), there is a sequence (ε_n) with $0 < \varepsilon_n \to 0$ and such that

$$\sum_{s=1}^{\infty} \varepsilon_s^{-1}[\phi(D_s) + \sum_{m=1}^{2^s-1} \|\beta\gamma_{s,m}\|^2 + \|\beta g_{2^s}\|^2 + \|h_s\|^2] < \varepsilon \tag{30}$$

By the Maximal Ergodic Lemma 2.2 (applied to the kernels α and id_M) there exists a projection $p \in M$ such that the following inequalities hold

$$\phi(1 - p) < 4\varepsilon$$

$$\|s_n h_k\|_p < 5\varepsilon_k^{1/2}; \qquad n,k = 1,2,\dots \tag{31}$$

$$\|pD_s p\|_\infty < 2\varepsilon_s; \qquad s = 1,2,\dots \tag{32}$$

$$\|\beta\gamma_{s,m}\|_p < 5\varepsilon_s^{1/2} \qquad m = 1,2,\dots 2^s-1, \quad s = 1,2,\dots \tag{33}$$

$$\|\beta q_{2^s}\|_p < 5\varepsilon_s^{1/2} \qquad s = 1,2,\dots \ . \tag{34}$$

Let us notice that by (23) and (32) we have

$$\|\beta(y_{s,m}\Omega)\|_p \lneqq \|\beta_o y_{s,m} p\|_\infty \le \|y_{s,m} p\|_\infty$$

$$= \|p|y_{s,m}|^2 p\|_\infty^{1/2} \le \|pD_s p\|_\infty^{1/2} < 2\varepsilon_s^{1/2}. \tag{35}$$

From (10), (19), (21), (23) and (32), (33), (34), (35) it follows that, for $2^s \le n < 2^{s+1}$, we have

$$\|\beta g_{2^s} - \beta g_n\|_p \le C\varepsilon_s^{1/2}. \tag{36}$$

Consequently,

$$\|\beta(\alpha^n\xi - s_n\xi)\|_p = \|\beta g_n\|_p \le \|\beta g_n - \beta g_{2^s}\| + \|\beta g_{2^s}\| < C\varepsilon_s^{1/2}, \qquad \text{for } 2^s \le n < 2^{s+1}. \tag{37}$$

Let us remark now that, for every $\eta \in H$, we have

$$\|\beta\eta\|_p \le \|\eta\|_p. \tag{38}$$

Indeed, β is generated by the kernel β_o which is the contraction in M. If $\eta = \sum_k x_k\Omega$ in H and the series $\sum_k x_k p$ converges in norm in M, then $\beta\eta = \sum_k \beta(x_k\Omega) = \sum_k (\beta_o x_k)\Omega$ and the series $\sum_k \beta_o x_k p = \beta_o \sum_k x_k p$ converges in M. Thus if $(x_k) \in S_{\eta,p}$ then $(\beta_o x_k) \in S_{\beta\eta,p}$ and $\|\sum_k \beta_o x_k p\| \le \|\sum_k x_k p\|$ which implies (38).

Thus, we have by (31), (28) and (38) that

$$\begin{aligned}\|\beta s_n\xi - \beta\bar{\xi}\|_p &= \|\beta s_n(\xi - \bar{\xi})\|_p \le \|\beta s_n h_k\|_p \\ &+ \|\beta s_n(\alpha_o y_k - y_k)\Omega\|_p \le \|s_n h_k\|_p \\ &+ \|\sigma_n(y_k - \alpha_o y_k)p\|_\infty \le 5\varepsilon_k^{1/2} \\ &+ \|\sigma_n(y_k - \alpha_o y_k)\|_\infty.\end{aligned} \tag{39}$$

From the estimations (37) and (39) it follows immediately that $\beta\alpha^n\xi - \beta\bar{\xi}$ almost surely as $n \to \infty$ (still under the assumption that the formula (27) holds). Thus it remains to show that

$$\sum_{n=1}^{\infty} \sum_{q=1}^{n} q^2 \sum_{j=1}^{2^q} \phi(|d_{n,q,j}|^2) < \infty, \tag{40}$$

where the operators $d_{n,q,j}$ are defined by formula (22).

We have

$$\phi(|d_{n,q,j}|^2) = \sum_{t=1}^{t_n} |R_{n,q,j}(z_t^{(n)})|^2 \phi(|x_{n,t}|^2)$$

$$+ \sum_{\substack{t,v=1\\ t\neq v}}^{t_n} R_{n,q,j}(z_t^{(n)}) R_{n,q,j}(z_v^{(n)}) \phi(x_{n,t}^* x_{n,v})$$

$$= A_{n,q,j} + B_{n,q,j}. \tag{41}$$

By (18), we easily get

$$\sum_{n=1}^{\infty} \sum_{q=1}^{n} q^2 \sum_{j=1}^{2^q} |B_{n,q,j}| < \infty, \tag{42}$$

so it remains to prove that

$$A = \sum_{n=1}^{\infty} \sum_{q=1}^{n} q^2 \sum_{j=1}^{2^q} |A_{n,q,j}| < \infty \tag{43}$$

holds. Using the estimation (18) and

$$\phi(|x_{n,t}|^2) = \|x_{n,t}\Omega\|^2 \leq 2\|\zeta_t^{(n)}\|^n + 2\|\zeta_t^{(n)}\|^n,$$

the orthogonality of vectors $\zeta_t^{(n)}$ and (17), we obtain after rather standard calculations that

$$A \leq C \sum_{n=1}^{\infty} \sum_{q=1}^{n} q^2 \sum_{j=1}^{2^q} \int_0^1 |R_{n,q,j}(z)|^2 F(dz) + C,$$

with some constant C, so it is enough to show that

$$\sum_{n=1}^{\infty} \sum_{q=1}^{n} q^2\, 2^q \max_{1\leq j\leq 2^q} \int_0^1 |R_{n,q,j}(z)|^2 F(dz) < \infty \tag{44}$$

holds. In order to do this we set

$$A_1(n,q) = \{1 - 2^{-n} < x < 1\}$$

$$A_2(n,q) = \{1 - 2^{-n+q} < x \leq 2^{-n}\} \tag{45}$$

$$A_3(n,q) = \{0 < x \leq 1 - 2^{-n+q}\},$$

$$a_k = F(\{1 - 2^{-k+1} \leq x < 1 - 2^{-k}\}) \tag{46}$$

and

$$B_s(n,q) = \max_{1\le j\le 2^q} \int_{A_s(n,q)} |R_{n,q,j}(x)|^2 F(dx). \qquad (47)$$

By (14), using the estimation

$$|k_n(x) - k_m(x)| \le C(n - m)(1 - x), \qquad \text{for } 0 < x\ 1,$$

we get

$$B_1(n,q) \le C\, 2^{2(n-q)} \sum_{k=n}^{\infty} 2^{-2k} a_k$$

(of course, the constant C is different in different formulae).
Consequently,

$$\sum_{n=1}^{\infty} \sum_{q=1}^{n} q^2 2^q B_1(n,q) \le C \sum_{n=1}^{\infty} \sum_{q=1}^{n} q^2 2^q \sum_{k=n}^{\infty} 2^{2(n-q)} 2^{-2k} a_k$$

$$\le C \sum_{q=1}^{\infty} q^2\, 2^{-q} \sum_{k=1}^{\infty} a_k 2^{-2k} \sum_{n=1}^{k} 2^{2n} \le C \sum_{k=1}^{\infty} a_k < \infty.$$

Using the estimation

$$|k_n(x) - k_m(x)| \le Cm^{-1}(n - m), \qquad \text{for } 0 < x < 1 \text{ and } n > m,$$

we easily obtain

$$B_2(n,q) \le C \sum_{k=n-q+1}^{n} 2^{2(n-q)} 2^{-2n} a_k.$$

Consequently,

$$\sum_{n=1}^{\infty} \sum_{q=1}^{n} q^2 2^q B_2(n,q) \le C \sum_{n=1}^{\infty} \sum_{q=1}^{n} q^2 2^q \sum_{k=n-q+1}^{n} 2^{2(n-q)} 2^{-2n} a_k$$

$$\le C \sum_{n=1}^{\infty} \sum_{k=1}^{n} a_k \sum_{q=n-k+1}^{n} q^2\, 2^{-q}$$

$$\le C \sum_{k=1}^{\infty} a_k \sum_{n=k}^{\infty} (n - k + 1)^2\, 2^{-(n-k)} < \infty.$$

Finally, using the inequality

$$|k_n(x)| \leq Cn^{-1}(1 - x)^{-1},$$

we get

$$B_3(n,q) \leq C \sum_{k=1}^{n-q} a_k 2^{-2n} 2^{2k}.$$

Thus

$$\sum_{n=1}^{\infty} \sum_{q=1}^{n} q^2 2^q B_3(n,q) \leq C \sum_{n=1}^{\infty} \sum_{q=1}^{n} q^2 2^q \sum_{k=1}^{n-q} a_k 2^{-2n} 2^{2k}$$

$$\leq C \sum_{n=1}^{\infty} 2^{-2n} \sum_{k=1}^{n} a_k \sum_{q=1}^{n-k} q^2 2^q 2^{2k}$$

$$\leq C \sum_{n=1}^{\infty} 2^{-2n} \sum_{k=1}^{n} a_k 2^{2k} (n - k)^2 2^{n-k}$$

$$\leq C \sum_{k=1}^{\infty} a_k \sum_{n=k}^{\infty} (n - k)^2 2^{-(n-k)} \leq C \sum_{k=1}^{\infty} a_k < \infty.$$

Summing up, we get (44) and, consequently (40) which completes the proof of our theorem.

Acknowledgement. The first draft of this paper was written while the author was attending the Symposium on Quantum Probability and Applications, Rome 1987. The author wishes to express his gratitude to professor Accardi for his warm hospitality and discussions. He is also indebted to Professor Parthasarathy for his suggestion of the problem solved in Theorem 3.2.

References

[1] Alexits, G.: Convergence problems of orthogonal series. New York-Oxford-Paris: Pergamon Press 1961

[2] Bratteli, O., Robinson, D.W.: Operator algebras and quantum statistical mechanics, I, New York-Heidelberg-Berlin: Springer 1979.

[3] Burkholder, D.L, and Chow, Y.S.: Iterates of conditional expectation operators. Proc. Amer. Math. Soc. 12, 490-495 (1961)

[4] Gaposhkin, V.F.; Criteria of the strong law of large numbers for some classes of stationary processes and homogeneous random fields (Russian). Theory Probab. Appl. 22, 295-319 (1977)

[5] Gaposhkin, V.F.: Individual ergodic theorem for normal operators in L_2 (Russian). Funkt. Anal. Appl. 15, 18-22 (1981)

[6] Goldstein, M.S.: Theorems in almost everywhere convergence in von Neumann algebras (Russian). J. Oper. Theory 6, 233--311 (1981)

[7] Hensz, E., Jajte, R.: Pointwise convergence theorems in L_2 over a von Neumann algebra. Math. Z. 193, 413-429 (1986)

[8] Jajte, R.: Strong limit theorems in non-commutative probability. Lect, Notes in Math., No 1110. Berlin-Heidelberg--New York: Springer 1985

[9] Jajte, R.: Ergodic theorem in von Neumann algebras. Semesterbericht Funktioneanalysis, Tübingen, Sommersemester, 135--144 (1986)

[10] Jajte, R.: Contraction semigroups in L_2 over a von Neumann algebra. Proc. Oberwolfach Conf. on Quantum Probability and Applications 1987. Lect. Notes in Math. 1303, 149--153 (1988)

[11] Jajte R.: Asymptotic formula for normal operators in non--commutative L_2-spaces. to appear

[12] Kümmerer, B.: A non-commutative individual ergodic theorem. Invent. Math. 46, 136-145 (1978)

[13] Lance, E.C.: Ergodic theorems for convex sets and operator algebras. Invent. Math. 37, 201-214 (1976)

[14] Lindblad, G.: On the generators of quantum dynamical semi-groups. Comm. Math. Phys. 48, 119-130 (1970)

[15] von Neumann, J.: Functional operators, vol. 2. Princeton: University Press 1950

[16] Petz, D.: Ergodic theorems in von Neumann. Acta Sci. Math. 46, 329-343 (1983)

[17] Petz, D.: Quasi-uniform ergodic theorems in von Neumann algebras. Bull.Lond. Mat. Soc. 16, 151-156 (1984)

[18] Sinai, Y.G., Aushelevich, V.V.: Some problems of non-commutative ergodic theory. Russian Math. Surv. 31, 157--174 (1976)

[19] Stein, E.M.: Topics in harmonic analysis. Ann. Math, Stud. No 63. Princeton: University Press 1980

[20] Takesaki, M.: Theory of operator algebras I. Berlin--Heidelberg-New York: Springer 1979

[21] Watanabe, S.: Ergodic theorems for dynamical semigroups on operator algebras. Hokkaido Math. J. 8, 176-190 (1979)

[22] Watanabe, S.: Asymptotic behaviour and eigenvalues of dynamical semi-groups on operator algebras. J. Math. Anal. Appl. 86, 411-424 (1982)

[23] Yeadon, F.J.: Ergodic theorems for semifinite von Neumann algebras I. J. Math Soc. 2 (16), 326-332 (1977)

[24] Yeadon, F.J.: Ergodic theorems for semifinite von Neumann algebras II. Math. Proc. Camb., Phil. Soc. 88, 135-147 (1980)

DUALITY TRANSFORM AS *-ALGEBRAIC ISOMORPHİSM

J.M.Lindsay and H.Maassen

It is well known that the symmetric Fock space over a real Hilbert space h and L^2- of the Gaussian space over h are naturally isomorphic (Wiener – Segal – Itô – $\cdots$). The purpose of this note is to demonstrate the algebraic character of this isomorphism in the case where h is L^2- of a non-atomic measure space.

Let Γ be the *symmetric measure space* of a σ-finite non-atomic measure space X ([Gui]), let $h = L^2_{\mathbf{R}}(X)$ and let $\mathcal{C}$ be the *Gaussian space over* h ([Sim]). Then $L^2(\Gamma)$ is a realisation of the (symmetric) Fock space over h. For $v \in h$ let $\pi_v \in L^2(\Gamma)$ be the *product function* $\sigma \mapsto \prod_{s\in\sigma} v(s)$ and let $\varepsilon_v \in L^2(\mathcal{C})$ be given by $\exp(\varphi_v - \frac{1}{2}\|v\|^2)$ where $\{\varphi_v : v \in h\}$ is the *Gaussian process indexed by* h ([Sim]). The (complex) linear spans Ψ and $\mathcal{E}$ of $\{\pi_v : v \in h\}$ and $\{\varepsilon_v : v \in h\}$ are dense respectively in $L^2(\Gamma)$ and $L^2(\mathcal{C})$, and the relation

$$< \pi_u, \pi_v > = \exp < u, v > = < \varepsilon_u, \varepsilon_v >,$$

for the respective inner products, is easily verified. Thus the isomorphism $\widehat{\ }: L^2(\Gamma) \to L^2(\mathcal{C})$ may be described as the continuous linear extension of the map $\pi_v \mapsto \varepsilon_v$.

Let us view the number operator: $Nf(\sigma) = \#\sigma\ f(\sigma)$ on $L^2(\Gamma)$ equally as an operator on $L^2(\mathcal{C})$ via the above isomorphism. The relation

$$a^N \pi_v = \pi_{av}, \tag{1}$$

for $a \in \mathbf{R}$ and $v \in h$ implies that

$$\Psi \subset \bigcap_{a>0} \mathcal{D}(a^N), \tag{2}$$

where $\mathcal{D}$ denotes domain. For $f, g \in \mathcal{D}(\sqrt{2}^N) \subset L^2(\Gamma)$, the *Wiener product* $f * g$ is the measurable function defined almost everywhere by

$$f * g : \sigma \mapsto \int_\Gamma \chi(\sigma, \omega) \sum_{\alpha\subset\sigma} f(\alpha \cup \omega) g(\omega \cup \overline{\alpha})\, d\omega,$$

where the sum is over partitions $\alpha \cup \overline{\alpha}$ of σ and χ is the indicator function of *some* product measurable set contained in $\Gamma^{(2)} := \{(\alpha, \beta) \in \Gamma^2 : \alpha \cap \beta = \emptyset\}$

which differs from $\Gamma^{(2)}$ by a null set (see [L 2]). (In this *non-atomic* case χ is almost everywhere equal to one.) $\tilde{}$ will denote (pointwise) complex conjugation on $L^2(\Gamma)$.

Ψ is invariant under $\tilde{}$, and also closed under $*$ since

$$\pi_u * \pi_v(\sigma) = \left(\sum_{\alpha \subset \sigma} \pi_u(\alpha)\pi_v(\overline{\alpha})\right) \int_\Gamma \pi_u(\omega)\pi_v(\omega)d\omega$$
$$= e^{<u,v>}\pi_{(u+v)},$$

for $u, v \in h$. Moreover, in view of the identity $\varepsilon_u \varepsilon_v = \exp < u, v > \ \varepsilon_{(u+v)}$, the isomorphism satisfies

$$\widehat{f * g} = \hat{f}\hat{g}, \tag{3}$$

for $f, g \in \Psi$. A straight-forward consequence of the $\int$–$\sum$ identity (equation (1.1) in [LM]; lemma 1.1 in [LP] these proceedings; [L 2]) is the inequality

$$\|f_1 * f_2\| \leq \|\sqrt{3}^N f_1\| \; \|\sqrt{3}^N f_2\|, \tag{4}$$

for $f_1, f_2 \in \mathcal{D}(\sqrt{3}^N)$. Thus for $F = \hat{f} \in \mathcal{E}$,

$$\begin{aligned} \|F\|_4^2 = \|\overline{F}F\|_2 &= \|\tilde{f} * f\| \\ &\leq \|\sqrt{3}^N f\|^2 = \|\sqrt{3}^N F\|_2^2. \end{aligned} \tag{5}$$

But, by (1), $\mathcal{E}$ is invariant under the operators a^N for $a \in \mathbf{R}$ so that

$$\|(1/\sqrt{3})^N G\|_4 \leq \|G\|_2, \tag{6}$$

for $G \in \mathcal{E}$. Since $\mathcal{E}$ is dense in $L^2(C)$, $(1/\sqrt{3})^N$ extends to a contraction from $L^2(C)$ to $L^4(C)$. Otherwise put:

$$\mathcal{D}(\sqrt{3}^N) = \mathcal{R}((1/\sqrt{3})^N) \subset L^4(C). \tag{7}$$

where $\mathcal{R}$ denotes range. With the relations

$$\widetilde{f_1 * f_2} \;=\; \tilde{f}_2 * \tilde{f}_1; \tag{8}$$
$$< f_1, f_2 * f_3 > \;=< \tilde{f}_2 * f_1, f_3 >, \tag{9}$$

which are easily verified from the $\int$–$\sum$ identity when $f_1, f_2, f_3 \in \mathcal{D}(\sqrt{3}^N)$, all the players are in position.

Proposition: *Let $f, g \in \mathcal{D}(\sqrt{3}^N) \subset L^2(\Gamma)$ then*

$$\widehat{f * g} = \hat{f}\hat{g}. \tag{10}$$

proof: We have already mentioned (eqn.(3)) that the identity holds on Ψ. Let $f_1, f_2 \in \mathcal{D}(\sqrt{3}^N)$ and $f_3 \in \Psi$ then, by (4),(9) and (7),

$$
\begin{aligned}
<\widehat{\tilde{f}_1 * f_2}, \hat{f}_3> &= < f_2, f_1 * f_3 > \\
&= < \hat{f}_2, \hat{f}_1 \hat{f}_3 > = < \hat{\tilde{f}}_1 \hat{f}_2, \hat{f}_3 >,
\end{aligned} \tag{11}
$$

as long as $f_1 \in \Psi$. By the density of Ψ, its invariance under $\tilde{}$ and the identity (9), the equality (10) is valid as long as either f or g belongs to Ψ. But this means that (11) holds without the restriction on f_1. By the invariance of $\mathcal{D}(\sqrt{3}^N)$ under $\tilde{}$ and the density of Ψ this implies that (10) holds without restriction and the proof is complete.

□

Remarks:

1. The proposition remains valid if $*$ is replaced by the *Clifford convolution* ([Mey]) given, in terms of a measurable total order on X, by

$$
f *_C g : \sigma \mapsto \sum_{\alpha \subset \sigma} \int_{\Gamma_X} \chi(\sigma, \omega)(-1)^{n(\alpha \cup \omega, \omega \cup \overline{\alpha})} f(\alpha \cup \omega) g(\omega \cup \overline{\alpha})\, d\omega,
$$

where $n(\alpha, \beta) = \#\{(a,b) \in \alpha \times \beta : a > b\}$ and $\hat{}$ is now the *fermionic duality transform* relating Fock space to the non-commutative L^2-space of the Clifford algebra over $L^2(X)$. The above proof has been written in such a way that it carries over well to this context. If Ψ is the *algebraic* span of $\delta_\emptyset$ and the one particle functions (supported by singletons), and $\tilde{}$ is defined by $\tilde{f}(\sigma) = (-1)^{n(\sigma,\sigma)}\overline{f(\sigma)}$, then (2)-(9) remain valid — (3) now being a consequence of the associativity of $*_C$. In this context the non-atomicity assumption on X is superfluous ([L 2]): all the (real) Clifford algebras with positive definite quadratic form are represented in this way. For further elucidation see [L 1].

2. The inequality (6) gives an easy proof of Nelson's hypercontractive estimate ([Ne 1], [Ne 2]) — see [L 1].

3. The proposition appeared in [LM] for the case when X is a compact real interval with Lebesgue measure, however the proof there was incomplete.

Acknowledgement: JML and HM are grateful for financial support from the Science and Engineering Research Council, U.K. and the Netherlands Organisation of Scientific Research (NWO) respectively.

References

[Gui] A.Guichardet, *Symmetric Hilbert Spaces and Related Topics.* Springer LNM **261** (1972), Heidelberg.

[L 1] J.M.Lindsay, *Gaussian hypercontractivity revisited.* Preprint 1989.

[L 2] J.M.Lindsay, *On set convolutions and integral-sum kernel operators.* In preparation.

[LM] J.M.Lindsay and H.Maassen, *An integral kernel approach to noise.* in Quantum Probability and Applications III, (L.Accardi and W.von Waldenfels eds.) Springer LNM **1303** (1988), 192-208.

[LP] J.M.Lindsay and K.R.Parthasarathy, *Rigidity of the Poisson convolution.* These Proceedings (1989).

[Mey] P-A.Meyer, *Eléments de probabilités quantiques I.* Springer LNM **1204** (1986) 186-312.

[Ne 1] E.Nelson, *A quartic interaction in two dimensions.* in Mathematical Theory of Elementary Particles, (R.Goodman and I.E.Segal eds.), M.I.T. Press, Cambridge, Mass. (1966), pp.69-73.

[Ne 2] E.Nelson, *Free Markov field.* J.Funct.Anal. **12** (1975), 211-227.

[Sim] B.Simon, *The $P(\phi)_2$ Euclidean (Quantum) Field Theory.* Princeton University Press, Princeton, NJ (1974).

Mathematics Department
King's College
Strand
London WC2R 2LS
U.K.

Mathematisch Instituut
Katholieke Universiteit
Toernooiveld
6525 ED Nijmegen
Netherlands

This paper is in final form and no similar paper has been or is being submitted elsewhere.

RIGIDITY OF THE POISSON CONVOLUTION

J.M.Lindsay and K.R.Parthasarathy

Introduction

In a recent article ([LP]) we classified algebraic deformations of the abstract Wiener convolution on Fock space. We were motivated by the appearance of a plethora of convolution products ([Mey], [LM 1], [LM 2]) that followed an observation of H.Maassen's ([Maa]). He noticed that if Guichardet's representation of Fock space ([Gui]) is employed then the pointwise product for Wiener functionals is transformed, via the natural isomorphism of Wiener space and Fock space, to a kind of convolution product for *Fock functionals.* P-A.Meyer observed that Clifford multiplication is similarly representable as a convolution of Fock functionals, now involving a multiplier, by means of the fermionic duality tranform. For further clarification of these remarks in an abstract context see [LM 3] (these proceedings). Lindsay and Maassen exploited a representation of Bose (CCR) multiplication as a *twisted* convolution in their treatment of the stochastic calculus for non-unit variance quantum Brownian motion. A similar representation of Fermi (CAR) multiplication may also be exploited. The convolutions investigated in [LP] were of the form

$$f *_p g : \sigma \mapsto \sum_{\alpha \subset \sigma} \int_\Gamma p(\omega, \alpha, \overline{\alpha}) f(\alpha \cup \tilde{\omega}) g(\omega \cup \overline{\alpha})\, d\mu(\omega), \qquad (0.1)$$

$\overline{\alpha}$ denoting the complement of α in σ and p being subject to the constraint that $*_p$ be associative. Detailed explanation of the terms in this formula are given in section 1. We gave a complete classification of associative products of the form (0.1), up to algebraic isomorphism through a multiplication operator, subject to p being both bounded and bounded away from zero, and a naturally arising sufficient condition for associativity.

In this article our point of departure is the less well-known isomorphism between the L^2-space of the standard Poisson process and Fock space, which also leads to a convolution of Fock functionals. We investigate deformations of this *Poisson convolution* — namely products of the form:

$$f \star_P g : \sigma \mapsto \sum_{|\underline{\alpha}| = \sigma} \int_\Gamma P(\omega, \underline{\alpha}) f(\alpha_1 \cup \alpha_2 \cup \tilde{\omega}) g(\omega \cup \alpha_2 \cup \alpha_3)\, d\mu(\omega),$$

where the sum is over all partitions of σ into three: $|\underline{\alpha}| = \alpha_1 \cup \alpha_2 \cup \alpha_3$, and again P satisfies a naturally arising sufficient condition for associativity. Our results are essentially negative. We show that all Poisson convolutions are isomorphic — in particular, no non-commutative ones exist.

We also give a proof that, under a separability assumption on the measure space, the sufficient condition on p for $*_p$ to be associative which was used in [LP] is also necessarily satisfied almost everywhere. We have been unable to fill this gap in the Poisson case. The question remains of whether one can construct interesting examples of convolutions for which the multiplier satisfies the above condition *almost everywhere* but p may not be modified on a null set so that it satisfies the condition *everywhere*. The analysis of [LP] and this paper is algebraic; the consideration of limits of these convolutions may lead to interesting further examples. The discrete case also merits elucidation.

§1. The symmetric functor.

For a set S, Γ_S denotes the collection $\{\sigma \subset S : \#\sigma < \infty\}$ and for a complex valued function φ on S, ε_φ denotes the *product function* $\sigma \in \Gamma_S \mapsto \prod_{s\in\sigma} \varphi(s)$ (in notation agreeing with [LP] but clashing with [LM 3]). If S carries a measurable structure $\mathcal{M}$ then Γ_S naturally inherits a measurable structure. Furthermore if $\mathcal{M}$ carries a measure m then this induces a measure μ on Γ_S, called the *symmetric measure of* m (see [Gui],[LP],[L]). We usually abbreviate $dm(s)$, $d\mu(\sigma)$ and $d\mu^d(\underline{\sigma})$ to ds, $d\sigma$ and $d\underline{\sigma}$ respectively. If $X = (S, \mathcal{M}, m)$ is σ-finite and non-atomic (as all our measure spaces will be) then $\Gamma_X = (\Gamma_S, \mu)$ is (σ-finite and) non-atomic also except that $\emptyset$ is an atom of measure unity; $\mu(\Gamma_S) = \exp\{m(S)\}$. More importantly the set $\Gamma_S^{(d)} = \{\underline{\alpha} \in (\Gamma_S)^d : \alpha_i \cap \alpha_j = \emptyset \text{ if } i \neq j\}$ differs from $(\Gamma_S)^d$ by a μ^d-null set. Since $\Gamma_S^{(d)}$ itself may not be measurable we fix, for each $d \geq 2$, a product measurable set $\Gamma_S^{[d]}$ of full μ^d- measure and contained in $\Gamma_S^{(d)}$. Since these sets only serve to make various integrands measurable their arbitrariness does not matter; let $\chi_{[d]}$ denote the indicator function of $\Gamma_S^{[d]}$. If $\varphi : X \to \mathbf{C}$ is measurable then $\int |\varepsilon_\varphi|^2 d\mu = \exp \int |\varphi|^2 dm$ and if $\psi = \varphi$, m-a.e. then $\varepsilon_\psi = \varepsilon_\varphi$ μ-a.e. A convention we adopt is that if Y and Z are measurable spaces then $f : Y \to Z$ automatically denotes a measurable function. The following identity is essential.

Lemma 1.1: ($\int$–$\sum$) *If $g : \Gamma_X^d \mapsto \mathbf{C}$ vanishes outside $\Gamma_X^{(d)}$ then the function $G : \sigma \mapsto \{\sum_{|\underline{\alpha}|=\sigma} g(\underline{\alpha})\}$ is measurable and*

$$\int_{\Gamma_X^d} g \, d\mu^d = \int_{\Gamma_X} G \, d\mu,$$

whenever either side is defined.

Proofs are given in [LM 2], [LP] and [L]. Thinking of G as the *complete symmetrisation* of g, the identity is purely combinatorial.

The set up here is a *twisted measure space* $Z = (S, \mathcal{M}, m, j, S_+)$ where $X = (S, \mathcal{M}, m)$ is a σ-finite non-atomic measure space, $j : X \to X$ is involutive ($j^2 = \mathrm{id}._X$) and $S_0 \cup S_+ \cup j(S_+)$ is a measurable partition of S (S_0 being the fixed point set of j) — in other words S_+ is a measurable section of $S \setminus S_0$ for j. Γ_Z is then Γ_X together with the involution defined by $\tilde{\sigma} = \{j(s) : s \in \sigma\}$. The symmetric measure of the measure $m^\dagger = m + \tilde{m}$ will be denoted $\mu^\dagger$. The convolutions considered in [LP] are defined on $\mathcal{K}(Z) := \{f : \Gamma_X \to \mathbf{C}\,\text{s.t.} \int a^{\#\sigma}|f(\sigma)|^2 d\mu^\dagger(\sigma) < \infty\, \forall a > 0\}$, and in view of the estimate

$$\|a^N(f *_p g)\|_{L^2(d\mu^\dagger)} \le \|p\|_\infty \left\| \sqrt{1+2a^2}^N f \right\|_{L^2(d\mu^\dagger)} \left\| \sqrt{1+2a^2}^N g \right\|_{L^2(d\mu^\dagger)},$$

which follows from a straight forward application of the $\int$–$\sum$ identity, $*_p$ descends to the quotient of $\mathcal{K}(Z)$ modulo $\mu^\dagger$-null functions.

§2. Microproduct.

As in the case of the Wiener convolution, the analysis of Poisson convolutions depends on that of a simpler product on functions on Γ_S. Let $\mathcal{F}(\Gamma_S)$ denote the linear space of complex valued functions on Γ_S. A *2-cocycle* on Γ_S is a map $q : \Gamma_S^{(2)} \to \mathbf{C}^\times := \mathbf{C} \setminus \{0\}$ which satisfies

$$q(\alpha, \beta \cup \gamma) q(\beta, \gamma) = q(\alpha, \beta) q(\alpha \cup \beta)$$

for all $(\alpha, \beta, \gamma) \in \Gamma_S^{(3)}$. q is a *product cocycle* if it is multiplicative in each argument: $q(\alpha \cup \beta, \gamma) = q(\alpha, \gamma) q(\beta, \gamma)$, $q(\alpha, \beta \cup \gamma) = q(\alpha, \beta) q(\alpha, \gamma)$, and it is *trivial* if $q(\alpha, \beta) = g(\alpha) g(\beta)/g(\alpha \cup \beta)$ for some $g : \Gamma \to \mathbf{C}^\times$ — write $q = \tau_g$. The only result from [LP] that we shall need is:

Proposition 2.1: *Let $q : \Gamma_S^{(2)} \to \mathbf{C}^\times$ be a 2-cocycle. Then there is a product cocycle π and a trivial cocycle τ_g such that $q = \pi\tau_g$.*

Proof: [LP] Proposition 2.3.

□

Let $p : \Gamma_S^{(3)} \to \mathbf{C}^\times$ then define a product $\circ_p$ on $\mathcal{F}(\Gamma_S)$ by

$$f \circ_p g(\sigma) = \sum_{|\underline{\alpha}| = \sigma} p(\underline{\alpha}) f(\alpha_1 \cup \alpha_2) g(\alpha_2 \cup \alpha_3)$$

— the sum being over partitions of $\sigma \in \Gamma_S$ into $\alpha_1, \alpha_2, \alpha_3$.

Lemma 2.2: *The product $\circ_p$ is associative if and only if p satisfies*

$$p(\omega_2 \cup \omega_3 \cup \omega_4, \omega_1 \cup \omega_0 \cup \omega_5, \omega_6)\, p(\omega_1 \cup \omega_2, \omega_0 \cup \omega_3, \omega_4 \cup \omega_5) = \\ p(\omega_2, \omega_3 \cup \omega_0 \cup \omega_1, \omega_4 \cup \omega_5 \cup \omega_6)\, p(\omega_3 \cup \omega_4, \omega_0 \cup \omega_5, \omega_6 \cup \omega_1) \tag{2.1}$$

for all $\underline{\omega} \in \Gamma_S^{(7)}$.

Proof: For $f, g, h \in \mathcal{F}(\Gamma_S)$,

$$(f \circ g) \circ h(\sigma) = \sum_{|\underline{\alpha}|=\sigma} p(\underline{\alpha}) f \circ g(\alpha_1 \cup \alpha_2) h(\alpha_3)$$

which, by dividing up $\underline{\alpha}$ as follows

$$\begin{array}{ccc} \alpha_1 & \alpha_2 & \alpha_3 \\ \hline \omega_2 & \omega_1 & \\ \omega_3 & \omega_0 & \omega_6 \\ \omega_4 & \omega_5 & \end{array},$$

equals

$$\sum_{|\underline{\omega}|=\sigma} p(\omega_2 \cup \omega_3 \cup \omega_4, \omega_1 \cup \omega_0 \cup \omega_5, \omega_6)\, p(\omega_1 \cup \omega_2, \omega_0 \cup \omega_3, \omega_4 \cup \omega_5) \times \\ f(\omega_0 \cup \omega_1 \cup \omega_2 \cup \omega_3)\, g(\omega_0 \cup \omega_3 \cup \omega_4 \cup \omega_5)\, h(\omega_5 \cup \omega_6 \cup \omega_1 \cup \omega_0).$$

On the other hand the following division of $\sigma = |\underline{\alpha}|$:

$$\begin{array}{ccc} \alpha_1 & \alpha_2 & \alpha_3 \\ \hline & \omega_3 & \omega_4 \\ \omega_2 & \omega_0 & \omega_5 \\ & \omega_1 & \omega_6 \end{array},$$

gives

$$\sum_{|\underline{\omega}|=\sigma} p(\omega_2, \omega_3 \cup \omega_0 \cup \omega_1, \omega_4 \cup \omega_5 \cup \omega_6)\, p(\omega_3 \cup \omega_4, \omega_0 \cup \omega_5, \omega_6 \cup \omega_1) \times \\ f(\omega_0 \cup \omega_1 \cup \omega_2 \cup \omega_3)\, g(\omega_0 \cup \omega_3 \cup \omega_4 \cup \omega_5)\, h(\omega_0 \cup \omega_5 \cup \omega_6 \cup \omega_1)$$

for $f \circ (g \circ h)(\sigma)$. Comparing coefficients gives the result. □

Notice the *trivial* (co-boundary like) solutions of (2.1):

$$\underline{\alpha} \mapsto \frac{f(\alpha_1 \cup \alpha_2) f(\alpha_2 \cup \alpha_3)}{f(\alpha_1 \cup \alpha_2 \cup \alpha_3)} \tag{2.2}$$

for $f : \Gamma_S \to \mathbf{C}^\times$.

Proposition 2.3: *All solutions of (2.1) are trivial.*

Proof: Let $p : \Gamma_S^{(3)} \to \mathbf{C}^\times$ be a solution of (2.1) which, without loss of generality, we suppose to be normalised: $p(\emptyset, \emptyset, \emptyset) = 1$. Putting $\omega_0 = \omega_1 = \omega_3 = \omega_5 = \emptyset$ in (2.1) gives

$$p(\omega_2 \cup \omega_4, \emptyset, \omega_6)\, p(\omega_2, \emptyset, \omega_4) = p(\omega_2, \emptyset, \omega_4 \cup \omega_6)\, p(\omega_4, \emptyset, \omega_6)$$

for all $\omega \in \Gamma_S^{(7)}$. In other words $q : (\alpha, \beta) \mapsto p(\alpha, \emptyset, \beta)$ is a 2-cocycle. By Proposition 2.1 there is a product cocycle π and a trivial cocycle τ_g such that

$$q(\alpha, \beta) = \frac{g(\alpha) g(\beta)}{g(\alpha \cup \beta)} \pi(\alpha, \beta). \tag{2.3}$$

Putting $\omega_1 = \omega_3 = \omega_4 = \omega_5 = \omega_6 = \emptyset$ in (2.1) gives

$$p(\omega_2, \omega_0, \emptyset) = p(\emptyset, \omega_0, \emptyset). \tag{2.4}$$

Let f' be the map $\alpha \mapsto p(\emptyset, \alpha, \emptyset)$ then putting $\omega_0 = \omega_1 = \omega_4 = \omega_5 = \emptyset$ in (2.1) gives

$$p(\omega_2 \cup \omega_3, \emptyset, \omega_6)\, p(\omega_2, \omega_3, \emptyset) = p(\omega_2, \omega_3, \omega_6)\, p(\omega_3, \emptyset, \omega_6),$$

in other words, using (2.3) and (2.4),

$$p(\alpha, \beta, \gamma) = \frac{q(\alpha \cup \beta, \gamma)}{q(\beta, \gamma)} f'(\beta) = h(\beta) \frac{g(\beta \cup \gamma) g(\alpha \cup \beta)}{g(\alpha \cup \beta \cup \gamma)} \pi(\alpha, \gamma), \tag{2.5}$$

where $h = f'/g$. Substituting (2.5) back into (2.1) with $\omega_0 = \omega_4 = \omega_5 = \emptyset$ gives

$$\pi(\omega_3, \omega_1) = \frac{h(\omega_1) h(\omega_3)}{h(\omega_3 \cup \omega_1) h(\emptyset)},$$

so that (2.5) now reads

$$p(\alpha,\beta,\gamma)=\frac{h(\alpha)h(\beta)h(\gamma)}{h(\alpha\cup\gamma)h(\emptyset)}\frac{g(\alpha\cup\beta)g(\beta\cup\gamma)}{g(\alpha\cup\beta\cup\gamma)}. \qquad (2.6)$$

Introducing this into (2.1) with $\omega_5=\omega_4=\omega_6=\emptyset$ gives the indentity

$$\frac{h(\omega_0\cup\omega_1)h(\omega_1\cup\omega_3)h(\omega_3\cup\omega_0)h(\emptyset)}{h(\omega_0\cup\omega_1\cup\omega_3)h(\omega_0)h(\omega_1)h(\omega_3)}=1.$$

(2.6) may therefore be written

$$p(\alpha,\beta,\gamma)=\frac{f(\alpha\cup\beta)f(\beta\cup\gamma)}{f(\alpha\cup\beta\cup\gamma)},$$

where $f=gh$, and the proposition is established.

□

Multiplication by f gives an isomorphism from $(\mathcal{F}(\Gamma_S),\circ_p)$ to $(\mathcal{F}(\Gamma_S),\circ)$ when p is determined by f as in (2.2).

§3. Poisson convolution.

In order to define the convolutions we need a couple of notations. First fix a twisted measure space Z. Let $\mathcal{F}_r(\Gamma^d)$ be the collection of *measurable* $\mathbf{C}^\times$-valued functions on Γ^d whose range is relatively compact, and for $f:\Gamma_X\to\mathbf{C}$ let $f^{[d]}:\Gamma^d\to\mathbf{C}$ be the (measurable) map $\underline{\alpha}\mapsto\chi_{[d]}(\underline{\alpha})f(|\underline{\alpha}|)$ where $|\underline{\alpha}|$ denotes the (disjoint) union $\alpha_1\cup\cdots\cup\alpha_d$. For $P\in\mathcal{F}_r(\Gamma^4)$, $f\star_P g$ is defined almost everywhere, for suitable $\mathbf{C}$-valued functions f and g, by

$$f\star_P g:\sigma\mapsto\sum_{|\underline{\alpha}|=\sigma}\int_\Gamma P(\omega,\underline{\alpha})f^{[3]}(\alpha_1,\alpha_2,\widetilde{\omega})\,g^{[3]}(\omega,\alpha_2,\alpha_3)d\omega. \qquad (3.1)$$

Although $f\star_P g$ is well defined if f and g lie in $\mathcal{K}(Z)$ (as long as m is totally finite) it is easy to see that in general $f\star_P g$ will not be square integrable, let alone belong to $\mathcal{K}(Z)$. For instance if f is a 1-particle function invariant under ~ (and P is identically equal to 1) then

$$f\star f=\int f^2dm\,\delta_\emptyset+f^2+2\varepsilon_f\chi_2,$$

where χ_2 is the indicator function of $\{\sigma\in\Gamma:\#\sigma=2\}$, so that (if f is real-valued)

$$\int|f\star f|^2\,d\mu=3\|f\|^4+\int|f|^4\,dm,$$

which in general is not finite. (D.Surgailis also noticed that there are elements of the second chaos whose *Poisson square* are not square integrable ([Mey], p.283).)

Since we are interested in algebraic structures we are thus forced to work with a more restrictive class than $\mathcal{K}(Z)$. We choose $\mathcal{K}_0(Z)$ defined as follows:

$$\left\{f : \Gamma_X \to \mathbf{C} \text{ s.t. } |f| < K\varepsilon_\varphi \text{ for some } K \in \mathbf{R}, \varphi \in \bigcap_{2\leq p<\infty} L^p(dm^\dagger)\right\}$$

(c.f. Maassen's original choice of domain). It is easy to see that $\mathcal{K}_0(Z)$ is closed under $\star_P$ (and also $\sim$). Given $(S, \mathcal{M}, j, S_+)$, $\mathcal{K}_0(Z')$ coincides with $\mathcal{K}_0(Z)$ if, and only if, the measures $m'^\dagger$ and $m^\dagger$ are mutually equivalent with Radon-Nikodym derivative in $\mathcal{F}_r(\Gamma)$ (up to a $m^\dagger$-null set). To emphasise this we write $\mathcal{K}_0[Z]$.
We again employ *proof by horoscope* to obtain sufficient conditions for associativity.

Lemma 3.1: *There are no associative products on $\mathcal{K}_0[Z]$ of the form (3.1) unless the twist is essentially the identity map on X. Suppose that $j = \mathrm{id}._X$ and $P \in \mathcal{F}_r(\Gamma^4)$ satisfies the identity*

$$Q_1(\underline{\omega}, \underline{\beta}) = Q_2(\underline{\omega}, \underline{\beta}) \qquad \text{for a.a. } (\underline{\omega}, \underline{\beta}) \in \Gamma_X^4 \times \Gamma_X^7, \tag{3.2}$$

where

$$\begin{aligned} Q_1(\underline{\omega}, \underline{\beta}) =& P(\omega_1 \cup \omega_2 \cup \omega_3, \beta_2 \cup \beta_3 \cup \beta_4, \beta_1 \cup \beta_0 \cup \beta_5, \beta_6) \times \\ & P(\omega_0, \omega_1 \cup \beta_1 \cup \beta_2, \omega_2 \cup \beta_0 \cup \beta_3, \omega_3 \cup \beta_4 \cup \beta_5); \\ Q_2(\underline{\omega}, \underline{\beta}) =& P(\omega_0 \cup \omega_1 \cup \omega_2, \beta_2, \beta_3 \cup \beta_0 \cup \beta_1, \beta_4 \cup \beta_5 \cup \beta_6) \times \\ & P(\omega_3, \omega_0 \cup \beta_3 \cup \beta_4, \omega_2 \cup \beta_0 \cup \beta_5, \omega_1 \cup \beta_6 \cup \beta_1), \end{aligned}$$

then $\star_P$ is associative.

Proof: Let $f, g, h \in \mathcal{K}_0(Z)$. For expressing $(f\star_P g)\star_P h(\sigma)$ and $f\star_P(g\star_P h)(\sigma)$ we use respectively the partitions

$$\begin{array}{cccc} \omega & \alpha_1 & \alpha_2 & \alpha_3 \\ \hline \omega_1 & \beta_2 & \beta_1 & \\ \omega_2 & \beta_3 & \beta_0 & \beta_6 \\ \omega_3 & \beta_4 & \beta_5 & \end{array} \qquad \text{and} \qquad \begin{array}{cccc} \omega & \alpha_1 & \alpha_2 & \alpha_3 \\ \hline \omega_0 & \beta_4 & \beta_3 & \\ \omega_2 & \beta_5 & \beta_0 & \beta_2 \\ \omega_1 & \beta_6 & \beta_1 & \end{array} .$$

Thus

$$
\begin{aligned}
&(f \star_P g) \star_P h(\sigma) \\
&= \int \sum_{|\underline{\alpha}|=\sigma} P(\omega, \alpha_1, \alpha_2, \alpha_3) f \star_P g(\alpha_1 \cup \alpha_2 \cup \tilde{\omega}) h(\omega \cup \alpha_2 \cup \alpha_3)\, d\omega \\
&= \iint \sum_{|\underline{\beta}|=\sigma, |\underline{\omega}|=\omega} P(\omega_1 \cup \omega_2 \cup \omega_3, \beta_2 \cup \beta_3 \cup \beta_4, \beta_1 \cup \beta_0 \cup \beta_5, \beta_6) \times \\
&\qquad P(\omega_0, \tilde{\omega}_1 \cup \beta_2 \cup \beta_1, \tilde{\omega}_2 \cup \beta_3 \cup \beta_0, \tilde{\omega}_3 \cup \beta_4 \cup \beta_5) \times \\
&\qquad f(\tilde{\omega}_1 \cup \tilde{\omega}_2 \cup \tilde{\omega}_0 \cup \beta_0 \cup \beta_1 \cup \beta_2 \cup \beta_3) \times \\
&\qquad g(\omega_0 \cup \tilde{\omega}_2 \cup \tilde{\omega}_3 \cup \beta_0 \cup \beta_3 \cup \beta_4 \cup \beta_5) \times \\
&\qquad h(\omega_1 \cup \omega_2 \cup \omega_3 \cup \beta_0 \cup \beta_1 \cup \beta_5 \cup \beta_6)\, d\omega\, d\omega_0,
\end{aligned}
$$

which, by an application of the $\int$–$\sum$ lemma, equals

$$
\iiiint \sum_{|\underline{\beta}|=\sigma} \text{(same integrand)}\, d\omega_0\, d\omega_1\, d\omega_2\, d\omega_3.
$$

On the other hand

$$
\begin{aligned}
&f \star_P (g \star_P h)(\sigma) \\
&= \int \sum_{|\underline{\alpha}|=\sigma} P(\omega, \alpha_1, \alpha_2, \alpha_3) f(\alpha_1 \cup \alpha_2 \cup \tilde{\omega})\, g \star_P h(\omega \cup \alpha_2 \cup \alpha_3) d\omega \\
&= \iiiint \sum_{|\underline{\beta}|=\omega} Q_2(\underline{\omega}, \underline{\beta}) f(\beta_0 \cup \beta_3 \cup \beta_1 \cup \beta_2 \cup \tilde{\omega}_0 \cup \tilde{\omega}_1 \cup \tilde{\omega}_2) \times \\
&\qquad g(\tilde{\omega}_3 \cup \omega_0 \cup \omega_2 \cup \beta_0 \cup \beta_3 \cup \beta_4 \cup \beta_5) \times \\
&\qquad h(\omega_1 \cup \omega_2 \cup \omega_3 \cup \beta_0 \cup \beta_5 \cup \beta_6 \cup \beta_1)\, d\omega_0\, d\omega_1\, d\omega_2\, d\omega_3.
\end{aligned}
$$

Now choose one-particle functions f, g and h such that the supports of f, g and $\tilde{h}$ lie in S_+. Then $f \star_P (g \star_P h)(\emptyset) = 0$ whereas

$$
(f \star_P g) \star_P h(\emptyset) = \int_{j(S_+)} P(\{s\}, \emptyset, \emptyset, \emptyset) P(\emptyset, \emptyset, \{\bar{s}\}, \emptyset) \bar{g}(s) \tilde{f}(s) h(s)\, ds.
$$

But P is nowhere zero so that $S_- := j(S_+)$ must be null. A similar argument shows that S_+ is also null, in other words for associativity to hold there can be no twist. If, on the other hand, there is no twist then the above expressions show that associativity follows from (3.2).

□

Functions $P \in \mathcal{F}_r(\Gamma^4)$ satisfying (3.2) for *all* $(\underline{\omega}, \underline{\beta}) \in \Gamma^{(11)}$ will be called *Poisson multipliers.* A class of Poisson multipliers is given by pairs $\varepsilon, g \in$

$\mathcal{F}_r(\Gamma)$, where $\varepsilon = \varepsilon_\varphi$ is multiplicative, through

$$P : (\omega, \alpha, \beta, \gamma) \mapsto \varepsilon(\omega) \frac{g(\alpha \cup \beta \cup \omega)\ g(\omega \cup \beta \cup \gamma)}{g(\alpha \cup \beta \cup \gamma)}.$$

These are *trivial multipliers* in the sense that M_g, the multiplication operator $k \mapsto gk$, is an isomorphism from $(\mathcal{K}_o[Z], m, \star_p)$ to $(\mathcal{K}_o[Z], m', \star)$ where $dm' = \varphi dm$. Note that φ is necessarily unit modulus-valued.

Proposition 3.2: *All Poisson multipliers are trivial.*

Proof: Let $P \in \mathcal{F}_r(\Gamma^4)$ be a Poisson multiplier. Notice that the map $(\alpha, \beta, \gamma) \mapsto (\emptyset, \alpha, \beta, \gamma)$ satisfies (2.1) and so is given by (2.2) for some $f \in \mathcal{F}_r(\Gamma)$. Putting $\omega_0 = \omega_1 = \omega_2 = \beta_0 = \beta_3 = \beta_4 = \beta_5 = \emptyset$ in (3.2) gives

$$P(\alpha, \beta, \gamma, \delta) = \frac{f(\alpha \cup \beta \cup \gamma) f(\gamma \cup \delta)}{f(\alpha) f(\beta \cup \gamma \cup \delta)} P(\alpha, \emptyset, \emptyset, \gamma \cup \delta). \tag{3.3}$$

Putting $|\underline{\beta}| = \omega_1 = \omega_2 = \emptyset$ in (3.2) and using (3.3) yields

$$P(\alpha, \emptyset, \emptyset, \beta) = \frac{f(\alpha \cup \beta)}{f(\beta)} g(\alpha),$$

where $g(\alpha) = f(\emptyset) f(\alpha)^{-1} P(\alpha, \emptyset, \emptyset, \emptyset)$. Substituting this into (3.3) now gives

$$P(\alpha, \beta, \gamma, \delta) = \frac{f(\alpha \cup \beta \cup \gamma) f(\alpha \cup \gamma \cup \delta)}{f(\beta \cup \gamma \cup \delta)} h(\alpha), \tag{3.4}$$

where $h = g/f$. Putting this back into (3.2) gives

$$h(\omega_0) h(\omega_1 \cup \omega_2 \cup \omega_3) = h(\omega_0 \cup \omega_1 \cup \omega_2) h(\omega_3).$$

In other words h is a constant times a multiplicative function. Absorbing the constant into f in (3.4) establishes the result.

□

§4. Associativity for $*_p$.

For this section we suppose that the σ-finite non-atomic measure space $X = (S, \mathcal{M}, m)$ is also *separable*, that is there is a countable sub-collection $\mathcal{R}$ of $\mathcal{M}$ (which we may take to be a ring all of whose sets have finite measure) which generates $\mathcal{M} : \mathcal{M} = \sigma(\mathcal{R})$. Let $\mathcal{D}$ be the complex-rational span of the indicator functions of sets from $\mathcal{R}$. For $\tau \subset S, f \in [\mathbf{Q} + i\mathbf{Q}]^\tau$, let $\mathcal{D}_{(\tau,f)}$ denote the functions in $\mathcal{D}$ which extend f. For almost all $\tau \in \Gamma_X$ $\mathcal{D}_{(\tau,f)}$ is dense in $L^2(X)$ for each f in $[\mathbf{Q}+i\mathbf{Q}]^\tau$. This follows since non-atomicity and σ- finiteness ensure that almost every element of X^d $(d \geq 2)$ lies in a rectangle with disjoint sides from $\mathcal{R}$ and that the rectangle may be chosen of arbitrarily small measure.

Proposition 4.1: *Let $Q : \Gamma_X^6 \to \mathbf{C}$ be bounded and satisfy*

$$\sum_{|\underline{\alpha}|=\tau} \iiint Q(\underline{\omega},\underline{\alpha})k_1^{[3]}(\alpha_1,\omega_2,\omega_3)k_2^{[3]}(\omega_1,\alpha_2,\omega_3)k_3^{[3]}(\omega_1,\omega_2,\alpha_3)\,d\underline{\omega} = 0, \tag{4.1}$$

for almost all τ in Γ_X, for each triple k_1, k_2, k_3 from $\mathcal{K}(X)$. Then $Q = 0$ almost everywhere.

The proof depends on the following lemma.

Lemma 4.2: *For almost all τ in Γ_X the family*

$$\{\varepsilon_{uv} \otimes \varepsilon_{vw} \otimes \varepsilon_{wv} : u \in \mathcal{D}_{(\tau,f)}, v \in \mathcal{D}_{(\tau,g)}; w \in \mathcal{D}_{(\tau,h)}\} \tag{4.2}$$

is total in $L^2(\Gamma_X^3)$ for each $f, g, h \in [\mathbf{Q}+i\mathbf{Q}]^\tau$.

Proof: Let $\mathcal{D}'$ be the collection of simple functions $\sum \lambda_i \chi_i$ in which the χ_i are indicator functions of disjoint sets from $\mathcal{R}$ but the λ_i are *squares* of complex rational numbers. For $x, y, z \in \mathcal{D}'$ with common support R let u, v and w be given by

$$u = z'x'/y'; \quad v = x'y'/z'; \quad w = z'y'/x'$$

on R and zero elsewhere, where x', y' and z' are respectively square roots of x, y and z. Then $\varepsilon_{uv} \otimes \varepsilon_{vw} \otimes \varepsilon_{wu} = \varepsilon_x \otimes \varepsilon_y \otimes \varepsilon_z$. Since $\{\varepsilon_x \otimes \varepsilon_y \otimes \varepsilon_z : x, y, z \in \mathcal{D}', \operatorname{supp} x = \operatorname{supp} y = \operatorname{supp} z\}$ is obviously total in $L^2(\Gamma_X^3)$ so also is $\{\varepsilon_{uv} \otimes \varepsilon_{vw} \otimes \varepsilon_{wu} : u, v, w \in \mathcal{D}\}$. The totality of (4.2) now follows by the same argument that establishes the density of $\mathcal{D}_{(\tau,f)}$.

□

Proof of Proposition: Putting $k_i = \varepsilon_{\varphi_i}$ in (4.1) shows that there is a null set $\mathcal{N}$ such that for $\tau \notin \mathcal{N}$

$$\iiint \varepsilon_{\varphi_1\varphi_3}(\omega_2)\varepsilon_{\varphi_1\varphi_2}(\omega_3)\varepsilon_{\varphi_2\varphi_3}(\omega_1) \left(\sum_{|\underline{\alpha}|=\tau} \varepsilon_{\varphi_1}(\alpha_1)\varepsilon_{\varphi_2}(\alpha_2)\varepsilon_{\varphi_3}(\alpha_3)Q(\underline{\omega},\underline{\alpha})\right)d\underline{\omega} = 0,$$

for all $\varphi_i \in \mathcal{D}$. In particular, for $\tau \notin \mathcal{N}$ and $f_i \in [\mathbf{Q}+i\mathbf{Q}]^\tau$,

$$\int \varepsilon_{\varphi_2\varphi_3} \otimes \varepsilon_{\varphi_1\varphi_3} \otimes \varepsilon_{\varphi_1\varphi_2}(\underline{\omega})\Big(\sum_{|\underline{\alpha}|=\tau} \varepsilon_{f_1}(\alpha_1)\varepsilon_{f_2}(\alpha_2)\varepsilon_{f_3}(\alpha_3)Q(\underline{\omega},\underline{\alpha})\Big)d\underline{\omega} = 0,$$

for all $\varphi_i \in \mathcal{D}_{(\tau_1 f_i)}$. By Lemma 4.2 we have, for almost all $(\tau,\underline{\omega}) \in \Gamma_X^4$,

$$\sum_{|\underline{\alpha}|=\tau} \varepsilon_{f_1}(\alpha_1)\varepsilon_{f_2}(\alpha_2)\varepsilon_{f_3}(\alpha_3)Q^{\underline{\omega}}(\underline{\alpha}) = 0.$$

Put f_i equal to the indicator function of β_i $(i = 1,2,3)$ where $\underline{\beta} \in 3^\tau$ (that is $\{\beta_i\}$ are disjoint and $|\underline{\beta}| = \tau$) then for almost all $(\tau,\underline{\omega}) \in \Gamma_X^4$, $P^{\underline{\omega}}(\underline{\beta}) = 0$ for all $\underline{\beta}$ such that $|\underline{\beta}| = \tau$. By the Fubini and $\int$–$\sum$ lemma $P = 0$ almost everywhere and the proposition is proved.

□

Since the proof of Proposition 3.3 in [LP] gives the expression

$$\sum_{|\underline{\alpha}|=\sigma} \iiint Q_p(\underline{\omega},\underline{\alpha})f^{[3]}(\alpha_1,\omega_2,\omega_3)g^{[3]}(\omega_1,\alpha_2,\omega_3)h(\omega_1,\omega_2,\alpha_3)d\omega_1\, d\omega_2\, d\omega_3$$

for the difference $f *_p (g *_p h) - (f *_p g) *_p h$ at σ, where

$$Q_p(\underline{\omega},\underline{\alpha}) = p(\omega_2 \cup \omega_3, \alpha_1, \alpha_2 \cup \alpha_3)\, p(\omega_1, \alpha_2 \cup \omega_3, \alpha_3 \cup \omega_2) - p(\omega_1 \cup \omega_2, \alpha_1 \cup \alpha_2, \alpha_3)\, p(\omega_3, \alpha_1 \cup \omega_2, \alpha_2 \cup \omega_1),$$

we have established

Proposition 4.3: *For $p \in \mathcal{F}_r(\Gamma^3)$, the vanishing of Q_p almost everywhere is necessary and sufficient for the associativity of $*_p$.*

Acknowledgement: This work was carried out whilst JML was visiting the Delhi Centre of the Indian Stastistical Institute. He would like to thank the research, secretarial and library staff for providing a stimulating and supportive environment. A research fellowship under the Royal Society – Indian National Science Academy exchange agreement is gratefully acknowledged.

References

[Gui] A.Guichardet, *Symmetric Hilbert Spaces and Related Topics*, Springer LNM **261** (1972) Heidelberg.

[L] J.M.Lindsay, *On set convolutions and integral-sum kernel operators* ,(In preparation).

[LM 1] J.M.Lindsay and H.Maassen, *An integral kernel approach to noise*, in Quantum Probability and Applications III, (L.Accardi and W.von Waldenfels eds.) Springer LNM **1303** (1988) 192-208.

[LM 2] J.M.Lindsay and H.Maassen, *The stochastic calculus of Bose noise*, Preprint (1988).

[LM 3] J.M.Lindsay and H.Maassen, *Duality transform as *-algebraic isomorphism*, These proceedings (1989).

[LP] J.M.Lindsay and K.R.Parthasarathy, *Cohomology of power sets with applications in quantum probability*, Commun. Math. Phys. **124** (1989) 337-364.

[Maa] H.Maassen, *Quantum Markov processes on Fock space described by integral kernels*, in Quantum Probability and Applications II, (L.Accardi and W.von Waldenfels eds.) Springer LNM **1136** (1985) 361-374.

[Mey] P-A.Meyer, *Eléments de probabilités quantiques I,II*, Springer LNM **1204, 1247** (1986) 186-312, 38-80.

Mathematics Department
King's College
Strand
London WC2R 2LS
U.K.

Indian Statistical Institute
(Delhi Centre)
7 SJS Sansanwal Marg
New Delhi 110016
INDIA

This paper is in final form and no similar paper has been or is being submitted elsewhere.

A discrete entropic uncertainty relation

H. Maassen*
University of Nijmegen,the Netherlands

Recently [MaU] a new class of 'generalised entropic' uncertainty relations for the probability distributions of non-commuting random variables was proved as a simple consequence of the Riesz-Thorin interpolation theorem. Here we shall give a quite explicit proof of the central inequality of this class, an 'entropic' uncertainty relation, which has been conjectured by Kraus [Kra].

We consider the following situation, not uncommon in quantum mechanics. Two observables of a physical system are represented by symmetric complex $n \times n$ matrices A and B, which we shall assume to have non-degenerate spectra. We can write A and B in the form

$$A = \sum_{i=1}^{n} \alpha_i P_i \quad \text{and} \quad B = \sum_{i=1}^{n} \beta_i Q_i,$$

where $P_1, \cdots, P_n$ and $Q_1, \cdots, Q_n$ are sequences of mutually orthogonal one-dimensional projections, and the sequences $\alpha_1, \cdots, \alpha_n$ and $\beta_1, \cdots, \beta_n$ consist of distinct real numbers, to be interpreted as the values which the observables can take. Each state ω on the algebra M_n of all complex $n \times n$ matrices then induces probability distributions on the spectra of A and B: $\omega(P_i)$ (or $\omega(Q_i)$) is the probability to find the value α_i (or β_i) when measuring the observable A (or B). One now defines the *uncertainty* $H(A,\omega)$ *of* A *in the state* ω as the Shannon entropy of this probability distribution:

$$H(A,\omega) = -\sum_{i=1}^{n} \omega(P_i) \log \omega(P_i).$$

The question was raised ([BBM], [Deu], [Kra]), what can be said about $H(A,\omega) + H(B,\omega)$, more in particular about its lower bound

$$d(A,B) = \inf_{\omega} (H(A,\omega) + H(B,\omega)).$$

One may regard this infimum as a "degree of incompatibility" of the observables A and B.

As a first reduction, let us note that $H(A,\omega)$ does not depend on the real numbers $\alpha_1, \cdots, \alpha_n$, but only on the projections $P_1, \cdots, P_n$, which are the minimal projections in the maximal abelian von Neumann algebra $\mathcal{A}$ generated by A. Let us therefore write $H(\mathcal{A},\omega), H(\mathcal{B},\omega)$ and $d(\mathcal{A},\mathcal{B})$ in what follows. When viewed in this way, d becomes a natural distance function between maximal abelian von Neumann algebras, comparable to the distance of point *sets* (not of points) in geometry: $d(\mathcal{A},\mathcal{B}) = 0$ if and only if $\mathcal{A}$ and $\mathcal{B}$ have a minimal projection in common.

*Supported by the Netherlands Organisation of Scientific Research NWO.

The latter observation suggests to consider the following easily computable functional on pairs of abelian von Neumann algebras in M_n:

$$m(\mathcal{A},\mathcal{B}) = \max\{\mathrm{tr}PQ | P \in \mathcal{A}, Q \in \mathcal{B} \text{ minimal projections}\}.$$

This definition amounts to

$$m(\mathcal{A},\mathcal{B}) = \max_{1\leq i,j\leq n} \mathrm{tr}(P_i Q_j) = \max_{1\leq i,j\leq n} |<e_i, f_j>|^2,$$

where e_i and f_j are unit vectors in the ranges of P_i and Q_j respectively. Note that

$$m(\mathcal{A},\mathcal{B}) \leq 1,$$

with equality if and only if $d(\mathcal{A},\mathcal{B}) = 0$. On the other hand, since for all j

$$\sum_{i=1}^{n} \mathrm{tr}(P_i Q_j) = \mathrm{tr} Q_j = 1,$$

we have

$$m(\mathcal{A},\mathcal{B}) \geq \frac{1}{n}.$$

It was observed by Kraus [Kra] that this lower bound is reached for 'complementary' observables A and B, which corresponds to e_j and f_j of the form

$$(e_j)_k = 1 \text{ if } j = k, 0 \text{ otherwise};$$

$$(f_j)_k = \frac{1}{\sqrt{n}} e^{\frac{2\pi i j k}{n}}.$$

Note that the algebras $\mathcal{A}$ and $\mathcal{B}$ then take the form:

$$\mathcal{A} = \{X \in M_n | X \text{ diagonal}\},$$

$$\mathcal{B} = \{X \in M_n | X_{i+k,j+k} = X_{i,j} \text{ for all } i,j,k\},$$

where the addition of indices is taken modulo n.

Kraus went on to conjecture that for such complementary observables (or algebras in our terminology)

$$d(\mathcal{A},\mathcal{B}) = \log n.$$

Indeed, the inequality $d(\mathcal{A},\mathcal{B}) \leq \log n$ is easily established: choose $\omega(X) = <e_1, Xe_1>$, so that $(\omega(P_1),\cdots,\omega(P_n)) = (1,0,\cdots,0)$ and $(\omega(Q_1),\cdots,\omega(Q_n)) = (\frac{1}{n},\frac{1}{n},\cdots,\frac{1}{n})$; then $H(\mathcal{A},\omega) = 0$ and

$$H(\mathcal{B},\omega) = -\sum_{i=1}^{n} \frac{1}{n} \log \frac{1}{n} = \log n.$$

Kraus' conjecture is therefore a consequence of the following theorem.

Theorem 1 *For all maximal abelian von Neumann subalgebras $\mathcal{A}$ and $\mathcal{B}$ of M_n one has*

$$d(\mathcal{A},\mathcal{B}) \geq -\log m(\mathcal{A},\mathcal{B}).$$

Proof. Let $\{P_1, \cdots, P_n\}$ and $\{Q_1, \cdots, Q_n\}$ be complete sets of minimal projections in $\mathcal{A}$ and $\mathcal{B}$ respectively, and let e_j and f_j be unit vectors in the ranges of P_j and Q_j respectively $(j = 1, \cdots, n)$. We may assume that $\{e_j\}$ is the canonical basis of $\mathbf{C}^n$. From the concavity of the function $\eta : [0,1] \to [0,\infty) : x \mapsto -x \log x$ (with $\eta(0) := 0$) it follows that the minimum of $H(\mathcal{A},\omega) + H(\mathcal{B},\omega) = \sum_{j=1}^n (\eta(\omega(P_j)) + \eta(\omega(Q_j)))$ is taken in a vector state $\omega(X) = <\psi, X\psi>$ on M_n. It therefore suffices to prove that for all $\psi \in \mathbf{C}^n$:

$$\sum_{j=1}^n (\eta(|<e_j,\psi>|^2) + \eta(|<f_j,\psi>|^2)) \geq -\log \max_{i,j} |<e_i, f_j>|^2. \tag{1}$$

Now let $m = m(\mathcal{A},\mathcal{B}) = \max_{i,j} |<e_i, f_j>|^2$ and let a unitary map $T : \mathbf{C}^n \to \mathbf{C}^n$ be defined by $Tf_j = e_j$. Then $\psi_i = <e_i, \psi>$ and $(T\psi)_i = <e_i, T\psi> = <T^{-1}e_i, \psi> = <f_i, \psi>$. If we now write $h(\psi)$ for $\sum_{j=1}^n \eta(|\psi_j|^2)$, then the inequality takes the form

$$h(\psi) + h(T\psi) \geq -\log m. \tag{2}$$

For $n \in \mathbf{N}$ and $p \in [1,\infty]$ let $l^p(n)$ denote the Banach space $\mathbf{C}^n$ with norm

$$\|\psi\|_p = \begin{cases} (\sum_{i=1}^n |\psi_i|^p)^{\frac{1}{p}} & \text{if } 1 \leq p < \infty \\ \max_{1 \leq i \leq n} |\psi_i| & \text{if } p = \infty. \end{cases}$$

We now make the following observation.

Lemma 2 *For all unit vectors ψ in $\mathbf{C}^n$*

$$\frac{d}{dp^{-1}} \|\psi\|_p \big|_{p=2} = h(\psi).$$

Proof. First we note that

$$\frac{d}{dp}\|\psi\|_p^p\big|_{p=2} = \frac{d}{dp}\sum_{j=1}^n |\psi_j|^p\big|_{p=2} =$$

$$\sum_{j=1}^n |\psi_j|^2 \log|\psi_j| = -\tfrac{1}{2}\sum_{j=1}^n \eta(|\psi_j|^2) = -\tfrac{1}{2}h(\psi).$$

Therefore, since $\|\psi\|_2 = 1$,

$$\frac{d}{dp^{-1}}\|\psi\|_p\big|_{p=2} = \frac{d}{dp^{-1}}(\|\psi\|_p^p)^{p^{-1}}\big|_{p=2} =$$

$$\log\|\psi\|_2^2 + \frac{1}{p}(\|\psi\|_p^p)^{\frac{1}{p}-1} \cdot \frac{dp}{dp^{-1}} \cdot \frac{d}{dp}\|\psi\|_p^p\big|_{p=2}$$

$$= \tfrac{1}{2} \cdot (-4)(-\tfrac{1}{2}h(\psi)) = h(\psi).$$

□

Let $\|T\|_p$ denote the norm of the linear map $T : \mathbf{C}^n \to \mathbf{C}^n$, viewed as an operator $l^p(n) \to l^q(n)$, where $\frac{1}{p} + \frac{1}{q} = 1$. (Here we make the usual convention that $\frac{1}{\infty} = 0$.) Then $\|T\|_2 = 1$ and $\|T\|_1 = \max_{j,k} |T_{jk}| = \max_{j,k} |<f_j, e_k>| = \sqrt{m}$.

Theorem 3 (Riesz-Thorin interpolation) *For a linear map $T : \mathbf{C}^n \to \mathbf{C}^n$ the function*

$$f_T : [0,1] \to \mathbf{R} : \frac{1}{p} \mapsto \log \|T\|_p$$

is convex.

Proof: [Rie]; see also [HLP]. □

It follows that f_T has a right derivative $f_T'(\frac{1}{2})$ at $\frac{1}{2}$, and that, since $f_T(\frac{1}{2}) = \log \|T\|_2 = 0$ and $f_T(1) = \log \|T\|_1 = \frac{1}{2} \log m$, we have

$$f_T'(\tfrac{1}{2}) \le \frac{f_T(1) - f_T(\frac{1}{2})}{1 - \frac{1}{2}} = \log m.$$

On the other hand, by the definition of the operator norm $\|T\|_p$, we have for all $p \in [1, \infty]$ and all unit vectors $\psi \in \mathbf{C}^n$:

$$\log \|T\|_p \ge \log \|T\psi\|_q - \log \|\psi\|_p,$$

where $\frac{1}{p} + \frac{1}{q} = 1$. Equality holds here for $p = 2$, hence we may differentiate with respect to $\frac{1}{p}$ at $\frac{1}{p} = \frac{1}{2}$:

$$f_T'(\tfrac{1}{2}) \ge -h(T\psi) - h(\psi).$$

It follows that $h(T\psi) + h(\psi) \ge -\log m$. □

The equality (2) is optimal if $|T_{ij}| = 1$ for some pair (i,j) and in the case of complementary observables, when

$$T_{jk} = \frac{1}{\sqrt{n}} e^{\frac{2\pi i jk}{n}}.$$

In general however, f_T will be strictly convex, so that $f_T'(\frac{1}{2}) < \log m$ and no ψ exists reaching equality in (2).

References

[BBM] I. Białynicki-Birula, J. Mycielsky: "Uncertainty Relations for Information Entropy in Wave Mechanics"; Commun. math. Phys. *44* (1975) 129-132.

[Deu] D. Deutsch: "Uncertainty in quantum measurements" Phys. Rev. Lett. *50* (1983) 631-633.

[HLP] G. Hardy, J.E. Littlewood, G. Pólya: Inequalities. Cambridge University Press 1934.

[Kra] K. Kraus: "Complementary observables and uncertainty relations"; Phys. Rev. D *35* (1987) 3070-3075.

[MaU] H. Maassen, J. Uffink: "Generalized Entropic Uncertainty Relations"; Phys. Rev. Lett. *60* (1988) 1103-1106.

[Rie] M. Riesz: "Sur les maxima des formes bilinéaires et sur les fonctionnelles linéaires"; Acta Math. *49* (1927) 465-497.

Working with Quantum Markov States and their classical analogues *†

B. Nachtergaele ‡
Instituut voor Theoretische Fysica
Universiteit Leuven, B-3030 Leuven, Belgium §

Abstract

We investigate in detail the structure of the set of Quantum Markov States (QMS), as defined by Accardi and Frigerio. A detailed study of their classical analogues reveals that they contain a lot more than just the Markovian measures. Furthermore we prove a canonical representation theorem and this result is then used to analyze the consequences of group invariance for a QMS. The theory is applied in a variational approximation for the ground state of the antiferromagnetic Heisenberg model. We thus show that the QMS might provide an interesting new tool to investigate ground state problems for 1-dimensional quantum spin systems.

I Introduction

The search for a natural notion of Markovianity for Quantum States, that is both tractable and general enough, has still not come to an end. For the moment, several candidates are put forward by different authors [1,2,3] and any proposal has specific advantages and disadvantages w.r.t. mathematical properties. To our opinion, the respective relevance of the different notions for the Quantum Processes which arise from physical problems, might serve as an additional useful argument in the discussion. In this contribution, we communicate preliminary results which were recently obtained about the Quantum Markov States (QMS) as defined by Accardi and Frigerio [4]. We investigate the applicability of the QMS in the study of ground-state properties of 1-dimensional Quantum Spin systems. We also study the classical analogue of QMS, partially for the sake of motivation, partially for their own interest. A more complete account of our study will be given in future papers [5].

We mainly deal with the following two questions:

1. How can we use the structure and the specific properties of QMS in calculations; how can they be classified and even parametrized in some important cases?

2. For what kind of physical problems do they have the right properties (e.g. can they coincide with or at least approximate ground states or equilibrium states corresponding to "physical" hamiltonians?).

---

*Talk held at the 4th Workshop on Quantum Probability and Applications, Heidelberg, September 1988
†The paper is in final form and no similar paper has been or is being submitted elsewhere
‡Onderzoeker I.I.K.W., Belgium
§Present address: Departamento de Física, Universidad de Chile, Casilla 487-3, Santiago de Chile.

As we already said, we only present preliminary results and this is true for both questions.

About 1. It seemed to be quite interesting to make a thorough study of the construction of Accardi and Frigerio (in a generalized form) in the case of abelian algebras (often called the "classical case"). In this case, one recovers not just the classical Markov processes, but a largely more general class of measures: the so called functions of Markov processes. We introduce a class of measures which we call "algebraic measures" because of their nice computational properties and it turns out that they provide an elegant tool to calculate with the probability measures corresponding to functions of Markov processes. Various types of correlation functions can be easily computed, and in [5] an expression for the mean entropy of such a measure is derived.

Both in the classical and in the quantum case, one can derive a kind of a representation theorem which will enable us to parametrize certain subclasses of the QMS, namely those which have an additional group invariance (translation invariance is always assumed). A problem which is under investigation now is the characterization of the pure states among the QMS.

About 2. One-dimensional quantum spin systems are presently studied with a big variety of approximation methods, not only aiming at approximate values for the ground-state energy but in many cases one is also interested in correlation functions etc. Quite often, these methods have the defect that one cannot be sure that the results are not in contradiction with the principles of Quantum Mechanics itself, i.e. the set of correlation functions that is obtained cannot necessarily be extended to a state of the system (in the sense of a normalized positive linear functional on the algebra of observables). An approximation method that is completely save in this respect is the so called Hartree-Fock or variational method. In this method, one minimizes a functional of the set of all states (which attains its absolute minimum in the true solution(s) of the problem) over a restricted set of well-described states. (Notice that in the literature the term "Hartree-Fock" approximation is often used to indicate a particular type of trial states, e.g. quasi-free states or Slater determinants or in other cases product states.) To solve this "restricted" variational problem, one needs detailed knowledge about the set of all trial states, and so there are not too many possibilities: product states (or quasi-free states in another context), spin waves, classical states,... and that is more or less it.

Now, if we succeed in analyzing the QMS in sufficient detail, we could apply them in such variational problems. An example given at the end of this paper shows that doing so, one can obtain considerably improved results compared with e.g. product-state approximations. Furthermore, we believe that there are special Hamiltonians that have QMS as their exact ground states.

The paper contains the following sections:

II. Definitions

III. Analysis of the abelian case

IV. A representation theorem

V. Quantum Markov States with a local symmetry

VI. Application to the Heisenberg model

VII. Conclusion

References

II Definitions

We are interested in translation-invariant states for quantum chains, i.e. one-dimensional quantum spin systems. Classical homogeneous stochastic processes with discrete time will be considered as a special case. To avoid any confusion about notations and definitions, we will spend some time here to explain what are the ingredients of the theory.

The observables at one site (or at a given moment in time in the case of classical stochastic processes) form a complex unital C^*-algebra $\mathcal{A}$, and we consider a copy of $\mathcal{A}$, attached at site $i \in \mathbb{Z}$, denoted by $\mathcal{A}_i$. For any finite subset $\Lambda \subset \mathbb{Z}$, the algebra of observables localized in the volume Λ is given by a tensor product of copies of $\mathcal{A}$, and denoted by $\mathcal{A}_\Lambda$:

$$\mathcal{A}_\Lambda = \otimes_{i\in\Lambda}\mathcal{A}_i.$$

By an inductive limit, we then define the C^*-algebra $\mathfrak{A}$ of quasi-local observables:

$$\mathfrak{A} = \overline{\cup_\Lambda \mathcal{A}_\Lambda}$$

where the bar denotes the C^*-norm closure. The embedding of $\mathcal{A}_\Lambda$ in $\mathcal{A}_{\Lambda'}$ for any $\Lambda \subset \Lambda' \subset \mathbb{Z}$ is the natural one. We consider states ω of $\mathfrak{A}$, i.e. positive normalized linear functionals of the algebra. On $\mathfrak{A}$, we define the translations as a group of automorphisms, forming a representation of $\mathbb{Z},+$; for any $a \in \mathbb{Z}$ there is a unique automorphism τ_a of $\mathfrak{A}$ satisfying for all $x = x_i \otimes x_{i+1} \otimes \cdots \otimes x_j \in \mathcal{A}_{[i,j]}$, $i \leq j$:

$$\tau_a(x) = y_{i+a} \otimes \cdots \otimes y_{j+a} \in \mathcal{A}_{[i+a,j+a]}$$

such that

$$y_{k+a} = x_k \qquad \text{for all } k,\ i \leq k \leq j.$$

In the theorems, we will be dealing only with translation-invariant states ω, i.e.

$$\omega(\tau_a(x)) = \omega(x) \qquad \text{for all } x \in \mathfrak{A}, a \in \mathbb{Z}.$$

Furthermore, we mostly assume we are in the simplest case where $\mathcal{A} = M_d$ (the $d \times d$ complex matrices) or $\mathcal{A} = C(I)$ (the continuous functions on I, where I is a compact space or even a inite set). Generalizations of the results are possible in several ways, but are beyond the scope of this paper.

The most convenient way to define a state ω of $\mathfrak{A}$ is by giving its local restrictions ω_Λ:

$$\omega_\Lambda = \omega|_{\mathcal{A}_\Lambda}$$

or all finite $\Lambda \subset \mathbb{Z}$. A family of states $\{\omega_\Lambda : \mathfrak{A}_\Lambda \to \mathbb{C} \mid \Lambda \subset \mathbb{Z}\}$ gives rise to a state ω of $\mathfrak{A}$ iff he compatibility conditions are satisfied, i.e.:

$$\omega_{\Lambda'}|_{\mathcal{A}_\Lambda} = \omega_\Lambda \qquad \text{for all } \Lambda \subset \Lambda' \subset \mathbb{Z}.$$

The basic ingredients in the construction of Accardi and Frigerio (for translation-invariant states) consist of a unity-preserving completely positive map (a C.P.U. map) E:

$$E : \mathcal{A} \otimes \mathcal{A} \to \mathcal{A}$$

and a state ϕ of $\mathcal{A}$, which satisfy the following relation:

$$\phi(E(\mathbb{1}\otimes x)) = \phi(x) \qquad \text{for all } x \in \mathcal{A}. \tag{1}$$

For any $\Lambda = [a,b] \equiv \{a, a+1, \ldots, b\} \subset \mathbb{Z}$ and $x_i \in \mathcal{A}$, $a \leq i \leq b$, one then defines

$$\begin{aligned} &\omega_\Lambda(x_a \otimes x_{a+1} \otimes \cdots \otimes x_b) \\ &= \phi(E(x_a \otimes E(x_{a+1} \otimes E(x_{a+2} \otimes \cdots \otimes E(x_b \otimes \mathbb{1})\cdots)))). \end{aligned} \tag{2}$$

We have the following construction theorem:

Theorem II.1 *For any unital C^*-algebra $\mathcal{A}$, C.P.U. map E and state ϕ satisfying (1), there is a unique family of compatible local states $\{\omega_\Lambda \mid \Lambda \subset \mathbb{Z}\}$ satisfying (2).*
Thus, due to (1), we have defined a translation-invariant state ω of $\mathfrak{A}$.

Proof: Because E is unity preserving, it is immediately clear that (2) will give rise to a state of $\mathfrak{A}$, if for any Λ, it can be linearly extended to a positive functional. Due to the translation-invariance (which is built in in the construction) and the fact that the relations (2) are compatible (by (1) and the fact that E is unity preserving), we can restrict our attention to intervals of the type $[0,n] \subset \mathbb{Z}$. To present the argument, it is convenient to introduce the following extensions of E, for all $m = 1, 2, \ldots$

$$\begin{aligned} E^{(m)}: \quad & \mathcal{A}_{[0,m]} \to \mathcal{A}_{[0,m-1]} \\ & x_0 \otimes \cdots \otimes x_m \to x_0 \otimes \cdots \otimes x_{m-2} \otimes E(x_{m-1} \otimes x_m). \end{aligned} \tag{3}$$

Clearly, all $E^{(m)}$ are (completely) positive, and one can now write $\omega_{[0,n]}$ as a composition of linear positive maps:

$$\begin{aligned} &\omega_{[0,n]}(x_0 \otimes \cdots \otimes x_n) \\ &= \phi \circ E^{(1)} \circ E^{(2)} \circ \cdots \circ E^{(n+1)}(x_0 \otimes \cdots x_n \otimes \mathbb{1}) \\ &= \phi \circ E^{(1)} \circ E^{(2)} \circ \cdots \circ E^{(n+1)} \circ i_{n+1}(x_0 \otimes \cdots x_n) \end{aligned}$$

where

$$i_{n+1} : \mathcal{A}_{[0,n]} \to \mathcal{A}_{[0,n+1]} : x \to x \otimes \mathbb{1}$$

is also a linear positive map. ∎

Remark that complete positivity of E is indeed required, namely to assure that the extensions $E^{(m)}$ defined in (3) are positive. In the classical case Theorem II.1 is nothing more than Kolmogorov's Extension Theorem.

III Analysis of the abelian case

We start from a slightly different way of presenting QMS. For simplicity, assume $\mathcal{A} \subseteq M_d$. For each $x \in \mathcal{A}$, define the following linear map E_x:

$$E_x : \mathcal{A} \to \mathcal{A} : y \to E(x \otimes y).$$

The normalization and translation invariance of the state ω can now be expressed in terms of $E_{\mathbb{1}}$ and its dual $E^*_{\mathbb{1}}$:

$$\text{normalization}: \quad E_{\mathbb{1}}(\mathbb{1}) = \mathbb{1} \tag{4}$$
$$\text{translation invariance}: E_{\mathbb{1}}^*(\phi) = \phi. \tag{5}$$

It will become yet more clear later that $E_{\mathbb{1}}$ fulfills somehow the role of a Markov operator (stochastic matrix) in the classical theory of Markov processes.

The state ω is expressed in terms of the E_x as follows:

$$\begin{aligned} &\omega(x_1 \otimes \cdots \otimes x_n) \\ &= \phi(E(x_1 \otimes E(x_2 \otimes \cdots \otimes E(x_n \otimes \mathbb{1})))) \\ &= \phi(E_{x_1} \circ E_{x_2} \circ \cdots \circ E_{x_n}(\mathbb{1})) \\ &= \langle \rho \mid E_{x_1} \cdots E_{x_n} \mathbb{1} \rangle \end{aligned}$$

where we now consider $\mathcal{A}$ as a Hilbert space with inner product $\langle \cdot \mid \cdot \rangle$ defined by

$$\langle A \mid B \rangle = \text{Tr } A^* B \qquad A, B \in \mathcal{A}$$

with Tr the unnormalized trace on $\mathcal{A} \subseteq M_d$.

It is easy to represent a classical Markovian measure on $I^{\mathbb{Z}}$ in this scheme. For later convenience, we take $I = \{0, \ldots, d-1\}$. Suppose μ is a translation-invariant Markovian measure on $I^{\mathbb{Z}}$ with local densities μ_Λ. Then there exist probability distributions ρ^n on I^n such that:

$$\mu_{[a,a+n-1]}(i_1, \ldots, i_n) = \frac{\rho^2(i_1, i_2)\rho^2(i_2, i_3) \cdots \rho^2(i_{n-1}, i_n)}{\rho^1(i_2) \cdots \rho^1(i_{n-1})}.$$

This kind of measures and their natural generalizations to the n-dependent Markovian case solve the DLR-equations for the finite-range Hamiltonians of classical one-dimensional lattice gases (see e.g. [6,7,8]). Unfortunately, this kind of analogy cannot be kept in the quantum case.

As $\mathcal{A} = C(I)$, the linear mappings E_x can be represented by $d \times d$ matrices; for all $i \in I$ define

$$(E_i)_{k,l} = \delta_{i,k} \frac{\rho^2(k, l)}{\rho^1(k)}.$$

The quotient is defined to be equal to zero whenever the denominator vanishes. So E_i differs from zero only on its ith row, which has as entries the conditional probabilities

$$c(i, j) = \frac{\rho^2(i, j)}{\rho^1(i)}.$$

The local densities μ_Λ can now be written in the general form for QMS:

$$\begin{aligned} \mu_\Lambda(i_a, \ldots i_b) &= \sum_{i,j} \rho^1(i)(E_{i_a} \cdots E_{i_b})_{ij} \\ &= \langle \rho^1 \mid E_{i_a} \cdots E_{i_b} \mathbb{1} \rangle \end{aligned} \tag{6}$$

where ρ^1 and $\mathbb{1}$ are considered as elements of $C(I)$. In terms of the algebra of observables $C(I^{\mathbb{Z}})$, (6) reads:

$$\int d\mu(i_a \cdots i_b) f_a(i_a) \cdots f_b(i_b) = \langle \rho^1 \mid E_{f_a} \cdots E_{f_b} \mathbb{1} \rangle$$

where for all $f \in C(I)$

$$E_f = \sum_{i \in I} f(i) E_i$$

and then $E_{\mathbb{1}}$ is a stochastic matrix satisfying

$$E_{\mathbb{1}}\mathbb{1} = \mathbb{1}$$
$$E^*_{\mathbb{1}}\rho^1 = \rho^1.$$

A first generalization of classical Markov processes arises immediately from this representation. Indeed, take a collection of $d \times d$ matrices $\{E_i \mid i \in I\}$ with nonnegative entries, such that $E_{\mathbb{1}} = \sum_i E_i$ leaves $\mathbb{1}$ invariant. Then there exists a nonnegative vector ρ satisfying

$$E^*_{\mathbb{1}}\rho = \rho$$

and one can define a measure on $I^{\mathbb{Z}}$ by relation (6). Thus one obtains a much larger class of measures than merely the Markovian ones. Let we call them Generalized Markovian Measures (GMM), just for later reference in this paper. One can generalize still further by taking the E_i to be positive elements in an abstract ordered algebra. This will be done in [5]. One of the applications is a nice characterization of the measures which are associated to functions of Markov processes. Because of their elegant computational properties, we call them algebraic measures. Also for the Quantum Markov States, a similar generalization can be made.

We conclude this section by studying the relation between the GMM and functions of Markov processes. We repeat that by a GMM, we mean a measure μ on $\{0,\ldots,d-1\}^{\mathbb{Z}}$ defined by a set of $d \times d$ matrices $\{E_i \mid i \in I\}$ with nonnegative entries as discussed above in (6). To formulate the result, it is convenient (certainly not necessary) to consider the stochastic process given by random variables $(X_i)_{i\in\mathbb{Z}}$ taking values in $I = \{0,\ldots,d-1\}$, which corresponds to the measure μ.

Theorem III.1 *Let $X = (X_i)_{i\in\mathbb{Z}}$ be a stochastic process with values in I, corresponding to a GMM μ on $I^{\mathbb{Z}}$, defined as above in (6).*

Then there exists a finite state space $\tilde{I}$ and a Markov process $\widetilde{X} = (\widetilde{X}_i)_{i\in\mathbb{Z}}$ taking values in $\tilde{I}$, and a function

$$F : \tilde{I} \to I$$

such that $X = F(\widetilde{X})$ i.e.:

$$(X_i)_{i\in\mathbb{Z}} = (F(\widetilde{X}_i))_{i\in\mathbb{Z}}.$$

One can always take $\tilde{I} = I^4$.

Proof: First define an auxiliary function f as:

$$f : I \times I \to I : (x_1, x_2) \to (x_1 + x_2) \bmod d$$

and a nonnegative $d^2 \times d^2$ matrix T by its matrix elements:

$$T_{(p_0,q_0),(p_1,q_1)} = \left(E_{f(q_0,q_1)}\right)_{p_0,p_1}.$$

Obviously, the function f has the following property: for each x and $p_1 \in I$, the equation $f(p_0, p_1) = x$ has one and only one solution. This enables us to write:

$$\begin{aligned}\mu(x_1,\ldots,x_n) &= \langle \rho \mid E_{x_1} \cdots E_{x_n}\mathbb{1}\rangle \\ &= \sum_{p_0,\ldots,p_n\in I} \rho_{p_0}(E_{x_1})_{p_0,p_1}(E_{x_2})_{p_1,p_2}\cdots(E_{x_n})_{p_{n-1},p_n}\end{aligned}$$

$$= \frac{1}{d} \sum_{p_0,\ldots,p_n \in I} \sum_{\substack{q_0,\ldots,q_n \in I \\ f(q_0,q_1)=x_1,\ldots,f(q_{n-1},q_n)=x_n}} \rho_{p_0} T_{(p_0,q_0),(p_1,q_1)} \cdots T_{(p_{n-1},q_{n-1}),(p_n,q_n)}.$$

Defining a vector $t \in \mathbb{R}^{d^2}$ by

$$t_{p,q} = \frac{1}{d}\rho_p \qquad p, q \in I$$

one checks immediately that

$$T\mathbb{1} = \mathbb{1}\ , \quad T^* t = t \quad \text{and} \quad \langle t \mid \mathbb{1} \rangle = 1$$

where in the above expressions $\mathbb{1}$ is now the constant vector $(1, 1, \ldots, 1) \in \mathbb{R}^{d^2}$. Due to the properties above, we can introduce a translation-invariant Markovian measure λ on $(I \times I)^{\mathbb{Z}}$. We call the corresponding random variables $\omega_i = (p_i, q_i)$. The values they take we also denote by $\omega_1, \omega_2, \ldots$:

$$\lambda(\omega_1, \ldots, \omega_n) = t_{\omega_1} T_{\omega_1,\omega_2} T_{\omega_2,\omega_3} \cdots T_{\omega_{n-1},\omega_n}.$$

Introducing a function g by

$$g(\omega_1, \omega_2) = f(q_1, q_2) \qquad \omega_1 = (p_1, q_1), \omega_2 = (p_2, q_2),$$

we have the following relation between λ and μ:

$$\mu(x_1, \ldots, x_n) = \sum_{\substack{\omega_0,\ldots,\omega_n \in I \times I \\ g(\omega_0,\omega_1)=x_1,\ldots,g(\omega_{n-1},\omega_n)=x_n}} \lambda(\omega_0, \ldots, \omega_n) \tag{7}$$

or at the level of the random variables:

$$X_i = g(\omega_{i-1}, \omega_i). \tag{8}$$

It is now clear that the variables $\widetilde{X}_i$ required in the theorem can be taken to be (ω_{i-1}, ω_i) and $\widetilde{I} = ((I \times I) \times (I \times I))$ and the function F as:

$$F : \widetilde{I} \to I : (\omega_0, \omega_1) = (p_0, q_0, p_1, q_1) \to g(\omega_0, \omega_1).$$

Then (8) can be rewritten as

$$X_i = F(\widetilde{X}_i).$$

On the level of the measures, we have to extend λ to a new translation-invariant Markovian measure ν as follows:

$$\begin{aligned} \nu(y_1, \ldots, y_n) &= \lambda(\omega_0, \ldots, \omega_n) \ \text{ if } y_j = (\omega_{j-1}, \omega_j) \ \ j = 1, \ldots, n \\ &= 0 \qquad\qquad\qquad\quad \text{otherwise.} \end{aligned}$$

(7) then becomes:

$$\mu(i_1, \ldots, i_n) = \sum_{\substack{y_1,\ldots,y_n \in \widetilde{I} \\ F(y_1)=i_1,\ldots,F(y_n)=i_n}} \nu(y_1, \ldots, y_n).$$

■

We still mention that, up to the statement $\tilde{I} = I^4$, the same theorem holds in the more general case where the dimension of the defining matrices E_i is not necessarily equal to $d = \#I$ (see [5]).

Remark III.2 *The QMS of an abelian algebra exactly coincide with the Generalized Markovian Measures discussed above, which in general are non-Markovian.*

It is also easy to see that a random function of a GMM, in particular of a Markov Process, has a description as a GMM. By a random function of a stochastic process we mean (formulated for the restricted case of a finite state space per value of the discrete time parameter) the process (Y_i) which arises from the original process (X_i) by putting:

$$Prob(Y_i = y \mid X_i = x) = F_{yx}$$

where Y_i is taking values $y \in J$ and where for any $x \in I, F_{yx}$ is a probability function on J. If the GMM μ defined by the matrices $\{E_x \mid x \in I\}$ and the vector ρ, describes (X_i), then (Y_i) is described by the GMM ν defined by $\{E'_y = \sum_{x \in I} F_{yx} E_x \mid y \in J\}$ and the same ρ. So the set of all GMM with given state space I is closed for "noisy observation".

As a corollary of Theorem III.1 we therefore have that any random function of a Markov Process can also be represented as a deterministic function of a (in general different) Markov Process.

We conclude this section with the remark that this treatment of functions of Markov Processes is very much related to the theory of Stochastic Modules (see [9] and references therein).

IV A representation theorem

In this section, we study QMS as defined in section II (see (2) and Theorem II.I). We will treat only the simplest case where $\mathcal{A} \subseteq M_d$. So we deal with a triple $(\rho, E, \mathbb{1})$ defining a state of $\mathfrak{A}$. This triple could also be represented as $(\phi, \tilde{E}, \psi)$ where ϕ and ψ are now vectors in $\mathcal{A}$ with the scalar product $\langle \phi \mid \psi \rangle = \mathrm{Tr}\, \phi^* \psi$ and where $\tilde{E}$ is a vectorspace representation

$$\tilde{E} : \mathcal{A} \to \mathcal{B}(\mathcal{A}) : x \to E_x$$

where as above

$$E_x : \mathcal{A} \to \mathcal{A} : y \to E(x \otimes y). \tag{9}$$

This kind of structure we will call a vectorrepresentation of the Quantum Markov State ω. This vectorrepresentation has the following obvious properties:

- ϕ and ψ are elements of an ordered vectorspace V with inner product $\langle \cdot \mid \cdot \rangle$ (here $V = \mathcal{A}$) normalized such that

$$\langle \phi \mid \psi \rangle = 1 \tag{10}$$

- compatibility and translation invariance:

$$\begin{aligned} E_{\mathbb{1}} \psi &= \psi & (11) \\ E^*_{\mathbb{1}} \phi &= \phi & (12) \end{aligned}$$

- positivity:

$$x_1 \otimes \cdots \otimes x_n \mapsto \langle \phi \mid E_{x_1} \cdots E_{x_n} \rangle \tag{13}$$

extends to a positive linear functional of $(\otimes\mathcal{A})^n$.

According to convenience we will use freely both the notations (ϕ, E, ψ) and (ϕ, π, ψ) to denote the vectorrepresentation of a QMS. In the first one we stress that there are linear transformations E_x of the representation space, representing the state (formula (13)). With the second one we indicate that we are in a situation where the linear transformations are of the form $\pi(E_x)$, where π is a representation of the algebra generated by the E_x.

Definition IV.1 *Two vectorrepresentations (ϕ, E, ψ) and (ϕ', E', ψ') on representation spaces V and V' respectively, are called equivalent iff there exists a non-singular linear mapping*

$$S : V \to V'$$

such that

$$\begin{aligned} \phi &= S^*\phi' \\ \psi &= S^{-1}\psi' \\ E_x &= S^{-1}E'_x S \end{aligned}$$

It is obvious that equivalent triples (ϕ, E, ψ) and $(\phi', E'\psi')$ necessarily define the same QMS. If we associate to the state ϕ its density matrix and take $\psi = \mathbb{1} \in \mathcal{A}$, working in this vectorrepresentation is only a matter of notation. It is clear however that any triple $(\phi, \tilde{E}, \psi)$ with the properties stated above, defines a state of $\mathfrak{A}$. What we would like to show now is that there exists a unique (up to equivalence) vectorrepresentation for any Quantum Markov State if we require additional cyclicity properties. This result will play an important role in the parametrization of certain subclasses of QMS.

Let $\mathfrak{G}$ be the algebra generated by $\{E_x \mid x \in \mathcal{A}\}$, then the theorem can be formulated as follows.

Theorem IV.2 *Let ω be a Quantum Markov State of $\mathfrak{A}(\mathcal{A})$ defined by the triple $(\rho, E, \mathbb{1})$, with $\mathcal{A}$ finite dimensional. Then there exists a vectorrepresentation of ω given by the triple (ϕ, π, ψ) on a vectorspace V with inner product $\langle\cdot \mid \cdot\rangle$, where $\phi, \psi \in V$ and where π is a representation of $\mathfrak{G}$ on V, more specifically, there exists a linear mapping $\tilde{E} : \mathcal{A} \otimes V \to V$ and a morphism $\pi : \mathfrak{G} \to \mathcal{B}(V)$ such that for all n and all $x_1, \ldots, x_n \in \mathcal{A}$*

$$\tilde{E}_{x_1} \cdots \tilde{E}_{x_2} = \pi(E_{x_1}) \cdots \pi(E_{x_n}) \in \mathcal{B}(V)$$

where the E_x and $\tilde{E}_x$ are defined as in (9), and which has the following properties:

1. $\langle \phi \mid \psi \rangle = 1$
2. $\pi(E_{\mathbb{1}})\psi = \psi$ *and* $\pi(E_{\mathbb{1}})^*\phi = \phi$
3. ψ *is cyclic for* $\pi(\mathfrak{G})$ *and* ϕ *is cyclic for* $\pi(\mathfrak{G})^*$ *(and hence* ψ *is separating for* $\pi(\mathfrak{G})^*$ *and* ϕ *is separating for* $\pi(\mathfrak{G})$*), i.e.*

$$\pi(\mathfrak{G})\psi = \pi(\mathfrak{G})^*\phi = V.$$

Moreover, any two such representations of ω with these properties are equivalent.

Proof: As a starting point for our construction, we take the vectorrepresentation $(\rho, E, \mathbb{1})$ which was discussed above.

Let $V_\rho = \mathfrak{G} \subseteq \mathcal{A}^*\rho$ and denote by P_ρ the orthogonal projection of $\mathfrak{G}$ on V_ρ. Then, for all n, m, all $x_1, \cdots x_m, y_1, \cdots y_n \in \mathfrak{G}$:

$$\begin{aligned}
&\langle E^*_{x_1} \cdots E^*_{x_m} \mid E_{y_1} \cdots E_{y_n} \mathbb{1}\rangle \\
&= \langle P_\rho E^*_{y_1} P_\rho E^*_{x_1} \cdots E^*_{x_m} \rho \mid E_{y_2} \cdots E_{y_n} \mathbb{1}\rangle \\
&= \langle P_\rho E^*_{y_n} P_\rho E^*_{y_{n-1}} P_\rho \cdots P_\rho E^*_{y_1} P_\rho E^*_{x_1} \cdots E^*_{x_m} \rho \mid \mathbb{1}\rangle \\
&= \langle E^*_{x_1} \cdots E^*_{x_m} \rho \mid (P_\rho E_{y_1} P_\rho)(P_\rho E_{y_2} P_\rho) \cdots (P_\rho E_{y_n} P_\rho) P_\rho \mathbb{1}\rangle \qquad (14)
\end{aligned}$$

We then define $V \subset V_\rho$ to be the vectorspace generated by the

$$(P_\rho E_{y_1} P_\rho) \cdots (P_\rho E_{y_n} P_\rho) P_\rho \mathbb{1},$$

and denote the corresponding orthogonal projection by P. Clearly $P P_\rho = P = P_\rho P$ and $P_\rho \mathbb{1} \in V$, and therefore

$$\begin{aligned}
&\langle E^*_{x_1} \cdots E^*_{x_m} \rho \mid E_{y_1} \cdots E_{y_m} \mathbb{1}\rangle \\
&= \langle \rho \mid P_\rho E_{x_m} P_\rho \cdots E_{x_1} P_\rho E_{y_1} \cdots P_\rho E_{y_n} P_\rho \mathbb{1}\rangle \\
&= \langle \rho \mid (P E_{x_m} P) \cdots (P E_{x_1} P)(P E_{y_1} P) \cdots (P E_{y_n} P) P \mathbb{1}\rangle \\
&= \langle (P E^*_{x_1} P) \cdots (P E^*_{x_m} P) P \rho \mid (P E_{y_1} P) \cdots (P E_{y_n} P) P \mathbb{1}\rangle \qquad (15)
\end{aligned}$$

It is now clear that we obtained a new vectorrepresentation of ω by the triple (ϕ, π, ψ) with the following definitions:

- $\phi = P\rho$
- $\psi = P\mathbb{1}$
- $\pi(E_x) = P E_x P \equiv \widetilde{E}_x$

Notice that $\pi(E^*_x) = \pi(E_x)^*$. Due to (15) property 1 of the theorem is automatically satisfied. The vector space spanned by the $\widetilde{E}_{y_1} \cdots \widetilde{E}_{y_n} \psi$ is the whole of V, by the definition of V, and as

$$\begin{aligned}
&\langle P_\rho E^*_{x_1} P_\rho \cdots P_\rho E^*_{x_m} P_\rho \rho \mid \widetilde{E}_{y_1} \cdots \widetilde{E}_{y_n} \psi\rangle \\
&= \langle P E^*_{x_1} P \cdots P E^*_{x_m} P \rho \mid \widetilde{E}_{y_1} \cdots \widetilde{E}_{y_n} \psi\rangle \qquad (16)
\end{aligned}$$

it follows that the vectorspace spanned by the vectors $P E^*_{x_1} P \cdots P E^*_{x_m} P\rho$ is at least V and so coincides with it. Indeed suppose that there is a vector $v \in V$ such that

$$\langle P E^*_{x_1} P \cdots P E^* x_m P \rho \mid v\rangle = 0$$

for all $x_1, \cdots, x_m \in \mathcal{A}$, then by (16) $\langle w \mid v\rangle = 0$ for all $w \in V_\rho \supseteq V$ and so $v = 0$. Hence also condition 3 is satisfied. The invariance of ϕ and ψ under $\widetilde{E}^*_{\mathbb{1}}$ and $\widetilde{E}_{\mathbb{1}}$ respectively, now follows from (15), 3. and the invariance of ρ and $\mathbb{1}$ under $\widetilde{E}^*_{\mathbb{1}}$ and $\widetilde{E}_{\mathbb{1}}$.

We now prove that such a vectorrepresentation is unique up to equivalence. Suppose that we are given two such representations of ω, both satisfying 1.-3: (ϕ, E, ψ) and (ϕ', E', ψ'), on Hilbert spaces $V, \langle\cdot \mid \cdot\rangle$ and $V', \langle\cdot \mid \cdot\rangle'$ respectively. We demonstrate the existence of a non-singular transformation $S : V \to V'$ such that

$$S^*\phi' = \phi,\ S^{-1}\psi' = \psi, \tag{17}$$

$$E_x = S^{-1}E_x S \tag{18}$$

for all $x \in \mathcal{A}$. Let

$$\{e_i = \sum_j E_{a^i_{1j}} \cdots E_{a^i_{nj}} \psi \mid i = 1, \cdots, p\}$$

be a basis of V, where obviously the a^i_{kj} are elements of $\mathcal{A}$, and p is the dimension of V. From the fact that both are vectorrepresentations of ω, we can conclude that

$$\{e'_i = \sum_j E'_{a^i_{1j}} \cdots E'_{a^i_{nj}} \psi' \mid i = 1, \cdots, p\}$$

is an independent set in V' and as we can always assume that $dim\, V' \leq dim\, V$, it is again a basis. (We do not consider infinite-dimensional vectorrepresentations of the state ω, as we know for sure that there is at least one finite-dimensional representation). The same basis e_i of V can be constructed starting from ϕ using elements $b^i_{kj} \in \mathcal{A}$:

$$e_i = \sum_j E^*_{b^i_{1j}} \cdots E^*_{b^i_{nj}} \phi$$

and there is an associated basis $e"_i$ of V' :

$$e"_i = \sum_j E'^*_{b^i_{1j}} \cdots E'^*_{b^i_{nj}} \phi'$$

Take S to be the non-singular mapping, relating these two bases with one another:

$$S\, e_i = e'_i$$

Then it immediately follows that:

$$\begin{aligned} \langle \phi \mid e_i \rangle &= \sum_j \omega(a^i_{1j} \otimes \cdots \otimes a^i_{nj}) \\ &= \langle \phi' \mid e'_i \rangle' \\ &= \langle \phi' \mid S\, e_i \rangle' \\ &= \langle S^*\, \phi' \mid e_i \rangle \end{aligned}$$

Hence

$$S^*\, \phi' = \phi$$

Denote by T the non-singular transformation of V' mapping $\{e'_i\}$ to $\{e"_i\}$:

$$T\, e'_i = e"_i$$

Then

$$\begin{aligned} \langle e_i \mid \psi \rangle &= \langle e"_i \mid \psi' \rangle' \\ &= \langle T\, S\, e_i \mid \psi' \rangle' \end{aligned}$$

$$= \langle e_i \mid S^* T^* \psi' \rangle$$

and so

$$\psi = S^* T^* \psi'$$

Analogously

$$\begin{aligned} \langle e_i \mid E_x e_j \rangle &= \langle e"_i \mid E'_x e'_j \rangle' \\ &= \langle T S e_i \mid E'_x S e_j \rangle' \\ &= \langle e_i \mid S^* T^* E'_x e_j \rangle \end{aligned}$$

and therefore

$$E_x = S^* T^* E'_x S$$

To complete the proof of (17) and (18) we just have to show that $T = S^{-1*} S^{-1}$ indeed:

$$\begin{aligned} \langle T e'_i \mid e'_j \rangle' &= \langle e"_i \mid e'_j \rangle' \\ &= \langle e_i \mid e_j \rangle \\ &= \langle S^{-1} e'_i \mid S^{-1} e'_j \rangle \\ &= \langle S^{-1*} S^{-1} e'_i \mid e'_j \rangle' \end{aligned}$$

This concludes the proof of equivalence. ■

In the sequel the vectorrepresentation of a QMS, guaranteed by this theorem, will be called the minimal representation of ω. Indeed, it can be shown that the dimension of a vectorrepresentation of ω (the dimension of the vector space) cannot be smaller than the dimension of the cyclic representation of Theorem IV.2.

Whenever two triples are equivalent (and hence define the same state) we write:

$$(\phi, E, \psi) \cong (\phi', E', \psi').$$

Remark IV.3 *The non-singular linear mapping which intertwines between two given minimal representations of a state ω, is unique.*

V Quantum Markov States with a local symmetry

Because of their importance in applications, we now want to investigate QMS with an additional symmetry (besides translation invariance). To be concrete, denote by ω a Quantum Markov State given in its minimal representation (ϕ, E, ψ), which was constructed in Theorem IV.1. Further we suppose that there is a group G, represented by *-automorphisms $\{\alpha_g \mid g \in G\}$ on the one-site algebra $\mathcal{A}$. Denote by ω_g the state rotated by the element $g \in G$: for all n, and all $x_1, \ldots, x_n \in \mathcal{A}$

$$\omega_g(x_1 \otimes \cdots \otimes x_n) = \omega(\alpha_g(x_1) \otimes \cdots \otimes \alpha_g(x_n)).$$

It is clear that a minimal representation of ω_g is given by (ϕ, E^g, ψ) where

$$E_x^g = E_{\alpha_g(x)}.$$

If E is C.P.U. then the E^g are C.P.U. because the automorphisms α_g are.

A state is called G-invariant if

$$\omega_g = \omega \qquad \text{for all } g \in G$$

The following theorem concerning G-invariant states is a straightforward application of Theorem IV.1.

Theorem V.1 *Let ω be a Quantum Markov State with minimal representation (ϕ, E, ψ). Then the following conditions are equivalent:*

1. *ω is G-invariant*
2. *$(\phi, E, \psi) \cong (\phi, E^g, \psi)$ for all $g \in G$*
3. *there exists a representation u_g of G on V such that for all $g \in G$*
   - $u_g E_x = E_{\alpha_g(x)} u_g$ (19)
   - $u_g \psi = \psi$ (20)
   - $u_g^* \phi = \phi$ (21)
4. *there exist well-defined linear mappings $v_g : V \to V$ such that for all $x_1, \ldots, x_n \in \mathcal{A}$*

$$v_g(E_{x_1} \cdots E_{x_n} \psi) = E_{\alpha_g(x_1)} \cdots E_{\alpha_g(x_n)} \psi$$

and such that $v_g^* \phi = \phi$ for all $g \in G$.

Proof:

1. $\Rightarrow$ 2. by Theorem IV.2.

2. $\Rightarrow$ 3. by Remark IV.3.

3. $\Rightarrow$ 4.: one actually has to take $v_g = u_g$; indeed, by applying (19) n times and finally (20), one gets:

$$\begin{aligned} u_g E_{x_1} \cdots E_{x_n} \psi &= E_{\alpha_g(x_1)} u_g E_{x_2} \cdots E_{x_n} \psi \\ &\;\;\vdots \\ &= E_{\alpha_g(x_1)} \cdots E_{\alpha_g(x_n)} u_g \psi \\ &= E_{\alpha_g(x_1)} \cdots E_{\alpha_g(x_n)} \psi. \end{aligned}$$

Property (21) then yields:

$$v_g^* \phi = u_g^* \phi = \phi.$$

4. $\Rightarrow$ 1. This implication follows by a straightforward computation: for all $g \in G$ and for all $x_1, \ldots, x_n \in \mathcal{A}$:

$$\begin{aligned} \omega(\alpha_g(x_1) \otimes \cdots \otimes \alpha_g(x_n)) &= \langle \phi \mid E_{\alpha_g(x_1)} \cdots E_{\alpha_g(x_n)} \psi \rangle \\ &= \langle \phi \mid v_g E_{x_1} \cdots E_{x_n} \psi \rangle \\ &= \langle v_g^* \phi \mid E_{x_1} \cdots E_{x_n} \psi \rangle \end{aligned}$$

$$
\begin{aligned}
&= \langle \phi \mid E_{x_1} \cdots E_{x_n} \psi \rangle \\
&= \omega(x_1 \otimes \cdots \otimes x_n).
\end{aligned}
$$

■

In view of Theorem V.1.3 it would be of great interest to characterize the mappings E that fulfill (19) and (11)-(13) for some representation u_g of G. By Remark V.2 below, we see that we only have to consider inequivalent representations, which can be chosen to be unitary representations if G is a compact group. We did not yet achieve the full solution of this problem. In the following we only treat a simplified case, but which is probably the most important one.

Remark V.2 *Let ω be a G– invariant Quantum Markov State and let u_g be the group representation of G which arises in Theorem V.1.3, with a given minimal representation (ϕ, E, ψ) of ω, on a vector space V. Let u'_g be a representation of G on V', which is equivalent to u_g. Then there exists a minimal representation (ϕ', E', ψ') on V' of ω such that Theorem V.1.3 yields the group representation u'_g. In particular, if G is a compact topological group we can always find a minimal representation such that the group representation generated on the representation space is unitary.*

Proof: The representation (ϕ', E', ψ') simply is the one obtained by applying the non-singular intertwining operator between u_g and u'_g to (ϕ, E, ψ). For compact G there is the well-known theorem that any representation is equivalent to a unitary representation. ■

Assumptions V.3 *1. $\mathcal{A} = V = M_d$*

2. for any $g \in G$: $u_g : \mathcal{A} \to \mathcal{A}$ is C.P.U.

About 1 we remark that in what follows we do not really need the minimal representations of the QMS. It is sufficient that we can work in a vectorrepresentation with $V = M_d$.

From assumption 2. it immediately follows that the u_g are automorphisms of $\mathcal{A}$. In this situation, we can work with two unitary representations ρ and σ of G on $\mathbb{C}^d$ such that:

$$\alpha_g(x) = \rho(g)^* x \rho(g) \tag{22}$$

$$u_g(x) = \sigma(g)^* x \sigma(g). \tag{23}$$

The C.P.U.-mappings satisfying (19) can now be characterized with the aid of the Choi-Effros representation of of completely positive maps $\Gamma : M_{d_1} \to M_{d_2}$, which goes as follows:
Take

$$\{e_{ij} \mid 1 \le i, j \le d_1\}$$

and

$$\{f_{kl} \mid 1 \le k, l \le d_2\}$$

bases of matrix units for M_{d_1} and M_{d_2} respectively. The linear maps Γ can then be represented by an element $\mathbb{\Gamma}$ in $M_{d_1} \otimes M_{d_2}$ such that

$$\Gamma(e_{i,j}) = \sum_{k,l} \mathbb{\Gamma}_{ik,jl} f_{kl}. \tag{24}$$

The theorem of Choi and Effros [10] states then that Γ is completely positive iff $\mathbb{\Gamma} \ge 0$ as an element of $M_{d_1} \otimes M_{d_2}$.

In our case we have $E : M_d \otimes M_d \to M_d$ and we choose a basis of matrix units

$$\{e_{ij} \mid 1 \leq i,j \leq d\}$$

in M_d and take the tensorproduct basis

$$\{e_{ij} \otimes e_{kl} \mid 1 \leq i,j,k,l \leq d\}$$

for $M_d \otimes M_d$. Denote by $\mathbb{E}$ the element in $M_d \otimes M_d \otimes M_d$ representing E. The following proposition then says that to satisfy (19), $\mathbb{E}$ has to commute with a certain tensorproduct representation of G.

Proposition V.4 *Let ρ and σ be two unitary representations of G on $\mathbb{C}^d$ and $E : M_d \otimes M_d \to M_d$ a C.P.U. map, characterized by $0 \leq \mathbb{E} \in M_d \otimes M_d \otimes M_d$, according to the Choi-Effros Theorem. Then*

$$\sigma(g)^* E(x \otimes y)\sigma(g) = E(\rho(g)^* x\rho(g) \otimes \sigma(g)^* y\sigma(g))$$

for all $g \in G$ and $x, y \in M_d$, iff

$$[\mathbb{E}, \overline{\rho}(g) \otimes \overline{\sigma}(g) \otimes \sigma(g)] = 0 \qquad \textit{for all } g \in G.$$

where $\overline{\rho}$ and $\overline{\sigma}$ denote the conjugate representation of ρ and σ (complex conjugation of the matrix elements).

The proof follows by a straightforward computation.

In principle, to obtain all E satisfying the third condition of Theorem V.1, we should consider all possible unitary representations σ of G on $\mathbb{C}^d$. The following remark, which is actually recalling Remark V.2, shows that the assumption of unitary group representation and the complete positivity of E are compatible.

Remark V.5 *Let σ and $\tilde{\sigma}$ two equivalent unitary representations of G on $\mathbb{C}^d$ with intertwining unitary w, and let ω be a G-invariant QMS, determined by (ϕ, E, ψ), with E C.P.U. and such that (19) is satisfied for $u_g = \sigma(g)^* \cdot \sigma(g)$ i.e.*

$$\sigma(g)^* E(x \otimes y)\sigma(g) = E(\alpha_g(x) \otimes \sigma(g)^* y\sigma(g)) \qquad g \in G, x \in M_d \tag{25}$$

then there exists a representation $(\tilde{\phi}, \tilde{E}, \tilde{\psi})$ of ω such that (25) holds with σ replaced by $\tilde{\sigma}$, and such that $\tilde{E}$ C.P.U.

Proof: Simply define

- $\tilde{E}(x \otimes y) = w^* E(x \otimes wyw^*)w$
- $\tilde{\phi} = w^* \phi w$
- $\tilde{\psi} = w^* \psi w$

then it is clear that $\tilde{E}$ is C.P.U. as it is a composition of C.P.U. maps. Further,

$$\widetilde{E_{\mathbb{1}}}(\tilde{\psi}) = w^* E(\mathbb{1} \otimes \psi)w = \tilde{\psi}$$

and also $\widetilde{E_{\mathbb{1}}}^*(\tilde{\phi}) = \tilde{\phi}$ since

$$\begin{aligned}
\langle \widetilde{E_{\mathbb{1}}}^*(\tilde{\phi}) \mid \xi \rangle &= \langle \tilde{\phi} \mid \widetilde{E_{\mathbb{1}}}(\xi) \rangle \\
&= \langle \tilde{\phi} \mid w^* E_{\mathbb{1}}(w\xi w^*)w \rangle \\
&= \langle \phi \mid E_{\mathbb{1}}(w\xi w^*) \rangle \\
&= \langle E_{\mathbb{1}}^* \phi \mid (w\xi w^*) \rangle
\end{aligned}$$

$$
\begin{aligned}
&= \langle \phi \mid (w\xi w^*) \rangle \\
&= \langle \tilde{\phi} \mid \xi \rangle.
\end{aligned}
$$

By definition, $(\phi, E, \psi) \cong (\tilde{\phi}, \tilde{E}, \tilde{\psi})$ and so they represent the same state. Finally, we check the covariance property w.r.t. $\tilde{\sigma}$:

$$
\begin{aligned}
\tilde{\sigma}(g)^* \tilde{E}(x \otimes y) \tilde{\sigma}(g) &= w^* \sigma(g) w w^* E(x \otimes wyw^*) w w^* \sigma(g) w \\
&= w^* E(\alpha_g(x) \otimes \sigma^*(g) wyw^* \sigma(g)) w \\
&= w^* E(\alpha_g(x) \otimes w\sigma^*(g) y \sigma(g) w^*) w \\
&= \tilde{E}(\alpha_g(x) \otimes \tilde{\sigma}(g)^* y \tilde{\sigma}(g))
\end{aligned}
$$

■

VI Application to the Heisenberg model

In this section, we want to study the ground-state behavior of the one-dimensional Heisenberg model for anti-ferromagnetism. In this model $\mathcal{A} = M_2$ and the local Hamiltonians can be written in terms of the Pauli matrices $S^\alpha \in M_2$, $\alpha \in \{x, y, z\}$. For all finite $\Lambda \subset \mathbb{Z}$,

$$
H_\Lambda = J \sum_{i \in \Lambda} \sum_\alpha S_i^\alpha S_{i+1}^\alpha \tag{26}
$$

where S_i^α denotes the embedding of S^α in $\mathfrak{A}$ at site $i \in \mathbb{Z}$. J is the coupling constant and is supposed to be strictly positive.

One can check that H_Λ is SU(2)-invariant, i.e. for all $u \in$SU(2):

$$
\sum_\alpha u^* S^\alpha u \otimes u^* S^\alpha u = \sum_\alpha S^\alpha \otimes S^\alpha. \tag{27}
$$

For any translation-invariant state ω of $\mathfrak{A}$, the following limit exists and its value is called the energy density $e(\omega)$ of ω:

$$
e(\omega) = \lim_{\Lambda \uparrow \mathbb{Z}} \frac{1}{|\Lambda|} \omega(H_\Lambda).
$$

Clearly,

$$
e(\omega) = J \sum_\alpha \omega(S^\alpha \otimes S^\alpha). \tag{28}
$$

Solving the ground-state problem for the model with Hamiltonians H_Λ means determining the minimal value e_G of e and the minimizing states ω_G. For the Heisenberg model (26), it is known [11] that

$$
e_G = (1 - 4 \ln 2) J \tag{29}
$$

but the state ω_G for which this value is attained is not known. Therefore, we try to approximate the true solution of the problem by the following "restricted" variational calculation:

$$
e_0 = \inf_{\omega \in \mathcal{S}} e(\omega)
$$

where $\mathcal{S}$ is the set of SU(2)-invariant QMS, having a vectorrepresentation (ϕ, E, ψ) with E a completely positive mapping satisfying (19) for some unitary representation σ as in Proposition V.3. This means that we work under the Assumption V.2.2. The restriction to SU(2)-invariant states is meaningful because the model is SU(2)-invariant (see (27)), which assures the existence of a SU(2)-invariant exact ground state. It is not clear however that including also non-invariant QMS would not yield a better (lower) value for e_0.

Due to Remark V.4, we only have to consider two possibilities for the representation σ of SU(2) occurring in the covariance condition (25).

1. The trivial representation: $\sigma(u) = \mathbb{1}$ for all $u \in \mathrm{SU}(2)$. In this case, (19) becomes:

$$E_{u^* x u} = E_x \qquad \text{for all } u \in \mathrm{SU}(2).$$

Denote by $d\lambda$ the normalized Haar measure on SU(2), then

$$\begin{aligned} E_x &= \int E_{u^* x u} d\lambda(u) \\ &= E_{\int u^* x u d\lambda(u)} \\ &= E_{\frac{1}{2}(\mathrm{Tr}x)\mathbb{1}} \\ &= \frac{1}{2}(\mathrm{Tr}x)E_{\mathbb{1}}. \end{aligned}$$

Therefore, the only SU(2)-invariant state obtained in this way is the tracial state:

$$\begin{aligned} \omega(x_1 \otimes \cdots \otimes x_n) &= \langle \rho \mid E_{x_1} \cdots E_{x_n} \mathbb{1} \rangle \\ &= \frac{1}{2^n} \mathrm{Tr}x_1 \mathrm{Tr}x_2 \cdots \mathrm{Tr}x_n. \end{aligned}$$

The energy density of this state is zero.

2. The two-dimensional irreducible representation $\sigma(u) = u$ for all $u \in \mathrm{SU}(2)$. In this case, according to Proposition V.3, we should find all positive complex 8×8 matrices $\mathbb{E}$ satisfying the covariance condition

$$[\mathbb{E}, \overline{u} \otimes \overline{u} \otimes u] = 0 \qquad \text{for all } u \in \mathrm{SU}(2)$$

and the normalization condition

$$\sum_{i,j} \mathbb{E}_{ijk,ijn} = \delta_{k,n}$$

which corresponds to $E(\mathbb{1} \otimes \mathbb{1}) = \mathbb{1}$.

This is done by decomposing the representation $\overline{u} \otimes \overline{u} \otimes u$ in its irreducible components. In this way, the commutant of the representation is easily determined. As the multiplicity of the irreducible components in this case does not exceed 2, the positivity conditions are simple quadratic inequalities. Skipping the boring calculations, we only give the general form of such a matrix $\mathbb{E}$, depending on three parameters λ, μ and ξ.

$$\mathbb{E} = W^* \begin{bmatrix} \lambda \mathbb{1}_2 & \nu \mathbb{1}_2 & 0 \\ \nu \mathbb{1}_2 & \mu \mathbb{1}_2 & 0 \\ 0 & 0 & \xi \mathbb{1}_4 \end{bmatrix} W$$

where W is the unitary in M_8 which reduces $\overline{u} \otimes \overline{u} \otimes u$. Therefore $\mathbb{E}$ is positive iff $\lambda, \mu, \xi \geq 0$, $\nu \in \mathbb{C}$, $|\nu^2| \leq \lambda\mu$ and satisfies the normalization condition if $2\lambda + 2\mu + 4\xi = 2$ i.e. if

$$\xi = \frac{1}{2}(1 - \lambda - \mu).$$

The energy density of the corresponding SU(2)-invariant state is now readily computed as

$$e = 12J\Big(-\frac{Re\nu}{\sqrt{3}} + \frac{\mu}{3} - \frac{\xi}{3}\Big)\Big(-\frac{\lambda}{2} + \frac{\mu}{6} + \frac{\xi}{3}\Big)$$

and the minimizing values of the parameters are

$$\lambda = \frac{3}{4} \qquad \mu = \frac{1}{4} \qquad \nu = -\sqrt{\lambda\mu} = -\frac{\sqrt{3}}{4} \qquad \xi = \frac{1}{2}(1 - \lambda - \mu) = 0$$

yielding as approximate ground-state energy density

$$e_0 = -\frac{4}{3}J$$

which is considerably close to the exact value $(1 - 4\ln 2 \approx 1.77)J$, compared with the product-state approximation which gives $e_0 = 0$.

Moreover, the correlation function can be computed for the minimizing state and it exhibits a nice anti-ferromagnetic behavior:

$$\omega(S_0^\alpha S_n^\beta) = \frac{4}{3}\delta_{\alpha,\beta}\Big(-\frac{1}{3}\Big)^n \qquad n \geq 1.$$

VII Conclusion

We can conclude that the Quantum Markov States are an interesting class of states for one-dimensional Quantum Spin Systems, although it is not clear that they are to be considered as the quantum analogue of Classical Markovian Measures.

It is possible to characterize and parametrize certain interesting subclasses of the QMS and to apply them as trial states in Hartree-Fock approximations for physical models.

The study of QMS for abelian algebras (giving rise to a class of measures we called Generalized Markovian Measures, just for the purpose of this paper, and which is a special case of the Algebraic Measures discussed elsewhere [5]) is a good guide to the analysis of the computational structure of QMS, but might also have interest on its own.

Finally, it must be said that a lot of interesting questions are still to be answered:

- some details related to the characterization of QMS with group-invariance,
- the identification of the pure states among the QMS, which might be related to the construction of GNS-representations,
- the study of the mean entropy and related quantities...

Furthermore, there are indications that the exact ground-states of some special Quantum Spin Systems are QMS. However, we do not know whether this is really the case in a non-trivial example. This point is still to be clarified [12].

Acknowledgment
The results presented in this paper were obtained in collaboration with M. Fannes and L. Slegers and the author takes the opportunity to express his deep gratitude to them.

References

[1] L. Accardi, Topics in Quantum Probability, Phys. Rep. **77**, 169-192 (1981)

[2] C.Cecchini, Stochastic couplings for von Neumann algebras, preprint and also in this volume.

[3] B. Kümmerer, *Survey on a Theory of Non-Commutative Stationary Markov Processes*, in *Quantum Probability III, Lecture Notes in Mathematics 1303*, Springer-Verlag Berlin 1988

[4] L. Accardi and A. Frigerio, *Markovian cocycles*, Proc. R. Ir. Acad. Vol. **831** 2, 251-269 (1983)

[5] M. Fannes, B. Nachtergaele and L. Slegers, *Functions of Markov Processes and Algebraic Measures*, Preprint-KUL-TF-89/10

[6] H.J. Brascamp, *Equilibrium states for a one-dimensional lattice gas*, Comm. Math. Phys. **21**, 56-70 (1971)

[7] M. Fannes and A. Verbeure, *On Solvable Models in Classical Lattice Systems*, Comm. Math. Phys. **96**, 115-124 (1984)

[8] B. Nachtergaele and L. Slegers, *Construction of equilibrium states for one-dimensional classical lattice systems*, Il Nuovo Cimento **100B**, 757-779 (1987)

[9] M. Rosenblatt, *Markov Processes. Structure and asymptotic behavior*, Springer-Verlag Berlin Heidelberg New-York 1971

[10] Choi and Effros, Ann. Math. **104**, 225 (1976)

[11] C. Domb and M.S. Green, in *Phase Transitions and Critical Phenomena, Volume 1: Exact Results*, Academic Press, London (1972)

[12] M. Fannes, B. Nachtergaele and R. Werner, Exact antiferromagnetic ground states for quantum chains, in preparation.

This paper is in final form and no similar paper has been or is being submitted elsewhere.

Dynamical Entropy, Quantum K-Systems and Clustering

Heide Narnhofer
Institut für Theoretische Physik, Universität Wien,
Boltzmanngasse 5, A-1090 Wien

Abstract

The two possibilities to define a quantum K-system, either using algebraic relations or using properties of the dynamical entropy, are compared. It is shown that under the additional assumption of strong asymptotic abelianess the algebraic relations imply the properties of the dynamical entropy.

1 Introduction

Ergodic theory is a highly developed field and has brought deep insight into the theory of classical dynamical systems. It is natural to try to translate successful concepts also to the theory of quantum dynamical systems. Also here some promising results have already been obtained. For instance, it was tried to find a quantum analogue to the classical system with the best mixing properties, the K-system [1,2]. The concept is fully based on algebraic relations between the development of subalgebras and though it gives a very transparent structure, some features are extremely sensitive against small perturbations of the dynamics or the observed subsystems. Therefore it seems necessary to find related quantities that serve to characterize a K-system but have better continuity properties. From classical theory we know that a K-system is completely characterized by properties of the dynamical entropy and since an analogue of the classical dynamical entropy can be constructed also for quantum systems [3], we suggest that a K-system in quantum theory should preferably be characterized in terms of its dynamical entropy [4]. The systems of [1,2] satisfy the desired properties under some additional assumption. It is an open problem whether [4] covers a larger range of systems and whether this assumption of asymptotic abelianess is stronger than necessary. In this note we want to concentrate on the continuity problem of the relevant quantities and show how they help to control the properties of the system.

2 K-Systems in the Sense of Emch and Schröder

Following [2] a K-system is defined as follows: Let $(\mathcal{A}, \omega, \alpha)$ be a von Neumann algebra $\mathcal{A}$ with automorphism α and invariant state $\omega \circ \alpha = \omega$. Let $\mathcal{A}_0$ be a subalgebra $\mathcal{A}_0 \subset \mathcal{A}$ such that

$$\alpha^n \mathcal{A}_0 \supset \mathcal{A}_0, \qquad n \geq 0,$$

$$\bigvee_{n=0}^{\infty} \alpha^n \mathcal{A}_0 = \mathcal{A}, \qquad \mathcal{A} \vee \mathcal{B} \equiv \{\mathcal{A} \cup \mathcal{B}\}'',$$
$$\bigwedge_{n=0}^{-\infty} \alpha^n \mathcal{A}_0 = \lambda 1, \qquad \mathcal{A} \wedge \mathcal{B} \equiv \{\mathcal{A}' \cup \mathcal{B}'\}'.$$

Then $(\mathcal{A}, \omega, \alpha, \mathcal{A}_0)$ is called a W^*-K-system. Evidently, the question arises, given $\mathcal{A}$, ω, α, is it possible to find such an $\mathcal{A}_0$ and how small changes in $\mathcal{A}_0$ spoil the picture. The theory of classical K-automorphisms tells us that we can start with an arbitrary finite-dimensional subalgebra $\mathcal{B}$, construct $\mathcal{B}_0 = \bigvee_{n=-\infty}^{0} \alpha^n \mathcal{B}$. If only $\bigwedge_{n=-\infty}^{0} \bigvee_{k=n}^{-\infty} \alpha^k \mathcal{B} = \lambda 1$, then $\bar{\mathcal{B}} = \bigvee_{n=-\infty}^{+\infty} \alpha^n \mathcal{B}_0$ is a K-system. So it only remains to check whether $\bar{\mathcal{B}} = \mathcal{A}$, i.e. if $\mathcal{B}_0$ resp. $\mathcal{B}$ was chosen sufficiently large.

Let us compare a typical quantum mechanical system: the Fermi algebra $\mathcal{A}$ of creation and annihilation operators $a(f)$, $f \in L^2(\mathbf{R}, dx)$. We consider the continuous automorphism group $\sigma_x a(f(y)) = a(f(x+y))$. Then $\mathcal{A}_0 = \mathcal{A}_{(-\infty,0]}$, built by the creation and annihilation operators $a(f)$ with $f \in L^2(\mathbf{R}_-, dx)$, serves the purpose. We can consider $\mathcal{A}_0$ to be $\bigcup_{x=0}^{-\infty} \alpha^x \mathcal{A}_{[-1,0]}$. So we can take for $\mathcal{B} = \mathcal{A}_{[-1,0]}$ to construct a K-system. However, if we take $\mathcal{B} = \{a(f), a^\dagger(f)\}''$ with $\tilde{f}(p) \neq 0 \; \forall p$ and $|\tilde{f}(p)| \leq e^{-\lambda|p|}$, so e.g. $\tilde{f}(p) = e^{-p^2}$, then $\mathcal{B}_0 = \bigcup_{x=0}^{-\infty} \alpha^x \mathcal{B} = \mathcal{A}$, and we failed to obtain a suitable $\mathcal{A}_0$. To demonstrate this claim all we have to show is that

$$\bigcup_{y \leq 0} f(x+y) = L^2(\mathbf{R}, dx).$$

Let $g \perp f(x+y) \; \forall y \leq 0$, then

$$\int \bar{\tilde{g}}(p) \tilde{f}(p) e^{ipx}\, dp = 0 \qquad \forall x \leq 0.$$

Consider

$$F(x+iy) = \int \bar{\tilde{g}}(p) \tilde{f}(p) e^{ipx - py}\, dp.$$

Due to our assumption on f this function is analytic for $|y| \leq \lambda$. Since it is $= 0$ for $x \leq 0$, $y = 0$, it has to be identical to zero, so also $\bar{\tilde{g}}(p)\tilde{f}(p) \equiv 0$ and this is only possible for $\tilde{g}(p) = 0$.

There is another shortcoming: we want to observe somehow the continuity of the automorphism group, i.e. it should be possible to discretize the steps of the automorphisms and to observe some limiting behaviour if the steps become sufficiently small. If we consider again $\mathcal{B} = \{a(f), a^\dagger(f)\}$, then $\{\bigvee_{k=0}^{n} \alpha^{(1/n)k} \mathcal{B}\}'' \equiv \mathcal{B}_n$ will have dimension 2^n, so we lose completely the fact that $\mathcal{B}_n$ and $\mathcal{B}_{n'}$ should be close to one another if $(n-n')/n \ll 1$.

3 The Dynamical Entropy and K-Systems

In classical theory the K-systems are exactly those for which the dynamical entropy of the automorphism with respect to any non-trivial subalgebra is strictly positive [5]. We repeat the necessary definitions:

Let $\mathcal{B}$ be a finite-dimensional subalgebra, built by the (in $\mathcal{B}$) minimal projections $P_1, \ldots P_n$. Then

$$H_\omega(\mathcal{B}) = -\sum \omega(P_i) \ln \omega(P_i).$$

Finally ($\omega = \omega \circ \alpha$),

$$h_\omega(\mathcal{B},\alpha) = \lim_{k\to\infty} \frac{1}{k} H_\omega(\mathcal{B},\ldots \alpha^k \mathcal{B}) = \inf_{k\to\infty} \frac{1}{k} H_\omega(\mathcal{B},\ldots \alpha^k \mathcal{B}).$$

We have to find corresponding definitions in the quantum situation. Now the von Neumann algebra $\mathcal{M}$ as well as the finite-dimensional subalgebra $\mathcal{B}$ are non-abelian. The minimal projections are not uniquely defined. We do not want to repeat the argumentation that led to the definition (see [3], [6]) but simply state ($S(\omega|\nu)$ is the relative entropy of the states ω and ν)

Definition 1:

$$H_{\omega,\mathcal{M}}(\mathcal{B}) = \sup_{\sum \lambda_i \omega_i = \omega} \sum [-\lambda_i \ln \lambda_i + S(\omega|\lambda_i\omega_i)_{\mathcal{B}}]$$

$$H_{\omega,\mathcal{M}}(\mathcal{B},\alpha\mathcal{B}) = \sup_{\sum \lambda_{ij}\omega_{ij}=\omega, \sum_j \lambda_{ij}\omega_{ij} = \lambda_i^{(1)}\omega_i^{(1)}, \sum_i \lambda_{ij}\omega_{ij} = \lambda_j^{(2)}\omega_j^{(2)}} \sum [-\lambda_{ij} \ln \lambda_{ij} +$$

$$+ S(\omega|\lambda_i^{(1)}\omega_i^{(1)})_{\mathcal{B}} + S(\omega|\lambda_j^{(2)}\omega_j^{(2)})_{\alpha\mathcal{B}}] =$$

$$= \sup \sum [-\lambda_{ij} \ln \lambda_{ij} + \lambda_i^{(1)} \ln \lambda_i^{(1)} + \lambda_j^{(2)} \ln \lambda_j^{(2)}] + \sum \lambda_i^{(1)} S(\omega|\omega_i^{(1)})_{\mathcal{B}} + \sum \lambda_j^{(2)} S(\omega|\omega_j^{(2)})_{\alpha\mathcal{B}}.$$

Generalization for k arguments is obvious. Finally, again

$$h_{\omega,\mathcal{M}}(\mathcal{B},\alpha) = \lim \frac{1}{k} H_{\omega,\mathcal{M}}(\mathcal{B},\alpha\mathcal{B},\ldots,\alpha^k\mathcal{B}).$$

Note that in the abelian case the optimal decomposition is given by

$$\omega(P_{i_1}\alpha P_{i_2}\ldots\alpha^k P_{i_k}\cdot) = \lambda_I \omega_I,$$

so in this case the definitions coincide.

In [4] we used the concept of dynamical entropy to define

Definition 2: A K-system is a quantum dynamical system $(\mathcal{M},\alpha,\omega)$ for which

$$\inf_{\mathcal{B}\neq c1} m_\omega(\mathcal{B},\alpha) \equiv \inf_{\mathcal{B}\neq c1, \dim \mathcal{B}<\infty} \lim_{n\to\infty} \frac{h_\omega(\mathcal{B},\alpha^n)}{H_\omega(\mathcal{B})} = 1.$$

The advantage of the definition is given by the following fact:

Theorem 1: Let $\mathcal{M}$ be a hyperfinite von Neumann algebra and $\mathcal{B}_k$ a sequence of finite-dimensional subalgebras, such that $\mathcal{M} = \overline{\lim}\, \mathcal{B}_k$ in the weak sense. Then

$$1 = \inf_{\mathcal{B}\neq c1} \lim_{n\to\infty} \frac{h_\omega(\mathcal{B},\alpha^n)}{H_\omega(\mathcal{B})}$$

iff

$$1 = \lim_{n\to\infty} \frac{h_\omega(\mathcal{A}_{d,k},\alpha^n)}{h_\omega(\mathcal{A}_{d,k})} \qquad \forall\, \mathcal{A}_{d,k} \subset \mathcal{B}_k, \dim \mathcal{A}_{d,k} = d \geq 2.$$

This means it suffices to check $(\cdot)$ for a countable set of subalgebras.

The theorem is a consequence of the following

Lemma 1: Monotonicity Let $\mathcal{B} \subset \mathcal{C}$. Then

$$H_\omega(\mathcal{B}, \alpha\mathcal{B}, \ldots, \alpha^n\mathcal{B}) \leq H_\omega(\mathcal{C}, \ldots, \alpha^n\mathcal{C}).$$

Proof: We take the optimal decomposition for $\alpha^k\mathcal{B}$ and use it for $\alpha^k\mathcal{C}$. The relative entropy $S(\omega|\varphi)_\mathcal{B} \leq S(\omega|\varphi)_\mathcal{C}$ is monotonic in $\mathcal{B}$ (see [7,8]).

Lemma 2: Continuity ([9]) in one subalgebra Let $\mathcal{B}$ be an algebra of dimension $d > 8$. Then

$$|H_\omega(\mathcal{B}) - H_{\bar{\omega}}(\mathcal{B})| \leq 12\varepsilon^{1/3} \ln \frac{2d}{\varepsilon} \qquad \text{with } \varepsilon = \|\omega - \bar{\omega}\|.$$

Remark: We do not claim that 12 is optimal. For taking into account smaller d, the number would change. But the essential interest lies in the way, how the continuity depends on the dimension for large algebras. The ε-dependence might also not be optimal and it is a challenge to improve it.

The estimates are based on the following facts:

$$\lambda S_\omega + (1-\lambda)S_{\bar{\omega}} \leq S_{\lambda\omega+(1-\lambda)\bar{\omega}} \leq \lambda S_\omega + (1-\lambda)S_{\bar{\omega}} - \lambda \ln \lambda - (1-\lambda)\ln(1-\lambda)$$

together with the scaling law

$$S_{\lambda\omega} = \lambda S_\omega - \omega(1)\lambda \ln \lambda$$

and

$$S_\omega(\mathcal{B}) \leq \ln d$$

leads to

$$|S_\omega - S_{\bar{\omega}}| \leq \varepsilon \ln \frac{2d}{\varepsilon}$$

using order relations between ω, $\bar{\omega}$ and $\nu = \dfrac{\omega + \bar{\omega}}{2} + \dfrac{|\omega - \bar{\omega}|}{2}$. To use this estimate for $H_\omega(\mathcal{B})$ we first observe that due to the continuity of S we may assume that we are arbitrarily close to the optimal decomposition with a finite decomposition $\omega = \sum \omega_i$, where $\omega_i(A) = \langle \Omega | x_i^* A x_i | \Omega \rangle$ with $x_i \in \mathcal{M}'$ and $\sum x_i^* x_i = 1$.

We use the same decomposition of unity to construct $\bar{\omega}_i = \langle \bar{\Omega} | x_i^* A x_i | \bar{\Omega} \rangle$. Since by assumption $\sum \|x_i|\Omega - \bar{\Omega}\rangle\|^2 = \|\Omega - \bar{\Omega}\|^2 < \varepsilon$, we split the x_i into those for which the normalized $\hat{\omega}_i$ and $\hat{\bar{\omega}}_i$ are close to one another (up to $\varepsilon^{2/3}$) and show that the rest only gives a contribution of $O(\varepsilon^{1/3})$. This gives the estimate of Lemma 2.

For $H_\omega(\mathcal{B})$ estimates using the norm difference of the state and the difference of two d-dimensional subalgebras (to every $b_i \in \mathcal{B}_i \, \exists \, b_j \in \mathcal{B}_j$ such that $\|(b_i - b_j)\Omega\| < \varepsilon$, $\|b_i^* - b_j^*)\Omega\| < \varepsilon$) are rather related. For an arbitrary number of subalgebras this does not hold anymore and in this case only a less explicit continuity property is available.

Lemma 3 [3]: Let $\mathcal{B}_n$ be a sequence of finite-dimensional subalgebras such that to every $b_{n_0} \in \mathcal{B}_{n_0}$ exists a sequence $b_n \in \mathcal{B}_n$, $b \in \mathcal{B}$, with st $\lim_{n\to\infty} b_n = b$. Then

$$\lim_{n\to\infty} \frac{1}{k} H_\omega(\mathcal{B}_n, \alpha\mathcal{B}_n, \ldots, \alpha^k\mathcal{B}_n) = \frac{1}{k} H_\omega(\mathcal{B}, \ldots \alpha^k\mathcal{B}).$$

Sketch of the proof: Due to the continuity of the entropy $S_\omega(\mathcal{B}_n)$ it suffices to consider only decompositions $\omega_{i_1,\dots,i_k}(\cdot) = \omega(x_{i_1,\dots,i_k}\cdot)$ with $i_k \le r(d,\varepsilon)$ to come within ε to the optimal result. By the same argument it suffices to take into account only those contributions where $\omega(x_{i_\ell}) > \varepsilon$ with $x_{i_\ell} = \sum_{I_k, i_\ell \, fixed} x_{I_k}$ with $I_k = \{i_1,\dots,i_k\}$. Now the convergence of the $\mathcal{B}_n$ gives

$$\lim_n \sup_{x,\omega(x)>\varepsilon} \left| \frac{\langle\Omega|xb_n|\Omega\rangle}{\langle\Omega|x|\Omega\rangle} - \frac{\langle\Omega|xb|\Omega\rangle}{\langle\Omega|x|\Omega\rangle} \right| = 0.$$

Lemma 4: Uppersemicontinuity Assume $\mathcal{B}_n \to \mathcal{B}$ in the above sense. Then

$$\lim_{n\to\infty} \lim_{k\to\infty} \frac{1}{k} H_\omega(\mathcal{B}_n, \alpha\mathcal{B}_n, \dots, \alpha^k\mathcal{B}_n) \le \lim_{k\to\infty} \frac{1}{k} H_\omega(\mathcal{B}, \dots, \alpha^k\mathcal{B}).$$

Proof:

$$H_\omega(\mathcal{B}_1,\dots,\mathcal{B}_\ell,\mathcal{B}_{\ell+1},\dots,\mathcal{B}_n) \le H_\omega(\mathcal{B}_1,\dots,\mathcal{B}_\ell) + H_\omega(\mathcal{B}_{\ell+1},\dots,\mathcal{B}_n),$$

which can be seen in the second form of Definition 1, if we notice that $\sum -\lambda_{ij}\ln\lambda_{ij} + \sum \lambda_i \ln \lambda_i + \lambda_j \ln \lambda_j \le 0$ due to the subadditivity of the entropy. Standard arguments tell us that $\lim_{k\to\infty}$ is in fact an infimum. Thus the continuity in Lemma 3 implies semicontinuity in Lemma 4.

Lemma 5: Let $\mathcal{B}_k$ be a sequence of finite-dimensional subalgebras with $\mathcal{M} = \overline{\lim \mathcal{B}_k}$. Let $\mathcal{B}$ be a finite-dimensional subalgebra. Then there exists for every $\varepsilon > 0$ some k such that

$$H_\omega(\mathcal{B}) \le H_\omega(\mathcal{B}_k) + \varepsilon,$$
$$\frac{1}{n} H_\omega(\mathcal{B},\dots\alpha^n\mathcal{B}) \le \frac{1}{n} H_\omega(\mathcal{B}_k,\dots\alpha^n\mathcal{B}_k) + \varepsilon.$$

Proof: Let E_k be the conditional expectation $\mathcal{M} \to \mathcal{B}_k$, τ_k the imbedding of $\mathcal{B}_k \to \mathcal{M}$ and γ the imbedding of $\mathcal{B}$ in $\mathcal{M}$. Then st lim $\tau_k E_k \gamma = \gamma$ and the continuity gives the desired result.

We combine the results to prove Theorem 1.

We may assume that $\inf \lim \dfrac{h_\omega(\mathcal{B},\alpha^n)}{H_\omega(\mathcal{B})}$ is attained within ε by some finite-dimensional $\mathcal{B}$. This $\mathcal{B}$ is the strong limit of $\mathcal{A}_{d,k}$. Therefore

$$1 = \lim_{k\to\infty} \lim_{n\to\infty} \frac{h_\omega(\mathcal{A}_{d,k},\alpha^n)}{H_\omega(\mathcal{A}_{d,k})} \le \lim \frac{h_\omega(\mathcal{B},\alpha^n)}{H_\omega(\mathcal{B})} \le 1$$

due to Lemmas 2 and 4.

It should be noted that it is not sufficient to calculate $m_\omega(\alpha,\mathcal{B}_n)$ only for an increasing set of subalgebras. Here we have the counterexample of the crossed product of an abelian von Neumann algebra with a Bernoulli shift [10,11] where it can be shown that $m_\omega(\alpha,\mathcal{B}_k) = 1$ for an appropriately chosen sequence of $\mathcal{B}_k$ (see [11]) but nevertheless there exist invariant elements. This is easily seen since the Bernoulli shift is an inner automorphism for the crossed product. Therefore there exist invariant subalgebras. For them $h_\omega(\alpha,\mathcal{A}) = 0$ and also $m_\omega(\alpha,\mathcal{A}) = 0$.

4 The Shift on the Quantum Lattice System

We consider a quantum lattice system $\mathcal{M} = \bigotimes_{x=-\infty}^{+\infty} \mathcal{A}_x$, where $\mathcal{A}_x$ are $d \times d$ matrix algebras. The K-automorphism maps $\alpha^n \mathcal{A}_x = \mathcal{A}_{x+n}$. It is obvious that it leads to a K-system in the sense of [1,2]. We consider some extremal invariant state ω (local correlations are permitted) and take

$$\mathcal{B}_0 = \bigotimes_{x=-\infty}^{0} \mathcal{A}_x.$$

Since ω is extremal invariant the state is clustering. Therefore w lim $\alpha^n b = \omega(b)$ for all $b \in \mathcal{M}$ and $\alpha^n b\Omega = b\Omega$ only holds for $b = \lambda \mathbf{1}$. Let P_n be the projection on $\alpha^{-n}\mathcal{B}_0|\Omega\rangle$. Then st $\lim_{n\to\infty} P_n = |\Omega\rangle\langle\Omega|$ and

$$\bigwedge_{n=0}^{-\infty} \alpha^n \mathcal{B}_0 = \lambda \mathbf{1}.$$

It was less trivial to show that $(\mathcal{M}, \omega, \alpha)$ is also a K-system in the sense of [4]. The argument runs as follows:

We know that for any finite-dimensional local algebra $\mathcal{B}$ a finite decomposition is sufficient. Without losing more than ε we can assume that this finite decomposition is given by $1 = \sum x_i$, $i = 1, \ldots, r$, where $x_i \in \bar{\mathcal{B}}$ is again a strictly local algebra. So with $\eta(\lambda) = -\lambda \ln \lambda$

$$H_\omega(\mathcal{B}) = \sum \eta(\omega(x_i)) + \sum S(\omega|\omega(x_i\cdot))_{|\mathcal{B}}.$$

If $\mathcal{B} \neq c\mathbf{1}$, there exists some x with $\omega(xb) \neq \omega(x)\omega(b)$ for $b \in \mathcal{B}$, so $H(\mathcal{B}) > 0$. For $\frac{1}{n} H_\omega(\mathcal{B}, \alpha^\ell \mathcal{B}, \ldots \alpha^{\ell n}\mathcal{B})$ we consider

$$\geq \sum \frac{1}{n} \eta(\omega(x_{I_n}^\ell) - \sum \eta(\omega(x_i)) + \sum \eta(\omega(x_i)) + \sum S(\omega|\omega(x_i\cdot))_{\mathcal{B}},$$

where $x_{I_n}^\ell = \prod_{k=0}^n \alpha^{\ell k} x_{i_k}$ is a positive operator if only ℓ is large enough so that $[x_{i_1}, \alpha^\ell x_{i_2}] = 0$. We can consider the index set I_n as an r-dimensional Bernoulli shift, $i \in \{1, \ldots, r\}$ and probabilities $\omega(x_i)$. So $\lim_{n\to\infty} \frac{1}{n} \sum \eta(\omega(x_I^\ell) = h_\omega(\alpha^\ell, \{1, \ldots, r\})$ and $\sum \eta(\omega(x_i)) = H_\omega(\{1, \ldots, r\})$. In the classical case we have strong subadditivity which gives

$$\lim_{n\to\infty} \frac{1}{n} H(\mathcal{B}, \ldots, \alpha^{n\ell}\mathcal{B}) = \lim_{n\to\infty} (H(\mathcal{B}, \ldots, \alpha^{n\ell}) - H(\alpha^\ell \mathcal{B}, \ldots, \alpha^{n\ell}\mathcal{B}) \equiv \lim_{n\to\infty} H(\mathcal{B} \setminus \bigvee_{k=1}^{n} \alpha^{k\ell}\mathcal{B}).$$

Now K-mixing implies the triviality of the tail [5]:

$$\lim_{\ell\to\infty} \lim_{n\to\infty} \bigvee_{k=\ell}^{n} \alpha^{k\ell}\mathcal{B} = \lambda \mathbf{1}.$$

Therefore we continue

$$\lim_{n\to\infty} \frac{1}{n} H(\mathcal{B}, \ldots, \alpha^{n\ell}\mathcal{B}) = H(\mathcal{B}).$$

In our case

$$\lim_{\ell\to\infty} \lim_{n\to\infty} \sum_{I_n} \frac{1}{n} \eta(\omega(x_{I_n}^\ell) - \sum_i \eta(\omega(x_i)) = 0$$

or

$$\lim h_\omega(\alpha^\ell, \mathcal{B}) = H_\omega(\mathcal{B}).$$

5 Clustering and K-Systems

In [2] it was shown that a K-system in the sense of [2] satisfies the following clustering property: Let x be given. Then to every $\varepsilon > 0 \; \exists \; n_0$ such that for all $n \geq n_0$ and all $y \in \mathcal{A}_0$

$$|\omega(x\alpha^{-n}y) - \omega(x)\omega(y)| \leq \varepsilon\|y\|.$$

It should be noted that this is the well-known clustering with respect to the shift given in [12]. We expect that this clustering is the tool to find a connection between K-systems in the sense of [2] and [4]. At least we have the following result:

Theorem: Let $(\mathcal{M}, \alpha, \mathcal{A}_0)$ be a K-system in the sense of [2]. Assume further that α is strongly asymptotically abelian. Let ω be an α invariant state which is KMS with respect to an automorphism τ. Then it is also a K-system in the sense of [4].

Remark: The clustering together with the KMS property already implies weak asymptotic abelianess, i.e. [13]

$$\lim \omega(a[b, \alpha^n c]d) = 0 \qquad \forall\, a, b, c, d.$$

We assume therefore in addition that

$$\lim_{n\to\infty} \|[b, \alpha^n c]d\Omega\| = 0.$$

Proof: More or less we have only to repeat the arguments of the preceding paragraph. We only have to be careful in the construction of x_I because we do not have strict abelianess. Therefore we choose

$$x^{\ell}_{I_n} = \sqrt{x_{i_0}}\alpha^{\ell}\sqrt{x_{i_1}} \ldots \alpha^{\ell n} x_{i_n} \ldots \sqrt{x_{i_0}} \in \mathcal{M}$$

which is obviously positive.

Take y_s a testelement of $\alpha^{\ell s}\mathcal{A}$. Then

$$\lim_{\ell\to\infty} \sum_{I_n, \widehat{i_s}} \omega(\sqrt{x_{i_0}} \ldots \alpha^{\ell s}\sqrt{x_{i_s}}\alpha^{\ell(s+1)}\sqrt{x_{i_{s+1}}} \ldots \alpha^{\ell s}\sqrt{x_{i_s}} \ldots \sqrt{x_{i_0}}\tau^{i/2}y_s) =$$

$$= \lim_{\ell\to\infty} \sum \omega(\alpha^{\ell(s+1)}\sqrt{x_{i_{s+1}}} \ldots \alpha^{\ell n}x_{i_n} \ldots \alpha^{\ell s}\sqrt{x_{i_s}} \ldots \tau^{i/2}y_s\tau^{i}(\sqrt{x_{i_0}} \ldots \alpha^{\ell s}\sqrt{x_{i_s}}))$$

due to the τ-KMS property. The operator $\alpha^{\ell(s+1)}\sqrt{x_{i_{s+1}}} \ldots \alpha^{\ell n}x_{i_n} \ldots \alpha^{\ell(s+1)}\sqrt{x_{i_{s+1}}}$ is separated from the rest by α^{ℓ}. So we use K clustering to obtain, again with the KMS condition,

$$= \lim_{\ell\to\infty} \sum \omega(\alpha^{\ell(s+1)}\sqrt{x_{i_{s+1}}} \ldots \alpha^{\ell(s+1)}\sqrt{x_{i_{s+1}}})\omega(\sqrt{x_{i_0}} \ldots \alpha^{\ell s}x_{i_s} \ldots \sqrt{x_{i_0}}\tau^{i/2}y_s) =$$

$$= \lim_{\ell\to\infty} \sum \omega(\cdot)\{\omega(\alpha^{\ell s}x_{i_s}\sqrt{x_{i_0}}\alpha^{\ell}\sqrt{x_{i_1}} \ldots \sqrt{x_{i_0}}\tau^{i/2}y_s) + \omega([\sqrt{x_{i_0}} \ldots, \alpha^{\ell s}x_{i_s}]\alpha^{\ell(s-1)}\sqrt{x_{i_{s-1}}} \ldots \tau^{i/}$$

The last term goes to zero, since we assume strong asymptotic abelianess. In the first term we can perform the summation and arrive at

$$= \omega(x_{i_s}\tau^{i/2}y_s)$$

which gives the optimal decomposition for $\alpha^{\ell s}\mathcal{A}$. In calculating $h_\omega(\alpha^\ell, \mathcal{A})$ we have therefore to show that

$$\lim_{\ell\to\infty}\lim_{n\to\infty}\frac{1}{n}\sum_{I_n}\eta(\omega(x^\ell_{I_n}))-\sum_i\eta(\omega(x_i))=0.$$

We can use again K clustering for

$$\lim_{\ell\to\infty}\omega(\sqrt{x_{i_0}}\alpha^\ell\sqrt{x_{i_1}}\ldots\alpha^\ell\sqrt{x_{i_1}}\sqrt{x_{i_0}})=\lim_{\ell\to\infty}\omega(x^\ell_{I_n})=$$

$$=\omega(x_{i_0})\omega(\sqrt{x_{i_1}}\ldots\sqrt{x_{i_1}})\equiv\omega(x_{i_0})\omega(x^\ell_{I_{n-1}}).$$

Notice that the limit is attained uniformly in n. Therefore

$$\lim_{\ell\to\infty}\lim_{n\to\infty}\frac{1}{n}[\sum\eta(\omega(x^\ell_{I_n})-n\sum\eta(\omega(x_{i_0})]=$$

$$=\lim_{\ell\to\infty}\lim_{n\to\infty}\frac{1}{n}\sum_{k=0}^{n}[\sum(\omega(x^\ell_{I_k})-\eta(\omega(x_{i_0}))-\eta(\omega(x^\ell_{I_{k-1}})].$$

We consider $\omega(x^\ell_{I_k})$ to be a state ω^k_ℓ on the abelian algebra $\mathcal{B}=\mathcal{B}_0\otimes\mathcal{B}^{k-1}$, $\mathcal{B}_0$ finite-dimensional, and K clustering tells us that $\lim_{\ell\to\infty}\|\omega^k_\ell-\omega_0\otimes\omega^{k-1}_\ell\|=0$ uniformly in k. It remains to show that this suffices to guarantee that

$$\lim_{\ell\to\infty}[S(\omega^k_\ell)-S(\omega_0)-S(\omega^{k-1}_\ell)]=0$$

uniformly in k.

ω^k_ℓ corresponds to a density matrix

$$\sum_{a,j}(\lambda_a\mu^\ell_j+\varepsilon^\ell_{aj})P_aQ_j=\sum\alpha^\ell_{aj}P_aQ_j,\qquad P_a\in\mathcal{B}_0,\qquad Q_j\in\mathcal{B}^{k-1},$$

with $a\in\{1,\ldots,n\}$ and $j\in\{1,\ldots\}$. Further $\sum_j\alpha^\ell_{aj}=\lambda_a$, $\sum_a\alpha^\ell_{aj}=\mu^\ell_j$, $\alpha^\ell_{aj}\geq 0$. Therefore

$$|\varepsilon^\ell_{aj}|\leq\max\{\lambda_a\mu^\ell_j,\alpha^\ell_{aj}\}\leq\max\{\lambda_a\mu^\ell_j,\mu^\ell_j\}\leq\mu^\ell_j$$

and also

$$|\varepsilon^\ell_{aj}|\leq\lambda_a.$$

Therefore we can write $\varepsilon^\ell_{aj}=\gamma^\ell_{aj}\mu^\ell_j$ with $|\gamma^\ell_{aj}|\leq 1$. We split the index set j into $I_a=\{j,|\gamma^\ell_{aj}|<\sqrt{\varepsilon}\}$ and $I^c_a=\{j,|\gamma^\ell_{i,j}|\geq\sqrt{\varepsilon}\}$. $\|\omega^k_\ell-\omega_0\otimes\omega^{k-1}_\ell\|<\varepsilon$ implies that $\sum|\varepsilon^\ell_{aj}|<\varepsilon$. It follows that

$$\sum_{j\in I^c_a}\mu^\ell_j<\sqrt{\varepsilon}.$$

Now we use

$$\begin{aligned}0&\leq S(\omega_0)+S(\omega^{k-1}_\ell)-S(\omega^k_\ell)=S(\omega_0\otimes\omega^{k-1}_\ell|\omega^k_\ell)=\\&=\sum_{a,j}\alpha^\ell_{aj}\ln\frac{\alpha^\ell_{aj}}{\lambda_a\mu^\ell_j}\leq\sum_{\gamma^\ell_{aj}\geq 0}(\lambda_a+\gamma^\ell_{aj})\mu^\ell_j\ln\frac{(\lambda_a+\gamma^\ell_{aj})}{\lambda_a}=\\&=\sum_{a,\gamma^\ell_{aj}\geq 0}\sum_{j,j\in I^c_a}+\sum_{a,\gamma^\ell_{aj}\geq 0}\sum_{j,j\in I_a}.\end{aligned}$$

λ_a runs through a finite set and $\neq 0$. Therefore

$$\ln(1+\frac{\gamma^{\ell}_{aj}}{\lambda_a}) \leq \ln(1+\frac{1}{\lambda_a})$$

is bounded and

$$\sum_{j\in I_a^c} \ldots \leq c_1\sqrt{\varepsilon}.$$

For $j \in I_a$,

$$\ln(1+\frac{\gamma^{\ell}_{aj}}{\lambda_a}) \leq \frac{\sqrt{\varepsilon}}{\lambda_a},$$

therefore

$$\sum_{j\in I_a} \ldots \leq c_2\sqrt{\varepsilon}.$$

All these estimates are independent of k, which proves the theorem.

Remark: It should be noted that a K-system in the sense of [2] need not be strongly asymptotically abelian. We have the example of the Fermi lattice system and the shift automorphism that is only weakly asymptotically abelian. On the other hand, it was only necessary that sufficiently many operators – those that are needed to construct optimal decompositions – asymptotically commute, and in fact this was necessary, otherwise we could not construct the positive x_I. So one can see that the C^*-algebra of fermions on a lattice is a K-system in the sense of [2], but only its even part a K-system in the sense of [4]. The next task that is under investigation [14] is to classify this class of operators that is needed for good decompositions for arbitrary decompositions for arbitrary subalgebras and so to see what kind of clustering is the consequence for a K-system in the sense of [4].

Acknowledgements

It is a pleasure to thank F. Benatti, T. Hudetz, and W. Thirring for critical remarks on the manuscript.

References

[1] G.G. Emch, Generalized K-flows, Commun. Math. Phys. 49 (1976) 191–215.

[2] W. Schröder, A hierarchy of mixing properties for non-commutative K-systems, in: Quantum Probability and Applications to the Quantum Theory of Irreversible Processes, ed. L. Accardi, A. Frigerio and V. Gorini, Springer Berlin, Heidelberg, New York, Tokyo, 1984, p. 340–351.

[3] A. Connes, H. Narnhofer, W. Thirring, Commun. Math. Phys. 112 (1987) 691–719.

[4] H. Narnhofer, W. Thirring, Quantum K-systems, Vienna preprint, UWThPh-1988-40.

[5] I.P. Cornfeld, S.V. Formin, Ya.G. Sinai, Ergodic Theory, Springer Berlin, Heidelberg, New York, 1982.

[6] H. Narnhofer, Dynamical Entropy in Quantum Theory, Vienna preprint UWThPh-1988-27, to be published in IAMP proceedings 1988.

[7] H. Kosaki, Commun. Math. Phys. 87 (1982) 315.

[8] H. Narnhofer, W. Thirring, Fizika 17 (1985) 257–265.

[9] H. Narnhofer, W. Thirring, Lett. Math. Phys. 15 (1988) 261–273.

[10] U. Quasthoff, On automorphisms of factors related to measure space transformations, Leipzig preprint 1988.

[11] H. Narnhofer, Beispiele von Algebren mit gleicher dynamischer Entropie, Vienna preprint, UWThPh-1988-42.

[12] R.T. Powers, Ann. of Math. 86 (1967) 138.

[13] H. Narnhofer, W. Thirring, Mixing properties of quantum systems, Vienna preprint, UWThPh-1988-18.

[14] F. Benatti, H. Narnhofer, W. Thirring, in preparation.

This paper is in final form and no similar paper has been or is being submitted elsewhere.

A continuous time version of Stinespring's theorem on completely positive maps

By

K.R.Parthasarathy
Indian Statistical Institute
7, S. J. S. Sansanwal Marg
New Delhi - 110016

The aim of this note is to prove the following version of Stinespring's theorem for a one parameter semigroup of completely positive maps [1], [2].

Theorem: Let A be W^* algebra of operators containing the identity and acting in a Hilbert space H_o and let $T_t : A \to A$ be a one parameter semigroup of completely positive linear maps such that T_o is identity and $T_t(1) \equiv 1$. Then there exists a family $\{H_t, t \geq 0\}$ of Hilbert spaces, W^* homomorphisms $J_t: A \to B(H_t)$, the algebra of all operators on H_t and isometries $V(s,t) : H_s \to H_t$, $0 \leq s < t < \infty$ such that

$$J_o(a) = a,$$

$$V^{\dagger}(s,t)\, J_t(a)\, V(s,t) = J_s(T_{t-s}(a)),$$

$$V(t,u)\, V(s,t) = V(s,u)$$

for all $0 < s < t < u < \infty$.

Proof: For any Hilbert space K we shall denote by $B(K)$ the W^* algebra of all operators on K. For any $a \in B(K)$, $a^{\dagger}$ will denote the adjoint of a. For any fixed t, applying Stinespring's theorem to the completely positive map T_t, construct the triple (K_t, h_t, V_t) where K_t is a Hilbert space, $h_t : A \to B(K_t)$ is a W^* homomorphism and $V_t : H_o \to K_t$ is an isometry such that

$$V_t^{\dagger}\, h_t(a)\, V_t = T_t(a) \quad \text{for all } a \in A,$$

nd the set $\{h_t(a)V_t u,\ a \in A,\ u \in H_o\}$ is total in K_t. We shall call K_t, h_t, V_t) the Stinespring triple associated with T_t. Now, for $\leq t_1 < t_2 < \infty$ consider the completely positive map $h_{t_1} \circ T_{t_2-t_1} : A \rightarrow B(K_{t_1})$ here o denotes composition. Applying Stinespring's theorem to this omposed completely positive map construct the Stinespring triple K_{t_1,t_2}, h_{t_1,t_2}, V_{t_1,t_2}) where K_{t_1,t_2} is a Hilbert space, $h_{t_1,t_2} : A \rightarrow B(K_{t_1,t_2})$ s a W* homomorphism and $V_{t_1,t_2} : K_{t_1} \rightarrow K_{t_1,t_2}$ is an isometry such that

$$V^{\dagger}_{t_1,t_2}\, h_{t_1,t_2}(a)\, V_{t_1,t_2} = h_{t_1} \circ T_{t_2-t_1}(a) \text{ for all } a \in A$$

nd $\{h_{t_1,t_2}(a)\, V_{t_1,t_2} u,\ a \in A,\ u \in K_{t_1}\}$ is total in K_{t_1,t_2}. Proceeding nductively we obtain a family of Stinespring triples $\{(K_{t_1,t_2,\ldots,t_n}$, $t_1,t_2,\ldots,t_n$, $V_{t_1,t_2,\ldots,t_n}$), $0 \leq t_1 < t_2 < \ldots < t_n < \infty$, $n = 1,2,\ldots\}$ where $t_1,t_2,\ldots,t_n : A \rightarrow B(K_{t_1,\ldots,t_n})$ is a W* homomorphism and $V_{t_1,t_2,\ldots,t_n}$: $t_1,t_2,\ldots,t_{n-1} \rightarrow K_{t_1,t_2,\ldots,t_n}$ is an isometry such that

$$V^{\dagger}_{t_1,t_2,\ldots,t_n}\, h_{t_1,t_2,\ldots,t_n}(a)\, V_{t_1,t_2,\ldots,t_n} = h_{t_1,t_2,\ldots,t_{n-1}} \circ T_{t_n-t_{n-1}}(a) \qquad (1)$$

nd $\{h_{t_1,t_2,\ldots,t_n}(a)\, V_{t_1,t_2,\ldots,t_n} u,\ u \in K_{t_1,t_2,\ldots,t_{n-1}},\ a \in A\}$ is total in $t_1,t_2,\ldots,t_n$.

A map $X : \mathbb{R}_+ \rightarrow A$ is said to have finite support if the set $(X) = \{t : X(t) \neq 1\}$ is finite. Let M denote the set of all such maps rom $\mathbb{R}_+$ into A with finite support and let $M_{t]} \subset M$ consist of those elements whose support $S(X)$ is contained in $[0,t]$. For any finite set $= \{t_1,t_2,\ldots,t_n\}$, $0 \leq t_1 < t_2 < \ldots < t_n < \infty$ and $X \in M$ define

$$X(F) = h_{t_1,t_2,\ldots,t_n}(X(t_n))V_{t_1,t_2,\ldots,t_n} h_{t_1,t_2,\ldots,t_{n-1}}(X(t_{n-1}))\ldots h_{t_1}(X(t_1))V_{t_1}. \qquad (2)$$

Thus X(F) is an operator from H_o into $K_{t_1,t_2,\ldots,t_n}$. From (1) and (2) we have for any $F = \{t_1,t_2,\ldots,t_n\}$, $0 \leq t_1 < t_2 < \ldots < t_n < \infty$, $X, Y \in M$ and $u, v \in H_o$

$$<X(F)u, Y(F)v>$$

$$= <u, T_{t_1}(X(t_1)^{\dagger}\ldots T_{t_{n-1}-t_{n-2}}(X(t_{n-1})^{\dagger} T_{t_n-t_{n-1}}(X(t_n)^{\dagger} Y(t_n))Y(t_{n-1}))\ldots Y(t_1))v> \tag{3}$$

We now make the important observation that

$$<X(F)u, Y(F)v> = <X(S(X)\cup S(Y))u,\ Y(S(X)\cup S(Y))v>$$

$$\text{whenever } F \supset S(X) \cup S(Y),\ u,v \in H_o \tag{4}$$

For any $u,v \in H_o$, $X,Y \in M$ define

$$K((u,X),(v,Y)) = <X(S(X)\cup S(Y))u,\ Y(S(X)\cup S(Y))v> \tag{5}$$

In view of (4), $K(.,.)$ is a positive definite kernel on $H_o \times M$. Hence by the GNS theorem there exists a Hilbert space H and a map $\delta: H_o \times M \to H$ such that

$$<\delta(u,X),\ \delta(v,Y)> = K((u,X),(v,Y)) \text{ for all } u,v \in H_o,\ X,Y \in M. \tag{6}$$

We denote by $H_{t]}$ the subspace spanned by $\{\delta(u,X),\ u \in H_o,\ X \in M_{t]}\}$.

For any $a \in A$, $X \in M$, $t \geq 0$ define $\ell_t(a)X \in M$ by

$$(\ell_t(a)X)(s) = X(s) \quad \text{if } s \neq t,$$
$$a\,X(t) \text{ if } s = t.$$

We now claim that for $X,Y \in M_{t]}$ and any unitary element $a \in A$

$$<\delta(u, \ell_t(a)X),\ \delta(v, \ell_t(a)Y)>$$
$$= <\delta(u,X),\ \delta(v,Y)>. \tag{7}$$

Indeed, let $S(X) \cup S(Y) = \{t_1,t_2,\ldots,t_n\} \subset [0,t]$. Then $S(\ell_t(a)X) \cup S(\ell_t(a)Y) \subset F$ where $F = \{t_1,t_2,\ldots,t_n,t\}$ if $t_n < t$ and $\{t_1,t_2,\ldots,t_n\}$ if $t_n = t$. By (3) we have

$$
\begin{aligned}
&<\delta(u,\ell_t(a)X),\ \delta(v,\ell_t(a)Y)> \\
&= <(\ell_t(a)X)(F)u,\ \ (\ell_t(a)Y)(F)v> \\
&= <u, T_{t_1}(X(t_1)^{\dagger}\ldots T_{t_n-t_{n-1}}(X(t_n)^{\dagger}a^{\dagger}a\,Y(t))Y(t_n))\ldots Y(t_1))v> \\
&= <X(F)u,\ Y(F)v> \\
&= <\delta(u,X),\ \delta(v,Y)>,
\end{aligned}
$$

which proves the claim (7). Hence for every unitary element $a \in A$ there exists a well defined isometric operator $J_t(a)$ in the Hilbert space $H_{t]}$ such that

$$J_t(a)\delta(u,X) = \delta(u,\ell_t(a)X) \text{ for all } u \in H_o,\ X \in M_{t]} \tag{8}$$

Furthermore for any two unitary elements a,b we have $J_t(ab)=J_t(a)J_t(b)$ and $J_t(1) = 1$. This implies that $J_t(a)$ defined by (8) is unitary. We now extend J_t to an arbitrary element as follows. Any $x \in A$ can be expressed as $x = \sum_{j=1}^{4} \alpha_j a_j$ where the α_j's are scalars and a_j's are unitary. Define $J_t(x) = \sum_{j=1}^{4} \alpha_j J_t(a_j)$. To show that $J_t(x)$ is well defined we have to only show that whenever $\sum \alpha_j a_j = 0$ the operator $\sum \alpha_j J_t(a_j)$ is also zero. By the definition of $H_{t]}$ and equations (3) - (6) and (8) it follows that whenever $\sum \alpha_j a_j = 0$ we have for any $X,Y \in M_{t]}$, $u,v \in H_o$

$$
\begin{aligned}
&<\delta(u,X),\ \textstyle\sum \alpha_j J_t(a_j)\ \delta(v,Y)> \\
&= <\delta(u,X),\ \delta(v,\ \ell_t(\textstyle\sum \alpha_j a_j)Y)> = 0.
\end{aligned}
$$

Since $\{\delta(u,X),\ u \in H_o,\ X \in M_{t]}\}$ span $H_{t]}$ it follows that $\sum \alpha_j J_t(a_j) = 0$. In other words we have shown that for every $x \in A$ there exists a well defined operator $J_t(x)$ on $H_{t]}$ such that

$$J_t(x)\delta(u,X) = \delta(u,\ell_t(x)X) \text{ for all } x \in A,\ u \in H_o,\ X \in M_{t]}. \tag{9}$$

This also shows that the map $x \to J_t(x)$ is a W* homomorphism for A into $B(H_{t]})$.

Let now $0 \leq s < t < \infty$, $X,Y \in M_{s]}$. We shall evaluate $\langle \delta(u,X), J_t(a)\delta(v,Y)\rangle$ for an arbitrary element $a \in A$. Let $S(X) \cup S(Y) = \{s_1,s_2,\ldots,s_n\}$ where $0 \leq s_1 < s_2 < \ldots < s_n \leq s$. Then

$$S(X) \cup S(\ell_t(a)Y) \subset \{s_1,s_2,\ldots,s_n,t\} = F, \quad \text{say.}$$

By (3) - (6) we have

$$\begin{aligned}
&\langle \delta(u,X), J_t(a)\delta(v,Y)\rangle \\
&= \langle \delta(u,X), \delta(v, \ell_t(a)Y)\rangle \\
&= \langle X(F)u, (\ell_t(a)Y)(F)v\rangle \\
&= \langle u, T_{s_1}(X(s_1)^\dagger \ldots T_{s_n-s_{n-1}}(X(s_n)^\dagger T_{t-s_n}(a)Y(s_n))\ldots Y(s_1))v\rangle \\
&= \langle u, T_{s_1}(X(s_1)^\dagger \ldots T_{s_n-s_{n-1}}(X(s_n)^\dagger T_{s-s_n}(X(s)^\dagger T_{t-s}(a)Y(s))Y(s_n))\ldots)v\rangle \\
&= \langle \delta(u,X), J_s(T_{t-s}(a))\delta(v,Y)\rangle.
\end{aligned}$$

If $V(s,t)$ denotes the identity map from $H_{s]}$ into $H_{t]}$ then the above equation implies that

$$V(s,t)^\dagger J_t(a) V(s,t) = J_s(T_{t-s}(a)),$$

For $X,Y \in M_{o]}$ we have

$$\langle \delta(u,X), \delta(v,Y)\rangle = \langle X(0)u, Y(0)v\rangle, \quad u,v \in H_o.$$

Thus the correspondence $X(0)u \to \delta(u,X)$, $u \in H_o$, $X \in M_{o]}$ is a unitary isomorphism between H_o and $H_{o]}$ and the algebras A and $M_{o]}$ are also isomorphic. Hence we may assume without loss of generality that $H_{o]} = H_o$. Putting $H_t = H_{t]}$ for all t the proof becomes complete.

References:

[1] E.B.Davies, Quantum Theory of Open Systems, Academic Press, New York (1976).

[2] W.F.Stinespring, Positive functions on C* algebras, Proc. Amer. Math. Soc. 6, 211-216 (1955).

This paper is in final form and no similar paper has been or is being submitted elsewhere.

THE TOPOLOGY OF THE CONVERGENCE IN PROBABILITY IN A W*-ALGEBRA IS NORMAL

Adam Paszkiewicz

Institute of Mathematics, Łódź University, ul. S. Banacha 22, 90-238 Łódź, Poland

1. Introduction and main result. Let M, M^h, $\mathrm{Proj}\,M$ and ρ denote, respectively, a W*-algebra acting in a Hilbert space H, the space of self-adjoint elements from M, the lattice of orthogonal projections from M and a faithful normal state on M. For any self-adjoint operator x in H, $e_x(\cdot)$ denotes its spectral measure, so $x = \int \lambda e_x(d\lambda)$. We say that an operator $x = x^*$, affiliated to M, is Segal measurable if there exists $N_0 > 0$ such that $e_x((-\infty,-N) \cup (N,\infty))$ is a finite projection of M for any $N > N_0$, cf. [4]. The space of all such operators will be denoted by $\tilde{M}$. In [2] we show that the sets

$$U(x,\varepsilon) = \{y \in \tilde{M};\ \exists\, e \in \mathrm{Proj}\,M : \rho(1-e) < \varepsilon \text{ and } \|(y-x)e\| < \varepsilon\} \quad \text{with } \varepsilon > 0$$

form the base of neighbourhoods of $x \in \tilde{M}$ for some Hausdorff topology τ in $\tilde{M}$, and that $x_n \to x$ in τ, $x_n x \in \tilde{M}$, if and only if $x_n \to x$ in probability (i.e. $\|(x_n - x)e_n\| \to 0$, $\rho(1-e_n) \to 0$ for some $e_n \in \mathrm{Proj}\,M$). The topology τ is determined by the condition $\bar{A} = \{x \in \tilde{M};\ x_n \to x$ in probability for some $x_n \in A,\ n \geq 1\}$ for any $A \subset \tilde{M}$. Now, we prove

1.1. THEOREM. The topology τ in $\tilde{M}$ is normal. In particular, τ is given by some uniform structure in $\tilde{M}$ [1].

As we suggest in [3], the results for M^h can often be extended to $\tilde{M}$ (in a rather standard way) by the following proposition, and this is true when we investigate the topology τ.

1.2. PROPOSITION. Let $e_n, f \in \mathrm{Proj}\,M$. If $\rho(1-e_n) \to 0$, and $1-f$ is a finite projection in M, then $\rho(f - e_n \wedge f) \to 0$.

As in [2, 3.1] we denote $P_F(K) = F - F \wedge (1 - K)$ for any projections $F, K \in \mathrm{Proj}\, M$. This notation gives, for example, a shorter formulation of the following

1.3. LEMMA [2, 4.1]. Let $h \leq f$, $h, f \in \mathrm{Proj}\, M$, let $1 - f$ be finite and let $\varepsilon > 0$. Then, for some $\delta > 0$, the inequality $\rho(1 - e) < \delta$, $e \in \mathrm{Proj}\, M$, implies the existence of $k \leq h$, $k \in \mathrm{Proj}\, M$, satisfying $\|k - P_{e \wedge f}(k)\| \leq \varepsilon$ and $\rho(h - k) < \varepsilon$.

We shall also use the following lemmas.

1.4. LEMMA [3, 7.1]. If $P \wedge (1 - K) = 0$, $K, P \in \mathrm{Proj}\, M$, then $\|K - P\| = \|K - PK\|$.

1.5. LEMMA (criterion of the convergence of projections to 0 [2, 4.4]). Let $c_m \in \mathrm{Proj}\, M$. If there exists projections $q_m \sim c_m$, $q_m \in \mathrm{Proj}\, M$, such that $\rho(q_m) \to 0$ and $q = \bigvee_{m \geq 1} q_m$ is a finite projection of M, then $\rho(c_m) \to 0$.

1.6. COROLLARY. For any finite projection $q \in \mathrm{Proj}\, M$ and for $\varepsilon > 0$, there exists a number $\delta > 0$ such that the conditions $q_1 \leq q$, $q_1 \sim c$, $\rho(q_1) < \delta$, $q_1, c \in \mathrm{Proj}\, M$, imply $\rho(c) < \varepsilon$.

1.7. LEMMA [3, 2.3]. Let a be any self-adjoint operator affiliated to M, $e \in \mathrm{Proj}\, M$ and let $eH \in D_a$ (the domain of a). For any $\varepsilon > 0$, there exists a projection $\tilde{e} \leq e$, $\tilde{e} \in \mathrm{Proj}\, M$, such that $\rho(e - \tilde{e}) < \varepsilon$ and $\|a e_a((-\infty, r) \cup (r, \infty))\tilde{e}\| \to 0$ as $r \to 0$, r is real and positive.

2. Construction of a subprojection almost commuting with a self-adjoint operator.

2.1. LEMMA. Let $K = \Sigma_i k_i$, $P = \bigvee_i p_i$, $k_i, p_i \in \mathrm{Proj}\, M$, $k_i k_j = 0$ for $i, j = 1, \ldots, m$, $i \neq j$, and let $0 < \varepsilon < 1/2$. If $\|k_i - p_i\| < \varepsilon/m$ for $1 \leq i \leq m$, then $\|K - P\| < \varepsilon$.

P r o o f. For a fixed vector $\xi = \Sigma_i \xi_i$, $\xi_i \in p_i H$, $\|\xi\| = 1$, denote $a = \max_i \|\xi_i\|$. Then $\|\xi - K\xi\| \leq \Sigma_i \|\xi_i - K\xi_i\| \leq \Sigma_i \|\xi_i - k_i \xi_i\| \leq \Sigma_i a\|p_i - k_i\| < a\varepsilon$. Moreover, $\|a\| < 2$. Indeed, for $\delta_i = \xi_i - k_i \xi_i$, we have

$$\|k_i\xi_j\| \leq \|\delta_j\| \leq a\|p_j - k_j\| < a\cdot\varepsilon/m \quad \text{if} \quad i \neq j,$$

and

$$1 = \|\xi\| \geq \max_i \|k_i\xi\| = \max_i \|k_i\xi_i + \Sigma_{j,j\neq i} k_i\xi_j\|$$

$$\geq \max_i (\|\xi_i\| - \|\delta_i\| - \Sigma_{j,j\neq i} \|\delta_j\|) \geq a - a\varepsilon > a/2.$$

We have proved, in fact, that $\|\xi - K\xi\| < 2\varepsilon < 1$ for any $\xi \in PH$, $\|\xi\| = 1$. Thus $P \wedge (1 - K) = 0$. Now, 1.4 can be used, and the inequality $\|K - P\| < \varepsilon$ is given by the following estimation: For $\zeta \in KH$, $\|\zeta\| = 1$, $\zeta_i = k_i\zeta$, the inequality $\|\zeta - P\zeta\| \leq \Sigma_i \|\zeta_i - P\zeta_i\| \leq \Sigma_i \|\zeta_i - p_i\zeta_i\| \leq \Sigma_i \|k_i - p_i\| < \varepsilon$ holds.

2.2. THEOREM. Let $x \in \tilde{M}$, $0 < \beta < 1$. There exists $\gamma > 0$ such that the inequality $\rho(1-E)<\gamma$, $E\in \mathrm{Proj}\, M$ implies $M \ni \overline{ex}$ (the closure of ex), $\|\overline{ex} - xe\| < \beta$, $\rho(1 - e) < \beta$ for some $e \leq E$, $e \in \mathrm{Proj}\, M$.

P r o o f. Let the number $N > 1$ be large enough for the inequality $\rho(1 - f) < \beta/4$ with $f = e_x(-N,N)$ being a finite projection in M, and let $h_i = e_x[i/n, (i+1)/n)$, $-Nn \leq i < Nn$, for some integer $n > 4/\beta$. By 1.3, we can find a number $\gamma > 0$ so small that the inequality $\rho(1 - E) < \gamma$ implies

$$\text{(1)} \qquad \|k_i - P_{E\wedge f}(k_i)\| < \beta/8nN^2, \quad \rho(h_i - k_i) < \beta/4nN, \quad -Nn \leq i < Nn,$$

for some projections $k_i \leq h_i$, $k_i \in \mathrm{Proj}\, M$. The projections k_i satisfy $\|\overline{k_i x} - xk_i\| = \|k_i x h_i - x h_i k_i\| \leq (i/n)\|k_i h_i - h_i k_i\| + 2\|x h_i - i h_i/n\| \leq 2/n < \beta/2$ for any i, $-Nn \leq i < Nn$. By 2.1 and (1), $\|\Sigma_i k_i - \vee_i P_{E\wedge f}(k_i)\| \leq \beta/4N$. Thus, for $e = \vee_i P_{E\wedge f}(k_i)$,

$$\|\overline{ex} - xe\| \leq 2(\beta/4N)\|xf\| + \|\overline{\Sigma_i k_i x} - x\Sigma_i k_i\| < \beta/2$$

$$+ \max\{\|\overline{k_i x} - xk_i\|;\ -Nn \leq i < Nn\} < \beta$$

and

$$\rho(1 - e) \leq \|e - \Sigma_i k_i\| + \rho(1 - \Sigma_i k_i) < \beta/4N + \rho(1 - f)$$

$$+ \Sigma_i \rho(h_i - k_i) < \beta \quad \text{and also} \quad e \leq E.$$

3. Proof of the normality of τ.

3.1. LEMMA. Let $t \geq 0$, $t,z \in M^h$, $0 < \alpha < \beta < 1$, $\gamma > 0$. If $\|tz - zt\| < \gamma$, then $\|e_t[\beta,\infty)ze_t[0,\alpha)\| < \gamma/(\beta - \alpha)$.

The proof is given by the inequality $\gamma > \|tz - zt\| \geq \|e_t[\beta,\infty)\cdot(tz - zt)e_t[0,\alpha)\| \geq \|e_t[\beta,\infty)tze_t[0,\alpha)\| - \|e_t[\beta,\infty)zte_t[0,\alpha)\| \geq (\beta - \alpha)\|e_t[\beta,\infty)ze_t[0,\alpha)\|$.

3.2. LEMMA. Let $0 \leq t \leq 1$, $t,z \in M^h$, $0 < \alpha < 1$, $\gamma > 0$. If $\|tz - zt\| < \gamma$, then $\|[(1 - t)^{1/2}z - z(1 - t)^{1/2}]e_t[0,\alpha)\| < \gamma/(1 - \alpha)$.

P r o o f. The series $1 - t/2 - \Sigma_{n\geq 2}(2n - 3)!!t^n/2^n n!$ converges in norm to the operator $(1 - t)^{1/2}$, and

$$\|[t^n z - zt^n]e_t[0,\alpha)\| = \|\sum_{k=0}^{n-1}[t^{n-k}zt^k - t^{n-k-1}zt^{k+1}]e_t[0,\alpha)\|$$

$$\leq \|tz - zt\|\sum_{k=0}^{n-1}\|t^k e_t[0,\alpha)\| < \gamma/(1 - \alpha).$$

The equality $1/2 + \Sigma_{n\geq 2}(2n - 3)!!/2^n n! = 1$ finishes the proof.

3.3. Any projections $e,f \in \mathrm{Proj}\,M$ can be written in the form

$$e = \begin{bmatrix} 1 & 0 \\ 0 & 0 \end{bmatrix} \oplus 1 \oplus 1 \oplus 0 \oplus 0, \qquad f = \begin{bmatrix} c^2 & sc \\ sc & s^2 \end{bmatrix} \oplus 1 \oplus 0 \oplus 1 \oplus 0,$$

$s,c \geq 0$, $c^2 + s^2 = 1$, $\ker c = \ker s = 0$, c,s act in $H_0 = H_1$ (the 2×2 matrices act in $H_0 \oplus H_1$), according to some decomposition $H = H_0 \oplus \ldots \oplus H_5$ (cf. [3, 7.1]). Then a unitary operator

$$(1) \qquad u = \begin{bmatrix} c & -s \\ s & c \end{bmatrix} \oplus 1 \oplus 1 \oplus 1 \oplus 1$$

can be used to show the equivalence of the projections $e - e \wedge (1 - f)$ and $f - (1 - e) \wedge f = u[e - e \wedge (1 - f)]u^*$. In the sequel, the following property of u will be crucial.

THEOREM. Let $e,f \in M$, $0 < \varepsilon < 1/2$ and let u be of the form (1). Then

(i) $\xi \in e_{efe}[1 - \varepsilon^2,1]H$, $\|\xi\| = 1$ imply $\|\xi - u\xi\| < 2\varepsilon$;

(ii) $\|ze - ez\| < \varepsilon^3$, $\|zf - fz\| < \varepsilon^3$, $z \in M^h$, $\xi \in e_{efe}[1 - \varepsilon^2, 1]H$, $\|\xi\| = 1$ imply $\|z(\xi - u\xi)\| < 5\varepsilon\|z\xi\| + 15\varepsilon$.

P r o o f. Obviously, e,f,z,u can also be represented by 6×6 matrices if a vector $\xi \in H$ is treated as the column-vector $\xi = [\xi_i]_{0 \le i \le 5}$, $\xi_i \in H_i$, $0 \le i \le 5$. Namely,

$$e = \begin{bmatrix} 1 &&&&& \\ & 0 &&&& \\ && 1 &&& \\ &&& 1 && \\ &&&& 0 & \\ &&&&& 0 \end{bmatrix}, \quad f = \begin{bmatrix} c^2 & sc &&&& \\ sc & s^2 &&&& \\ && 1 &&& \\ &&& 0 && \\ &&&& 1 & \\ &&&&& 0 \end{bmatrix}, \quad z = [z_{ij}]_{0 \le i, j \le 5},$$

$z_{ij} = z_{ji}^*$, $z_{ij} : H_j \to H_i$,

$$u = \begin{bmatrix} c & -s &&&& \\ s & c &&&& \\ && 1 &&& \\ &&& 1 && \\ &&&& 1 & \\ &&&&& 1 \end{bmatrix}$$

(the elements which are not typed vanish).

Denote $ez - ze = [\varepsilon_{ij}]$, $fz - zf = [\delta_{ij}]$, $\varepsilon_{ij}, \delta_{ij} : H_j \to H_i$, $0 \le i,j \le 5$. Then $\|\varepsilon_{ij}\|, \|\delta_{ij}\| < \varepsilon^3$. Observe that some elements ε_{ij} are equal to $\pm z_{ij}$. Namely,

$$[z_{ij}] = \left[\begin{array}{cc|cccc} \bullet & \varepsilon_{01} & \bullet & \bullet & \varepsilon_{04} & \varepsilon_{05} \\ -\varepsilon_{10} & \bullet & -\varepsilon_{12} & -\varepsilon_{13} & \bullet & \bullet \\ \hline \bullet & \varepsilon_{21} & \bullet & \bullet & \varepsilon_{24} & \varepsilon_{25} \\ \bullet & \varepsilon_{31} & \bullet & \bullet & \varepsilon_{34} & \varepsilon_{35} \\ -\varepsilon_{40} & \bullet & -\varepsilon_{42} & -\varepsilon_{43} & \bullet & \bullet \\ -\varepsilon_{50} & \bullet & -\varepsilon_{52} & -\varepsilon_{53} & \bullet & \bullet \end{array}\right].$$

Then some elements δ_{ij} can easily be calculated:

$$[\delta_{ij}] = \begin{bmatrix} c^2z_{00}-z_{00}c^2, & \dots & \\ scz_{00}-z_{11}cs, & \dots & \\ z_{20}s^2, & -z_{20}sc, & \dots \\ -z_{30}c^2, & \dots & \\ -z_{41}sc, & \dots & \\ -z_{51}sc, & \dots & \end{bmatrix} + [\gamma_{ij}]$$

with some operators $\gamma_{ij} : H_j \to H_i$, $\|\gamma_{ij}\| < 2\varepsilon^3$, $0 \leq i,j \leq 5$. Then

$$\|c^2z_{00} - z_{00}c^2\|,\ \|csz_{00} - z_{11}cs\|,\ \|z_{20}s^2\|,\ \|z_{20}cs\|, \tag{2}$$
$$\|z_{41}cs\|,\ \|z_{51}cs\| < 3\varepsilon^3 < \varepsilon.$$

For $\xi \in e_{efe}[1 - \varepsilon^2,1]H$, $\|\xi\| = 1$, we have

$$\xi = \begin{matrix} \zeta \\ 0 \\ \eta \\ 0 \\ 0 \\ 0 \end{matrix}, \quad \xi - u\xi = \begin{matrix} (1-c)\zeta \\ -s\zeta \\ 0 \\ 0 \\ 0 \\ 0 \end{matrix}, \quad z(\xi - u\xi) = \begin{matrix} z_{00}(1-c)\zeta+\nu_0 & \\ \mu_1 & - z_{11}s\zeta \\ z_{20}(1-c)\zeta+\nu_2 & \\ z_{30}(1-c)\zeta+\nu_3 & \\ \mu_4 & - z_{41}s\zeta \\ \mu_5 & - z_{51}s\zeta \end{matrix}$$

with some vectors $\eta \in H_2$, $\zeta \in e_{c^2}[1 - \varepsilon^2,1]H_0 = e_s[0,\varepsilon]H_0$, $\|\eta\|,\|\zeta\|<1$, and with $\mu_i = - \varepsilon_{i0}(1-c)\zeta$, $\nu_i = -\varepsilon_{i1}s\zeta$, $\|\mu_i\|,\|\nu_i\| < \varepsilon^3 < \varepsilon/4$. It remains to estimate the norms of the entries of $\xi - u\xi$ and of $z(\xi - u\xi)$ with the use of inequalities (2). Obviously, $\|(1-c)\zeta\| \leq \|(1-c)e_c[1 - \varepsilon^2,1]\| = 1 - (1 - \varepsilon^2)^{1/2} < \varepsilon$, $\|s\zeta\| \leq \|se_s[0,\varepsilon]\| = \varepsilon$, thus (i).

It is also easy to obtain that

$$\|z_{20}(1-c)\zeta\| \leq \|(1-c)z_{02}\| \leq \|s^2z_{02}\| = \|z_{20}s^2\| < 3\varepsilon^3 < \varepsilon, \tag{3}$$

and similarly (as $\zeta \in h = e_c[3^{1/2}/2,1]$),

(4) $$\|z_{30}(1-c)\zeta\| \le \|z_{30}(1-c)h\| \le \|z_{30}c^2\| < 3\varepsilon^3 < \varepsilon,$$

(5) $$\|z_{i1}s\zeta\| \le \|z_{i1}sh\| \le (2/3^{1/2})\|z_{i1}cs\| < (2/3^{1/2})3\varepsilon^3 < \varepsilon,$$

$i = 4,5$.

The estimations of $\|z_{00}(1-c)\zeta\|$, $\|z_{11}s\zeta\|$ are more complicated. At first, we obtain some inequality for $\|sz_{00}\zeta\|$. By (2), $\|s^2z_{00} - z_{00}s^2\| = \|c^2z_{00} - z_{00}c^2\| < 3\varepsilon^3$ and, by lemma 3.1,

$$\|e_s[2\varepsilon,1]z_{00}e_s[0,\varepsilon]\| = \|e_{s^2}[4\varepsilon^2,1]z_{00}e_{s^2}[0,\varepsilon^2]\| \le 3\varepsilon^3/3\varepsilon^2 = \varepsilon.$$

Let $z\xi$ have entries $(z\xi)_i \in H_i$, $0 \le i \le 5$. Then

$$\|z_{00}\zeta + z_{02}\eta\| = \|(z\xi)_0\| \le \|z\xi\|.$$

In consequence,

$$\begin{aligned}\|sz_{00}\zeta\| &\le \|sz_{00}\zeta + se_s[0,2\varepsilon)z_{02}\eta\| + \|se_s[0,2\varepsilon)z_{02}\eta\|\\ &\le \|se_s[0,2\varepsilon)(z_{00}\zeta + z_{02}\eta)\| + \|se_s[2\varepsilon,1]z_{00}\zeta\|\\ &+ \|(sc + s^2)z_{02}\| \le 2\varepsilon\|z_{00}\zeta + z_{02}\eta\|\\ &+ \|e_s[2\varepsilon,1]z_{00}e_s[0,\varepsilon]\| + \|z_{20}sc\| + \|z_{20}s^2\|\\ &< 2\varepsilon\|z\xi\| + 3\varepsilon.\end{aligned}$$

Analogously to (4), (5), we can now write

(6) $$\begin{aligned}\|z_{11}s\zeta\| &\le \|z_{11}sh\| \le (2/3^{1/2})\|z_{11}sc\|\\ &\le (2/3^{1/2})(\|csz_{00}\| + \|csz_{00} - z_{11}cs\|)\\ &\le (2/3^{1/2})(\|sz_{00}\| + \varepsilon) \quad \text{(by (2))}\\ &\le (2/3^{1/2})(2\varepsilon\|z\xi\| + 4\varepsilon) < 3\varepsilon\|z\xi\| + 5\varepsilon.\end{aligned}$$

Moreover,

$$\begin{aligned}\|(cz_{00} - z_{00}c)e_{s^2}[0,\varepsilon^2]\| &\le \|s^2z_{00} - z_{00}s^2\|/(1-\varepsilon^2)\\ &< 3\varepsilon^3/(1 - \varepsilon^2) < \varepsilon\end{aligned}$$

by lemma 3.2 with $t = s^2$ and by (2). Thus

$$(7)\qquad \|z_{00}(1-c)\zeta\| \le \|(1-c)z_{00}\zeta\| + \|(cz_{00} - z_{00}c)\zeta\|$$
$$\le \|sz_{00}\| + \|(cz_{00} - z_{00}c)e_{s^2}[0,\varepsilon^2]\|$$
$$\le 2\varepsilon\|z\xi\| + 4\varepsilon.$$

Inequalities (3),...,(7) and $\|\mu_i\|, \|\nu_i\| < \varepsilon/4$ imply

$$\|z(\xi - u\xi)\| \le \Sigma_{0\le i\le 5} \|(z(\xi - u\xi))_i\| < 5\varepsilon\|z\xi\| + 15\varepsilon.$$

3.4. LEMMA. For $\|x_n\| \le 1$, $x_n \in M^h$, the conditions $\rho(1-x_n)\to 0$ and $x_n \to 1$ in the strong operator topology are equivalent. In particular, there exists a function $\beta : [0,1] \to [0,1]$ such that $\rho(1-e) < \beta(\delta)$, $\rho(1-f) < \beta(\delta)$, $e,f \in \operatorname{Proj} M$, $0 < \delta < 1$ imply $\rho(1-efe) < \delta$ (and $\rho(1 - e_{efe}[1-\varepsilon^2,1]) < \delta/\varepsilon^2$ for any $\varepsilon > 0$).

3.5. Proof of the normality of the topology τ. Let $A=\bar{A}$, $B=\bar{B}$ be any disjoint subsets of the space $\tilde{M}$, closed in τ. For a fixed $x \in A$ ($x \in B$), let us denote $\alpha(x) = \inf\{\alpha;\ \exists\, y \in B$ ($y \in A$, respectively), $e \in \operatorname{Proj} M : \|(y-x)e\| < \alpha,\ \rho(1-e) < \alpha\}$. By the definition of τ, $0 < \alpha(x) \le 1$. Using 1.6, 3.4 and 2.2, respectively, we can define numbers $N(x) \ge 1$, $\beta(x) > 0$, $\gamma(x) > 0$ satisfying, for $c,e,E,f \in \operatorname{Proj} M$,

(1) $\quad c \le e_x((-\infty,-N(x)) \cup (N(x),\infty))$ implies $\rho(c) < \alpha(x)/4$;

(2) $\quad \rho(1-e) < \beta(x)$, $\rho(1-f) < \beta(x)$ imply

$$\rho(1 - e_{efe}[1-\varepsilon^2(x),1]) < \alpha(x)/4,$$

with

(3) $\quad \varepsilon(x) = \alpha(x)/(10N(x) + 40)$;

(4) $\quad \beta(x) < \varepsilon^3(x)/3$;

(5) $\quad \rho(1-E) < \gamma(x)$ implies $\|\overline{ex} - xe\| < \beta(x)$,

$\rho(1-e) < \beta(x)$ for some $e \le E$, $e \in \operatorname{Proj} M$;

(6) $\quad \gamma(x) < \beta(x)$.

If we show that $U(x,\gamma(x)) \cap U(y,\gamma(y)) = \emptyset$ for any operators $x \in A$, $y \in B$, then the existence of the required disjoint open sets

$V, W \subset \tilde{M}$ containing A and B, respectively, will be proved. Suppose that $z \in U(x,\gamma(x)) \cap U(y,\gamma(y))$; $x \in A$, $y \in B$. By 1.7, we can assume that $z \in M^h$. Changing the notation, if necessary, we can also obtain $\beta(x) > \beta(y)$. By (5) and the definition of $U(x,\gamma(x))$, there exist projections $e \leq E$, $f \leq F$, $e,E,f,F \in \text{Proj } M$, satisfying

$$\|(z - x)E\| < \gamma(x) < \varepsilon^3(x)/3 \qquad \text{(cf. (6), (4))},$$

$$\|\overline{ex} - xe\| < \varepsilon^3(x)/3, \qquad \rho(1 - e) < \beta(x),$$

and also

$$\|(z - y)F\| < \gamma(y) < \beta(y) < \beta(x) < \varepsilon^3(x)/3,$$

$$\|\overline{fy} - yf\| < \varepsilon^3(x)/3, \qquad \rho(1 - f) < \beta(x).$$

Then

$$\|ez - ze\| \leq \|\overline{ex} - xe\| + \|(z - x)e\| + \|\overline{e(z - x)}\|$$
$$= \|\overline{ex} - xe\| + 2\|(z - x)e\| \leq \|\overline{ex} - xe\|$$
$$+ 2\|(z - x)E\| < \varepsilon^3(x),$$

$$\|fz - zf\| < \varepsilon^3(x).$$

Let the isometry u be given by 3.3 (1) and let $e_1 = e_{efe}[1 - \varepsilon^2(x),1]$, $e_2 = e_1 \wedge e_x[-N(x),N(x)]$, $f_1 = ue_2u^*$, $f_2 = f_1 \wedge e_x[-N(x),N(x)]$, $e_3 = u^*f_2u$. Then $\rho(1 - e_1) < \alpha(x)/4$ by (2); $e_1 - e_2 \leq e_x((-\infty,-N(x)),(N(x),\infty))$ and $\rho(e_1 - e_2) < \alpha(x)/4$ by (1); $f_1H = ue_2H$ and $\|f_1 - e_2\| \leq \|(u - 1)e_2\| < 2\varepsilon(x) < \alpha(x)/4$ by theorem 3.3. (i); $f_1 - f_2 \leq e_x((-\infty,N(x)) \cup (N(x),\infty)$ and $\rho(f_1 - f_2) < \alpha(x)/4$ by (1). In consequence, $\rho(1 - f_2) < \alpha(x)$ and $f_2H \subset ue_2 H$. It only remains to estimate $\|(x - y)f_2\|$. Obviously,

$$(7) \qquad \|(z - y)f_2\| \leq \|(z - y)F\| < \varepsilon^3(x)/3 < \alpha(x)/10 \qquad \text{(by (3))}.$$

For any vector $\tilde{\xi} \in f_2H$, $\|\tilde{\xi}\| = 1$, we have $\tilde{\xi} = u\xi$ with some $\xi \in e_2H \subset e_{efe}[1 - \varepsilon^2(x),1]H$, $\|\xi\| = 1$, and $\|z\xi\| \leq \|xe_2\| + \|(z - x)e\| \leq \|xe_x(-N(x),N(x))\| + \|(z - x)e\| < N(x) + 1$. By virtue of lemma 3.3,

$$\|z(\xi - \tilde{\xi})\| < (5N(x) + 20)\varepsilon(x) \leq \alpha(x)/2$$

(cf. (3)). Obviously,

$$\|(z - x)\xi\| \leq \|(z - x)E\| < e^3(x)/3 < \alpha(x)/10,$$

$$\|x(\xi - \tilde{\xi})\| \le \|x(e_2 \vee f_2)\| \, \|\xi - u\xi\| < 2N(x)\varepsilon(x) < \alpha(x)/5$$

(cf. (3)), and, by (7), we finally have

$$\|(y - x)\tilde{\xi}\| \le \|(y - z)\tilde{\xi}\| + \|z(\tilde{\xi} - \xi)\| + \|(z - x)\xi\| + \|x(\xi - \tilde{\xi})\| < 9\alpha(x)/10.$$

In fact, we have proved that $y \in U(x,\alpha(x) - \delta)$ with some $\delta > 0$, and this contradicts the definition of $\alpha(x)$.

REFERENCES

[1] R. Engelking, General topology, PWN, Warszawa 1977.

[2] A. Paszkiewicz, A limit in probability in a W*-algebra is unique, to appear.

[3] A. Paszkiewicz, Convergences in W*-algebras - their strange behaviour and tools for their investigation, to appear in Proc. of Symposium on Quantum Prob. & Appl., Rome, October ´86 - July ´87, in Lecture Notes Math.

[4] I.E. Segal, A Non-Commutative Extension of Abstract Integration, Ann. Math. 57 (1953), 401-457.

This paper is in final form and no similar paper has been or is being submitted elsewhere.

First steps towards a Donsker and Varadhan theory in operator algebras

Dénes Petz
Instituut voor Theoretische Fysika
Universiteit Leuven
B-3030 Leuven, Belgium

1.Introduction and motivation

Let $\xi_1, \xi_2, \ldots$ be a sequence of independent identically distributed random variables with mean $m = E(\xi_1)$. If $m < a < b$ then the law of large numbers states that

$$P\left(\frac{\xi_1 + \xi_2 + \cdots + \xi_n}{n} \in [a,b]\right) \to 0 \quad \text{as} \quad n \to \infty .$$

However, much more is true, namely, that the convergence is exponentially fast. There is $\lambda(a,b) > 0$ such that

$$P\left(\frac{\xi_1 + \xi_2 + \cdots + \xi_n}{n} \in [a,b]\right) \le C\, e^{-n\,\lambda(a,b)}$$

with some constant C, and in fact, $\lambda(a,b)$ depends on $a-m$. The *large deviation principle* (LDP) is gathered from this example.

Let $\eta_1, \eta_2, \ldots$ be a sequence of random variables and $I : \mathbb{R} \to \mathbb{R}^+ \cup \{+\infty\}$. (η_n) is said to satisfy the LDP with rate function I if

(i) $\limsup_{n\to\infty} \frac{1}{n} \log P(\eta_n \in F) \le -\inf\{I(x) : x \in F\}$ for every closed set $F \subset \mathbb{R}$,

(ii) $\liminf_{n\to\infty} \frac{1}{n} \log P(\eta_n \in G) \ge -\inf\{I(x) : x \in G\}$ for every open set $G \subset \mathbb{R}$,

(iii) $I \not\equiv +\infty$ and $\{t \in \mathbb{R} : I(t) \le L\}$ is compact for every $L \in \mathbb{R}$.

Cramér essentially proved in the 30's the following.

Theorem 1. *If $\xi_1, \xi_2, \ldots$ are identically distributed independent random variables and*

$$L(u) = E(\exp u\xi_1) < +\infty \qquad (u \in \mathbb{R})$$

then $\eta_n = (\xi_1 + \xi_2 + \cdots + \xi_n)/n$ satisfies the LDP with rate function

$$I(x) = \sup\{u\,x - \log L(u)\ :\ u \in \mathbb{R}\}.$$

If you are pushed by the strong desire of noncommutativization everything possible then Theorem 1 does not give much chance. A reformulation due to Varadhan is more suitable.

Theorem 2. *Under the assumption of Theorem 1 the relation*

$$\lim_{n\to\infty} \frac{1}{n} \log E\,(\exp n f(\eta_n)) = \sup\{f(n) - I(u) : u \in \mathbb{R}\}$$

holds for every continuous bounded function $f : \mathbb{R} \to \mathbb{R}$.

In Theorem 2 probability, open and closed sets do not appear and a more functional analytic view point seems to be possible.

Personally I met LDP first when I visited Luigi Accardi at the 2nd University of Rome in 1984. We formulated Theorem 2 on the language of operator algebras as follows. Let $\mathcal{B}_n$ be a copy of a finite dimensional C^*-algebra $\mathcal{B}$ and take $\mathcal{A} = \otimes_{n\in\mathbb{Z}} \mathcal{B}_n$. Assume that φ is a product state on $\mathcal{A}$. Then φ and the embeddings $i_n : \mathcal{B}_n \to \mathcal{A}$ play the role of a Bernoulli sequence. For $b \in \mathcal{B}$ one can ask the existence of the limit

$$\lim_{n\to\infty} \frac{1}{n} \log\varphi\left(\exp\left(n f\left(\frac{i_1(b) + i_2(b) + \cdots + i_n(b)}{n}\right)\right)\right)$$

(where f is a continuous function $\mathbb{R} \to \mathbb{R}$) and we observed that

$$\liminf_{n\to\infty} \frac{1}{n} \log\varphi\left(\exp\left(n f\left(\frac{i_1(b) + i_2(b) + \cdots + i_n(b)}{n}\right)\right)\right) \geq \omega(i_1(b)) - S_M(\varphi,\omega) \qquad (1)$$

whenever ω is a stationary state on $\mathcal{A}$ and $S_M(\varphi,\omega)$ is the mean relative entropy (see below). If you take simply $f = id$ then the lim inf in (1) is a limit with value $\log\varphi(\exp i_1(b))$. It can not be the sup of the right hand side (ω runs over stationary states), since the latter is a convex function of b. (Remember that $A \mapsto \log\varphi(\exp A)$ is not convex in general, being the exp function not operator convex). Hence the noncommutative generalization must go in another direction.

Over the pure noncommutativization of the LDP (or more precisely, Theorem 2), a motivation arises from quantum statistical mechanics. I briefly recall the notion of a quantum lattice system. To each $n \in \mathbb{Z}$ a copy $\mathcal{B}_n$ of the 2×2 matrices is associated. If $I \subset \mathbb{Z}$ then $\mathcal{A}_I$ denotes $\otimes_{n\in I} \mathcal{B}_n$ and I write $\mathcal{A}$ for $\mathcal{A}_{\mathbb{Z}}$. Having fixed some selfadjoint matrices x and h, one defines the local Hamiltonians by the formula

$$H_{[m,n]} = \sum_{i=m}^{n} h_i + \frac{1}{n-m} \sum_{i,j=m}^{n} x_i x_j\,.$$

(This interaction is called mean field [4].) The mean free energy at the inverse temperature β is given as

$$F_n(\beta) = \frac{1}{n} \log \operatorname{Tr} \exp\left(\sum_{i=1}^{n} \log D_i - n\beta\left(\frac{x_1 + x_2 + \cdots + x_n}{n}\right)^2\right) \qquad (2)$$

where $D_i = \exp(-\beta h_i)/\operatorname{Tr}\exp(-\beta h_i)$. The similarity of (1) and (2) is striking if f is the square function. (For the sake of simplicity, you may take $\beta = -1$.) On the other hand, statistical mechanics suggests the correction of (1) in order to get an equality (which is, in fact, the Gibbs variational principle).

This lecture is based on the joint work [7] with Raggio and Verbeure and benefited also from from the generalization [9] of Raggio and Werner. In the latter paper the interested reader may find more physical interpretations. Concerning large deviations and (classical) statistical mechanics I refer to the book [3].

2. Preliminaries

Let $\mathcal{A}$ be a C^*-algebra with a state φ. Performing the GNS-construction we arrive at a cyclic vector $\Phi \in \mathcal{H}_\varphi$ and a representation $\pi_\varphi : \mathcal{A} \to B(\mathcal{H}_\varphi)$. If Φ is separating for the von Neumann algebra $\pi_\varphi(\mathcal{A})''$ then we say that φ is separating. (Typical examples of separating states are the KMS-states.) Having a cyclic and separating vector Φ (with respect to $\pi_\varphi(\mathcal{A})''$), the corresponding modular operator Δ is at our disposal. The formula

$$\Phi^h = \exp\left(\frac{1}{2}(\log\Delta + \pi_\varphi(h)\right)\Phi$$

defines the perturbed vector for $h = h^* \in \mathcal{A}$ and

$$\varphi^h(a) =< \pi_\varphi(a)\,\Phi^h\,,\Phi^h > \qquad (a \in \mathcal{A})$$

is the (unnormalized) perturbed functional. In the next mainly the quantity $\varphi^h(1) = \|\Phi^h\|^2$ will occur. It is known that $h \mapsto \log\varphi^h(I)$ is a convex continuous function on $\mathcal{A}^{sa}$ [1].

Whenever F is a convex function on a normed space X then its conjugate (or Legendre-Fenchel transform, see [10]) is defined on X^* as

$$F^*(X^*) = \sup\{X^*(x) - F(x) \,:\, x \in X\}.$$

The conjugate of $F \,:\, h \mapsto \log\varphi^h(I)$ is the relative entropy.

Proposition 1. *If $\omega \in \mathcal{A}_h^*$ then*

$$F^*(\omega) = \begin{cases} S(\varphi,\omega): & \text{if } \omega \text{ is a state} \\ +\infty & \text{otherwise.}\end{cases}$$

For the sake of a simple presentation I prove only that $F^*(\omega) = +\infty$ if ω is not a state. You may take the other part of the proposition as the definition of the relative entropy. (See [8] for the details; concerning the definition and general properties of the relative entropy I refer to the survey papers [2] and [5].)

For $h = t\,I$ we have

$$\omega(h) - F(h) = (\omega(I) - 1)t$$

and in the case of $\omega(I) \neq 1$, $F^*(\omega)$ must be $+\infty$. If $h_0 \in \mathcal{A}_0 \subset \mathcal{A}$ then the monotonicity theorem ([6]) tells that

$$\omega(h_0) - \log\varphi^{h_0}(I) \geq \omega(h_0) - \log(\varphi|\mathcal{A}_0)^{h_0}(I)\,.$$

When ω is not positive, there exists $0 \leq h_0$ with $\omega(h_0) < 0$. Let $\mathcal{A}_0$ be the commutative subalgebra generalized by $\{h_0\}$. Then

$$\omega(t\,h_0) - \log(\varphi|\mathcal{A}_0)^{t h_0}(I) = t\,\omega(h_0) - \log\varphi(e^{t h_0}) \to +\infty$$

if $t \to -\infty$.

Let $\mathcal{A}_1$ be a C^*-algebra with a separating state ρ. I write $\mathcal{A}_n$ for the n-fold minimal C^*-tensor product of $\mathcal{A}_1$ with itself and $\mathcal{A}$ for the inductive limit C^*-algebra. φ stands for the product state $\rho \otimes \rho \otimes \dots$ of $\mathcal{A}$. Note that φ is also separating. For a shift invariant state ψ of $\mathcal{A}$ the limit of

$$\frac{1}{n} S(\varphi|\mathcal{A}_n\,,\,\psi|\mathcal{A}_n)$$

exists and is called the mean relative entropy of φ and ψ; in notation $S_M(\varphi, \psi)$. In fact,

$$S_M(\varphi,\psi) = \sup\left\{\frac{1}{n} S(\varphi|\mathcal{A}_n\,,\,\psi|\mathcal{A}_n) \,:\, n \in \mathbb{N}\right\},$$

therefore $\psi \mapsto S_M(\varphi, \psi)$ is weak* lower semicontinuous and convex ([8]).

It is worthwhile to state in this section a theorem of Størmer which will play an important role below. The result says that if ψ is a state of A invariant under all finite permutations of the factors of the infinite tensorproduct then ψ is an average (i.e., integral) of product states [11].

3. The perturbational limit principle

Let $\mathcal{A}$ be a C^*-algebra with a separating state φ. Assume that for every $n \in \mathbb{N}$ a completely positive unital mapping α_n of a C^*-algebra $\mathcal{B}$ into $\mathcal{A}$ is given and the invariance $\varphi \circ \alpha_n = \varphi \circ \alpha_m$ $(n, m \in \mathbb{N})$ holds. Motivated by Section 1, I say that (α_n) satisfies the *perturbational limit principle* (PLP) if there exists a lower semicontinuous function I from the state space $\mathcal{S}(\mathcal{B})$ of $\mathcal{B}$ into $\mathbb{R}^+ \cup \{+\infty\}$ such that

$$\lim_{n\to\infty} \frac{1}{n} \log \varphi^{n f(\alpha_n(a))}(I) = \sup\left\{f\left(\nu(a)\right) - I(\nu) \,:\, \nu \in \mathcal{S}(\mathcal{B})\right\} \qquad (3)$$

holds for every continuous function $f : \mathbb{R} \to \mathbb{R}$ and for every $a = a^* \in \mathcal{B}$.

Fix $a = a^* \in \mathcal{B}$ and assume that $|f(t) - g(t)| < \varepsilon$ for $|t| \le \|a\|$. Then $\|f(\alpha_n(a)) - g(\alpha_n(a))\| \le \varepsilon$ and

$$\varphi^{n f(\alpha_n(a))}(I) \le \varphi^{n g(\alpha_n(a))}(I) \cdot e^{n\varepsilon}.$$

It follows that the left hand side of (3) is continuous in f. Since the right hand side is obviously continuous, due to the Weierstrass approximation theorem (3) holds for every continuous function f whenever it holds for all polynomials.

Proposition 2. *If (3) holds for every polynomial f and for every element a of a norm dense set $\mathcal{D}$ in $\mathcal{B}^{sa}$ then the PLP holds.*

It should be showed that (3) is true for a polynomial f and an arbitrary $b \in \mathcal{B}^{sa}$. The proof is an application of the Golden-Thompson-Araki inequality [1]. If $\varepsilon > 0$ is given then for a small $\delta > 0$ then $\|b_1 - b_2\| < \delta$ implies $\|f(\alpha_n(b_1)) - f(\alpha_n(b_2))\| < \varepsilon$ and we have

$$\left|\limsup_{n\to\infty} \frac{1}{n} \log \varphi^{n f(\alpha_n(b_1))}(I) - \liminf_{n\to\infty} \frac{1}{n} \log \varphi^{n f(\alpha_n(b_2))}(I)\right| \le \varepsilon.$$

Hence the uniform continuity of the functional calculus was used and I state it for the sake of completeness as a separate lemma.

Lemma 1. *Let f be a polynomial. Then for every $\varepsilon > 0$ and $K > 0$ there exists a $\delta > 0$ such that for every selfadjoint operator A, B with $\|A\|, \|B\| < K$ and $\|A - B\| < \delta$ the estimate*

$$\|f(A) - f(B)\| \leq \varepsilon$$

holds.

The argument that proved Proposition 2 yields also the following.

Proposition 3. *Let $\alpha_n, \beta_n : \mathcal{B} \to \mathcal{A}$ be completely positive unital mappings. Assume that $\varphi \circ \alpha_n = \varphi \circ \beta_m$ $(n, m \in \mathbb{N})$ and $\|\alpha_n(a) - \beta_n(a)\| \to 0$ for every $a \in \mathcal{B}^{sa}$. Then (α_n) satisfies the PLP if and only if (β_n) does so.*

Proposition 3 supplies us immediately with some trivial examples. Let $\mathcal{A} = \mathcal{B}$ and $\alpha_n = \frac{1}{n}(a + (n-1)\varphi(a)I)$. Then the PLP holds with the rate function

$$I(\nu) = \begin{cases} 0 & \nu = \varphi \\ +\infty & \text{otherwise.} \end{cases}$$

I note that taking $a = I$ it follows from (3) that $I(\nu) \geq 0$ and $I \not\equiv +\infty$. It is also clear that if the PLP holds then the rate function is uniquely determined. Set

$$F(a) = \lim_{n\to\infty} \frac{1}{n} \log \varphi^{n\,\alpha_n(a)}(I).$$

Then F is a lower semicontinuous convex function and I is the restriction of the conjugate of F to $S(\mathcal{B})$.

In the rest of the lecture I consider only $\mathcal{A} = \mathcal{A}_1 \otimes \mathcal{A}_1 \otimes \ldots$ and $\varphi = \rho \otimes \rho \otimes \ldots$.

3. Perturbational limit theorems

Let $S_n \subset \mathrm{Aut}\,\mathcal{A}$ be the group of permutations of the first n factors in $\mathcal{A}_0 \otimes \mathcal{A}_s \otimes \ldots$. A state ψ of $\mathcal{A}$ is called symmetric if it is invariant under $\cup_{n=1}^{\infty} S_n$; the set of all symmetric states is denoted by $I(\mathcal{A})$. The mapping $E_n : \mathcal{A} \to \mathcal{A}$ defined by

$$E_n(a) = \frac{1}{n!} \sum_{\alpha \in S_n} \alpha(a) \qquad (a \in \mathcal{A})$$

is a projection of norm one onto the fixed point algebra $\{a \in \mathcal{A}_{n-1} : \alpha(a) = a$ for every $\alpha \in S_n\}$. When a happens to be in $\mathcal{A}_k$ and $n > k$ then we have also

$$E_n(a) = \frac{(n-k)!}{n!} \sum_{\alpha \in S_n/S_k} \alpha(a).$$

Note that for $a \in \mathcal{A}_0$ we have

$$E_1(a) = \frac{1}{n}\left(a + \gamma(a) + \cdots + \gamma^{n-1}(a)\right)$$

where γ is the right shift endomorphism of $\mathcal{A}$.

Lemma 2. *Let $a \in \mathcal{A}_k$, $l \in \mathbb{N}$, $k < n \in \mathbb{N}$ and $\psi \in I(\mathcal{A})$. Then*

$$\left|\psi\left(E_n(a)^l\right) - \psi\left(a\,\gamma^{k+1}(a)\cdots\gamma^{l(k+1)}(a)\right)\right| \le \frac{c(l,k)}{n}\|a\|^l.$$

The proof is elementary combinatorics. One should consider the multinomial expansion of

$$\left(\sum_{\alpha \in S_n|S_{k+1}} \alpha(a)\right)^2,$$

in which the typical term is $\alpha_{\pi_1}(a)\,\alpha_{\pi_2}(a)\ldots\alpha_{\pi_l}(a)$ where π_i's are such permutations of $\{0,1,\ldots,n-1\}$ such that

$$\pi_i(\{0,1,\ldots,k\}) \cap \pi_j(\{0,1,2,\ldots,k\}) = \emptyset$$

for every $i \neq j$. Due to the symmetry condition the state ψ takes at such terms the value $\psi\left(a\,\gamma^{k+1}(a)\ldots\gamma^{l(k+1)}(a)\right)$.

It follows from Lemma 2 that $F_f^a(\psi) = \lim_{n\to\infty} \psi(f(E_n(a)))$ exists for all polynomial f and $a \in \mathcal{A}_k^{sa}$. By approximation the existence of $F_f^a(\psi)$ extends to every continuous f and $a \in \mathcal{A}^{sa}$.

Lemma 3. *We have for every continuous $f : \mathbb{R} \to \mathbb{R}$, $a \in \mathcal{A}^{sa}$ and $\psi \in I(\mathcal{A})$ the following relation*

$$\liminf_{n\to\infty} \frac{1}{n} \log \varphi^{n\,f(E_n(a))}(I) \ge F_f^a(\psi) - S_M(\varphi,\psi)$$

holds.

According to Proposition 1

$$\log \varphi^{n\,f(E_n(a))}(I) \ge \psi(n\,f(E_n(a)) - S(\varphi|\mathcal{A}_{n-1}, \psi|\mathcal{A}_{n-1})$$

holds. Devided by n and letting $n \to \infty$ the assertion is concluded.

Let us denote by $\overline{\omega}_n$ the state $\varphi^{n\,f(E_n(a))}/\varphi^{n\,f(E_n(a))}(I)$ restricted to $\mathcal{A}_{n-1}$. We write $\tilde{\omega}_n$ for the periodical state $\overline{\omega}_n \otimes \overline{\omega}_n \otimes \ldots$ of $\mathcal{A}$. $\tilde{\omega}$ is not stationary but S_m-invariant if $n > m$. Since $\tilde{\omega}_n$ is invariant under γ^n, the state

$$\omega_n = \frac{1}{n}\left(\tilde{\omega}_n + \tilde{\omega}_n \circ \gamma + \cdots + \tilde{\omega}_n\,\gamma^{n-1}\right)$$

will be γ invariant. It is easy to see that the sequences (ω_n) and $(\tilde{\omega}_n)$ have the same weak* limit points, and all of them are symmetric.

Lemma 4. *Let $N_0 \subset \mathbb{N}$ be an infinite subset, ψ a weak* limit point of $\{\omega_n : n \in \mathbb{N}_0\}$, $\varepsilon > 0$ and f a polynomial. Then*

$$\left|E_f^a(\psi) - \overline{\omega}_n(f(E_n(a))\right| < \varepsilon$$

is valid for infinitely many n in N_0.

Let $f(u) = \sum_{t=0}^{l} c_t\, u^t$. Remember that due to Lemma 2

$$E_f^a(\psi) = \sum_{t=0}^{l} c_t\, \psi\left(a\, \gamma^{k+1}(a) \ldots \gamma^{t(k+1)}(a)\right)$$

and ψ is also a limit point of $\{\tilde{\omega}_n : n \in N_0\}$. Therefore

$$\left|\sum_{t=0}^{l} c_t \left(\psi\left(a\, \gamma^{k+1}(a) \ldots \gamma^{t(k+1)}(a)\right) - \tilde{\omega}_n\left(a\, \gamma^{k+1}(a) \ldots \gamma^{t(k+1)}(a)\right)\right)\right| < \delta$$

for infinitely many $n \in N_0$. Since for n big enough

$$(\tilde{\omega}_n - \overline{\omega}_n)\left(a\, \gamma^{k+1}(a) \ldots \gamma^{t(k+1)}(a)\right) = 0$$

and thanks to Lemma 2

$$\left|\tilde{\omega}_n\left(a\, \gamma^{k+1}(a) \ldots \gamma^{t(k+1)}(a)\right) - \tilde{\omega}_n\left(E_n(a)^t\right)\right| < \delta;$$

the proof is completed.

Theorem 3. *Let $f : \mathbb{R} \to \mathbb{R}$ be continuous and $a \in \mathcal{A}^{sa}$. Then*

$$\lim_{n\to\infty} \frac{1}{n} \log \varphi^{n\, f(E_n(a))}(I) = \sup\left\{F_f^a(\psi) - S_M(\varphi, \psi) : \psi \in I(\mathcal{A})\right\}.$$

A part of the theorem is contained in Lemma 3. Set

$$C \equiv \limsup_{n\to\infty} \frac{1}{n} \log \varphi^{n\, f(E_n(a))}(I).$$

There is an infinite subset N_0 of the integers such that

$$\overline{\omega}_n\left(f(E_n(a))\right) - \frac{1}{n} S\left(\varphi|\mathcal{A}_{n-1}, \overline{\omega}_n\right) > C - \varepsilon$$

for every $n \in N_0$. Let ψ be a weak* limit point of $\{\omega_n : n \in \mathbb{N}_0\}$. According to Lemma 4 there is an infinite subset N_1 of N_0 such that

$$E_f^a(\psi) \geq \overline{\omega}_n\left(f(E_n(a)\right) - \varepsilon$$

for every $n \in N_1$. By the lower semicontinuity of $\omega \mapsto S_M(\varphi, \omega)$

$$S_M(\varphi, \psi) \leq S_M(\varphi, \omega_m) + \varepsilon$$

for some $m \in N_1$. Since $S_M(\varphi, \omega_m) = \frac{1}{m} S\left(\varphi|\mathcal{A}_{n-1}, \overline{\omega}_m\right)$ we estimate as

$$\begin{aligned} E_f^a(\psi) \quad - \quad S_M(\varphi, \psi) &\geq \overline{\omega}\left(f(E_m(a))\right) - S_M(\varphi, \omega_m) - 2\varepsilon = \\ &= \overline{\omega}_m\left(f(E_m(a))\right) - \frac{1}{m} S\left(\varphi(\mathcal{A}_{m-1}, \overline{\omega}_m\right) \geq C - 3\varepsilon. \end{aligned}$$

The proof is complete.

Theorem 3 in the case $a \in \mathcal{A}_0^{sa}$ was obtained in [7]. Essentially the same argument worked here in the more general situation.

Corollary. *Let $a \in \mathcal{A}_k^{sa}$. Then*

$$\lim_{n\to\infty} \frac{1}{n} \log \varphi^{n f(E_n(a))}(I) =$$

$$\sup\left\{ f(\nu(a)) - \frac{1}{k+1} S(\varphi|\mathcal{A}_k, \nu) : \nu = \rho \otimes \ldots \otimes \rho \in \mathcal{A}_k^* \text{ with some } \rho \in \mathcal{S}(\mathcal{A}_0)\right\}.$$

If μ is a Radon measure on $\mathcal{S}(\mathcal{A}_0)$ then

$$\omega = \int (\rho \otimes \rho \otimes \ldots) d\mu(\rho) \in I(\mathcal{A})$$

and due to Størmer theorem we get all elements of $I(\mathcal{A})$ this way. We may assume again that f is a polynomial. If so then

$$E_f^a(\omega) = \lim_{n\to\infty} \omega(f(E_n(a)) = \int f(\rho \otimes \ldots \otimes \rho(a)) d\mu(\rho)$$

(The tensorproduct is of $(k+1)$-fold.) On the other hand,

$$S_M(\varphi, \omega) = \int S(\varphi|\mathcal{A}_0, \rho) d\mu(\rho) = \frac{1}{k+1} \int S(\varphi|\mathcal{A}_k, \rho \otimes \ldots \otimes \rho) d\mu(\rho).$$

Therefore,

$$E_f^a(\omega) - S_M(\varphi, \omega) \leq$$

$$\sup\left\{ f(\rho \otimes \ldots \otimes \rho(a)) - \frac{1}{k+1} S(\varphi) A_k, \rho \otimes \ldots \rho) : \rho \in \mathcal{S}(\mathcal{A}_0)\right\}$$

and the converse inequality is obvious.

The corollary tells that $\alpha_n : \mathcal{A}_0 \to \mathcal{A}$, $\alpha_n = E_n|\mathcal{A}_0$ satisfies the PLP and the rate function is the relative entropy. I note that

$$\sup\{f(\rho(a)) - S(\varphi|\mathcal{A}_0, \rho) : \rho \in \mathcal{S}(\mathcal{A}_0)\} = \sup\{f(u) - \mathcal{F}(u) : u \in \mathbb{R}\}$$

where $\mathcal{F} : \mathbb{R} \to \mathbb{R}^+ \cup \{+\infty\}$ is the Legendre transform of the function $t \mapsto \log \varphi^{ta}(I)$. This is important from physical point of view (see [7,9] and references therein).

References

1. H. Araki, Golden-Thompson and Peierls-Bogoliubov inequalities for a general von Neumann algebra, Commun. Math. Phys. **44**, 1-7 (1975).

2. H. Araki, Recent progress on entropy and relative entropy, Proceedings of the VIIIth Congress on Mathematical Physics, (Eds. M. Mebkhout, R. Senor), 354-365, World Sci. Publ., 1987, Singapore.

3. R.S. Ellis, Entropy, large deviations and statistical mechanics, Springer, 1985, New York, Berlin, Heidelberg, Tokyo.

4. M. Fannes, H. Spohn and A. Verbeure, Equilibrium states of mean field models, J. Math. Phys. **21**, 355-358 (1980).

5. D. Petz, Properties of quantum entropy, Lecture Notes in Math. **1136**, 428-441 (1985).

6. D. Petz, A variational expression for the relative entropy, Commun. Math. Phys. **114**, 345-349 (1988).

7. D. Petz, G.P. Raggio and A. Verbeure, Asymptotics of Varadhan-type and the Gibbs variational principle, Commun. Math. Phys., to appear.

8. D. Petz, On certain properties of the relative entropy of states of operator algebras, Preprint KUL-TF-88/19, Leuven, 1988.

9. G. Raggio and R.F. Werner, Quantum statistical mechanics of general mean field systems, Preprint DIAS-STP-88-49, Dublin, 1988.

10. R.T. Rockefeller, Convex analysis, Princeton University Press, 1970, Princeton.

11. E. Størmer, Symmetric states of infinite tensor products of C^*-algebras, J. Functional Anal. **3**, 48-68 (1969).

QUANTUM CONDITIONAL PROBABILITY SPACES

Sylvia Pulmannová
Mathematics Institute, Slovak Academy of Sciences
Obrancov mieru 49, 814 73 Bratislava, Czechoslovakia

According to Rényi [1],[2], a classical conditional probability space $(S,\mathcal{A},\mathcal{B},P)$ is defined as follows : S is an abstract space, $\mathcal{A}$ is a σ-algebra of subsets of S, $\mathcal{B}$ is a nonempty subset of $\mathcal{A}$ and $P = P(A/B)$ is a set-function of two set variables defined for all $A \in \mathcal{A}$ and $B \in \mathcal{B}$. The elements of $\mathcal{A}$ are interpreted as events and $P(A/B)$ as the conditional probability of the event A with respect to the condition B. It is supposed that the following axioms are fulfilled:

A. $P(./B)$ is for every fixed $B \in \mathcal{B}$ a measure (i.e. a nonnegative and countably additive set function on $\mathcal{A}$) and $P(B/B) = 1$ for any $B \in \mathcal{B}$.

B. For any $A \in \mathcal{A}$, $B \in \mathcal{B}$ and $C \in \mathcal{B}$, for which $B \subset C$ we have $P(A \cap B/C) = P(A/B)\,P(B/C)$.

Theory of probability, based on the above axioms, should be considered as a natural generalization of the probability theory of Kolmogorov. Theory of conditional probability spaces has been further developed by Császár [3] and Krauss [4]. Attempts to generalize the Rényi's approach to the frame of quantum logics have been made by Kalmár [5] and Nánásiov [6]. In the present paper, a more natural generalization is presented. In particular, quantum conditional probability spaces based on the logics of projections in von Neumann algebras are studied. We arrive at a "representation theorem" generalizing the "classical" theorem of Krauss [4].

Throughout this paper L will be a (quantum) logic, i.e. an orthomodular lattice and we use the symbols $0,1,\vee,\wedge,-,{}',\leq$ with their ordinary meaning in L (see [7], [8] ,[9] for the details). Two elements $a,b \in L$ are orthogonal ($a \perp b$) if $a \leq b'$, and $a,b \in L$ are compatible ($a \leftrightarrow b$) if $a = a \wedge b \vee a \wedge b'$. A measure on L is a map $m:L \to R$, which is finitely or σ-additive on orthogonal elements. A measure m is called a state if $m(1) = 1$. A measure on L is a Jauch - Piron (JP) measure if $m(a) = 0 = m(b)$ imply $m(a \cup b) = 0$ $(a,b \in L)$. We say that L is a JP-logic if every state on L is a JP-state. An ideal in L is a subset $P \subset L$ such that (i) $a \leq b$, $b \in P$ imply $a \in P$, and (ii) $a \in P$ and $b \in P$ imply $a \vee b \in P$. By a measure on P we mean a map $m: P \to R$ which is finitely additive or σ-additive on orthogonal elements of P (i.e. $m(\vee a_i) = \sum m(a_i)$ for every sequence of pairwise orthogonal elements of P such that $\vee a_i$ belongs to P in the σ-additive case). Let m be a measure on an ideal P of L. An element $b \in L$ is called a support of m if $m(a) = 0$ iff $a \perp b$. For $a,b \in L$ we put $a*b = (a \vee b') \wedge b$.

DEFINITION 1. Let L be a logic . Let L_c be any subset of $L \sim \{0\}$. A conditional probability is a function $p: L \times L_c \to R$ such that :

(a) For any fixed $b \in L_c$, $p(./b)$ is a measure on L and $p(b/b) = 1$.

(b) For any $a,b,c \in L$ such that $a \leq b \leq c$ and $b,c \in L_c$ we have $p(a/c) = p(a/b)\,p(b/c)$.

A quantum conditional probability space is a triple (L,L_c,p), where p satisfies axioms (a), (b).

It is easy to check that if L is a Boolean algebra our definition is equivalent to the Rényi's definition, in general case our definition is weaker.

Now we shall consider a conditional probability space (L,L_c,p), where L is a JP-logic and $L_c = L-\{0\}$. We note that the JP property implies that for every $x,y \in L$ such that $x \vee y \neq 0$ it is $p(x/x \vee y) + p(y/x \vee y) > 0$. Following Rényi [2] and Krauss [4], we define

$$x \overset{<}{=} y \quad \text{if} \quad p(y/x \vee y) > 0$$
$$x < y \quad \text{if} \quad p(x/x \vee y) = 0.$$

Further for every $x \in L-\{0\}$ we put

$$P(x) = \{y \in L-\{0\} \mid y \overset{<}{=} x\} \cup \{0\},$$

and for every $x \in L-\{0\}$ and $y \in P(x)$ define

$$m_x(y) = \frac{p(y/x \vee y)}{p(x/x \vee y)}.$$

The following theorem is a quantum analogue of the Krauss representation theorem.

THEOREM 1. Let L be a JP-logic and let $(L,L-\{0\},p)$ be a conditional probability space. Then there exists a family $\{(P_i,m_i) \mid i \in I\}$ such that

(i) for every $i \in I$, P_i is an ideal in L, and the family $\{P_i \mid i \in I\}$ is linearly ordered by set inclusion,

(ii) for every $x \in L-\{0\}$ there exists $i \in I$ such that $x \in P_i - \bigcup\{P_j \mid P_j \subset P_i\}$, where $\subset$ means a proper inclusion,

(iii) for every $i \in I$, m_i is a measure on P_i such that for every $x \in P_i$, $m_i(x) = 0$ iff $x \in \bigcup\{P_j \mid P_j \subset P_i\} \cup \{0\}$. Moreover if p is σ-additive, then m_i is σ-additive,

(iv) for every $i \in I$ and every $x,y \in P_i$ such that $x \leftrightarrow y$ $m_i(x \wedge y) = p(x/y)m_i(y)$,

(iv') if $s(m_i)$ is a support of m_i and $m_i(x) > 0$, then the support of $p(./x)$ exists and is equal to $s(m_i) * x$.

Proof. For every $i \in I$, $P_i = P(x)$ for some $x \in L-\{0\}$ and $m_i = m_x$. The proof of (i)-(iv) can be obtained by an appropriate modification of the technics used in [4] Th.2.6. We shall prove (iv'). Put $s(m_i) = b$ and let $x \in P_i$ be such that $m_i(x) > 0$. (1) Suppose that $y \perp b*x$, $y \in L$. Then $p(y/x) \leq p(b' \wedge x/x) + p(x'/x) = 0$, since by (iv) we have $p(b' \wedge x/x) = \dfrac{m_i(b' \wedge x)}{m_i(x)} = 0$. (2) Let $p(y/x) = 0$ $(y \in L)$. Then the JP property implies that $p(y \vee x'/x) = 0$, and hence $p(y*x/x) = 0$. By (iv), $m_i(y*x) = p(y*x/x)\, m_i(x) = 0$. Hence $b \perp (y*x)$ i.e $b*x \leq ((y' \wedge x) \vee x') \wedge x = y' \wedge x \leq y'$. We have shown that $p(y/x) = 0$ iff $y \perp b*x$. It is easy to show that $b*x$ is independent on the choice of the support b of m_i.

In analogy with the classical case, a family $\{(P_i,m_i) \mid i \in I\}$ satisfying conditions (i) - (iv) of Theorem 1 will be called a chain representation of p. The next theorem shows that the chain representation is essentially unique.

THEOREM 2. If $\{(P_i,m_i) \mid i \in I\}$ is a chain represntation of p then $\{P_i \mid i \in I\} = \{P(x) \mid x \in L - \{0\}\}$, and for every $i \in I$ and $x \in L - \{0\}$, if

$P_i = P(x)$, then there exists $r > 0$ such that $m_i = r.m_x$.

Proof. Is analogous to [4] Th.2.7, if we take into account that $x \leftrightarrow x \vee y$.

In general, our representation theorem is considerably weaker than the Krauss' theorem for classical case. Given a chain representation $\{(P_i,m_i) | i \in I\}$, we are not able to construct the corresponding conditional probabilities in terms of m_i. In what follows we shall show that it is possible to obtain a stronger result if L is the projection lattice of a von Neumann algebra.

Let W be any von Neumann algebra of operators acting on a complex, separable, infinite dimensional Hilbert space H. Denote by $L(W)$ the lattice of all orthogonal projections belonging to W. A partial ordering is introduced in $L(W)$ defining $E \leq F$ when $EF = FE = E$, and an orthocomplementation is defined via $E \mapsto E' = I-E$. We shall need the following results.

PROPOSITION 1. Let W be a von Neumann algebra not containg a factor of type I_2 as direct summand. Then every finite measure on $L(W)$ is the restriction of a positive functional on W to $L(W)$, and every σ-additive measure is the restriction of a normal functional on W ([12], [13], [14], 15], [16], [17]).

Prop.1 implies that every σ-additive measure on $L(W)$ is a JP-measure. For a measure m on $L(W)$ we denote by $\tilde{m}$ the functional on W extending m.

PROPOSITION 2. Let W be a von Neumann algebra with no type I_2 summand. Let m be a σ-aditive state on $L(W)$) and let $E \in L(W)$ be such that $m(E) \neq 0$. Then $F \mapsto p_m(F/E) := \frac{\tilde{m}(EFE)}{m(E)}$ is the unique functional on $L(W)$) satisfying the following conditions

(i) $p\ (./E))$ is a σ-additive state on $L(W)$) and

(ii) for all $F \in L(W)$ such that $F \leq E$, $p_m(F/E) = \frac{m(F)}{m(E)}$ ([18]).

Proposition 2 enables us to construct examples of quantum conditional probability spaces.

THEOREM 3. (i) Let W be a von Neumann algebra and let $\tilde{m}$ be a state on W. Put $L = L(W)$, $L_c = \{E \in L(W) | \tilde{m}(E) \neq 0\}$ and $p(F/E) = \frac{\tilde{m}(EFE)}{\tilde{m}(E)}$ for $F \in L(W)$ and $E \in L_c$. Then (L,L_c,p) is a quantum conditional probability space. If $\tilde{m}$ is normal, then p is σ-additive.

(ii) Conversely, let $(L(W),L_c,p)$ be a σ-additive conditional probability space, where $L(W)$ is the logic of a von Neumann algebra W with no type I_2 summand, $1 \in L_c$ and $L_c \subset \{E \in L(W) | p(E/1) > 0\}$. Then $p(F/E) = \frac{\tilde{m}(EFE)}{\tilde{m}(E)}$ for some normal state $\tilde{m}$ on W.

Proof. (i) The first point is immediate. To prove the second part (ii), put $m(E) = p(E/1)$. By axiom (a) of Def.2.2, m is a σ-additive state on $L(W)$. Let $\tilde{m}$ be the normal state on W which extends m by Prop.1. Let $E \in L_c$ and $F \in L(W)$, $F \subseteq E$. Axiom (b) of Def.1 gives us $p(F/E) = \frac{p(F/1)}{p(E/1)} = \frac{m(F)}{m(E)}$. Therefore, applying Prop.2 we obtain that $p(G/E) =$

$\frac{\tilde{m}(EGE)}{m(E)}$, $G \in L(W)$.

REMARK Von Neumann [10] introduced so-called transition probabilities on a continuous geometry. His transition probabilities enjoied, among many others, the following properties: (i) for every $b \in L$, $b > 0$, the map $a \mapsto p(a/b)$ is a σ-additive state on L and $p(a/b) = 1$ iff $b \le a$, (ii) if $a \le b$ then $p(a/b) = \frac{D(a)}{D(b)}$, where D is the dimension function on L. It is easy to see that $(L, L-\{0\}, p)$ is a conditional probability space in the sense of Def.1.

A measure m on an ideal P of a logic L is said to be bounded if $\sup\{m(a) \mid a \in P\} < \infty$.

We shal say that a family $\{(P_i, m_i \mid i \in I\}$, where P_i is an ideal in L and m_i is a mesure on P is σ-additive (bounded, extendable) if m_i is σ-additive (bounded, extendable to a measure on the whole L) for every $i \in I$.

The following proposition shows that there is a wide family of extendable measures on the ideals in $L(W)$.

PROPOSITION 3. Let W be a von Neumann algebra with no type I_2 summand. Let $P \subset L(W)$ be such that $p \in P$, $q \in L(W)$, $q \le p \Rightarrow q \in P$ and $p,q \in P$, $|pq| < 1 \Rightarrow p \vee q \in P$. Then every bounded measure on P extends to a measure on the whole $L(W)$ ([19]).

THEOREM 4. Let W be a von Neumann algebra with no type I_2 summand. Let $(L, L-\{0\}, p)$ be a σ-additive conditional probability space such that its chain representation $\{(P_i, m_i) \mid i \in I\}$ is extendable. Then

(iv'') for every $i \in I$ and every $y \in P_i$, $x \in L$ we have $\tilde{m}_i(yxy) = p(x/y) m_i(y)$ where $\tilde{m}_i$ is the extension of m_i to a normal functional on W, which exists by Prop. 1.

Conversely, every extendable, σ-additive family $\{(P_i, m_i) \mid i \in I\}$ satisfying the conditions (i) - (iii) of Theorem 1 and the condition

(v) for every $i, j \in I$, if $P_i = P_j$ then there is $r > 0$ such that $m_i = r.m_j$

uniquely determines a σ-additive conditional probability p on $L(W)$ such that (iv'') holds.

Proof. Let $x \in L-\{0\}$ and let $\tilde{m}_x$ be of the form $\tilde{m}_x(y) = \sum_{n=1}^{\infty} (\phi_n, y\phi_n)$, $\phi_n \in H$, $n \in N$ (see [11]). Let $z \in P(x)$, and let $m_x(z) = 0$. Then $z\phi_n = 0$ for all $n \in N$. Therefore $zyz\phi_n = 0$, and hence $\tilde{m}_x(zyz) = 0$, i.e. (iv'') holds.

Now let $m_x(z) > 0$, $z \in P(x)$. By Theorem 1 (iv), for every $y \le z$, $p(y/z) = \frac{m_x(y)}{m_x(z)}$. This shows that the functional $p(./z)$ on $L(W)$ satisfies the conditions of Prop.2, and therefore $p(y/z) = \frac{\tilde{m}_x(zyz)}{m_x(z)}$ for every $y \in L(W)$.

To prove the converse statement, let an extendable, σ-additive family $\{(P_i,m_i) | i \in I\}$ satisfy the conditions (i) - (iii) of Th.1 and (v). Consider $x,y \in L(W)$, $y \neq 0$. By (ii) there is $i \in I$ such that $y \in P_i - \bigcup \{P_j | P_j \subset P_i\}$. By (iii), $m_i(y) > 0$. By the suppositions, m_i extends to a normal functional $\tilde{m}_i$ on W. Thus define $p(x/y) = \dfrac{\tilde{m}_i(yxy)}{m_i(y)}$ for all $x \in L(W)$. By (v), p is well-defined and it is easy to check that p is the desired conditional probability on $L(W)$. Uniqueness of p follows from Prop.2.

A simple example of a conditional probability space with $L_c = L - \{0\}$ can be constructed as follows. Let $0 < a_1 < \ldots < a_n < a_{n+1} = 1$ be a finite sequence in $L(W)$. Put $P_j = [0,a_j]$, $j \leq n$, $P_{n+1} = L(W)$. Then $P_1 \subset P_2 \subset \ldots \subset P_n \subset P_{n+1}$ is a linearly ordered set of ideals of $L(W)$. Moreover, for every x there is $i \leq n+1$ such that $x \in P_i - P_{i-1}$. Let $m_1, m_2, \ldots, m_n, m_{n+1}$ be σ-additive measures on $L(W)$ with the supports $s(m_j) = a_j \wedge a'_{j-1}$, $j \leq n+1$. If $x \in P_j$, then $m_j(x) = 0$ iff $x \perp s(m_j)$ i.e. iff $x \leq (a'_j \vee a_{j-1}) \wedge a_j = a_{j-1}$, i.e. iff $x \in P_{j-1}$. This shows that (i)-(iii) of Th.1. are satisfied. By Th.4, $p(x/y) = \dfrac{\tilde{m}_i(yxy)}{m_i(y)}$, where $i \leq n+1$ is such that $m_i(y) > 0$, defines a conditional probability on $L(W)$.

The "top-down" construction of Krauss can also be adapted (see [4]). We construct an ordinal $\alpha > 0$ and a sequence $\{(P_\xi, m_\xi) | \xi < \alpha\}$ such that for each $\xi < \alpha$, P_ξ is an ideal in $L(W)$ and m_ξ is a nontrivial bounded σ-additive measure on P_ξ, as follows: Let $P_0 = L(W)$, and let m_0 be a σ-additive state on $L(W)$. Suppose $\eta > 0$ and $\{(P_\xi, m_\xi) | \xi < \eta\}$ has been constructed. If $\eta = \xi + 1$ and $\{x \in P_\xi | m_\xi(x) = 0\} = \{0\}$, let $\alpha = \eta$. If $\{x \in P_\xi | m_\xi(x) = 0\} \neq \{0\}$ let $P_\eta = \{x \in P_\xi | m_\xi(x) = 0\}$ and let m_η be a σ-additive measure on P_η (which is a restriction of a σ-additive state on $L(W)$ to P_η). If $\eta = \bigcup_{\xi<\eta} \xi$ and $\bigcap_{\xi<\eta} P_\xi = \{0\}$, let $\alpha = \eta$. If $\bigcap_{\xi<\eta} P_\xi \neq 0$, let $P_\eta = \bigcap_{\xi<\eta} P_\xi$ and let m_η be a restriction of a σ-additive state on $L(W)$ to P_η, which is nontrivial on P_η. By a cardinality argument this construction has to terminate at some ordinal α. By the construction, the sequence $\{(P_\xi, m_\xi) | \xi < \alpha\}$ satisfies conditions (i) and (iii) of Th.1. Let $x \in L(W) - \{0\}$. If $x \in \bigcap_{\xi<\alpha} P_\xi$, then for some $\xi < \alpha$, $\alpha = \xi + 1$ and condition (ii) of Th.1 is satisfied. If $x \notin \bigcap_{\xi<\alpha} P_\xi$, let $\xi < \alpha$ be the first ordinal such that $x \notin P_\xi$. Then for some $\eta < \xi$, $\xi = \eta + 1$ and condition (ii) of Th.1 is again satisfied. Then by Th.3 there is a unique conditional probability p satisfying (iv).

Theorem 4 describes a wide class of conditional probabilities on the logics of von Neumann algebras. It remains an open question if it is possible to find a similar representation for not necessarily bounded conditional probabilities on von Neumann algebras. Another open problem is to describe the class of quantum logics on which conditional probabilities may exist.

REFERENCES

1. RÉNYI, A.: On a new axiomatic theory of probability, Acta Math. Acad. Sci.Hung.6, (1955),285-335.
2. RÉNYI, A.: On conditional probability spaces generated by a dimensionally ordered set of measures, Teorija verojatnostej i jeje primenenija 1, (1956), 61-71.
3. CSÁSZÁR, A.: Sur la structure des éspaces de probabilité conditionelle, Acta Math. Acad. Sci. Hung.6, (1955), 337-361.
4. KRAUSS, P.H.: Representation of conditional probability measures on Boolean algebras, Acta Math. Acad. Sci. Hung.19, (1965), 229-241.
5. KALMÁR, J.G.: Conditional probability measures on propositional systems. Publ. Math. Debrecen 30, (1983), 101-115.
6. NÁNÁSIOVÁ, O.: Conditional probability on a quantum logic, Int. J. Theor. Phys. 25, (1986), 1155-1162.
7. BELTRAMETTI, E.G., CASSINELLI,G.: The logic of quantum mechanics, Addison-Wesley, Reading, Mass. 1981.
8. VARADARAJAN, V.S.: Geometry of quantum theory, Springer, New York 1985.
9. PIRON, C.: Foundations of quantum physics, Benjamin, Reading Mass. 1976.
10. Von NEUMANN, J.: Continuous geometries with a transition probability, Mem. Amer. Math. Soc. 252, (1981).
11. DIXMIER, J.: Les algébres d'opérateurs dans l'espace Hilbertien, Gauthier-Villars, Paris 1957.
12. GLEASON,A.M.: Measures on the closed subspace of a Hilbert space, J. Math. Mech.6, (1957), 447-452.
13. GUNSON, J.: Physical states on quantum logics. Ann. Inst. H. Poincaré 17, (1972), 295-311.
14. MATVEICHUK, M..: Odna teorema o sostojanii na kvantovych logikach, Teoret. i matem. fiz.45, (1980), 244-250.
15. PASKIEWICZ, A.: Measures on projections of von Neumann algebras, J. Funct. Anal.62, (1985) 87-117.
16. CHRISTENSEN, E.: Measures on projections and physical states, Commun. Math. Phys.86, (1982), 529-538.
17. YEADON, F.: Measures on projections in W^*-algebras of type II_1, Bull. London Math. Soc.15, (1983), 139-145.
18. CASSINELLI, G., ZANGHI, N.: Conditional probabilities in quantum mechanics, I,II, Nuovo Cimento 73 B, (1983), 237-245, ibid. 79 B, (1984), 141-145.
19. MATVEICHUK, M.S.: Verojatnostnaja mera na ideale projektorov. Verojatnostnyje metody i kibernetika, Kazan univ. 19, (1983) 51-55.

This paper is in final form and no similar paper has been or is being submitted elsewhere.

Quantum Diffusions on the Rotation Algebras and the Quantum Hall effect

Paul Robinson
Mathematisch Instituut, Katholieke Universiteit
Toernooiveld, 6525 ED Nijmegen

Acknowledgements

This work is based on my PhD thesis. I would like to thank Robin Hudson for his invaluable help and the SERC for financial support.

§1. Introduction

Quantum diffusions on a C^*-algebra A are quantum stochastic processes on A in the sense of Accardi, Lewis and Frigerio [1], constructed by solving quantum stochastic differential equations (quantum s.d.e.'s) defined on norm-dense 'smooth' subalgebras A^∞ of A.

We will take A to be a unital C^*-subalgebra of $B(H_0)$, the bounded operators on a Hilbert space H_0; and a process $j = (j_t : t \geq 0)$ on A will be a family of identity preserving $*$-homomorphisms from A into $B(H_0 \otimes H)$, where H is the Boson Fock space over $L^2(\mathbf{R}_+)$. The process j will be assumed to be adapted in the sense that for each $x \in A$ and $t \geq 0$, $j_t(x) \in B(H_0 \otimes H_t) \otimes \mathbf{1}$, where H_t is Boson Fock space over $L^2([0,t])$.

To construct stochastic differential equations we will use the calculus of Hudson and Parthasarathy [2]. Define time indexed creation and annihilation operators on the algebraic tensor product of H_0 with the span of the exponential vectors $\psi(f)$, $f \in L^2(\mathbf{R}_+)$ in H by the respective actions,

$$A^\dagger(t)u \otimes \psi(f) = \frac{d}{d\varepsilon} u \otimes \psi(f + \varepsilon\chi_{[0,t]})\Big|_{\varepsilon=0}$$

$$A(t)u \otimes \psi(f) = \int_o^t f u \otimes \psi(f)$$

for each $u \in H_0$, $f \in L^2(\mathbf{R}_+)$.

The Hudson and Parthasarathy calculus now gives meaning to quantum stochastic differential equations,

$$dj_t(x) = j_t(\alpha(x))dA^\dagger(t) + j_t(\alpha^\dagger(x))dA(t) + j_t(\tau(x))dt, \;\; j_0(x) = x \otimes \mathbf{1} \tag{1.1}$$

where $\alpha, \alpha^\dagger, \tau : A^\infty \to A$ and for each $x \in A^\infty$.

The maps $\alpha, \alpha^\dagger$ and τ cannot be freely chosen [3], [4]. Since each j_t is to be a $*$-homomorphism then it must be contractive; thus for each $x, y \in A^\infty$ the stochastic differentials $dj_t(x)$ and $dj_t(y)$ satisfy the condition ensuring validity of the quantum Ito product formula of [2] and necessarily,

$$dj_t(xy) = dj_t(x)j_t(y) + j_t(x)dj_t(y) + dj_t(x)dj_t(y)$$

evaluated by bilinear extension of the rule $dAdA^\dagger = dt$, all other products of differentials vanishing.

Following the analysis of [3] we find that for each $x, y \in A^\infty$,

$$\alpha(xy) = x\alpha(y) + \alpha(x)y \tag{1.2}$$
$$\alpha^\dagger(x) = \alpha(x^*)^*$$
$$\tau(xy) = x\tau(y) + \tau(x)y + \alpha^+(x)\alpha(y) \tag{1.3}$$

Thus α is a l-cocycle taking values in A and the map $\eta_\alpha : (x, y) \mapsto -\alpha^\dagger(x)\alpha(y)$ is a Hochschild 2-coboundary on A^∞ taking values in A [4].

In §2 we construct an example of a pair (A^∞, A) where η_α is not a 2-coboundary, thus exhibiting a cohomological obstruction to the existence of a diffusion. This obstruction is then 'removed' in a certain sense by constructing a diffusion on a C^*-algebra extension of A. This work essentially clarifies the results of [5].

In §3 a speculative application of this obstruction/extension result to a problem in solid state physics is discussed.

§2. Quantum Diffusion on the Rotation algebras

Let $\chi = \{U(m) : m \in \mathbf{Z}^2\}$ denote the set of characters on the 2-torus $\mathbf{T}^2$ i.e. for each $m = (m_1, m_2) \in \mathbf{Z}^2$ and $s = (s_1, s_2) \in \mathbf{T}^2$,

$$U(m)(s) = \exp 2\pi i(m_1 s_1 + m_2 s_2)$$

Let $\theta \in [0, 1]$ and denote by A_θ^F the vector space of finite sums of elements of χ over $\mathbf{C}$; then under linear extension of the multiplication,

$$U(m) \times_\theta U(n) = \exp \pi i\theta(m \wedge n)U(m + n) \tag{2.1}$$

where $m \wedge n = m_1 n_2 - n_1 m_2$, and the involution,

$$(aU(m))^* = \bar{a}U(-m), \qquad a \in \mathbf{C} \tag{2.2}$$

A_θ^F becomes a $*$-algebra.

We can easily define a representation of A_θ^F on $L^2(\mathbf{T}^2)$ via the following easy lemma,

Lemma 2.1.

Fix $x \in A_\theta^F$ and define an action of x on A_θ^F by,

$$(x)y = x \times_\theta y \tag{2.3}$$

for each $y \in A_\theta^F$. Then this action is bounded in L^2-norm. ◇

In Lemma 2.1 the vector space A_θ^F is viewed as a dense subset of $L^2(\mathbf{T}^2)$. It is easy to show that the action (2.3) extends to a faithful $*$-representation of A_θ^F on $L^2(\mathbf{T}^2)$, which we also denote by A_θ^F.

Let A_θ denote the C^*-algebra completion of A_θ^F in the prenorm,

$$\|x\|_\theta = \sup_{\|f\|_2=1} \|(x)f\|_2 \tag{2.4}$$

where $x \in A_\theta^F$ and $\| \quad \|_2$ denotes L^2-norm.

Define unbounded derivations α_1^θ and α_2^θ on A_θ^F by linear extension of the actions,

$$(\alpha_i^\theta(U(n)))f = 2\pi i \delta_{ij} n_j (U(n))f \tag{2.5}$$

$F \in L^2(\mathbf{T}^2)$, $n = (n_1, n_2)$.

Finally, let 1 be the identity function in $L^2(\mathbf{T}^2)$, then it can be easily shown that $w_\theta : A_\theta \to \mathbf{C}$ defined by,

$$w_\theta(x) = \langle 1, (x)1 \rangle \tag{2.6}$$

is a tracial state on A_θ.

For $\theta \in [0,1]$ the C^*-algebra A_θ is a representation of the rotation algebra [6].

The question of existence of a diffusion is now the following - given a derivation $\alpha^\theta : A_\theta^F \to A_\theta$, can we find a $\tau^\theta : A_\theta^F \to A_\theta$ satisfying,

$$\tau^\theta(xy) = x\tau^\theta(y) + \tau^\theta(x)y + \alpha^{\theta^\dagger}(x)\alpha^\theta(y) \tag{2.7}$$

Choosing $\alpha^\theta = \alpha_1^\theta + i\alpha_2^\theta$ we find,

Theorem 2.2.

Let $\alpha^\theta = \alpha_1^\theta + i\alpha_2^\theta$. Then there exists no map $\tau^\theta : A_\theta^F \to A_\theta$ satisfying (2.7).

Proof:

Write $U = U(1,0)$ and $V = U(0,1)$ and suppose there exists a map $\tau^\theta : A_\theta^F \to A_\theta$ satisfying (2.7). Using (2.1) and (2.5) it is easy to sec that,

$$e^{2\pi i\theta} V\tau^\theta(U) - \tau^\theta(U)V + e^{2\pi i\theta}\tau^\theta(V)U - U\tau^\theta(V) = -4\pi^2 iUV \tag{1}$$

Multiplying through in (1) on the left by U^* and on the right by V^*, applying the state w_θ and noting that U and V are unitary we see that,

$$w_\theta(VU^*\tau^\theta(U)V^* - U^*\tau^\theta(U) + U^*\tau^\theta(V)V^*U - \tau^\theta(V)V^*) = -4\pi^2 i$$

where we have used (2.1) several times.

But now w_θ is tricial on A_θ which gives,

$$0 = -4\pi^2 i$$

and no such map τ^θ can exist. ◇

We will now look at an 'extension' of A_θ on which, in some sense, a solution of (2.7) exists.

Let $f \in L^2([0,1], L^2(\mathbf{T}^2))$, the $L^2(\mathbf{T}^2)$ valued square integrable functions on $[0,1]$ and for each $m, n \in \mathbf{Z}^2$ define operators $W(m)$ and $Z(m,n)$ on $H_0 = L^2([0,1], L^2(\mathbf{T}^2))$ by

$$\left.\begin{aligned}(W(m)f)(\theta) &= U(m)(f(\theta))\\ (Z(m,n)f)(\theta) &= \exp 2\pi i(m \wedge n)f(\theta)\end{aligned}\right\} \tag{2.8}$$

using (2.1) we see that,

$$W(m)W(n) = Z(m,n)W(n)W(m), \quad [W(m), Z(n,p)] = 0 \tag{2.9}$$

for all $m, n, p \in Z^2$.

Let A^F denote the collection of finite polynomials in the $W(m)$'s and $Z(n,p)$'s and define a homomorphism $\rho_\theta : A^F \to A_\theta$ by linear extension of the rule,

$$\rho_\theta(W(m)) = U(m), \quad \rho_\theta(Z(m,n)) = \exp 2\pi i(m \wedge n)\mathbf{1} \tag{2.10}$$

wehere $\mathbf{1}$ is the identity in A_θ.

Let $x \in A^F$ and define an involution on A^F by,

$$(x^* f)(\theta) = \rho_\theta(x)^* f(\theta) \tag{2.11}$$

From (2.9) and (2.11) A^F forms a $*$-algebra and ρ_θ is a $*$-homomorphism. Clearly, ρ_θ maps A^F onto A_θ^F (though certainly not injectively).

Define a C^*-prenorm on A^F by,

$$\|x\| = \sup_{\theta \in [0,1]} \|\rho_\theta(x)\|_\theta \tag{2.12}$$

and let A be the C^*-algebra completion of A^F. Then,

Lemma 2.3.

The $*$-homomorphism ρ_θ extends uniquely to a $*$-homomorphism of A onto A_θ, for each $\theta \in [0,1]$.

Proof:

This follows easily from the fact that ρ_θ is contractive, maps A^F onto A_θ^F and that A_θ^F is dense in A_θ. ◇

In this sense we regard A as an extension of A_θ.

We can define unbounded derivations α_1 and α_2 on A^F via the prescription,

$$\rho_\theta \circ \alpha_j = \alpha_j^\theta \circ \rho_\theta, \quad j = 1,2 \tag{2.13}$$

where the α_j^θ's are defined by (2.5).

On this extension A, can we now find a $\tau : A^F \to A$ satisfying,

$$\tau(xy) = x\tau(y) + \tau(x)y + \alpha^\dagger(x)\alpha(y), \quad x, y \in A^F$$

where $\alpha = \alpha_1 + i\alpha_2$? In fact we can, and solutions of the diffusion equation,

$$dj_t(x) = j_t(\alpha(x))dA^\dagger + j_t(\alpha^\dagger(x))dA + j_t(\tau(x))dt \tag{2.14}$$

are given explicitely by,

$$\left.\begin{aligned} j_t(W(m)) &= W(m) \otimes e^{2\pi i m_1 Q(t)} e^{2\pi i m_2 P(t)} \\ j_t(Z(m,n)) &= Z(m,n) \otimes e^{8\pi^2 it(m\wedge n)} \end{aligned}\right\} \tag{2.15}$$

where $m = (m_1, m_2)$ and the maps $Q(t) = A^\dagger(t) + A^+(t)$ and $P(t) = -i(A^\dagger(t) - A(t))$ are non-commuting realisations of classical Brownian motion on Fock space [7].

Then τ satisfies,

$$\tau(W(m)Z(n,p)) = 8\pi^2\left(\frac{1}{2}(im_1 m_2 + \frac{1}{4}m_1^2 + \frac{1}{4}m_2^2 + i(n \wedge p)\right)W(m)Z(n,p) \tag{2.16}$$

for each $m, n, p \in \mathbf{Z}^2$.

Note: In contrast to (2.13) we see that there is no map τ^θ mapping A_θ^F into A_θ such that

$$\rho_\theta \circ \tau = \tau^\theta \circ \rho_\theta.$$

§3. A speculative application of this result in Solid State Physics

The abstract C^*-algebras A_θ and A defined by the relations,

$$U(m)U(n) = \exp 2\pi i\theta(m \wedge n)U(n)U(m) \tag{3.1}$$

and,

$$W(m)W(n) = Z(m,n)W(n)W(m), \quad [W(m), Z(n,p)] = 0 \tag{3.2}$$

respectively, have been used by Bellissard [8] to compute the integer valued quantum Hall voltage. The algebras arise as follows,

(i) For an electron moving in a periodic 2-dimensional lattice and subjected to a uniform magnetic field B, perpendicular to the lattice, every observable of interest is a self-adjoint element of A_θ. Here, $\theta = BS/(h/e)$ where h is Planck's constant, e is the electron charge and S is the area of a unit cell of the lattice.

(ii) If we want to investigate the 'evolution' of observables as we change the field B (or the parameter θ) then we must also make the field an observable. Thus we 'envelope' all of the algebras A_θ in A, in the sense that A is an extension of all of them.

Bellissard now defines an 'Ito like' derivation in A [8]. Let x be in A^F and satisfy,

$$\rho_\theta(x) = \sum a(m;\theta)U(m) \in A_\theta^F \tag{3.3}$$

and define $\partial : A^F \to A^F$ by

$$\rho_\theta(\partial x) = \sum \frac{\partial}{\partial\theta}(a(m;\theta))U(m) \tag{3.4}$$

then for each $x, y \in A^F$

$$\partial(xy) = \partial(x)y + x\partial(y) + \frac{i}{4\pi}(\alpha_1(x)\alpha_2(y) - \alpha_2(x)\alpha_1(y)) \tag{3.5}$$

We will now see how this might arise through the Quantum Ito formula. We make the following speculative proposition; as we change the magnetic field strength B the evolution of observables is governed by the process $j = (j_t : t \geq 0)$ satisfying,

$$dj_t(x) = j_t(\alpha_1(x))dQ(t) + j_t(\alpha_2(x))dP(t) + j_t(\tau(x))dt, \quad j_0(x) = x \otimes 1 \tag{3.6}$$

on a suitably smooth domain in A, where t is the variable $t = BS/(h/e)$ and $Q(t), P(t)$ satisfy the commutation relations,

$$[P(BS/(h/e)), Q(BS/(h/e))] = -2iBS/(h/e) \tag{3.7}$$

Equation (3.6) is essentially (2.14). If such a Bose field satisfying (3.7) exists, it must certainly be related to the fact that at low temperature (which is the realm in which the quantum Hall effect occurs) the applied field must be quantised [9].

We will see how far this view takes us. The differentials $dQ(t)$ and $dP(t)$ satisfy,

	$dQ(t)$	$dP(t)$
$dQ(t)$	dt	$-idt$
$dP(t)$	idt	dt

(3.8)

Writing,

$$\tau(xy) = x\tau(y) + \tau(x)y + \alpha_1(x)\alpha_1(y) + \alpha_2(x)\alpha_2(y) + i(\alpha_1(x)\alpha_2(y) - \alpha_2(x)\alpha_1(y)) \quad (3.9)$$

and comparing with (3.8) we see that the term $i(\alpha_1(x)\alpha_2(y) - \alpha_2(x)\alpha_1(y))$ comes from the $dPdQ$ and $dQdP$ terms in the Quantum Ito formula. Put $4\pi\partial = -\tau + \frac{1}{2}\alpha_1 \circ \alpha_1 + \frac{1}{2}\alpha_2 \circ \alpha_2$ in (3.9) then,

$$\partial(xy) = \partial(x)y + x\partial(y) + \frac{i}{4\pi}(\alpha_1(x)\alpha_2(y) - \alpha_2(x)\alpha_1(y))$$

which is precisely Bellissards Ito derivation! Moreover, the correction term arising from the $dPdQ$ and $dQdP$ terms in the Quantum Ito formula is thus a purely quantum effect.

Noting from (2.15) that $j_t(x) \in A \otimes B(H)$, define the vacuum conditional expectation $\mathbf{E}_0$ from $A \otimes B(H)$ to A, by continuous linear extension of,

$$\mathbf{E}_0(x \otimes T) = <\psi_0, T\psi_o>x, \quad x \in A, \quad T \in B(H)$$

where ψ_0 is the Fock vacuum.

Then,

Theorem 3.1.

Let $\theta, t \in \mathbf{R}_+$ then,

(i) for each $x \in A$,

$$w_{\theta+t}\rho_{\theta+t}(x) = w_\theta\rho_\theta(\mathbf{E}_0(j_{t/4\pi})(x))) \quad (3.10)$$

where $j = (j_t : t \geq 0)$ is the process defined by (2.15)

(ii) for each $x \in A^F$,

$$\frac{\partial}{\partial\theta} w_\theta\rho_\theta(x) = \frac{1}{4\pi} w_\theta\rho_\theta(\tau(x)) \quad (3.11)$$

where $\tau : A^F \to A$ satisfies (3.9).

Proof:

(i) Simple varification on a monomial $U(m)Z(n,p)$, $m, n, p \in \mathbf{Z}^2$. Then extend to A by continuity from the dense subalgebra A^F.

(ii) For $x \in A^F$ compute,

$$\frac{\partial}{\partial\theta} w_\theta\rho_\theta(x) = \lim_{t\to0} \frac{w_{\theta+t}\rho_{\theta+t}(x) - w_\theta\rho_\theta(x)}{t}$$

using (3.10). ◇

Formula (3.11) will extend to a larger class of smooth elements $A^\infty \subset A$ [10]. The algebra A_θ also has a smooth subalgebra containing A^F, $\rho_\theta(A^\infty) = A_\theta^\infty$, and we can obtain Streda's formula [11], [8],

Corollary 3.2.

Put $\theta = BS/(h/e)$ in (3.11). Let $x \in A^\infty$ be such that for B in some interval (B_1, B_2), $\rho_\theta(x)$ is a projection in A_θ^∞. Then in this range,

$$\frac{e}{S}\frac{\partial}{\partial B}w_\theta\rho_\theta(x) = \frac{e^2}{h}\cdot Ch(\rho_\theta(x)) \tag{3.12}$$

where,

$$Ch(\rho_\theta(x)) = (\frac{1}{2\pi i})w_\theta[\rho_\theta(x)(\alpha_1^\theta(\rho_\theta(x))\alpha_2^\theta(\rho_\theta(x)) - \alpha_2^\theta(\rho_\theta(x))\alpha_1^\theta(\rho_\theta(x)))] \tag{3.13}$$

Proof:

The argument is essentially Bellisards in [8]. The maps τ and ∂ differ by a factor $\frac{1}{2}\alpha_1 \circ \alpha_1 + \frac{1}{2}\alpha_2 \circ \alpha_2$ which gives rise to an extra term on the right hand side of (3.12),

$$\frac{e^2}{h}(\frac{1}{4\pi})w_\theta[(1-2\rho_\theta(x))(\alpha_1^\theta(\rho_\theta(x))^2 + \alpha_2^\theta(\rho_\theta(x))^2)]$$

but it is easy to show that this vanishes if $\rho_\theta(x)$ is a projection. ◇

The map Ch defined by (3.13) is the chern character of A_θ, and for $e \in A_\theta$ a projection, Connes [12] has shown that $Ch(e)$ is an integer.

The expression $\frac{1}{S}w_\theta\rho_\theta(x)$ is called the integrated density of states (per unit volume) of the system, when $\rho_\theta(x)$ is the projection onto eigenstates of the hamiltonian of energy less than the Fermi energy. Streda's formula then says that the Hall voltage is e times the derivative with respect to B of the integrated density of states.

Finally, we can compute a version of Streda's formula(3.13) at temperatures which are greater than zero. Without worrying about details, which can be found in [13], a finite temperature Bose noise satisfying (3.7) is one for which we have,

	$dQ(t)$	$dP(t)$
$dQ(t)$	$\sigma^2 dt$	$-idt$
$dP(t)$	idt	$\sigma^2 dt$

where $\sigma^2 = (1 + e^{-T_0/kt})/(1 - e^{-T_0/kT})$, T_0 is a constant, k is Boltzmann's constant and T is the temperature of the field.

We assume that our system is in thermal equilibrium with this Bose field; thus the system plus field is now described as being at temperature T.

Repeating Theorem 3.1 and Corollary 3.2 with this field and denoting the reciprocal of the Hall voltage by R_H (the Hall resistivity) we find that, at low temperature,

$$R_H \alpha (1 + 2e^{-T_0/kT})$$

Temperature variation of R_H at low temperatures has been found experimentally to vary from the zero temperature resistivity by a factor $e^{-T_0/kT}$ [14].

References

[1] Accardi, L., Frigerio, A and Lewis, J.T., Quantum Stochastic Processes, Proc. Res. Inst. Math. Sci. Kyoto 18, 94-133 (1982)

[2] Hudson, R.L. and Parthasarathy, K.R., Quantum Ito's Formula and Stochastic Evolutions, Commun. Math. Phys. 93, 301-323 (1984).

[3] Hudson, R.L., Algebraic Theory of Quantum Diffusions, Stochastic Mechanics and Stochastic Processes, Swansea Proceedings, eds. I. Davies, A. Truman (1986).

[4] Evans, M.P. and Hudson, R.L., Multidimensional Quantum Diffusions, Quantum Probability and Applications III, proceedings Oberwolfach, eds. Accardi, L. and von Waldenfels, W., 1303, Springer-Verlag (1987).

[5] Hudson, R.L., Robinson, P., Quantum Diffusions and the Noncommutative Torus, Lett. Math. Phys. 15, 47-53 (1988).

[6] Rieffel, M.A., C^*-algebras Associated with Irrational Rotations, Pacific J. Math. 93, 415-429 (1981).

[7] Hudson, R.L., Quantum Stochastic Calculus in Fock space: a review, Fundamental Aspects of Quantum Theory, eds. Frigerio, A. and Gorini, V., Plenum Press, New York and London (1986).

[8] Bellisard, J., C^*-algebras in Solid State Physics, in D. Evans (ed.), Proc. Anglo-American Conf. on Operator ALgebras, Warwick (1987), to appear.

[9] Srivastava, Y., Beyond the Hall Effect: Practical Engineering form Relativistic Quantum Field Theory, Fundamental Aspects of Quantum Theory, eds. Frigerio, A., and Gorini, V., Plenum Press, New York and London (1986).

[10] Robinson, P., Nottingham PhD thesis (1988).

[11] Streda, P., Theory of quantised Hall conductivity in two dimensions, J. Phys. C. 15, L717-L723 (1982).

[12] Connes, A., A survey of Foliations and Operator Algebras, chapter 13, Proc. Symposia in Pure Mathematics, Operator Algebras and Applications, American Math. Soc. 38 (part 1), 616-628 (1982).

[13] Hudson, R.L., Lindsay, J.M., Uses for non-Fock Quantum Brownian motion and a quantum Martingale representation theorem, Proceedings of the 2nd workshop on Quantum Probability and its applications, eds. Accardi, L. and von Waldenfels, W., 1136, Springer-Verlag (1985).

[14] Prange, R.E., Girvin, S.M. (eds.), Quantum Hall Effect, Graduate Texts in Contemporary Physics, New York, Springer (1987).

This paper is in final form and no similar paper has been or is being submitted elsewhere.

QUANTUM DIRICHLET FORMS, DIFFERENTIAL CALCULUS AND SEMIGROUPS

Jean-Luc Sauvageot
Laboratoire de Probabilités
Université P. & M. Curie
Tour 56 4 place Jussieu F-75252 Paris Cedex 05

Introduction

This work is a continuation, and a kind of reciprocal, of a previous work ([6]) which pointed out the relationship between quantum semigroups and non commutative differential geometry.

In [6], to any quantum semigroup was associated a differential calculus of order one. Here we start from a differential calculus on a C^*-algebra $\mathbb{A}$ with a faithful l.s.c. trace τ, satisfying some natural domain and symmetry assumptions, and show that it is actually the differential calculus associated with a quantum semigroup on the von Neumann algebra $\pi_\tau(\mathbb{A})''$. The method is most classical, an obvious non abelian extension of the concept of Dirichlet form (cf. [2]), and provides a very general method for getting semigroups of completely positive contractions on C^*- and W^*-algebras.

The conclusion is that there are many quantum semigroups, and that a quantum semigroup is a geometrical object. In order to illustrate this last idea, we give an example which is basically a non commutative geometrical Riemannian structure: the canonical construction of a quantum semigroup (*the transverse heat semigroup)* on a Riemannian foliation W^*-algebra (a next paper will show that it is actually a semigroup on the foliation C^*-algebra).

The first section is just the definition of Dirichlet forms as closable positive quadratic forms on the Hilbert space $L^2(\mathbb{M},\tau)$ associated with a trace τ on a von Neumann algebra $\mathbb{M}$, which satisfy some functional calculus inequalities.

Section 2 is the proof of the fundamental property of a Dirichlet form: if Q is the self adjoint positive operator on $L^2(\mathbb{M},\tau)$ associated with its closure, then the semigroup $\{e^{-tQ}\}_{t\geq 0}$ in $\mathcal{L}(L^2(\mathbb{M},\tau))$ extends to a pointwise σ-weakly continuous semigroup of positive normal contractions on the von

Neumann algebra $\mathbb{M}$.

In section 3 we establish a relationship between differential calculus and Dirichlet forms: we start with a C^*-algebra $\mathbb{A}$ (with a trace τ) and a first order differential calculus, that is a densely defined derivation ∂ from $\mathbb{A}$ into an $\mathbb{A}$-$\mathbb{A}$ C^*-Hilbert bimodule, which is closable in the L^2 sense (with some other simple assumptions), and show that the quadratic form corresponding to the self adjoint positive operator $\partial^*\partial$ is a Dirichlet form: from which we conclude that the operator $\Delta = -\partial^*\partial$ is the generator of a σ-weakly pointwise continuous semigroup of *completely positive* contractions of the von Neumann algebra $\pi_\tau(\mathbb{A})''$.

The last section is a presentation of our example: the transverse heat semigroup on a Riemannian foliation W^*-algebra.

1. Dirichlet forms on von Neumann algebras

1.1. Notations.

1.1.1. Throughout this section and the next one, $\mathbb{M}$ denotes a fixed von Neumann algebra (with separable predual), and τ a faithful normal semifinite trace τ on $\mathbb{M}$.

$\mathbb{M}_+$ is the cone of positive elements of $\mathbb{M}$, and $\mathbb{M}_{sa}$ the real vector space of self adjoint elements of $\mathbb{M}$.

1.1.2. To the pair $(\mathbb{M},\tau)$ are classically associated:

- the hereditary cone $\mathcal{M}_\tau^+ = \{ x\in\mathbb{M}_+ / \tau(x) < \infty \}$
- the bilateral ideal $\mathcal{M}_\tau$ = linear span of $\mathcal{M}_\tau^+$ $[\mathcal{M}_\tau = \mathcal{N}_\tau^2]$
- the bilateral ideal $\mathcal{N}_\tau = \{ x\in\mathbb{M} / \tau(x^*x) < \infty \}$, whose self adjoint part is $\mathcal{N}_\tau^{sa} = \mathcal{N}_\tau \cap \mathbb{M}_{sa}$; $\mathcal{N}_\tau$ and $\mathcal{N}_\tau^{sa}$ appear respectively as complex and real pre-hilbert spaces for the scalar product

$$< y , x >_\tau = \tau(y^*x) ;$$

- the (respectively complex and real) Hilbert spaces $L^2(\mathbb{M},\tau)$ and $L^2(\mathbb{M}_{sa},\tau)$ obtained by completion of $\mathcal{N}_\tau$ and $\mathcal{N}_\tau^{sa}$ respectively.

$L^2(\mathbb{M},\tau)$ is the Hilbert space complexification of $L^2(\mathbb{M}_{sa},\tau)$, and the corresponding isometric involution J $[\xi+i\eta \to \xi-i\eta,\ \xi,\eta \in L^2(\mathbb{M}_{sa},\tau)]$ extends the involution $x \to x^*$ of $\mathcal{N}_\tau$.

1.1.3. There is an obvious one to one correspondance between

- closed or closable positive quadratic forms on $L^2(\mathbb{M}_{sa},\tau)$
- closed or closable J-invariant quadratic forms on $L^2(\mathbb{M},\tau)$

- self-adjoint positive operators on $L^2(\mathbb{M}_{sa},\tau)$
- self-adjoint positive operators on $L^2(\mathbb{M},\tau)$ commuting with J .

[To a positive closed quadratic form q on $L^2(\mathbb{M},\tau)$ corresponds the self adjoint positive operator Q satisfying $\langle\xi,Q\xi\rangle=q(\xi)$, $\forall\xi\in\mathcal{D}(Q^{1/2})=\mathcal{D}(q)$. One has $JQJ = Q$ iff $q(J\xi) = q(\xi)$, $\forall\xi\in\mathcal{D}(q)=J\mathcal{D}(q)$. etc.]

Then, for any positive Borel function f on $\mathbb{R}_+$, $f(Q)$ maps $L^2(\mathbb{M}_{sa},\tau)$ onto itself and can be viewed as an operator on $L^2(\mathbb{M}_{sa},\tau)$.

1.2. Functional calculus:

1.2.1. An element ξ of $L^2(\mathbb{M}_{sa},\tau)$ can be understood as a self adjoint (possibly unbounded) operator in $L^2(\mathbb{M}_{sa},\tau)$ affiliated with the von Neumann algebra $\mathbb{M}$ (cf. **[3]**); the notation spec(ξ) will refer to the spectrum of ξ as an element of $\mathbb{M}$, or affiliated to $\mathbb{M}$.

For any Borel function f: $\mathbb{R}\to\mathbb{R}$, or spec(ξ) $\to\mathbb{R}$, the operator $f(\xi)$ exists as an operator affiliated with $\mathbb{M}$; it belongs to $\mathbb{M}$ whenever f is bounded; it belongs to $L^2(\mathbb{M}_{sa},\tau)$ whenever f satisfies $|f(t)|\le k|t|$ for a k in $\mathbb{R}_+$, e.g. if f is Lipschitz and $f(0) = 0$.

From now on, we follow the presentation of [2], §1.3:

1.2.2. For any $\varepsilon>0$, let us fix a C^∞ function f_ε: $\mathbb{R}\to\mathbb{R}$ satisfying the three conditions

(i) $-\varepsilon \le f_\varepsilon(t) \le 1+\varepsilon$, $\forall\ t \in \mathbb{R}$

(ii) $0 \le f'_\varepsilon(t) \le 1$, $\forall\ t \in \mathbb{R}$

(iii) $f_\varepsilon(t) = t$, $\forall\ t \in [0,\ 1]$.

Notice that for any ξ in $L^2(\mathbb{M}_{sa},\tau)$, $f_\varepsilon(\xi)$ belongs to $\mathcal{N}_\tau$, with spectrum in $[-\varepsilon,\ 1+\varepsilon]$.

Moreover, one can check easily $|f_\varepsilon(s)-t| \le |s-t|$, $\forall\ s \in \mathbb{R}$, $\forall\ t \in [0,1]$.

1.3. Definition of a Dirichlet form

We shall call *Dirichlet form* on $\mathbb{M}$ a densely defined, closable, positive quadratic form q on $L^2(\mathbb{M}_{sa},\tau)$, with closure $\bar{q}$, which satisfies the condition:

$$\forall\ \xi \in \mathcal{D}(q),\ \forall\ \varepsilon > 0,\ \text{then}\ f_\varepsilon(\xi) \in \mathcal{D}(\bar{q})\ \text{and}\ \bar{q}(f_\varepsilon(\xi)) \le q(\xi)\ .$$

2. Quantum semigroup associated with a Dirichlet form

This section is devoted to the proof of the following property: the semigroup $\{e^{-tQ}\}_{t\ge0}$ of contractions of $L^2(\mathbb{M},\tau)$, where Q is the positive self adjoint operator on $L^2(\mathbb{M},\tau)$ associated with the closure of a Dirichlet form q, extends to a semigroup of positive normal contractions of $\mathbb{M}$.

2.1. Notations

Let q be a Dirichlet form on $\mathbb{M}$, $\bar{q}$ its closure, and Q be the self-adjoint positive operator on $L^2(\mathbb{M}_{sa},\tau)$ associated with it (cf.1.1.3).

Fix ξ in $\mathcal{N}_\tau^+$ with norm in $\mathbb{M}$ less than one, i.e. spec(ξ) $\subset$ [0, 1], and set

$$\eta = \frac{1}{1+Q}\,\xi \quad \text{[which is defined as an element of } L^2(\mathbb{M}_{sa},\tau)\text{]}.$$

We want to show that η belongs to $\mathcal{N}_\tau$, with spectrum in [0, 1]

Notice first that $\eta \in \mathcal{D}(Q) \subset \mathcal{D}(Q^{1/2}) = \mathcal{D}(\bar{q})$.

Then define a closed form ψ on $L^2(\mathbb{M}_{sa},\tau)$, $\mathcal{D}(\psi) = \mathcal{D}(\bar{q})$, by

$$\psi(\zeta) = \bar{q}(\zeta) + \|\zeta-\xi\|^2 \ , \ \forall\ \zeta \in \mathcal{D}(\bar{q})$$

2.2. Lemma

For any ζ in $\mathcal{D}(\bar{q})$, one has $\quad \psi(\zeta) = \psi(\eta) + \bar{q}(\zeta-\eta) + \|\zeta-\eta\|^2$.

Proof:

One has

$$\bar{q}(\zeta-\eta) = \bar{q}(\zeta) + \bar{q}(\eta) -2\langle Q\eta,\zeta\rangle = \bar{q}(\zeta) + \bar{q}(\eta) -2\langle\xi-\eta,\zeta\rangle$$

thus

$$\begin{aligned}\bar{q}(\zeta-\eta) + \|\zeta-\eta\|^2 &= \bar{q}(\zeta) +\|\zeta\|^2 + \bar{q}(\eta) + \|\eta\|^2 - 2\langle\xi,\zeta\rangle \\ &= \psi(\zeta) + \bar{q}(\eta) + \|\eta\|^2 - \|\xi\|^2 \\ &= \psi(\zeta) - \psi(\eta) +2[\ \bar{q}(\eta) + \|\eta\|^2 - \langle\eta,\xi\rangle\ ] \\ &= \psi(\zeta) - \psi(\eta) +2[\ \langle\eta,Q\eta+\eta\rangle - \langle\eta,\xi\rangle\ ] \\ &= \psi(\zeta) - \psi(\eta)\ .\end{aligned}$$

■

2.3. Lemma

For any $\varepsilon>0$ and f_ε the function defined in 1.2.3, for any ζ in $\mathcal{D}(q)$, one has

$$\psi(f_\varepsilon(\zeta)) \le \psi(\zeta)\ .$$

Proof:

By definition of a Dirichlet form one has $\bar{q}(f_\varepsilon(\zeta)) \le \bar{q}(\zeta)$. Hence it is enough to show that $\|f_\varepsilon(\zeta)-\xi\| \le \|\zeta-\xi\|$.

As a general property for normal traces on von Neumann algebras, there exists a positive Radon measure ν on $\mathbb{R}\times\mathbb{R}$, with support in spec($\zeta$) x spec($\xi$) and such that $f(\zeta)g(\xi)$ belongs to $\mathcal{M}_\tau$ iff the function $f(s)g(t)$ belongs to $L^1(\nu)$, with

$$\tau[f(\zeta)g(\xi)] = \int f(s)g(t)d\nu(s,t)\ .$$

We get then

$$\tau[(f_\varepsilon(\zeta)-\xi)^2] = \int(f_\varepsilon(s)-t)^2 d\nu(s,t)$$

and, because $\mathrm{spec}(\xi) \subset [0, 1]$ and $|f_\varepsilon(s)-t| \le |s-t|$ whenever $0 \le t \le 1$,

$$\tau[(f_\varepsilon(\zeta)-\xi)^2] \le \int (s-t)^2 d\nu(s,t) = \tau[(\zeta-\xi)^2] = \|\zeta-\xi\|^2 . \quad \blacksquare$$

2.4. Lemma

η belongs to $\mathcal{N}_\tau^{sa}$ and, as an operator in $\mathbb{M}$, is positive with norm less than one.

Proof:

Let $\{\zeta_n\}_{n>0}$ be a sequence in $\mathcal{D}(q)$ which converges to η in $L^2(\mathbb{M}_{sa},\tau)$ with $\lim_{n\to\infty} q(\zeta_n) = \bar{q}(\eta)$: hence $\lim_{n\to\infty}\psi(\zeta_n) = \psi(\eta)$.

Lemmas 2.2 and 2.3 imply $\psi(\eta) \le \psi(f_\varepsilon(\zeta_n)) \le \psi(\zeta_n)$, hence $\lim_{n\to\infty}\psi(f_\varepsilon(\zeta_n)) = \psi(\eta)$ for any $\varepsilon>0$; which, by lemma 2.2, imply that, for any $\varepsilon>0$, the sequence $\{f_\varepsilon(\zeta_n)\}_{n>0}$ converges to η in $L^2(\mathbb{M}_{sa},\tau)$; which in turn implies that η belongs to $\mathcal{N}_\tau^{sa}$ with spectrum in $[-\varepsilon, 1+\varepsilon]$, for any $\varepsilon>0$; hence the result. $\blacksquare$

What we have got with those three lemmas is the following result:

If Q is the positive self adjoint operator associated with the closure of a Dirichlet form in $L^2(\mathbb{M}_{sa},\tau)$, then the operator $(1+Q)^{-1}$ maps $\mathcal{N}_\tau^{sa}$ into itself and extends to a positive normal contraction of the von Neumann algebra $\mathbb{M}$.

2.5. Proposition

Let Q be the self adjoint positive operator associated with the closure of a Dirichlet form on the von Neumann algebra $\mathbb{M}$. Then, for any $t \ge 0$, the operator e^{-tQ} of $L^2(\mathbb{M},\tau)$ maps $\mathcal{N}_\tau$ onto itself and extends to a normal positive contraction of $\mathbb{M}$.

In other words, the operator $(-Q)$ appears as the generator of a σ-weakly pointwise continuous semigroup of normal positive contractions of the von Neumann algebra $\mathbb{M}$.

Proof:

For any $\lambda > 0$, λq is a Dirichlet form, and $(1+\lambda Q)^{-1}$ maps $\mathcal{N}_\tau$ into itself, extending to a normal positive contraction of $\mathbb{M}$.

Write

$$e^{-tQ} = \lim_{\lambda\to 0} e^{-\left(\frac{tQ}{1+\lambda Q}\right)} = \lim_{\lambda\to 0} e^{-t/\lambda} \sum_{n=0}^{\infty} \frac{t^n}{\lambda^n n!} (1+\lambda Q)^{-n}$$

(cf. [2],1.3.2) to get the result.

2.6. Remark

The positivity of the semigroup $\{e^{-tQ}\}$ obtained in proposition 2.5. is only a partial conclusion: what we are looking for is a *quantum semigroup*, that is a semigroup of *completely positive* normal contractions of $\mathbb{M}$. One could of course define, in a more or less obvious way, something like a "completely Dirichlet form" which would lead to such a semigroup. However, proposition 2.5 is enough for most concrete examples, for instance the forthcoming one (the Dirichlet form associated with a differential calculus) where the step from positivity to complete positivity does not raise any difficulty.

3. Differential calculus and Dirichlet forms

3.1. Terminology and notations

We start with a C^*-algebra $\mathbb{A}$ and a l.s.c. faithful trace τ on $\mathbb{A}$. The ideals $\mathcal{M}_\tau$ and $\mathcal{N}_\tau$, the real and complex Hilbert spaces $L^2(\mathbb{A}_{sa},\tau)$ and $L^2(\mathbb{A},\tau)$, and the involution J are defined as in 1.1.2.

Let π_τ be the representation of $\mathbb{A}$ in $\mathcal{L}(L^2(\mathbb{A},\tau))$ extending the left ideal structure of $\mathcal{N}_\tau$, and $\pi_\tau(\mathbb{A})''$ the von Neumann algebra which is the σ-weak closure of $\pi_\tau(\mathbb{A})$ in $\mathcal{L}(L^2(\mathbb{A},\tau))$. Then τ extends to a faithful trace $\bar{\tau}$ on $\pi_\tau(\mathbb{A})''$.

We assume to be given an $\mathbb{A}$-$\mathbb{A}$ C^*-bimodule $\mathcal{E}$ and a derivation ∂ from a dense $*$-algebra $\mathbb{A}_\infty$ of $\mathbb{A}$ into $\mathcal{E}$, that is a linear map from $\mathbb{A}_\infty$ into $\mathcal{E}$ satisfying the Leibnitz rule

$$\partial(ab) = a\partial(b) + \partial(a)b, \ \forall\ a,b \in \mathbb{A}_\infty .$$

By definition of a C^*-bimodule, $\mathcal{E}$ is equipped with an $\mathbb{A}$-valued right-$\mathbb{A}$-linear positive definite scalar product $[\eta,\xi \in \mathcal{E} \to \langle\eta,\xi\rangle_{\mathbb{A}} \in \mathbb{A}]$, and $\mathcal{N}_{\mathcal{E}} = \{ \xi \in \mathcal{E} \ /\ \tau(\langle\xi,\xi\rangle_{\mathbb{A}}) < \infty \}$ is a sub $\mathbb{A}$-$\mathbb{A}$-bimodule of $\mathcal{E}$, dense whenever $\mathcal{N}_\tau$ is dense in $\mathbb{A}$, and a pre-hilbert space for the scalar product

$$\langle\eta,\xi\rangle_\tau = \tau(\langle\eta,\xi\rangle_{\mathbb{A}}),\ \ \eta,\xi \in \mathcal{N}_{\mathcal{E}} .$$

The completed Hilbert space will be denoted $L^2(\mathcal{E},\tau)$ and is again an $\mathbb{A}$-$\mathbb{A}$-bimodule.

3.2. Assumptions

The first assumption is that $\mathbb{A}_\infty$ is included in $\mathcal{N}_\tau$, and $\partial(\mathbb{A}_\infty)$ in $\mathcal{N}_{\mathcal{E}}$: for any a in $\mathbb{A}_\infty$, one has $\tau(a^*a) < \infty$ and $\tau(\langle\partial a,\partial a\rangle_{\mathbb{A}}) < \infty$.

The second and main assumption is that ∂, viewed as a densely defined linear operator from $L^2(\mathbb{A},\tau)$ into $L^2(\mathcal{E},\tau)$, is closable. Or, equivalently, that the quadratic form q densely defined on $L^2(\mathbb{A},\tau)$, with domain $\mathbb{A}_\infty$, by

$$q(a) = \tau(\langle\partial a,\partial a\rangle_{\mathbb{A}}) = \langle\partial a,\partial a\rangle_\tau,\ \ a \in \mathbb{A}_\infty ,$$

is closable.

[We shall from now on denote again by ∂ the closed operator from $L^2(\mathbb{A},\tau)$ into $L^2(\mathcal{E},\tau)$; ∂^* will be its adjoint, as a closed operator from $L^2(\mathcal{E},\tau)$ into $L^2(\mathbb{A},\tau)$.]

The third assumption is that the quadratic form q (whose domain is J-invariant) is real, that is $q(J\xi) = q(\xi)$, $\forall\ \xi\in\mathcal{D}(q)$, which reads

$$\tau(\ \langle\partial(a^*),\partial(a^*)\rangle_{\mathbb{A}}) = \tau(\ \langle\partial a,\partial a\rangle_{\mathbb{A}}),\ \forall\ a \in \mathbb{A}_\infty .$$

Thus, q appears as a closable quadratic form on $L^2(\mathbb{A}_{sa},\tau)$, and our purpose is to show that it is actually a Dirichlet form, and that the operator $\Delta = -\partial^*\partial$ is the generator of a quantum semigroup on the von Neumann algebra $\pi_\tau(\mathbb{A})''$.

3.3. Derivations and functional calculus

3.3.1 Notations

Fix a self adjoint element a of $\mathbb{A}_\infty$; let $X \subset \mathbb{R}$ be the spectrum of a: then one gets two representations λ and ρ from C(X) into $\mathcal{L}(L^2(\mathcal{E},\tau))$ defined respectively by

$$\lambda(f).\xi = f(a).\xi \qquad \text{and} \quad \rho(f).\xi = \xi.f(a), \qquad \forall f\in C(X),\ \forall\xi\in L^2(\mathcal{E},\tau),$$

where f(a) is well defined (by functional calculus) as an element of $\tilde{\mathbb{A}}$, the C^*-algebra deduced from $\mathbb{A}$ by adjoining a unit.

From the two commuting representations λ and ρ of C(X) one deduces a representation $\lambda\otimes\rho$ from $C(X)\otimes C(X) = C(X\times X)$ into $\mathcal{L}(L^2(\mathcal{E},\tau))$.

3.3.2. Notation

Let f be a C^1-function defined in a neighbourhood of X: then $\tilde{f}$ will denote the function on XxX defined by

$$\tilde{f}(s,t) = \begin{cases} f'(s) & \text{if } s=t \\ \dfrac{f(s)-f(t)}{s-t} & \text{if } s\neq t \end{cases} .$$

3.3.3. Lemma

With the above notations, if f is a C^1-function on $\mathbb{R}$ with f(0)=0, then f(a) belongs to the domain of ∂, and

$$\partial(f(a)) = \lambda\otimes\rho(\tilde{f}).\partial a .$$

Proof:

As noticed in 1.2.1, for f as above, f(a) belongs to $\mathcal{N}_\tau$. By density of polynomials in the C^1-functional calculus on compact subsets of $\mathbb{R}$, and closability of ∂, it is enough to check the lemma for f a monomial of non zero degree, that is $f(t) = t^n$, $n > 0$.

We get then

$$\partial(a^n) = \textstyle\sum_{p=0}^{n-1} a^p.\partial a.a^{n-1-p} = \lambda\otimes\rho(\phi).\partial a ,$$

where ϕ is the function $[(s,t) \to \sum_{p=0}^{n-1} s^p t^{n-1-p}]$ on $\mathbb{R}\times\mathbb{R}$, which is obviously equal to $\tilde{f}$. ∎

3.4. Proposition

With the notations and assumptions made in 3.1 and 3.2 above, the quadratic form q is a Dirichlet form.

Proof:

For a in $\mathbb{A}_\infty$, $a=a^*$, and f_ε defined in 1.2.2, then, by lemma 3.3.3, $f_\varepsilon(a)$ belongs to $\mathcal{N}_\tau^{sa}$ and its image in $L^2(\mathbb{A}_{sa},\tau)$ belongs to $\mathcal{D}(\bar{q})$, with

$$\bar{q}(f_\varepsilon(a)) = \langle\lambda\otimes\rho(\tilde{f}_\varepsilon).\partial a, \lambda\otimes\rho(\tilde{f}_\varepsilon).\partial a\rangle \le \langle\partial a, \partial a\rangle_\tau = q(a)$$

because of $|\tilde{f}_\varepsilon(s,t)| \le 1$, $\forall\, s,t \in \mathbb{R}$ [the scalar products being taken in $L^2(\mathcal{E},\tau)$]. ∎

3.5. Corollary

For any $t\ge0$, the operator $e^{t\Delta}$ on $L^2(\mathbb{A},\tau)$ [with $\Delta = -\partial^*\partial$] maps the ideal $\mathcal{N}_{\bar{\tau}}$ of $\pi_\tau(\mathbb{A})''$ onto itself, and extends to a completely positive normal contraction of $\pi_\tau(\mathbb{A})''$.

In other words, Δ appears as the infinitesimal generator of a pointwise σ-weakly continuous semigroup of completely positive normal contractions of the von Neumann algebra $\pi_\tau(\mathbb{A})''$.

Proof:

By proposition 2.5, $e^{t\Delta}$ maps $\mathcal{N}_{\bar{\tau}}$ into itself and extends to a positive normal contraction of the von Neumann algebra $\pi_\tau(\mathbb{A})''$. To get complete positivity, apply this result to the algebra $\mathbb{A}\otimes\mathbb{F}_n$ and the derivation $\partial\otimes i_n$, where $\mathbb{F}_n$ is the algebra of nxn matrices with complex entries, and i_n the identity map of $\mathbb{F}_n$. ∎

4. Example: the heat semigroup of a Riemannian foliation C^*-algebra

Our presentation will be the simplest one, though not the most intrinsic. We omit the proofs, which will be given in a forthcoming paper.

4.1. Foliated Riemannian manifolds

4.1.1. Bundles of a foliated Riemannian manifold

We start with a compact foliated manifold $(V,\mathbb{F})$ equipped with a Riemannian structure (cf [1],[5]). That is

- V is a compact C^∞ n-dimensional Riemannian manifold;
- $\mathbb{F} = \{\mathbb{F}_x\}_{x\in V}$ is a C^∞ p-dimensional integrable subbundle of the tangent bundle T(V); the leaves of the foliations are the maximal connected integral

manifolds of $\mathbb{F}$;

- the transverse bundle is $\mathcal{T} = T(V)/\mathbb{F}$ [quotient bundle], of dimension q=n-p, and the cotransverse bundle $\mathcal{T}^*$ is a subbundle of the cotangent bundle $T^*(V)$; identifying $\mathcal{T}$ with the orthogonal of $\mathbb{F}$ (through the euclidian structure of $T(V)$), then $T(V)$ splits as the direct orthogonal sum of $\mathbb{F}$ and $\mathcal{T}$, $T(V) = \mathbb{F} \oplus \mathcal{T}$; similarly $T^*(V) = \mathbb{F}^* \oplus \mathcal{T}^*$. Sections of the bundle $\mathbb{F}$ are the *longitudinal vector fields;* sections of $\mathcal{T}^*$ are the *transverse differential forms* .

4.1.2. Local coordinates

By convention, a system $(x_1,\ldots,x_n)$ of local coordinates on V will always be implicitely assumed adapted to the foliation, where the first coordinates $(x_1,\ldots,x_p)$ are the longitudinal coordinates, and the last ones $(x_{p+1},\ldots,x_n)$ are the transverse coordinates.

Then $(\partial/\partial x_1,\ldots,\partial/\partial x_p)$ appears as a local field of basis for $\mathbb{F}$, while $(dx_{p+1},\ldots,dx_n)$ appears as a local field of basis for $\mathcal{T}^*$.

4.1.3. Transverse Riemannian structure (cf. [5])

The Riemannian structure of V induces a transverse Riemannian structure on the foliated manifold $(V,\mathbb{F})$, i.e. a structure of Riemannian foliation, if the following property holds:

If γ is a path with origin x and extremity y, tangent to the bundle $\mathbb{F}$ (i.e. $\gamma'(t)\in\mathbb{F}_{\gamma(t)}$, $\forall$ t), then the infinitesimal holonomy $dh(\gamma)$ of γ (cf. [7]) is an isometry from $\mathcal{T}_x$ onto $\mathcal{T}_y$.

In local coordinates (cf. convention 4.1.2), this properties reads:

for j,k > p, the scalar product $(dx_j|dx_k)$ in $\mathcal{T}^*$ depends only on the transverse coordinates $(x_{p+1},\ldots,x_n)$.

4.2. The foliation C^*-algebra

4.2.1. The holonomy graph of the foliation (cf. [1],[7])

Let γ be a path drawn on the leaf manifold, that is a path tangent to th foliation bundle $\mathbb{F}$: its origin will be $x=s(\gamma)$, its extremity $y=r(\gamma)$, whic will be written shortly $\gamma:x\to y$.

Let $\mathcal{G}$ be the graph of the foliation , i.e. the groupoid of equivalen classes of paths tangent to the foliation bundle $\mathbb{F}$, where two paths ar identified if they have the same origin, the same extremity, and induce th same holonomy map.

$\mathcal{G}$ is a n+p dimensional manifold and the map (s,r) from $\mathcal{G}$ into VxV is local submersion, through which the tangent bundle of $\mathcal{G}$ at $\gamma:x\to y$, i identified with $\mathbb{F}_x \oplus \mathbb{F}_y \oplus \mathcal{T}_\gamma$, where $\mathcal{T}_\gamma$ is isomorphic both to $\mathcal{T}_x$ and $\mathcal{T}_y$, thos two spaces being in turn identified through the infinitesimal holonomy dh(γ

of γ. So that $\mathcal{G}$ is equipped with a Riemannian structure inherited from the euclidian structures of $\mathbb{F}$ and $\mathcal{T}$.

Similarly, the cotangent bundle $T^*(\mathcal{G})$ splits into three orthogonal subbundles: $T^*_\gamma(\mathcal{G}) = \mathbb{F}^*_x \oplus \mathbb{F}^*_y \oplus \mathcal{T}^*_\gamma$, where the euclidian space $\mathcal{T}^*_\gamma$ can be identified indiferently with $\mathcal{T}^*_x$ or $\mathcal{T}^*_y$.

4.2.2 Foliation algebras (cf.**[1]**)

Let $\mathbb{A}_\infty$ be the space of complex valued C^∞ functions on $\mathcal{G}$ with compact support, equipped with the following structure of *-algebra:

the involution is defined by

$$f^*(\gamma) = \overline{f(\gamma^{-1})}$$

the product is defined by

$$f*g\ (\gamma) = \int_{\mathcal{G}^y} f(\gamma')g(\gamma'^{-1}\gamma)d\nu^y(\gamma')$$

where ν^y is the Riemannian measure on the submanifold $\mathcal{G}^y=\{\gamma'/r(\gamma')=y\}$ of $\mathcal{G}$.

$\mathbb{A}_\infty$ is equipped with a trace τ defined by

$$\tau(f) = \int_V f(\tilde{x})d\mu(x)$$

where μ is the Riemannian measure on V and, for $x \in V$, $\tilde{x}$ is the trivial path from x to x. The associated Hilbert space $L^2(\mathbb{A}_\infty,\tau)$ is canonically identified with the L^2-space of the Riemannian manifold $\mathcal{G}$, and the left multiplication of $\mathbb{A}_\infty$ onto itself extends to a *-representation π_τ of $\mathbb{A}_\infty$ into $\mathcal{L}(L^2(\mathbb{A}_\infty,\tau))$.

The C^* algebra $\mathbb{A} = C^*(V,\mathbb{F})$ of the foliation is the norm closure of $\pi_\tau(\mathbb{A}_\infty)$ in $\mathcal{L}(L^2(\mathbb{A}_\infty,\tau))$.

The von Neumann algebra $\mathbb{M} = W^*(V,\mathbb{F})$ of the foliation is its σ-weak closure, i.e. $\mathbb{M} = \pi_\tau(\mathbb{A}_\infty)''$.

4.3. The cotangent bimodule

Let $\mathcal{E}_\infty$ be the complexified space of C^∞ sections with compact support of the bundle $\mathcal{T}^*(\mathcal{G}) = \{\mathcal{T}^*_\gamma\}_{\gamma\in\mathcal{G}}$ [transverse differential forms with compact support on $\mathcal{G}$]. As an element of $\mathcal{T}^*_\gamma$ (γ:x→y) can be understood as belonging either to $\mathcal{T}^*_x$ or to $\mathcal{T}^*_y$ (cf.4.2.1), $\mathcal{E}_\infty$ is equipped with an obvious structure of $\mathbb{A}_\infty$-$\mathbb{A}_\infty$ bimodule, where left and right multiplication are defined by

$$\omega*f\ (\gamma) = \int_{\mathcal{G}^y} \omega(\gamma')f(\gamma'^{-1}\gamma)d\nu^y(\gamma') \quad [\omega(\gamma') \in \mathcal{T}^*_{y'},\ \forall\gamma'\in\mathcal{G}^y]$$

and γ:x→y

$$f*\omega\ (\gamma) = \int_{\mathcal{G}^y} f(\gamma')\omega(\gamma'^{-1}\gamma)d\nu^y(\gamma') \quad [\omega(\gamma'^{-1}\gamma) \in \mathcal{T}^*_{x'},\ \forall\gamma'\in\mathcal{G}^y].$$

Moreover, $\mathcal{E}_\infty$ is equipped with an $\mathbb{A}_\infty$-valued $\mathbb{A}_\infty$-right linear scalar product, where, for ω', ω belonging to $\mathcal{E}_\infty$, $\langle\omega', \omega\rangle_{\mathbb{A}_\infty}$ is the function

$$\gamma \to \langle\omega', \omega\rangle_{\mathbb{A}_\infty}(\gamma) = \int_{\mathcal{G}_y} (\omega'(\gamma')|\omega(\gamma'\gamma))d\nu_y(\gamma') ;$$

where ν_y is the Riemannian measure on the submanifold $\mathcal{G}_y=\{\gamma'/s(\gamma')=y\}$ of $\mathcal{G}$, and where the scalar product $(\omega'(\gamma')|\omega(\gamma'\gamma))$ is evaluated in $\mathcal{T}^*_{r(\gamma')}$.

$\mathcal{E}_\infty$ is a Hilbert pre C^*-$\mathbb{A}$-$\mathbb{A}$ bimodule (cf. **[4]**) and the completed Hilbert C^*-$\mathbb{A}$-$\mathbb{A}$ bimodule $\mathcal{E}$ is what can be called *the transverse cotangent bimodule* of the C^*-algebra $\mathbb{A}$. The next step is to build a derivation ∂ from $\mathbb{A}_\infty$ into $\mathcal{E}_\infty$ in order to get a differential calculus, to which will be associated in turn a transverse divergence $(-\partial^*)$, a transverse Laplacian $\Delta^{\mathcal{T}} = -\partial^*\partial$, and a transverse heat semigroup $\{e^{t\Delta^{\mathcal{T}}}\}_{t\geq 0}$.

4.4. Transverse differentiation

4.4.1. Notation

Let f be a C^∞ function on V or $\mathcal{G}$; then $d^{\mathcal{T}}f$ will denote the orthogonal projection of the differential df of f onto the cotransverse bundle $\mathcal{T}^*$ [$d^{\mathcal{T}}f$ is thus a transverse differential form].

4.4.2. Lemma

There exists a global C^∞ transverse differential form θ on V which, in any system $(x_1,\dots,x_n)$ of local coordinates (cf. 4.1.2), writes locally

$$\theta = \sum_{i=1}^{p} \mathcal{L}_{\partial/\partial x_i}(d^{\mathcal{T}}x_i) + \frac{1}{2}\, d^{\mathcal{T}}\mathrm{Log}|G_{11}| ,$$

where $|G_{11}|$ is the determinant of the leaf Riemannian matrix $G_{11} = \left((\partial/\partial x_i|\ \partial/\partial x_j)\right)_{i,j=1..p}$ and $\mathcal{L}_{\partial/\partial x_i}(d^{\mathcal{T}}x_i)$ is the Lie derivative of the transverse differential form $d^{\mathcal{T}}x_i$ with respect to the longitudinal vector field $\partial/\partial x_i$ (and is again a transverse differential form).

In some account, θ measures the non integrability of the transverse bundle $\mathcal{T}$, normal to the foliation.

4.4.3. Notation

To the transverse differential form θ on V are naturally associated two transverse differential forms $s^*\theta$ and $r^*\theta$ on $\mathcal{G}$, where, for $\gamma: x \to y$,

$s^*\theta\ (\gamma)$ is equal to θ(x) [identifying $\mathcal{T}^*_\gamma$ with $\mathcal{T}^*_x$]

$r^*\theta\ (\gamma)$ is equal to θ(y) [identifying $\mathcal{T}^*_\gamma$ with $\mathcal{T}^*_y$] .

Define a transverse differential form Θ on $\mathcal{G}$ by

$$\Theta = \frac{1}{2}(s^*\theta + r^*\theta) .$$

4.4.4. Proposition

1/ The linear map ∂ from $\mathbb{A}_\infty$ into $\mathcal{E}_\infty$ defined by

$$\partial f = d^{\mathcal{T}}f + f.\Theta ,\ f \in \mathbb{A}_\infty ,$$

is a derivation, i.e. satisfies the Leibnitz rule

$$\partial(f*g) = (\partial f)*g + f*(\partial g),\ \forall\ f,g \in \mathbb{A}_\infty .$$

2/ Wiewed as a densely defined operator from $L^2(\mathbb{A}_\infty,\tau)$ into $L^2(\mathcal{E}_\infty,\tau)$, ∂ is closable; $\mathcal{E}_\infty$ is an essential domain for ∂^*, on which it satisfies

$$\partial^*\omega = d^*\omega + (\ \Theta \mid \omega\ ),\ \forall\ \omega \in \mathcal{E}_\infty .$$

3/ The operator $\Delta^{\mathcal{T}} = -\partial^*\partial$ admits $\mathbb{A}_\infty$ as an essential domain, on which it writes as a diffusion operator:

$$\Delta^{\mathcal{T}}(f) = d^*d^{\mathcal{T}}f + f.\partial^*\Theta\ ,\ \forall\ f \in \mathbb{A}_\infty .$$

As a corollary of proposition 2.5, $\Delta^{\mathcal{T}}$ is the generator of a semigroup $\{e^{t\Delta^{\mathcal{T}}}\}_{t\geq 0}$ [the *transverse heat semigroup*] of completely positive normal contractions of the Riemannian foliation von Neumann algebra $\mathbb{M} = W^*(V,\mathbb{F})$.

4.5. Concluding remarks

Proposition 4.4.4. is only a first approach to the transverse Laplacian and transverse heat semigroup, which deserve further attention. Here are some questions about this transverse heat semigroup which will be studied in a forthcoming paper:

- is it Markov (i.e. $e^{t\Delta^{\mathcal{T}}}(1_{\mathbb{M}}) = 1_{\mathbb{M}},\ \forall t\geq 0$) ? is it a C^*-semigroup (i.e. $e^{t\Delta^{\mathcal{T}}}(C^*(V,\mathbb{F})) \subset C^*(V,\mathbb{F})\ \forall t\geq 0$) ? [to those two questions a positive answer will be given]
- has this transverse heat semigroup some regularizing property ? does any $e^{t\Delta^{\mathcal{T}}}$ map $W^*(V,\mathbb{F})$ into the multipliers of $C^*(V,\mathbb{F})$? etc.

Notice that the Laplacian $\Delta^{\mathcal{T}}$ and the heat semigroup depend only on the transverse structure (i.e. the Euclidian structure on the bundle $\mathcal{T}$ and the projection from T(V) on $\mathcal{T}$), and could be defined intrinsically, by use of half density bundles, without any reference to an Euclidian structure on the bundle $\mathbb{F}$.

Remark: While this paper was waiting for publication, the author developped the investigation of the foliation semigroup above, and obtained the result that the tranverse heat semigroup is a Markov C^*-semigroup whenever the riemannian foliation is C^∞ and the manifold compact; the reader will find a better and more intrinsic exposition of this semigroup in a note already published in the *Comptes-Rendus de l'Académie des Sciences* (série I, tome 310 n°7, 1990, p.531-537), and the complete proofs in the corresponding detailed preprint.

References

[1] Connes A., A survey of foliations and von Neumann algebras, *Proc.Symp.in Pure Maths., A.M.S.*, vol.38 (1982), p.521-628

[2] Fukushima M., *Dirichlet forms and Markov processes*, North Holland Mathematical Library n°23 (1980)

[3] Haagerup U., Operator valued weights in von Neumann algebras, I, *J.Func.Anal.* 32 (1979) n°2, p.175-206.

[4] Kasparov G.G., Hilbert C^*-modules, *J.Op.Th.*4 (1980), p.133-150

[5] Molino P., *Feuilletages riemanniens*, Université du Languedoc, Montpellier, 1983

[6] Sauvageot J.-L., Tangent bimodule and locality for dissipative operators on C^*-algebras, *Quantum Probability and Appl. IV*, *Lecture notes in Maths.* n°1396, 1989, p.322-338.

[7] Winkelnkemper H.E., The graph of a foliation, *Ann.Glob.Anal.Geom.* 1 (1983) n°3, p.51-75

This paper is in final form and no similar paper has been or is being submitted elsewhere.

Gaussian States on Bialgebras

Michael Schürmann
Institut für Angewandte Mathematik
Universität Heidelberg
Im Neuenheimer Feld 294
D-6900 Heidelberg
Federal Republic of Germany

1. Introduction. A stochastic increment process taking values in a semi-group G is a stochastic process

$$X_{st} : \Omega \to G$$

defined on a probability space $(\Omega, \mathcal{E}, P)$, indexed by pairs $(s,t) \in \mathbb{R}_+^2$, $s \leq t$, such that

$$X_{rs} X_{st} = X_{rt} \text{ for } r \leq s \leq t \tag{1}$$

and

$$X_{tt} = e, \; e \text{ the unit element of } G. \tag{2}$$

(If G is a group and $(X_t)_{t \in \mathbb{R}_+}$ is a G-valued stochastic process an example of an increment process is obtained by setting $X_{st} = (X_s)^{-1} X_t$.) To explain the passage to non-commutativity, we apply a 'Fourier transformation' by, rather than G itself, considering the algebra $\mathrm{R}(G)$ of complex-valued functions on G formed by the coefficients of finite-dimensional representations of G. This algebra is a *-algebra with complex conjugation of functions as the involution. The *-algebra $\mathrm{R}(G)$ reflects the correct positivity structure, but it does not reflect the semi-group structure of G. Therefore, some additional structure is needed. This is build in in the *-bialgebra $\mathrm{R}(G)$ which is the *-algebra $\mathrm{R}(G)$ together with the mappings

$$\Delta : \mathrm{R}(G) \to \mathrm{R}(G) \otimes \mathrm{R}(G)$$

and

$$\delta : \mathrm{R}(G) \to \mathbb{C}$$

given by

$$\Delta f(x,y) = f(xy)$$

and

$$\delta f = f(e)$$

where we identified $\mathrm{R}(G) \otimes \mathrm{R}(G)$ with its image under the injective linear mapping from $\mathrm{R}(G) \otimes \mathrm{R}(G)$ to $\mathrm{R}(G \times G)$ which maps $f \otimes g$ to the function $(x,y) \mapsto f(x)g(y)$. Both Δ and δ are *-algebra homomorphisms. The general notion of a *-bialgebra comprises a *-algebra $\mathcal{B}$ together with *-algebra homomorphisms

$$\Delta : \mathcal{B} \to \mathcal{B} \otimes \mathcal{B}$$

and

$$\delta : \mathcal{B} \to \mathbb{C}$$

satisfying the coassociativity rule

$$(\Delta \otimes \mathrm{id}) \circ \Delta = (\mathrm{id} \otimes \Delta) \circ \Delta$$

and the counit conditions

$$(\delta \otimes \mathrm{id}) \circ \Delta = \mathrm{id} = (\mathrm{id} \otimes \delta) \circ \Delta.$$

For an algebra $\mathcal{A}$ we can turn the vector space $\mathrm{L}(\mathcal{B}, \mathcal{A})$ of linear operators from $\mathcal{B}$ to $\mathcal{A}$ into an algebra by defining the product by

$$R \star S = \mathrm{M} \circ (R \otimes S) \circ \Delta$$

where $\mathrm{M} : \mathcal{A} \otimes \mathcal{A} \to \mathcal{A}$ denotes the multiplication on $\mathcal{A}$. A unit of $\mathrm{L}(\mathcal{B}, \mathcal{A})$ is given by $a \mapsto \delta(a)\mathbf{1}$. In particular, the algebraic dual space $\mathcal{B}^*$ of $\mathcal{B}$ becomes an algebra with unit δ. A *-bialgebra carries all the structure needed to make sense of a quantum stochastic independent stationary increment process over $\mathcal{B}$; see [2]. The latter is a family j_{st} of *-algebra homomorphisms from $\mathcal{B}$ to an algebra $\mathcal{A}$ where the index set is again the set of pairs $(s, t) \in \mathbb{R}_+^2$, $s \leq t$. Moreover, there is also given a state Φ on $\mathcal{A}$ and the following conditions are required to hold

(a) (increment property)

$$j_{rs} \star j_{st} = j_{rt} \text{ for } r \leq s \leq t$$
$$j_{tt} = \delta \mathbf{1}$$

(b) (independence and stationarity of increments)

(b1)

$$\Phi(j_{t_1 t_2}(a_1) \ldots j_{t_n t_{n+1}}(a_n)) = \varphi_{t_2 - t_1}(a_1) \ldots \varphi_{t_{n+1} - t_n}(a_n)$$

for $n \in \mathbb{N}$, $t_l \in \mathbb{R}_+$, $t_1 < \cdots < t_{n+1}$, $a_1, \ldots, a_n \in \mathcal{B}$, where φ_t, $t \in \mathbb{R}_+$, denotes the state $\Phi \circ j_{0t}$ on $\mathcal{B}$.

(b2)

$$[j_{st}(a), j_{s't'}(a')] = 0$$

for all $a, a' \in \mathcal{B}$ and all s, t, s', t' such that the open intervals (s, t) and (s', t') are disjoint.

(c) (weak continuity)

$$\lim_{t \downarrow 0} \varphi_t(a) = \delta(a) \text{ for all } a \in \mathcal{B}.$$

We say that two processes $j_{st}^{(1)}$ and $j_{st}^{(2)}$ are equivalent if

$$\Phi^{(1)}(j_{s_1 t_1}^{(1)}(a_1) \ldots j_{s_n t_n}^{(1)}(a_n)) = \Phi^{(2)}(j_{s_1 t_1}^{(2)}(a_1) \ldots j_{s_n t_n}^{(2)}(a_n))$$

for all choices of $n \in \mathbb{N}$, $(s_l, t_l) \in \mathbb{R}_+^2$, $s_l \leq t_l$, and $a_l \in \mathcal{B}$.

In the case of a G-valued stochastic increment process X_{st} (from now on called the 'classical' case) we set

$$\mathcal{B} = \mathrm{R}(G) \cap L^\infty(G),$$
$$\mathcal{A} = L^\infty(\Omega, \mathcal{E}, P),$$
$$\Phi = P,$$
$$j_{st}(f) = f \circ X_{st}.$$

In the classical case condition (a) is just the translation of (1) and (2). Condition (b1) is the independence and stationarity of the increments X_{st} in the usual sense of calssical probability theory whereas (b2) becomes trivial. Condition (c) guarantees the continuity of the convolution semi-group φ_t of probability measures on G and, hence, the existence of an infinitesimal generator. Our notion of equivalence of processes reduces to stochastic equivalence of classical stochastic processes, i.e. two classical stochastic processes are equivalent if their finite-dimensional distributions agree.

An important fact about an independent stationary increment process is that such a process is, up to equivalence, determined by its infinitesimal generator, which is a linear functional ψ on the underlying *-bialgebra $\mathcal{B}$. It is given as the pointwise derivative of the convolution semi-group φ_t at 0, that is

$$\psi(a) = \frac{\mathrm{d}}{\mathrm{d}t}\varphi_t(a)|_{t=0}.$$

The derivative exists as a consequence of condition (c) and of the fundamental theorem on coalgebras; see [1, 16]. Moreover, we have

$$\varphi_t = \exp_\star t\psi$$

where $\exp_\star$ denotes the convolution exponential series which exists as a pointwise limit for all elements in $\mathcal{B}^*$; see [13]. The set of generators of independent stationary increment processes coincides with the set of conditionally positive hermitian linear functionals on $\mathcal{B}$, where we say that $\psi \in \mathcal{B}^*$ is conditionally positive if it is positive on the kernel K^1 of the counit δ (i.e. $\psi(a^*a) \geq 0$ for all $a \in \mathcal{B}$ with $\delta(a) = 0$) and if $\psi(1) = 0$, and where we say that $\psi \in \mathcal{B}^*$ is hermitian if $\psi(a^*) = \overline{\psi(a)}$ for all $a \in \mathcal{B}$.

Let G be a compact real Lie group and denote by $\mathrm{K}(G)$ the coefficient algebra of G, that is the algebra formed by the coefficients of continuous irreducible representations of G; see [7]. $\mathrm{K}(G)$ is a sub-*-bialgebra of $\mathrm{R}(G)$. By the Stone-Weierstrass theorem probability measures on G are determined by their values on elements of $\mathrm{K}(G)$. Furthermore, the weakly continuous convolution semigroups φ_t of probability measures on G are, if restricted to $\mathrm{K}(G)$, precisely the convolution semi-groups of the form $\varphi_t = \exp_\star t\psi$ with ψ conditionally positive and hermitian. Thus, for a compact Lie group the classical theory is completely covered by our theory of independent, stationary increment processes over *-bialgebras. As a consequence of Hunt's formula [9] a generator ψ can always be decomposed into a 'maximal quadratic' part and a part with no 'quadratic component' in the following sense. One has

$$\psi(f) = \sum_{k=1}^{d} \alpha_k \mathrm{X}_k(f)(e) + \sum_{k,l=1}^{d} \alpha_{kl} \mathrm{X}_k \mathrm{X}_l(f)(e) + r(f)$$

where $X_1, \ldots, X_d$ is a basis of the Lie algebra $\mathcal{G}$ of G, $\alpha_1, \ldots, \alpha_d$ are real numbers, $(\alpha_{kl})_{k,l=1,\ldots,d}$ is a positive semi-definite real $d \times d$-matrix and r is another generator which *cannot* be decomposed into

$$f \mapsto Y^2(f)(e) + r'(f)$$

with $Y \in \mathcal{G}$, $Y \neq 0$, and r' a generator. In the sense of the present paper, $\psi - r$ is a maximal quadratic component of ψ. It is uniquely determined by ψ up to adding a term of the form $f \mapsto Y(f)(e)$ with $Y \in \mathcal{G}$.

In this paper, we generalize this result. First we say what we mean by a quadratic linear functional on a *-bialgebra. In Section 2 we give the motivation for our general definition. In Section 3 we find a cohomological condition on the *-bialgebra which guarantees the existence of maximal quadratic components. We show that commutative *-bialgebras satisfy this condition which, from the point of view of *-bialgebras, explains why Hunt's theorem holds. Next we prove that the so-called 'non-commutative analogue of the coefficient algebra of the unitary group' (see [5]) has this property, too. This is remarkable, because this *-bialgebra is neither commutative nor cocommutative. In Section 4 we apply the fact that free algebras always satisfy our condition (see also Theorem 4.3. of [2]) to prove that an infinitely divisible state on the formal algebra of the canonical commutation relations can be written, in a unique way, as the convolution of a gaussian (i.e. quasi free) state and a 'classical' state with no gaussian component.

2. Motivation. We begin with some examples.

2.1. Let γ_Q be a gaussian distribution on $\mathbb{R}^d$ with covariance matrix Q. Denote the restriction of γ_Q to the *-algebra $\mathbb{C}[d] = \mathbb{C}[x_1, \ldots, x_d] \subset L^1(\mathbb{R}^d, \gamma_Q)$ of polynomial functions on $\mathbb{R}^d$ with complex coefficients again by γ_Q. Then γ_Q is the convolution exponential of the linear functional g_Q on $\mathbb{C}[d]$ defined by $g_Q(x_k x_l) = Q_{kl}$ and $g_Q(M) = 0$ for a monomial M not of length 2.

2.2. A non-commutative generalization of 2.1. is obtained as follows. Let V be a complex vector space with a conjugation $v \mapsto v^*$, i.e. an antilinear mapping on V satisfying $(v^*)^* = v$. The conjugation can be extended to an involution of the tensor algebra

$$\mathrm{T}(V) = \mathbb{C} \oplus V \oplus (V \otimes V) \oplus \ldots$$

over V in a unique way. $\mathrm{T}(V)$ is turned into a *-bialgebra by extending Δ and δ with

$$\Delta(v) = v \otimes \mathbf{1} + \mathbf{1} \otimes v$$

and

$$\delta(v) = 0,$$

$v \in V$, to the whole of $\mathrm{T}(V)$ as *-algebra homomorphisms. A gaussian state γ_Q on $\mathrm{T}(V)$ is then a state of the form $\exp_\star g_Q$ where Q is a positive hermitian form on V and g_Q is defined by $g_Q(vw) = Q(v^*, w)$ for $v, w \in V$ and $g_Q(M) = 0$ for a monomial M not of length 2.

2.3. Let f_Q be the Fourier transform of the gaussian distribution γ_Q on $\mathbb{R}^d$. Then $f_Q(x) = \exp g_Q(x)$ with $g_Q(x) = -\frac{1}{2}\langle x, Qx\rangle$. Let G be a group. Functions on G can be identified with linear functionals on the group algebra $\mathbb{C}G$ of G in the obvious way. The pointwise product of functions comes from the bialgebra structure of $\mathbb{C}G$ given by

$\Delta(x) = x \otimes x$ and $\delta(x) = 1$, $x \in G$. If we extend $x \mapsto x^{-1}$ antilinearly $\mathbb{C}G$ becomes a *-bialgebra. In this sense, f_Q is the convolution exponential of g_Q.
2.4. Let G be a compact real Lie group. Gaussian distributions on G can be characterized as the states on the coefficient algebra $\mathrm{K}(G)$ of the form

$$\exp_\star(\sum \psi_j)$$

where the sum in the exponent is finite and each ψ_j is a linear functional on $\mathrm{K}(G)$ of the form

$$f \mapsto (\mathrm{D}^2 f)(e) = (\delta \circ \mathrm{D}^2)(f)$$

with D a left invariant *-derivation on G (i.e. D an element of $\mathcal{G}$).
2.5. We do not know how to construct a non-commutative analogue of $\mathrm{K}(G)$ for an arbitrary G, but for special cases we do. For example, in the case of the group $\mathcal{U}_d$ of unitary $d \times d$-matrices the coefficient algebra $\mathrm{K}[d] = \mathrm{K}(\mathcal{U}_d)$ of $\mathcal{U}_d$ can be described as the *commutative* *-algebra generated by indeterminates x_{kl}, $k, l = 1, \ldots, d$, with the relations

$$\sum_{n=1}^{d} x_{kn} x_{ln}^* = \delta_{kl} \tag{3}$$

$$\sum_{n=1}^{d} x_{nk}^* x_{nl} = \delta_{kl}$$

where we set $(x_{kl})^* = x_{kl}^*$. A non-commutative analogue $\mathrm{K}\langle d\rangle$ of $\mathrm{K}[d]$ is then the *-algebra generated by indeterminates x_{kl} satisfying the relations (3). $\mathrm{K}\langle d\rangle$ is a *-bialgebra if we define Δ and δ by extension of

$$\Delta(x_{kl}) = \sum_{n=1}^{d} x_{kn} \otimes x_{nl}$$

and

$$\delta(x_{kl}) = \delta_{kl}.$$

In [5] the *-bialgebra $\mathrm{K}\langle d\rangle$ was called the non-commutative analogue of the coefficient algebra of the unitary group. It may not be clear what a gaussian state on $\mathrm{K}\langle d\rangle$ should be.

In the cases 2.1.-2.4. gaussian states could be characterized as convolution exponentials of certain linear functionals. It can easily be seen that, up to a derivational component, the convolution logarithm ψ of a gaussian state is characterized by the conditions that ψ is conditionally positive and hermitian, and that ψ vanishes on all elements of the form abc with $a, b, c \in K^1$. (In 2.3. we actually have to make the additional asssumption that g_Q is continuous.)

We establish the connection to the quantum stochastic calculus of Hudson and Parthasarathy; see [8]. Given a conditionally positive linear functional ψ on a *-bialgebra $\mathcal{B}$ we define a hermitian form on $\mathcal{B}$ by

$$(a, b)_\psi = \psi((a - \delta(a)\mathbf{1})^*(b - \delta(b)\mathbf{1}));$$

see [12]. Denote by D_ψ the pre-Hilbert space obtained from $\mathcal{B}$ by dividing by the null space of the hermitian form, and let $\eta_\psi : \mathcal{B} \to D_\psi$ be the canonical mapping. A *-representation ρ_ψ of $\mathcal{B}$ on the pre-Hilbert space D_ψ is obtained by setting

$$\rho_\psi(a)\eta_\psi(b) = \eta_\psi(a(b - \delta(b)\mathbf{1})) = \eta_\psi(ab) - \eta_\psi(a)\delta(b).$$

Now we assume that ψ is also hermitian, so that ψ is the generator of some quantum independent, stationary increment process. Consider the quantum stochastic integral equation

$$j_{st}(a) = \delta(a)\mathrm{id} + \int_s^t (j_{s\tau} \star \mathrm{d}I_\tau^\psi)(a) \tag{4}$$

where

$$I_t^\psi(a) = A_t^\dagger(\eta(a)) + \Lambda_t(\rho(a) - \delta(a)\mathrm{id}) + A_t(\eta(a^*)) + \psi(a)t$$

and $A_t^\dagger$, $\Lambda_t(B)$ and $A_t(\xi)$ for a vector ξ in D_ψ and a linear operator B on D_ψ denote the creation, preservation and annihilation processes on the Bose Fock space $\mathcal{F}_\psi$ over $L^2(\mathbb{R}_+, H_\psi)$ with H_ψ the completion of D_ψ. It was proved in [4] that a solution of (4) always exists on a dense linear subspace of $\mathcal{F}$ spanned by certain exponential vectors. In a forthcoming paper it will be shown that the (in general unbounded but closable) operators $j_{st}(a)$ can be extended to an independent stationary increment process with generator ψ. The special cases $\mathrm{T}(V)$, $\mathbb{C}G$ and $\mathrm{K}\langle d\rangle$ have already been treated rigorously; see [13, 14, 4, 12]. We write down equation (4) for these cases. In the case of $\mathrm{T}(V)$ it becomes, for $v \in \mathrm{T}(V)$,

$$j_{st}(v) = A_{st}^\dagger(\eta(v)) + \Lambda_{st}(\rho(v)) + A_{st}(\eta(v)) + \psi(v)(t-s)$$

and $j_{st}(v)$ are combinations of creation, preservation, annihilation and scalar processes. In the case of $\mathbb{C}G$ we obtain for $x \in G$, using differential notation,

$$\begin{aligned} \mathrm{d}j_{st}(x) &= j_{st}(x)(\mathrm{d}A_t^\dagger(\eta(x)) + \mathrm{d}\Lambda_t(\rho(x) - \mathrm{id}) + \mathrm{d}A_t(\eta(x^{-1})) + \psi(x)\mathrm{d}t), \\ j_{tt}(a) &= \delta(a)\mathrm{id}. \end{aligned}$$

It is not difficult to check that the unitary operators on $\mathcal{F}$ given by

$$\begin{aligned} j_{st}(x)\mathrm{E}(f) = \exp((t-s)\psi(x) - \int_s^t \langle \eta(x), \rho(x) f(\tau)\rangle \mathrm{d}\tau) \\ \mathrm{E}(f\chi_{[0,s)} + (\rho(x) f\chi_{[s,t]} + \eta(x)\chi_{[s,t]}) + f\chi_{(t,\infty)}), \end{aligned}$$

where $\mathrm{E}(f)$ denotes the exponential vector of $f \in L^2(\mathbb{R}_+, H)$, satisfy these equations; see [6, 4]. For $\mathrm{K}\langle d\rangle$ we combine the operators $j_{st}(x_{kl})$ on $\mathcal{F}$ to an operator U_{st} on $\mathbb{C}^d \otimes \mathcal{F}$. For $\mathrm{N} \in M_d \otimes H$, $\mathrm{V} \in M_d \otimes B(H) = B(\mathbb{C}^d \otimes H)$ we set

$$\begin{aligned} (\mathrm{d}A^\dagger(\mathrm{N}))_{kl} &= \mathrm{d}A^\dagger(\mathrm{N}_{kl}), \\ (\mathrm{d}A(\mathrm{N}))_{kl} &= \mathrm{d}A(\mathrm{N}_{lk}), \\ (\mathrm{d}\Lambda(\mathrm{V}))_{kl} &= \mathrm{d}\Lambda(\mathrm{V}_{kl}). \end{aligned}$$

Equation (4) becomes

$$dU_{st} = U_{st}(dA_t(\mathrm{L}) + d\Lambda_t(\mathrm{W}-\mathbf{1}) - dA_t^\dagger(\mathrm{WL}) + (i\mathrm{H} - \frac{1}{2}\langle \mathrm{L},\mathrm{L}\rangle)dt), \qquad (5)$$
$$U_{tt} = \mathrm{id},$$

where $\mathrm{L} \in M_d \otimes H$, $\mathrm{W} \in M_d \otimes B(H)$ is unitary, and $\mathrm{H} \in M_d$ is selfadjoint with

$$\mathrm{L}_{kl} = \eta(x_{lk}{}^*),$$
$$\mathrm{W}_{kl} = \rho(x_{kl}),$$
$$(\langle \mathrm{L},\mathrm{L}\rangle)_{kl} = \sum_{n=1}^{d} \langle \mathrm{L}_{nk}, \mathrm{L}_{nl}\rangle,$$
$$\mathrm{H}_{kl} = -i(\psi(x_{kl}) + \frac{1}{2}(\langle \mathrm{L},\mathrm{L}\rangle)_{kl}).$$

It can be shown that $U_t = U_{0t}$ extend in a unique way to unitary operators on $\mathbb{C}^d \otimes \mathcal{F}$ and that $U_{st} = (U_s)^\dagger U_t$. In the special case $\dim H = 1$ the matrices L, W and H are all in M_d and (5) becomes

$$dU_{st} = U_{st}(\mathrm{L}^*\, dA_t - \mathrm{WL}\, dA_t^\dagger + (\mathrm{W}-\mathbf{1})\, d\Lambda_t + (i\mathrm{H} - \frac{1}{2}\mathrm{L}^*\mathrm{L})\, dt),$$
$$U_{tt} = \mathrm{id}.$$

Going back to the general formula (4), it seems reasonable to call a j_{st} a Gaussian process over $\mathcal{B}$ if the preservation integrator term $d\Lambda_r$ does not appear. This is the case if and only if $\rho(a) = \delta(a)\mathrm{id}$ for all $a \in K^1$.

3. Maximal quadratic components. As we do not need the comultiplication for most of this section, we start from a more general situation, namely a pair $(\mathcal{B},\delta)$ consisting of a *-algebra $\mathcal{B}$ and a *-algebra homomorphism δ from $\mathcal{B}$ to the field of complex numbers. The notion of conditional positivity and the construction of D_ψ, η_ψ and ρ_ψ out of a conditionally positive linear functional ψ on $\mathcal{B}$ are the same as before. For $n \in \mathbb{N}$ denote by K^n the linear span of the set $\{a_1 \ldots a_n : a_j \in K^1, j = 1,\ldots,n\}$. We call a conditionally positive hermitian linear functional on $\mathcal{B}$ *quadratic* if it satisfies one of the conditions of the following proposition.

3.1. PROPOSITION. *For a conditionally positive hermitian linear functional ψ on $\mathcal{B}$ the following are equivalent*

(i) $\psi \lceil K^3 = 0$
(ii) $\psi(a^*a) = 0$ *for all* $a \in K^2$
(iii)

$$\psi(abc) = \psi(ab)\delta(c) + \psi(ac)\delta(b) + \delta(a)\psi(bc) - \delta(a)\delta(b)\psi(c) - \delta(a)\psi(b)\delta(c) - \psi(a)\delta(b)\delta(c)$$

for all $a,b,c \in \mathcal{B}$.
(iv) $\rho_\psi \lceil K^1 = 0$
(v) $\rho_\psi(a) = \delta(a)\mathrm{id}$ *for all* $a \in \mathcal{B}$
(vi) $\eta_\psi \lceil K^2 = 0$

(vii) $\eta_\psi(ab) = \delta(a)\eta_\psi(b) + \eta_\psi(a)\delta(b)$ *for all* $a, b \in \mathcal{B}$.

PROOF: Straightforward.◇

We say that ψ is *degenerate* quadratic if ψ is hermitian and vanishes on K^2.

For convenience we use the following cohomological notation. Given a *-representation ρ of $\mathcal{B}$ on a pre-Hilbert space D we turn D into a $\mathcal{B}$-bimodule by setting

$$aub = \rho(a)(u)\delta(b)$$

for $a, b \in \mathcal{B}$ and $u \in D$. We denote by $Z^n(D,\rho)$ the n-cocycles and by $B^n(D,\rho)$ the n-coboundaries of the corresponding Hochschildt cohomology theory. Condition (vii) of Proposition 2.1. says that $\eta_\psi \in Z^1(D_\psi, \delta)$. It is clear that the bilinear form $-\mathcal{L}(\psi)$ on $\mathcal{B}$ with

$$\mathcal{L}(\psi)(a,b) = \langle \eta_\psi(a^*), \eta_\psi(b)\rangle = (a^*, b)_\psi$$

is the coboundary of ψ (we write $\partial\psi = -\mathcal{L}(\psi)$) which yields $\mathcal{L}(\psi) \in B^2(\mathbb{C}, \delta)$. If $\mathcal{L}$ is a bilinear form on $\mathcal{B}$ satisfying $\mathcal{L}(a,b) = \overline{\mathcal{L}(b^*, a^*)}$ then we have that $\partial\psi = \mathcal{L}$ implies $\partial\psi^* = \mathcal{L}$ where $\psi^* \in \mathcal{B}^*$ with $\psi^*(a) = \overline{\psi(a^*)}$. It follows that $\partial(\frac{1}{2}(\psi + \psi^*)) = \mathcal{L}$ which shows that a solution ψ of $\partial\psi = \mathcal{L}$ can be assumed to be hermitian without loss of generality. By the following simple proposition the study of quadratic linear functionals can be reduced to the study of elements of $Z^1(\mathbb{C}, \delta)$ in many cases.

3.2. PROPOSITION. *Let $\mathcal{B}$ be finitely generated and let η be in $Z^1(D, \delta)$. Then there exist $n \in \mathbb{N}$ and $d_1, \ldots, d_n$ in $Z^1(\mathbb{C}, \delta)$ such that*

$$\langle \eta(a), \eta(b)\rangle = \sum_{i=1}^{n} \overline{d_i(a)} d_i(b) \tag{6}$$

for all $a, b \in \mathcal{B}$.

PROOF: We claim that $\eta(\mathcal{B})$ is finite-dimensional. For let $\{a_0, a_1, \ldots, a_m\}$ be a set of generators of $\mathcal{B}$. We can assume that $a_0 = \mathbf{1}$ and $a_j \in K^1$ for $j \geq 1$, otherwise we pass to $a_j - \delta(a_j)\mathbf{1}$. Since $\eta \in Z^1(D, \delta)$ means $\eta(a) = 0$ for $a \in K^2$ and since $\eta(\mathbf{1}) = 0$, we have $\eta(\mathcal{B}) = \text{Span}\{\eta(a_1), \ldots, \eta(a_m)\}$. Now choose an orthonormal basis $\{e_1, \ldots, e_n\}$ of $\eta(\mathcal{B})$. Then $d_i \in Z^1(\mathbb{C}, \delta)$ where $d_i(a) = \langle e_i, \eta(a)\rangle$, and (6) holds.◇

3.3. PROPOSITION. *Let $\mathcal{B}$ be finitely generated and let ψ be a quadratic linear functional on $\mathcal{B}$. Then there exist $n \in \mathbb{N}$ and $d_1, \ldots, d_n$ in $Z^1(\mathbb{C}, \delta)$ such that*

$$\psi(ab) = \psi(a)\delta(b) + \delta(a)\psi(b) + \sum_{i=1}^{n} (d_i)^*(a) d_i(b)$$

for all $a, b \in \mathcal{B}$. If $\mathcal{B}$ is commutative the d_i can be chosen to be hermitian.

PROOF: The first part is only an application of Proposition 3.2. to $\eta_\psi \in Z^1(D_\psi, \delta)$.- For the second part one only has to remark that, if $\mathcal{B}$ is commutative,

$$\begin{aligned}\psi(ab) &= \psi(a)\delta(b) + \delta(a)\psi(b)\\ &+ \sum_{i=1}^{n} (d_i^{(1)}(a) d_i^{(1)}(b) + d_i^{(2)}(a) d_i^{(2)}(b))\end{aligned}$$

with

$$\mathrm{d}_i^{(1)} = \frac{1}{2}(\mathrm{d}_i + (\mathrm{d}_i)^*),$$
$$\mathrm{d}_i^{(2)} = \frac{i}{2}(\mathrm{d}_i - (\mathrm{d}_i)^*).\diamond$$

Assume for a moment that $\mathcal{B}$ has a comultiplication Δ and a counit δ turning $\mathcal{B}$ into a *-bialgebra. A *left invariant derivation* D on $\mathcal{B}$ is a derivation on the algebra $\mathcal{B}$ such that

$$\mathrm{D} = (\mathrm{id} \otimes (\delta \circ \mathrm{D})) \circ \Delta.$$

If D is a derivation then $\delta \circ \mathrm{D} \in Z^1(\mathbb{C}, \delta)$. If, on the other hand, $\mathrm{d} \in Z^1(\mathbb{C}, \delta)$ then $\mathrm{D} = (\mathrm{id} \otimes \mathrm{d}) \circ \Delta$ is a left invariant derivation. This establishes a one-to-one correspondence between left invariant derivations on $\mathcal{B}$ and elements in $Z^1(\mathbb{C}, \delta)$. The hermitian elements in $Z^1(\mathbb{C}, \delta)$ are precisely the degenerate quadratic linear functionals.

If $\mathcal{B}$ is commutative and finitely generated as an algebra then, by Proposition 3.3., for a quadratic ψ there exist $n \in \mathbb{N}$ and hermitian left invariant derivations $\mathrm{D}_1, \ldots, \mathrm{D}_n$ on $\mathcal{B}$ such that

$$\psi - \frac{1}{2}\sum_{i-1}^{n}(\delta \circ (\mathrm{D}_i)^2)$$

is a hermitian left invariant derivation. The proof of the following is again straightforward.

3.4. PROPOSITION. *For a linear functional ψ on $\mathcal{B}$ the following are equivalent*

(i) $\psi \in Z^1(\mathbb{C}, \delta)$
(ii) $\psi \lceil K^2 = 0$
(iii) $\psi(a^* a) = 0$ *for all* $a \in K^1$
(iv) ψ *is conditionally positive and* $D_\psi = \{0\}.\diamond$

Next we investigate under which circumstances we can extract a maximal quadratic component from a given hermitian conditionally positive linear functional. All is based on the following simple considerations. Let ρ be a *-representation of $\mathcal{B}$ on a pre-Hilbert space D and denote by H the completion of D. We form the closed linear subspace

$$H_1 = \{u \in H : \langle u, \rho(a)v\rangle = \langle u, v\rangle \delta(a) \text{ for all } a \in \mathcal{B} \text{ and for all } v \in D\}$$

of H. We denote by P_1 the orthogonal projection onto H_1 and we set $H_2 = (H_1)^\perp$, $P_2 = \mathrm{id} - P_1$, $D_1 = P_1 D$ and $D_2 = P_2 D$.

3.5. LEMMA. *For a *-representation ρ of $\mathcal{B}$ the equation*

$$\rho_2(a)P_2 v = \rho(a)v - \delta(a)P_1 v,$$

$v \in D$, $a \in \mathcal{B}$, *defines a *-representation of $\mathcal{B}$ on D_2, and*

$$\rho_2(a)P_2 v = P_2 \rho(a) v.$$

*Moreover, $\delta\,\mathrm{id} \oplus \rho_2$ is a *-representation of $\mathcal{B}$ which is an extension of ρ, i.e.*

$$D \subset D_1 \oplus D_2$$

and

$$(\delta \mathrm{id} \oplus \rho_2)\lceil D = \rho.$$

Given $\eta \in Z^1(D,\rho)$ we have that

$$P_1 \circ \eta \in Z^1(D_1,\delta)$$

and

$$P_2 \circ \eta \in Z^1(D_2,\rho_2).$$

PROOF: First we must prove that $\rho_2(a) : D_2 \to H$ is a well-defined linear mapping. So assume $P_2 v = 0$. Then

$$\begin{aligned} \rho_2 P_2 v &= \rho(a)v - \delta(a)P_1 v \\ &= \rho(a)P_1 v - \delta(a)P_1 v \\ &= 0, \end{aligned}$$

because for $u \in D$

$$\langle \rho(a)P_1 v - \delta(a)P_1 v, u\rangle = \langle P_1 v, \rho(a^*)u - \delta(a^*)u\rangle = 0.$$

Next we have

$$\langle u, \rho_2(a)P_2 v - P_2\rho(a)v\rangle = \langle u, P_1\rho(a)v - \delta(a)P_1 v\rangle = 0$$

which proves $\rho_2(a)P_2 v = P_2\rho(a)v$. Thus $\rho_2(a)$ maps D_2 to D_2 and the rest is immediate.◇

In the sequel, for a given $\eta \in Z^1(D,\rho)$ we put $\eta_1 = \eta \circ P_1$ and $\eta_2 = \eta \circ P_2$. Notice that H_2 is the closure of $\eta(K^2)$.

Let η be in $Z^1(D,\rho)$. A pair (E,ϑ) with E a pre-Hilbert space and $\vartheta \in Z^1(E,\delta)$ such that $\vartheta(\mathcal{B}) = E$ is called a *derivation component* of η if

$$\|\vartheta(a)\| \le \|\eta(a)\|$$

for all $a \in \mathcal{B}$. A derivation component (E,ϑ) of η is called *maximal* if

$$\|\vartheta'(a)\| \le \|\vartheta(a)\|$$

holds for all derivation components (E',ϑ') of η.

3.6. PROPOSITION. *Two maximal derivation components $(E^{(i)},\vartheta^{(i)})$, $i = 1,2$, of η are equivalent, i.e. there is a unitary operator*

$$\mathcal{U} : F^{(1)} \to F^{(2)},$$

$F^{(i)}$ the completion of $E^{(i)}$, such that

$$\mathcal{U}E^{(1)} = E^{(2)}$$

and

$$\vartheta^{(2)} = \mathcal{U} \circ \vartheta^{(1)}.$$

PROOF: Since $\|\vartheta^{(1)}(a)\| = \|\vartheta^{(2)}(a)\|$ for all $a \in \mathcal{B}$ it follows that $\mathcal{U}\vartheta^{(1)}(a) = \vartheta^{(2)}(a)$ gives the desired $\mathcal{U}$. ◇

3.7. THEOREM. *The pair (D_1, η_1) is a maximal derivation component of η.*

PROOF: We already know that $\eta_1 \in Z^1(D_1, \delta)$. We have

$$\|\eta_1(a)\| = \|P_1\eta(a)\| \leq \|\eta(a)\|$$

which shows that η_1 is a derivation component of η.- Proof of maximality: Suppose that (E, ϑ) is a derivation component of η. Since $\|\vartheta(a)\| \leq \|\eta(a)\|$ for all $a \in \mathcal{B}$ the Cauchy-Schwartz inequality yields

$$|\langle\vartheta(a), \vartheta(b)\rangle| \leq \|\vartheta(a)\|\|\vartheta(b)\| \leq \|\eta(a)\|\|\eta(b)\|.$$

Thus

$$(\eta(a), \eta(b))_\vartheta = \langle\vartheta(a), \vartheta(b)\rangle$$

defines a hermitian form on D with $|(u, v)_\vartheta| \leq \|u\|\|v\|$ for all $u, v \in D$. Therefore, there is a linear operator C on H, $0 \leq C \leq \text{id}$, such that $(u, v)_\vartheta = \langle Cu, v\rangle$. We have for $a, b \in \mathcal{B}$, $b \in K^2$

$$\langle C\eta(a), \eta(b)\rangle = \langle\vartheta(a), \vartheta(b)\rangle = 0,$$

because $\vartheta \in Z^1(E, \delta)$. But this means $C\eta(a) \in H_1$ for all $a \in \mathcal{B}$, so that $C = P_1C$, and, since C is selfadjoint, $P_1CP_1 = C$. Finally, we have for $a \in \mathcal{B}$

$$\begin{aligned}\|\vartheta(a)\|^2 &= \langle C\eta(a), \eta(a)\rangle \\ &= \langle CP_1\eta(a), P_1\eta(a)\rangle \\ &\leq \|\eta_1(a)\|^2. \diamond\end{aligned}$$

We say that a quadratic g is a *quadratic component* of the conditionally positive hermitian linear functional ψ if $\psi - \mathrm{g}$ is conditionally positive. Then we can write ψ as a sum of a quadratic linear functional and another conditionally positive hermitian linear functional. We say that a quadratic component g of ψ is maximal if $\mathrm{g} - \mathrm{g}'$ is conditionally positive for all quadratic components g' of ψ.

3.8. PROPOSITION. *The difference of two maximal quadratic components of ψ is a hermitian left invariant derivation.*

PROOF: Let g_1 and g_2 both be maximal quadratic components of ψ. Then

$$(\mathrm{g}_1 - \mathrm{g}_2)(a^*a) = 0 \text{ for all } a \in K^1.$$

Since g_1 and g_2 are hermitian, $\mathrm{g}_1 - \mathrm{g}_2$ is hermitian, too. ◇

For a linear mapping η from $\mathcal{B}$ to a pre-Hilbert space D we denote by $\mathcal{L}(\eta)$ the bilinear form om $\mathcal{B}$ with $\mathcal{L}(\eta)(a, b) = \langle\eta(a^*), \eta(b)\rangle$. We say that $\mathcal{B}$ has the property (C) if

(i) $\mathcal{L}((\eta_\psi)_1) \in B^2(\mathbb{C}, \delta)$ for all conditionally positive hermitian linear functionals ψ.

Other forms of the same condition (C) are

(ii) For all conditionally positive hermitian linear functionals ψ there is a hermitian linear functional g such that $\partial g = -\mathcal{L}((\eta_\psi)_1)$.

(iii) For all conditionally positive hermitian linear functionals ψ there is a hermitian linear functional g such that

$$g(ab) + g(a)\delta(b) - \delta(a)g(b) = \langle(\eta_\psi)_1(a^*), (\eta_\psi)_1(b)\rangle.$$

A stronger condition than (C) is given by

$$\mathcal{L}(\vartheta) \in B^2(\mathbb{C}, \delta) \text{ for all } \vartheta \in Z^1(D, \delta). \qquad \text{(C')}$$

If $\mathcal{B}$ is finitely generated (C') is equivalent to

$$\mathcal{L}(\mathrm{d}) \in B^2(\mathbb{C}, \delta) \text{ for all left invariant derivations d.}$$

An apparently even stronger condition is

$$\mathcal{L}(\eta) \in B^2(\mathbb{C}, \delta) \text{ for all } \eta \in Z^1(D, \rho). \qquad \text{(D)}$$

3.9. THEOREM. *Let ψ be a conditionally positive hermitian linear functional on $\mathcal{B}$. A hermitian linear functional g satisfying*

$$\partial g = -\mathcal{L}((\eta_\psi)_1)$$

is a maximal quadratic component of ψ. It follows that, if $\mathcal{B}$ has property (C), then each ψ has a maximal quadratic component.

PROOF: let g be hermitian and $\partial g = -\mathcal{L}((\eta_\psi)_1)$. Then for $a \in K^1$

$$g(a^*a) = \|\eta_1(a)\|^2 \le \|\eta(a)\|^2 = \psi(a^*a)$$

where we put $\eta_\psi = \eta$. If g' is quadratic with $g'(a^*a) \le \psi(a^*)$ it follows that

$$g'(a^*a) = \|\eta_{g'}(a)\|^2 \le \|\eta_1(a)\|^2 = g(a^*a).\diamond$$

Our next aim is to prove that each *commutative* *-algebra satisfies condition (C). We need the following lemma which is also useful for Section 4.

3.10. LEMMA. *Let ρ be a *-representation of $\mathcal{B}$ on the pre-Hilbert space D and let η be in $Z^1(D, \rho)$. Then the following are equivalent*

(i)

$$\langle\eta(a), \eta(b)\rangle = \langle\eta(b^*), \eta(a^*)\rangle$$

for all $a, b \in \mathcal{B}$ with $b \in K^2$

(ii)

$$\langle\eta_2(a), \eta_2(b)\rangle = \langle\eta_2(b^*), \eta_2(a^*)\rangle$$

for all $a, b \in \mathcal{B}$.

PROOF: (ii) $\Rightarrow$ (i) is clear, because $\eta\lceil K^2 = \eta_2\lceil K^2$.- (i) $\Rightarrow$ (ii): Let η satisfy (i). For $a, b \in K^2$ it follows that

$$\langle\eta(a), \eta(b)\rangle = \langle\eta(b^*), \eta(a^*)\rangle.$$

Thus we can define a conjugation $u \mapsto u^*$ on H_2 by

$$\eta(a)^* = \eta(a^*),$$

$a \in K^2$. Moreover, again by (i), we have for $a \in \mathcal{B}$, $b \in K^2$,

$$\begin{aligned}\langle P_2\eta(a), \eta(b)\rangle &= \langle\eta(a), \eta(b)\rangle \\ &= \langle\eta(b^*), P_2\eta(a^*)\rangle \\ &= \langle(P_2\eta(a^*))^*, \eta(b)\rangle\end{aligned}$$

which means

$$P_2\eta(a^*) = (P_2\eta(a))^*$$

for all $a \in \mathcal{B}$, and, hence, for $a, b \in \mathcal{B}$

$$\begin{aligned}\langle\eta_2(a), \eta_2(b)\rangle &= \langle P_2\eta(a), P_2\eta(b)\rangle \\ &= \langle(P_2\eta(a^*))^*, P_2\eta(b)\rangle \\ &= \langle P_2\eta(b^*), P_2\eta(a^*)\rangle \\ &= \langle\eta_2(b^*), \eta_2(a^*)\rangle. \diamond\end{aligned}$$

3.11. PROPOSITION. *Let $\mathcal{B}$ be a *-bialgebra. If $\vartheta \in Z^1(D, \delta)$ is such that*

$$\langle\vartheta(a), \vartheta(b)\rangle = \langle\vartheta(b^*), \vartheta(a^*)\rangle$$

for all $a, b \in \mathcal{B}$, then

$$g = -\frac{1}{2}(\mathcal{L}(\vartheta) \circ \Delta)$$

satisfies

$$\partial g = -\mathcal{L}(\vartheta).$$

PROOF: Let $a, b \in K^1$. Then

$$\begin{aligned}&\frac{1}{2}(\mathcal{L}(\vartheta) \circ \Delta)(ab) \\ &= \frac{1}{2}\sum_{i,j}\langle\vartheta((b_{1i})^*(a_{1j})^*), \vartheta(a_{2j}b_{2i})\rangle \\ &= \frac{1}{2}\sum_{i,j}(\delta(b_{1i})\langle\vartheta((a_{1j})^*), \vartheta(a_{2j})\rangle\delta(b_{2i}) + \delta(b_{1i})\langle\vartheta((a_{1j})^*), \vartheta(b_{2i})\rangle\delta(a_{2j}) \\ &\quad + \delta(a_{1j})\langle\vartheta((b_{1i})^*), \vartheta(b_{2i})\rangle\delta(a_{2j}) + \delta(a_{1j})\langle\vartheta((b_{1i})^*), \vartheta(a_{2j})\rangle\delta(b_{2i})) \\ &= \frac{1}{2}(\langle\vartheta(a^*), \vartheta(b)\rangle + \langle\vartheta(b^*), \vartheta(a)\rangle) \\ &= \langle\vartheta(a^*), \vartheta(b)\rangle. \diamond\end{aligned}$$

3.12. THEOREM. *A commutative *-bialgebra satisfies condition (C).*

PROOF: Let ψ be a conditionally positive hermitian linear functional on the commutative *-bialgebra $\mathcal{B}$. For $a, b \in K^1$ we have

$$\psi(ab) = \langle \eta(a^*), \eta(b) \rangle = \psi(ba) = \langle \eta(b^*), \eta(a) \rangle, \tag{7}$$

and (i) of Lemma 3.10. holds. Thus we have (ii) of Lemma 3.10. Combining this with (7) we obtain

$$\langle \eta_1(a), \eta_1(b) \rangle = \langle \eta_1(b^*), \eta_1(a^*) \rangle$$

for all $a, b \in \mathcal{B}$. Now the theorem follows from Proposition 3.11.◇

3.12. THEOREM.

(i) *The non-commutative coefficient algebra* $\mathrm{K}\langle d \rangle$ *has the property (D). For* $\eta \in Z^1(D, \rho)$ *a hermitian solution* ψ *of the problem*

$$\partial \psi = -\mathcal{L}(\eta) \tag{8}$$

is determined by the equations

$$\psi(\mathbf{1}) = 0$$
$$\psi(x_{kl}) = -\frac{1}{2} \sum_{n=1}^{d} \langle \eta(x_{kn}^*), \eta(x_{ln}^*) \rangle.$$

(ii) *For* $L \in M_d$, *a left invariant derivation on* $\mathrm{K}\langle d \rangle$ *is determined by the equations*

$$\mathrm{d}_L(x_{kl}) = L_{kl} \tag{9}$$
$$\mathrm{d}_L(x_{kl}^*) = -L_{lk}. \tag{10}$$

This establishes a 1-1-correspondence between M_d *and* $Z^1(\mathbb{C}, \delta)$. *A left invariant derivation* d_L *is hermitian if and only if* L *is skew-hermitian.*

(iii) *For* $L \in M_d$ *a hermitian solution* g *of the problem*

$$\partial \mathrm{g} = -\mathcal{L}(\mathrm{d}_L)$$

is determined by the equations

$$\mathrm{g}(\mathbf{1}) = 0$$
$$\mathrm{g}(x_{kl}) = -\frac{1}{2}(L^* L)_{kl}.$$

PROOF: (i): The statement follows from Theorem 5.1. of [12]. We give an elementary proof without using quantum stochastic calculus. Define a hermitian linear functional

$\tilde{\psi}$ on the free *-algebra $\mathrm{F}\langle d\rangle$ generated by indeterminates x_{kl}, $k,l = 1,\ldots d$, by

$$\tilde{\psi}(1) = 0,$$

$$\tilde{\psi}(x_{kl}) = -\frac{1}{2}\sum_{n=1}^{d}\langle\eta(x_{kn}^*),\eta(x_{ln}^*)\rangle,$$

$$\tilde{\psi}(ab) = \tilde{\psi}(a)\delta(b) + \delta(a)\tilde{\psi}(b) + \langle\eta(a^*),\eta(b)\rangle.$$

We have

$$\tilde{\psi}(\sum_{n=1}^{d} x_{kn}x_{ln}^*) = \tilde{\psi}(x_{kl}) + \overline{\tilde{\psi}(x_{lk})} + \sum_{n=1}^{d}\langle\eta(x_{kl}^*),\eta(x_{ln}^*)\rangle = 0.$$

Since η respects the relations in $\mathrm{K}\langle d\rangle$ it follows that

$$\begin{aligned}\eta(\sum_{n=1}^{d} x_{kn}x_{ln}^*) &= \sum_{n=1}^{d}(\rho(x_{kn})\eta(x_{ln}^*) + \eta(x_{kn})\delta_{ln})\\ &= \sum_{n=1}^{d}\rho(x_{kn})\eta(x_{ln}^*) + \eta(x_{kl})\\ &= 0\end{aligned}$$

which implies

$$\eta(x_{kl}) = -\sum_{n=1}^{d}\rho(x_{kn})\eta(x_{ln}^*).$$

As ρ respects the relations in $\mathrm{K}\langle d\rangle$, we have

$$\begin{aligned}&\tilde{\psi}(\sum_{n=1}^{d} x_{nk}^*x_{nl})\\ &= \overline{\tilde{\psi}(x_{lk})} + \tilde{\psi}(x_{kl}) + \sum_{n=1}^{d}\langle\eta(x_{nk}),\eta(x_{nl})\rangle\\ &= -\sum_{n=1}^{d}\langle\eta(x_{kn}^*),\eta(x_{ln}^*)\rangle + \sum_{n,m,r=1}^{d}\langle\rho(x_{nm})\eta(x_{mk}^*),\rho(x_{nr})\eta(x_{rl}^*)\rangle\\ &= -\sum_{n=1}^{d}\langle\eta(x_{kn}^*),\eta(x_{ln}^*)\rangle + \sum_{n=1}^{d}\langle\sum_{m,r=1}^{d}\rho(x_{nr}^*)\rho(x_{nm})\eta(x_{mk}^*),\eta(x_{rl}^*)\rangle\\ &= 0.\end{aligned}$$

Using again the definition of $\tilde{\psi}$, it follows that $\tilde{\psi}$ respects the relations in $\mathrm{K}\langle d\rangle$ and gives rise to a hermitian linear functional on $\mathrm{K}\langle d\rangle$ which by construction is a solution of (8).- (ii): Define the left invariant derivation $\tilde{d}_L$ on $F\langle d\rangle$ by $\tilde{d}_L(x_{kl}) = L_{kl}$ and $\tilde{d}_L(x^*_{kl}) = -L_{lk}$. Then

$$\begin{aligned}\tilde{d}_L(\sum_{n=1}^{d} x_{kn}x^*_{ln}) &= \sum_{n=1}^{d}(L_{kn}\delta_{ln} + \delta_{kn}(-L_{nl}))\\ &= 0\\ &= \tilde{d}_L(\sum_{n=1}^{d} x^*_{nk}x_{nl})\end{aligned}$$

which shows that $\tilde{d}_L$ respects the relations in $K\langle d\rangle$. Thus it gives rise to a left invariant derivation d_L on $K\langle d\rangle$ satisfying (9) and (10). On the other hand, a left invariant derivation d on $K\langle d\rangle$ must satisfy $d(x_{kl}) = -d(x_{lk})$. (iii): This is just (i) for the special case $\eta = d_L$.◇

4. Infinitely divisible states on the Weyl algebra. Let $T(V)$ be a tensor *-bialgebra (Example 2.2. of Section 2). The left invariant derivations on $T(V)$ can be identified with the linear functionals α on V via $d_\alpha \lceil V = \alpha$. A left invariant derivation d_α is hermitian if and only if α is hermitian. The quadratic linear functionals on $T(V)$ vanishing on V can be identified with the positive semi-definite sesquilinear forms Q on V via $g_Q(vw) = Q(v^*, w)$ for $v, w \in V$. Since there are no relations in $T(V)$ it is clear that $T(V)$ satisfies condition (D). For, if η is in $Z^1(D,\rho)$ a solution of $\partial\psi = -\mathcal{L}(\eta)$ is simply given by

$$\begin{aligned}\psi(\mathbf{1}) &= 0\\ \psi(v) &= 0\\ \psi(v_1 \dots v_n) &= \langle \eta(v_1^*), \eta(v_2 \dots v_n)\rangle,\end{aligned}$$

$v, v_1 \dots, v_n \in V$, $n \geq 2$. It follows that all conditionally positive hermitian linear functionals on $T(V)$ have maximal quadratic components. This result was already obtained in [2].

Let C be a bilinear form on V satisfying

$$C(v,w) = -C(w,v) = \overline{C(w^*,v^*)}.$$

We denote by $I(C)$ the ideal in $T(V)$ generated by the elements $[v,w] - C(v,w)$, $v, w \in V$. This makes sense, because $C(v,w) = S([v,w])$ with $S : V \otimes V \to \mathbb{C}$ the linear mapping given by $S(v \otimes w) = \frac{1}{2}C(v,w)$. Moreover, $I(C)$ is a *-ideal. The *-algebra $T(V)/I(C)$ is called the Weyl algebra $W_C = W_C(V)$ associated to C. If $C \neq 0$ the coalgebra structure of $T(V)$ cannot be transferred to W_C because $I(C)$ is not a coideal. A state φ on $T(V)$ is called *infinitely divisible* if for each $n \in \mathbb{N}$ there is a state φ_n on $T(V)$ such that $\varphi = (\varphi_n)^{\star n}$. A state φ on W_C is called infinitely divisible if the state $\tilde{\varphi}$ on $T(V)$ induced by φ via the canonical mapping from $T(V)$ to W_C is infinitely divisible. In other words, the infinitely divisible states on W_C are those infinitely divisible states on $T(V)$ vanishing on $I(C)$. We show in this section that an infinitely divisible state φ on W_C must be of the form $\varphi = g_Q \star \mu$ where μ vanishes on $W_0(V)$ (i.e. μ is 'classical') and Q is such that $Q(v^*,w) - Q(w^*,v) = C(v,w)$. This is the infinitesimal version of a result of Araki (see [3]) for the Heisenberg group.

We begin with a simple but useful proposition; see also [15].

4.1. PROPOSITION. *For a normalized linear functional φ on $\mathrm{T}(V)$ and $n \in \mathbb{N}$ there is a uniquely determined normalized linear functional φ_n on $\mathrm{T}(V)$ such that $\varphi = (\varphi_n)^{\star n}$.*

PROOF: We define φ_n by the recursion formula

$$n\varphi_n(v_1 \ldots v_m) = \varphi(v_1 \ldots v_m) - \sum_{A_1,\ldots,A_n} \varphi_n(v_{A_1}) \ldots \varphi_n(v_{A_n}) \tag{11}$$

($m \in \mathbb{N}$ the running parameter) where the sum is extended over all ordered disjoint partitions $(A_1, \ldots, A_n)$ of the set $\{1, \ldots, m\}$ with $A_j \neq \{1, \ldots, m\}$ (but $A_j = \emptyset$ is allowed), and where for $A = \{j_1 < \cdots < j_k\} \subset \{1, \ldots, m\}$ the symbol v_A denotes the monomial $v_{j_1} \ldots v_{j_k}$ (we put $v_\emptyset = \mathbf{1}$!). It is clear that $(\varphi_n)^{\star n} = \varphi$ and that, on the other hand, a normalized φ_n with $(\varphi_n)^{\star n} = \varphi$ has to satisfy (11).◇

If the linear functional ψ on $\mathrm{T}(V)$ vanishes at $\mathbf{1}$ we have $\psi^{\star n}(M) = 0$ for all monomials of length less than n, Therefore, the series

$$\exp_\star \psi = \sum_{n=0}^{\infty} \frac{\psi^{\star n}}{n!}$$

can be defined pointwise as a normalized linear functional on $\mathrm{T}(V)$. For the same reasons,

$$\ln_\star \varphi = \sum_{n=0}^{\infty} (-1)^{n+1} \frac{(\varphi - \delta)^{\star n}}{n}$$

makes sense for a normalized φ and is a linear functional vanishing at $\mathbf{1}$. Moreover, $\ln_\star$ is the inverse of $\exp_\star$.

4.2. THEOREM. *A state φ on $\mathrm{T}(V)$ is infinitely divisible if and only if there is a conditionally positive hermitian linear functional ψ on $\mathrm{T}(V)$ such that $\varphi = \exp_\star \psi$.*

PROOF: Let φ be infinitely divisible. We must prove that $\ln_\star \varphi$ is conditionally positive and hermitian. But we have

$$(\exp_\star(\frac{1}{n} \ln_\star \varphi))^{\star n} = \varphi_n$$

which by Proposition 4.1. yields

$$\exp_\star(\frac{1}{n} \ln_\star \varphi) = \varphi_n$$

and $n(\varphi_n - \delta)$ converges to $\ln_\star \varphi$ pointwise. Since $n(\varphi_n - \delta)$ are all conditionally positive and hermitian the same holds for $\ln_\star \varphi$.- The converse follows from the fact that $\exp_\star t\psi$ is a state for all $t \geq 0$ iff ψ is conditionally positive and hermitian; see [11].◇

Next we find necessary and sufficient conditions on ψ for $\exp_\star \psi$ to vanish on $I(C)$. We need the following lemma.

4.3. LEMMA. *Let $\varphi, \varphi_1, \dots, \varphi_n$ be normalized linear functionals on* $T(V)$.

(i) *If φ_j vanishes on $I(C_j)$ then $\varphi_1 \star \dots \star \varphi_n$ vanishes on $I(C_1 + \dots + C_n)$.*

(ii) *If $\varphi = (\varphi_n)^{\star n}$ and φ vanishes on $I(C)$ then φ_n vanishes on $I(\frac{1}{n}C)$.*

PROOF: (i): We have for $a, b \in T(V)$

$$\begin{aligned}
&(\varphi_1 \star \dots \star \varphi_n)(a([v,w] - \sum_{j=1}^{n} C_j(v,w))b) \\
&= (\varphi_1 \otimes \dots \varphi_n)(\Delta_n(a)(([v,w] - C - 1(v,w)) \otimes \mathbf{1} \otimes \dots \otimes \mathbf{1} \\
&\quad + \mathbf{1} \otimes (([v,w] - C_2(v,w)) \otimes \mathbf{1} \otimes \dots \otimes \mathbf{1} + \dots \\
&\quad \dots + \mathbf{1} \otimes \dots \otimes \mathbf{1} \otimes ([v,w] - C_n(v,w)))\Delta_n(b)) \\
&= \sum_{i,j} (\varphi_1(a_{1i}([v,w] - C_n(v,w))b_{1j})\varphi_2(a_{2i}b_{2j}) \dots \varphi_n(a_{ni}b_{nj}) + \dots \\
&\quad \dots + \varphi_1(a_{1i}b_{1j}) \dots \varphi_{n-1}(a_{(n-1),i}b_{(n-1),j})\varphi_n(a_{ni}([v,w] - C_n(v,w))b_{nj}) \\
&= 0
\end{aligned}$$

if φ_j vanishes on $I(C_j)$.- (ii): Suppose φ vanishes on $I(C)$ and

$$\varphi_n(M([v,w] - \frac{1}{n}C(v,w))N) = 0$$

for all $v, w \in V$ and all monomials M, N with the length of MN less than m. Then for monomials M, N with the length of MN equal to m we have

$$\begin{aligned}
0 &= \varphi(M([v,w] - C(v,w))N) \\
&= (\varphi_n)^{\star n}(M([v,w] - C(v,w))N) \\
&= n\varphi_n(M([v,w] - \frac{1}{n}C(v,w))N). \diamond
\end{aligned}$$

4.4. THEOREM. *Let φ be an infinitely divisible state on $T(V)$. Then the following are equivalent*

(i) *φ vanishes on $I(C)$.*

(ii)

$$\ln_\star \varphi([v,w]) = C(v,w) \tag{12}$$

and

$$\ln_\star \varphi(a[v,w]b) = 0 \tag{13}$$

for all $v, w \in V$ and $a, b \in T(V)$ with $ab \in K^1$.

PROOF: (i) $\Rightarrow$ (ii): We have

$$\ln_\star \varphi([v,w]) = \lim_{n \to \infty} n\varphi_n([v,w]) = C(v,w).$$

For $a, b \in \mathrm{T}(V)$ with $ab \in K^1$

$$\begin{aligned}\ln_\star \varphi(a[v,w]b) &= \lim_{n\to\infty} n\varphi_n(a[v,w]b)\\ &= \lim_{n\to\infty} \varphi_n(ab)C(v,w)\\ &= 0\end{aligned}$$

because

$$\begin{aligned}\lim_{n\to\infty} \varphi_n(a) &= \lim_{n\to\infty} \frac{1}{n}(n\varphi_n - \delta)(a)\\ &= \lim_{n\to\infty} \frac{1}{n}\psi(a)\\ &= 0\end{aligned}$$

for $a \in K^1$.- (ii) $\Rightarrow$ (i): For $a, b \in \mathrm{T}(V)$ and $v, w \in V$ we have

$$\begin{aligned}&(\ln_\star \varphi)^{\star(n+1)}(a([v,w] - C(v,w))b)\\ &= (\ln_\star \varphi)^{\otimes(n+1)}(\Delta_{n+1}(a)([v,w] \otimes \mathbf{1} \otimes \cdots \otimes \mathbf{1} + \ldots\\ &\qquad \cdots + \mathbf{1} \otimes \cdots \otimes \mathbf{1} \otimes [v,w])\Delta_{n+1}(b))\\ &\qquad - C(v,w)(\ln_\star \varphi)^{\star(n+1)}(ab)\\ &= (n+1)C(v,w)(\ln_\star \varphi)^{\star n}(ab) - C(v,w)(\ln_\star \varphi)^{\star(n+1)}(ab)\end{aligned}$$

so that

$$\begin{aligned}&\varphi(a([v,w] - C(v,w))b)\\ &= \sum_{n=0}^{\infty} \frac{(\ln_\star \varphi)^{\star n}}{n!}(a([v,w] - C(v,w))b)\\ &= -\delta(a)\delta(b)C(v,w) + \sum_{n=0}^{\infty} \frac{(\ln_\star \varphi)^{\star(n+1)}}{(n+1)!}(a[v,w] - C(v,w))b)\\ &= -\delta(a)\delta(b)C(v,w) + (\sum_{n=0}^{\infty} \frac{(n+1)(\ln_\star \varphi)^{\star n}}{(n+1)!}(ab)\\ &\qquad - \sum_{n=0}^{\infty} \frac{(\ln_\star \varphi)^{\star(n+1)}}{(n+1)!}(ab))C(v,w)\\ &= 0. \diamond\end{aligned}$$

We say that a quadratic g on $\mathrm{T}(V)$ is centralized if it vanishes on V. Clearly, each conditionally positive hermitian linear functional on $\mathrm{T}(V)$ has a unique centralized maximal quadratic component.

4.5. THEOREM. *Let ψ be a conditionally positive hermitian linear functional on* $\mathrm{T}(V)$ *satisfying the conditions (12) and (13) of Theorem 4.4. (ii). Let*

$$\psi = \mathrm{g}_Q + \mu$$

be the unique decomposition of ψ *into a centralized maximal quadratic component* g_Q *and a conditionally positive hermitian linear functional* μ *with degenerate maximal quadratic component. Then*

(i) $Q(v^*,w) - Q(w^*,v) = C(v,w)$.

(ii) μ *vanishes on* $I(0)$.

PROOF: It follows from (13) that $\psi(abc) = \psi(cab)$ for $a,b,c \in K^1$. But this means

$$\langle \eta(ab), \eta(c)\rangle = \langle \eta(c^*), \eta(b^* a^*)\rangle$$

for $a,b \in K^1$, $c \in \mathrm{T}(V)$. By Lemma 3.10. this implies

$$\langle \eta_2(a), \eta_2(b)\rangle = \langle \eta_2(b^*), \eta_2(a^*)\rangle$$

for $a,b \in \mathrm{T}(V)$, or

$$\mu\,(ab) = \mu(ba) \tag{14}$$

for $a,b \in K^1$. Moreover, μ agrees with ψ on elements of the form $a[v,w]b$ with $ab \in K^1$, $v,w \in V$. Together with (14) this implies (ii). By (12) we have

$$\psi([v,w]) = C(v,w) = g_Q([v,w]) = Q(v^*,w) - Q(w^*,v)$$

which is (i).◇

4.6. COROLLARY. *The infinitely divisible states on* $\mathrm{W}_C(V)$ *are precisely the states on* $\mathrm{T}(V)$ *of the form*

$$\gamma_Q \star \mu$$

where

$$Q(v^*,w) - Q(w^*,v) = C(v,w) \tag{15}$$

and μ *is an infinitely divisible state on* $\mathrm{W}_0(V)$ *which cannot be decomposed into a convolution product of a non-degenerate gaussian state and another infinitely divisible state on* $\mathrm{W}_0(V)$.◇

REMARKS: 1) If $C \neq 0$ it follows from the relation (15) for the covariance matrix of the maximal gaussian component that an infinitely divisible state on $\mathrm{W}_0(V)$ has a non-degenerate maximal gaussian component.

2) A graded version of Theorem 4.5. would include the result of Streater (see [10]) that an infinitely divisible state on the algebra of the canonical anti-commutation relations must be gaussian.

Acknowledgements

This work was supported by the SERC and by the SFB 123 'Stochastische Mathematische Modelle'. I have had useful discussions with P. Glockner and R.L. Hudson.

References

[1] Abe, E.: Hopf algebras. Cambridge: Cambridge University Press 1980

[2] Accardi, L., Schürmann, M., Waldenfels, W.v.: Quantum independent increment processes on superalgebras. Math. Z. 198, 451-477 (1988)

[3] Araki, H.: Factorizable representation of current algebra. Publ. Res. Inst. Math. Sci. 5, 361-422 (1970)

[4] Glockner, P.: *-Bialgebren in der Quantenstochastik. Dissertation, Heidelberg 1989

[5] Glockner, P., Waldenfels, W.v.: The relations of the non-commutative coefficient algebra of the unitary group. SFB-Preprint Nr. 460, Heidelberg 1988

[6] Guichardet, A.: Symmetric Hilbert spaces and related topics. (Lect. Notes Math. vol. 261). Berlin Heidelberg New York: Springer 1972

[7] Heyer, H.: Probability measures on locally compact groups. Berlin Heidelberg New York: Springer 1977

[8] Hudson, R.L., Parthasarathy, K.R.: Quantum Ito's formula and stochastic evolutions. Commun. Math. Phys. 93, 301-323 (1984)

[9] Hunt, G.A.: Semi-groups of measures on Lie groups. Trans. Amer. Math. Soc. 81, 264-293 (1956)

[10] Mathon, D., Streater, R.F.: Infinitely divisible representations of Clifford algebras. Z. Wahrscheinlichkeitstheorie verw. Geb. 20, 308-316 (1971)

[11] Schürmann, M.: Positive and conditionally positive linear functionals on coalgebras. In: Accardi, L., Waldenfels, W.v. (eds) Quantum Probability and Applications II. Proceedings, Heidelberg 1984. (Lect. Notes Math., vol. 1136). Berlin Heidelberg New York: Springer 1985

[12] Schürmann, M.: Noncommutative stochastic processes with independent and stationary increments satisfy quantum stochastic differential equations. To appear in Probab. Th. Rel. Fields

[13] Schürmann, M: A class of representations of involutive bialgebras. To appear in Math. Proc. Cambridge Philos. Soc.

[14] Schürmann, M.: Quantum stochastic processes with independent additive increments. Preprint, Heidelberg 1989

[15] Streater, R.F.: Infinitely divisible representations of Lie algebras. Z. Wahrscheinlichkeitstheorie verw. Geb. 42, 67-80 (1971)

[16] Sweedler, M.E.: Hopf algebras. New York: Benjamin 1969

The paper is in final form and no similar paper has been or is being submitted elsewhere.

QUANTUM MACROSTATISTICS AND IRREVERSIBLE THERMODYNAMIC.

by Geoffrey L. Sewell

Department of Physics, Queen Mary College, London E1 4NS

ABSTRACT

We formulate a mathematical framework for the non-equilibrium statistical thermodynamics of processes of the diffusive type. Both the classical phenomenological laws and the quantum mechanical ones for the fluctuations of the relevant macro-observables are expressed in terms of tempered distributions, and the connection between these laws is formulated on the basis of very general 'axioms' c physical origin, which incorporate the essential ideas of the Onsager theory [1]. On this basis, we show that the fluctuation dynamics of the quantum macro-observables reduces to a classical Markov process in a certain large-scale, hydrodynamical limit, and that the phenomenological transport coefficients satisfy the Onsager reciprocity relations.

1. INTRODUCTION.

The purpose of this talk is to present an approach to the statistical thermodynamics of irreversible processes, an old subject, which still lacks a general mathematical framework. Let me start by describing the situation in very general terms. We have a many (infinite)-particle system, S, occupying a space $\mathbf{X} = \mathbf{R}^d$. . the microscopic level, S is a quantum system, which can be formulated in standard algebraic terms [2-5]. At the macroscopic level, on the other hand, it correspond to a classical model, i. e. one whose variables are intercommuting c-numbers. Thus, its macroscopic equilibrium properties are given by classical thermodynamics, which may be based on the entropy function, s, of a certain set of variables $\mathbf{q} = (q_1,\ldots, q_n)$, that are global densities of extensive conserved quantities, e. g. energy, particle number, etc. These variables correspond to global intensive observables, i.e. ' observables at infinity' [6], $\hat{\mathbf{q}}$, in the quantum picture of S; and, in order th they provide a complete thermodynamic description of the system, the equilibrium states of S must be uniquely labelled by the expectation values $\mathbf{q}$ that they yield for $\hat{\mathbf{q}}$ [5, Ch. 4]. As regards non-equilibrium thermodynamics, this pertains to situations where the different spatial regions of S are not in equilibrium with one another, and is concerned, at the macroscopic level, with the dynamics of the local values of the same variables **q** that provided the equilibrium description. It is an empirical fact that these time-dependent local densities q_t generally evolve irreversibly, according to a self-contained law of the form

$$\dot{q}_t(x) = F(q_t; x) \qquad (1.1)$$

as in hydrodynamics or diffusion: in general, the functional **F** is non-linear. The task of non-equilibrium statistical thermodynamics is to relate the law (1.1) to the underlying quantum mechanics of **S**, and thus to determine the general constraints imposed on the form of **F** by microphysics.

In this work, I shall formulate a mathematical treatment of the subject for the relatively simple case where the macroscopic law (1.1) is of the linear diffusive type, i. e.

$$\partial q^{(j)}_t / \partial t = \sum_{k=1}^{n} L_{jk} \Delta q^{(k)}_t , \; j = 1, 2,.. , n, \qquad (1.2)$$

the L' s being real constants and the real parts of the latent roots of $[L_{jk}]$ positive, corresponding to a dissipative process. Such linear equations generally pertain to situations where the system is sufficiently close to (stable) equilibrium: a corresponding treatment of the non-linear case has been sketched in Ref.[7] on the basis of a certain assumption of local equilibrium.

The essential strategy adopted here is to interrelate the classical phenomenological law and the quantum stochastic one for the fluctuations of the relevant macro-observables, on the basis of very general 'axioms' of physical origin, which incorporate the essential ideas of the Onsager theory [1]. On this basis, we show that the fluctuation dynamics of the quantum macro-observables reduces to a classical Markov process, under rather general conditions and in a certain large-scale, hydrodynamical limit; and that the phenomenological transport coefficients satisfy the Onsager reciprocity relations.

The work will be set out as follows. In Section 2, the assumed macroscopic law (1.2) will be recast in distribution-theoretic terms, as the smoothing of the field **q** against suitable test-functions is natural for continuum mechanics. Section 3 will be devoted to the algebraic formulation of the quantum model of **S**. In Section 4, the macroscopic fluctuation observables will be formulated as operator-valued distributions, as in quantum field theory [8], and and it will be shown that classical properties of these emerge under rather general conditions (Theorem 1). In Section 5, the Onsager reciprocity relations will be derived for the transport coefficients L_{jk} of equation (1.2) (Theorem 3). In Section 6, the fluctuations of the quantum macro-observables will be shown to conform, under suitable conditions and in a large-scale limit, to a classical Markov process (Theorem 4)

2. THE PHENOMENOLOGICAL PICTURE.

Let $S(X)$ be the Schwartz space of smooth, fast-decreasing, real-valued functions on X ($= \mathbf{R}^d$ for some +ve d), and let $S'(X)$ be its dual, i. e. the space of real tempered distributions over X. We formulate the phenomenological law (1.2) as an equation in $S'(X)$. Thus, we take $t \to q^{(j)}{}_t$ to be a $C^{(1)}$-class mapping of $\mathbf{R}_+$, the non-negative reals, into $S'(X)$, and we define $\mathbf{q}_t$ to be the element of $S'(X) \otimes \mathbf{R}^n$ given by

$$\mathbf{q}_t = \sum_{j=1}^{n} q^{(j)}{}_t \otimes e_j , \qquad (2.1)$$

where $\{e_1, \dots, e_n\}$ is an orthonormal basis set in $\mathbf{R}^n$. Thus, $\mathbf{q}_t$ is also $C^{(1)}$-class w.r.t. t. We recast the assumed phenomenological law (1.2) as the following equation in $S'(X) \otimes \mathbf{R}^n$:

$$\partial \mathbf{q}_t / \partial t = (\Delta \otimes L)\mathbf{q}_t , \forall t \in \mathbf{R}_+, \text{ with } L = [L_{jk}] \qquad (2.2)$$

Assuming that the equilibrium, i. e. the terminal, value for $\mathbf{q}$ is space-translationally invariant, we may, without loss of generality, take this to be zero. The equation (2.2) may then be integrated to yield

$$\mathbf{q}_t = T(t)\mathbf{q} , \forall t \in \mathbf{R}_+, \text{ with } \mathbf{q} = \mathbf{q}_0 , \qquad (2.3)$$

where $\{T(t) \mid t \in \mathbf{R}_+\}$ is the one-parameter semigroup of linear transformations of $S'(X) \otimes \mathbf{R}^n$ given by

$$T(t) = \exp(\Delta \otimes L)t \qquad (2.4)$$

Hence, by the reflexivity of S-spaces,

$$\mathbf{q}_t(f) = \mathbf{q}(T^*(t)f) \quad \forall f \in S(X) \otimes \mathbf{R}^n, t \in \mathbf{R}_+ \qquad (2.5)$$

where $T^*(\mathbf{R}_+)$ is the one-parameter semigroup of linear transformations of $S(X) \otimes \mathbf{R}^n$, dual to $T(\mathbf{R}_+)$, i. e.

$$T^*(t) = \exp(\Delta \otimes L^*)t \qquad (2.6)$$

L^* being the adjoint of the matrix L.

3. THE QUANTUM PICTURE.

We formulate the quantum description of the system S in standard algebraic terms [2-5]. Thus we take the bounded observables for each bounded open region F of X to be the self-adjoint elements of a W^*-algebra A(F), that is C^*-isotonic w.r.t. F and satisfies the requirement that A(F) and A(F') intercommute if F and F' are disjoint. The local algebras A(F) are constructed as for finite systems of the given species, confined to the regions F. The algebra of observables **A** for **S** is taken to be the norm completion of the union of the local algebras A(F). We assume that space-translations are represented by a weakly continuous representation y of the additive group X in Aut(**A**), such that y(x)A(F) = A(F + x).

The states of the system are the positive normalised linear functionals on **A**, satisfying the condition of local normality, i. e. that their restrictions to the local algebras A(F) are normal. The dynamics of S, as given by a 'natural' limit of that of a finite system of the same species, is generally supported only by certain representations R, and in each of these it corresponds to a one-parameter group of automorphisms of R(**A**)'' [9, 10]. In particular, if W is an equilibrium state and (H, R, w) is its GNS triple, the dynamics of the states in the normal folium of W is given by the modular automorphism group [11] z(**R**) of R(**A**)'', corresponding to the canonical extension $\hat{W}$ of W to this algebra, which we shall henceforth denote by $\hat{A}$. Thus, in units for which the inverse temperature is unity, $\hat{W}$ satisfies the KMS condition

$$\int dt\, \hat{W}((z(t)a) b) h(t) = \int dt\, \hat{W}(b\, z(t)a) h(t - i) \qquad (3.1)$$

$$\forall\, a, b \in \hat{A} \quad \text{and} \quad h \in \hat{D}(\mathbf{R}),$$

$\hat{D}(\mathbf{R})$ being the Fourier transform of the Schwartz space D(**R**) of infinitely differentiable functions with compact support. Furthermore, defining v(**R**) to be the one-parameter continuous unitary group v(**R**) in H, that implements z(**R**) according to the formula

$$z(t)a = v(t)av(-t) \quad \text{and} \quad v(t)w = w, \qquad (3.2)$$

we see that

$$\hat{W}((z(t)a) b) = (a^* w, v(-t)bw) \qquad (3.3\,a)$$

$$\text{and} \quad \hat{W}(b\, z(t)a) = (b^* w, v(t)aw) \qquad (3.3\,b)$$

The state $\hat{W}$ is said to be weakly asymptotically abelian with respect to time-translations if [2] the commutator [$z(t)a$, b] tends weakly to 0, as $t \to \pm\infty$, for all a, b in $\hat{A}$, i.e if

$$\lim_{t \to \pm\infty} \left((z(t)a^* w_1 , b w_2) - (b^* w_1 , z(t) a w_2) \right) = 0, \qquad \forall a, b \in \hat{A} ; w_1 , w_2 \in H \tag{3.4}$$

This is a property of many-particle systems that is frequently ,though not universally, satisfied. We shall state explicitly when we are assuming this, or even some closely related dynamical property.

We shall henceforth assume that the state $\hat{W}$ under consideration is an extremal KMS state, corresponding to a pure phase [6, 12], and that it is translationally invariant. The latter property implies that the space-translational automorphisms $y(X)$ are implemented in H by a continuous unitary representation u of X and extend there to automorphisms $\hat{y}(X)$ of $\hat{A}$, with

$$\hat{y}(x)a = u(x)au(-x) \text{ and } u(x)w = w \tag{3.5}$$

Thus,

$$\hat{W}(a\,\hat{y}(x)b) = (a^* w, u(x)bw) \tag{3.6}$$

Further, the assumption that $\hat{W}$ is an extremal KMS state implies that it possesses the clustering, i. e. short range correlation property [6, 12]

$$(a^* w, u(x)bw) \to (a^* w, w)(w, bw) \text{ as } |x| \to \infty \tag{3.7}$$

Since we shall be dealing with unbounded observables, we note that the set of these for a finite region F consists of the unbounded self-adjoint operators affiliated to $A(F)$ (i. e . commuting weakly with its commutant) in the irreducible representation space of that algebra. Further, the local normality of W ensures [13] that its GNS representation has a canonical extension to the unbounded observables for all the local regions F, this extension being affiliated to $\hat{A}$. Suppose now that a and b are representations in H of unbounded observables, whose domains contain w. Then the r.h.s.' s of equations (3.3), (3.5), (3.6) are well-defined and [13] the KMS condition (3.1), with time-correlation functions defined by (3.3), the asymptotic abelian condition (3.4) (with w_1 , w_2 in the domains of $z(t)a$ and b) and the clustering property (3.7) all extend to these observables.

4. THE MACROSTATISTICS.

In order to specify the macroscopic observables, we recall from Section 1 (cf. also [5, Ch. 4]) that the thermodynamical variables $q^{(j)}$ are expectation values of 'observables at infinity' $\hat{q}^{(j)}$, given by global densities of extensive conserved quantities. We assume that, in the GNS representation of the equilibrium state W, the $\hat{q}^{(j)}$'s are space-averages of densities $\hat{q}^{(j)}(x)$, which are operator-valued tempered distributions, such that, if $h \in S(X)$, the 'smeared fields' $\hat{q}^{(j)}(h) = \int dx\, \hat{q}^{(j)}(x)\, h(x)$ are self-adjoint operators affiliated to $\hat{A}$: in many cases, such as those where $\hat{q}^{(j)}$ is the density of energy or particle number of a continuous system, these operators will be unbounded. We assume that the fields $\hat{q}^{(j)}$ transform covariantly under space translations, so that

$$\hat{y}(x)\hat{q}^{(j)}(h) = u(x)\hat{q}^{(j)}(h)u(-x) = \hat{q}^{(j)}(h_x),$$

$$\text{where } h_x(x') = h(x'-x) \qquad (4.1)$$

The time evolute of $\hat{q}^{(j)}$ is given, in view of (3.2), by

$$\hat{q}^{(j)}(h,t) = z(t)\hat{q}^{(j)}(h) = v(t)\hat{q}^{(j)}(h)v(-t) \qquad (4.2)$$

Since $z(t)$ is an automorphism of $\hat{A}'$, as well as of $\hat{A}$ [11], this equation implies that $\hat{q}^{(j)}(h,t)$ commutes weakly with the former algebra and is therefore affiliated to $\hat{A}$ for all t.

Corresponding to our definition (2.1) of $\mathbf{q}_t$ as an element of $\mathbf{S}'(X) \otimes \mathbf{R}^n$, we define $\hat{\mathbf{q}}(\,.\,,t)$ as an operator-valued element of that space by the analogous formula:

$$\hat{\mathbf{q}}(\,.\,,t) = \sum_{j=1}^{n} \hat{q}^{(j)}(\,.\,,t) \otimes e_j \qquad (4.3)$$

We assume that the monomials in $\{\hat{\mathbf{q}}(f,t) \mid f \in \mathbf{S}(X) \otimes \mathbf{R}^n;\ t \in \mathbf{R}\}$ have a common domain containing w, as is neccessary if the field $\hat{\mathbf{q}}$ and its fluctuations are to remain bounded in the state $\hat{W}$. In view of the translational invariance of this state, we may assume, without loss of generality, that its expectation value for $\hat{\mathbf{q}}$ is zero, i. e. that

$$(w, \hat{\mathbf{q}}(f)w) = 0 \quad \forall f \in \mathbf{S}(X) \otimes \mathbf{R}^n \qquad (4.4)$$

corresponding to the assumption that the equilibrium value of $\mathbf{q}$ in the phenomenological theory was zero. We further assume that the vectors $\hat{\mathbf{q}}(f_1, t_1) \ldots \hat{\mathbf{q}}(f_r, t_r)w$ are continuous w.r.t. the f' s : it follows from (4.2) and the continuity of $v(t)$ in t that this vector is also continuous in the t' s.

We assume that the properties of fluctuation field $\hat{q}$ pertinent to the phenomenological picture of the system are those that emerge on a 'large' length scale. Accordingly, we reformulate this field on a length scale L, which will eventually tend to infinity. We shall assume, following Onsager [1], that, in a sense that will be made precise in Section 5, the large-scale regressions of the fluctuations are governed by the phenomenological law (1.2). Thus, as that is invariant under $x \to kx$, $t \to k^2 t$, we formulate the fluctuation dynamics on the time-scale L^2, defining the rescaled smeared fluctuation field to be

$$\hat{q}_L(f, t) = \hat{q}(f_L, L^2 t) \; \forall f \in S(X) \otimes \mathbf{R}^n, \; t \in \mathbf{R}$$

$$\text{where } f_L(x) = L^{-d/2} f(x / L), \tag{4.5}$$

the normalisation factor $L^{-d/2}$ being the standard one for statistical fluctuations with short-range correlations.

Since, as assumed above, w lies in the domain of the monomials in $\hat{q}(f, t)$, and thus in $\hat{q}_L(f, t)$, we may define

$$K_L(f_1, t_1; \ldots; f_r, t_r) = (w, \hat{q}_L(f_1, t_1) \ldots \hat{q}_L(f_r, t_r) w) \tag{4.6}$$

Thus, by (4.5) and the continuity properties assumed following (4.4), K_L is continuous in the f' s and t's.

We assume that K_L tends to a limit, K, as $L \to \infty$, our hydrodynamical limit, i.e.

$$\lim_{L \to \infty} K_L(f_1, t_1; \ldots; f_r, t_r) = K(f_1, t_1; \ldots; f_r, t_r) \tag{4.7}$$

The macrostatistical properties of S are thus represented by K. To show that, under rather general conditions, these properties are of a classical nature, we introduce a concept corresponding to a macroscopic version of the asymptotically abelian condition (3.4). The essential idea behind it is that a non-zero time t on the macroscopic scale corresponds to an infinite time ($= \lim_{L \to \infty} L^2 t$) on the microscopic one. Thus, the following definition provides a macroscopic counterpart to the asymptotic abelian condition (3.4).

<u>Def. 1.</u> S is termed M-abelian if $K(f_1, t_1; \ldots; f_k, t_k; f_{k+1}, t_{k+1}; \ldots; f_r, t_r)$ is invariant under $(f_k, t_k) \rightleftharpoons (f_{k+1}, t_{k+1})$ when $t_k \neq t_{k+1}$.

Next, we note that, in view of the completeness of S'-spaces, it follows from (4.7) and the above-stated continuity properties of K_L that K is continuous in the f' and measureable in the t' s. It therefore corresponds, by nuclearity, to an element $\underline{K}$ of $S'((X\otimes R)^r)\otimes R^{nr}$, according to the formula

$$\underline{K}(f_1\otimes h_1, \dots, f_r\otimes h_r) =$$

$$\int dt_1 \dots dt_r\, K(f_1, t_1; \dots; f_r, t_r)\, h_1(t_1) \dots h_r(t_r) \quad \forall h_1, \dots h_r \in S(R) \tag{4.8}$$

$\underline{K}$ extends to a state of the tensor algebra $\underline{S} = \sum_{r=0}^{\infty}{}^{\oplus} S((X\otimes R)^r)\otimes R^{nr}$ [14], with

$$\underline{K}(F_1\otimes F_2 \dots \otimes F_r) = \underline{K}(F_1, F_2, \dots, F_r) \ \forall F_1, \dots, F_r \in S(X\otimes R)\otimes R^n \tag{4.9}$$

The positivity and space-time translational invariance of $\underline{K}$ follows from the corresponding properties of the state $\hat{W}$. Thus, $\underline{K}$ is a space-time translationally invariant state on the tensor algebra $\underline{S}$, and ($\underline{S}$, $\underline{K}$) corresponds to a field theory [14].

Def. 2. The system ($\underline{S}$, $\underline{K}$) is termed a classical field if $\underline{K}(F_1, \dots, F_r)$ is invariant under permutations of the F' s ($\in S(X\otimes R)\otimes R^n$) for all positive r. It is termed classical at the two-point level if it possesses this property for r = 2.

Theorem 1. (a) Assuming the KMS property for $\hat{W}$, ($\underline{S}$, $\underline{K}$) is classical at the two-point level.

(b) If, further, the (measureable)function $t\ (\in R) \to K(f, t; g, 0)$ is continuous, for all $f, g \in S(X)\otimes R^n$, then

$$K(f, t; g, 0) = K(g, 0; f, t) \ \forall t \in R \tag{4.10}$$

(c) ($\underline{S}$, $\underline{K}$) is a classical field if S is M-abelian.

(d) If, further, $\underline{K}$ satisfies the technical condition that

$$|\underline{K}(F^{(r)})| < A(F)^r, \tag{4.11}$$

where $F^{(r)}$ is the r-fold tensor product $F\otimes F \dots \otimes F$ and $A \in S'(X\otimes R)\otimes R^n$, then this field represents a classical stochastic process, Q, indexed by $S(X\otimes R)\otimes R^n$, with characteristic function

$$C(F) \ (= E(\exp(iQ(F)))) = \sum_{r=0}^{\infty} \underline{K}(F^{(r)})/r! \quad \forall F \in S(X\otimes R)\otimes R^n \tag{4.12}$$

Proof of Theorem 1. (a) Since the cyclic vector w representing the state $\hat{W}$ lies in the domain of the operators $\hat{q}(f)$, the KMS condition (3.1) is applicable with $a = \hat{q}(f)$ and $b = \hat{q}(g)$, and therefore, by (4.5) and (4.6),

$$\int dt\, K_L(f, t; g, 0)\, h(t) = \int dt\, K_L(g, 0; f, t)\, h(t - iL^{-2}) \qquad (4.13)$$

$$\forall\, f, g \in S(X) \otimes \mathbf{R}^n, \quad h \in \hat{D}(\mathbf{R})$$

Further, it follows from (3.2), (4.2), (4.3) and (4.6) that K_L satisfies the Schwartz inequalities

$$|K_L(f, t; g, 0)| \text{ and } |K_L(g, 0; f, t)|$$

$$\leq |K_L(f, 0; f, 0)|^{\frac{1}{2}} |K_L(g, 0; g, 0)|^{\frac{1}{2}} \qquad (4.14)$$

and (4.7) implies that the r.h.s. of this formula tends to a finite limit as $L \to \infty$. Hence, by the dominated convergence theorem, we may use (4.7) to pass to the following limiting form of (4.13):

$$\int dt\, K(f, t; g, 0)\, h(t) = \int dt\, K(g, 0; f, t)\, h(t) \quad \forall h \in \hat{D}(\mathbf{R})$$

Moreover, this formula extends to S-class test-functions h, since, by (4.7) and (4.14), the moduli of the K-functions in it are uniformly bounded by $|K(f, 0; f, 0)\, K(g, 0; g, 0)|^{\frac{1}{2}}$. Hence, the functions $t\ (\in \mathbf{R}) \longrightarrow K(f, t; g, 0)$ and $K(g, 0; f, t)$ are equal, as tempered distributions. Therefore, by the time-translational invariance of K, the same is true of the functions $(t, u)\ (\in \mathbf{R}^2) \to K(f, t; g, u)$ and $K(g, u; f, t)$. Consequently, by (4.8), $\underline{K}(F, G) = \underline{K}(G, F)$ for $F, G \in S(X \otimes \mathbf{R}) \otimes \mathbf{R}^n$, which means that ($\underline{S}$, $\underline{K}$) is classic at the two-point level.

(b) Since $K(f, t; g, 0) = K(f, 0; g, -t)$, by the time-translational invarianceof $\hat{W}$, and thus of K, it follows that the assumed continuity of $K(f, . ; g, 0)$, for all $f, g \in S(X) \otimes \mathbf{R}^n$, implies that of $K(f, 0; g, .)$ and thus of $K(g, 0; f, .)$. Therefore since, as shown in the proof of (a), the functions $K(f, . ; g, 0)$ and $K(g, 0; f, .)$ are equal, as tempered distributions, their continuity ensures that they are pointwise equal.

(c) follows from equation (4.8) and Def.' s 1 and 2.

(d) To prove this part of the theorem, it suffices to show that the function C, defined by the r.h.s. of(4.12), fulfills the conditions of Bochner' s theorem for a classical characteristic function. Thus, since it follows immediately from that equation that $C(0) = 1$, we have just to prove that C is continuous and that

$$\sum_{j,k=1}^{p} a_j^* a_k \, C(F_j - F_k) \geq 0 \quad \forall a_j \in \mathbf{C},\ F_j \in S(X \otimes \mathbf{R}) \otimes \mathbf{R}^n,\ j = 1,\ldots,p, \qquad (4.15)$$

for finite p. To establish these properties, we first note that, as $F^{(r)}$ is the r-fold tensor product of F, it follows from the classicality of ($\underline{S}$, $\underline{K}$) that

$$\underline{K}((F+G)^{(m)}) = \sum_{r=0}^{\infty} \binom{m}{r} \underline{K}(F^{(r)} \otimes G^{(m-r)}) \qquad (4.16)$$

while the condition (4.10) and the Schwartz inequality ensuing from the positivity of $\underline{K}$, ensure that

$$|\underline{K}(F^{(r)} \otimes G^{(s)})| < A(F)^r A(G)^s \qquad (4.17)$$

We infer the continuity of C from the fact that, by (4.10), (4.17) and the continuity of A, it follows from (4.16) that, if $\{G_n\}$ is a sequence converging to 0 in $S(X \otimes \mathbf{R}) \otimes \mathbf{R}^n$, then $C(F + G_n) \to 0$ as $n \to \infty$.

To prove (4.15), we note that the positivity of $\underline{K}$ implies that

$$\underline{K}(E^* E) \geq 0, \text{ with } E = \sum_{j=1}^{p} \sum_{r=0}^{N\oplus} a_j (-i)^r F_j^{(r)} / r! \quad \text{and N finite,}$$

i. e.,
$$\sum_{j,k=1}^{p} a_j^* a_k \sum_{r,s=0}^{N\oplus} i^{(r+s)} \underline{K}(F_j^{(r)} \otimes (-F_k)^{(s)}) / r!\, s! \geq 0 \qquad (4.18)$$

for finite N. Using the inequalities (4.10) and (4.17), it is a simple matter to pass to the limit of (4.18) as $N \to \infty$, and then to see that, by (4.10) and (4.16), the resultant formula is (4.15), as required.

□

5. THE ONSAGER RELATIONS.

The theory of this Section will be based on the properties of the two-point functions $K(f, t; g, u)$. We start with the following assumption, to the effect that, in the limit $L \to \infty$, the fluctuation observables, $\{\hat{q}_L(f)\}$, for $t = 0$, have the classical property that the expectation value of their commutators vanish.

$$\text{(I)} \qquad \lim_{L \to \infty} \left((\hat{q}_L(f)w, \hat{q}_L(g)w) - (\hat{q}_L(g)w, \hat{q}_L(f)w) \right) = 0,$$

i. e., by (4.6) and (4.7),

$$K(f, 0; g, 0) = K(g, 0; f, 0) \qquad (5.1)$$

Note. Since this equation is implied by (4.10), it follows that assumption (I) is weaker than that of Theorem 1(b). Nevertheless, (I) is not universally valid. For example, it fails if the observables comprising $\hat{q}$ include different components of magnetic polarisation.

Our next assumption is a macroscopic version of the cluster property (3.7). The observation behind it is that, if the supports of f and g are disjoint, then those of f_L and g_L are separated by a distance that tends to infinity with L; and our assumption is that $\hat{q}_L(f)$ and $\hat{q}_L(g)$ become uncorrelated in that limit. Thus, in view of (4.5)-(4.7), we formulate this as

(II) $K(f, 0; g, 0) = 0$ if f, g have disjoint supports. (5.2)

Theorem 2. Under the assumptions (I) and (II), the static two-point function $K(f, 0; g, 0)$ takes the form

$$K(f, 0; g, 0) = \langle f, (I \otimes B)g \rangle \qquad (5.3)$$

where the angular brackets denote the inner product in the real Hilbert space $L^2(X) \otimes \mathbf{R}^n$ and B is a Hermitian operator in $\mathbf{R}^n$.

Note. The matrix B is closely related to the equilibrium thermodynamical properties of the system, since, under mild supplementary conditions, its inverse is the second derivative of the specific entropy, s, w.r.t. **q** [7].

Proof of Theorem 2. Let a, b be vector in $\mathbf{R}^n$. Then it follows from our definition of K that the functional $(h, k) (\in S(X) \times S(X)) \to K(h\otimes a, 0; k\otimes b, 0)$ is a translationally invariant tempered distribution. It therefore corresponds canonically to an element G_{ab} of $S'(X)$, i. e.,

$$K(h\otimes a, 0 ; k\otimes b, 0) = \int dx\, G_{ab}(h_x) k(-x),$$

where h_x is defined in (4.1). Further, assumption (II) implies that G_{ab} has support at the origin and therefore consists of a finite linear combination of δ, the Dirac distribution supported by $\{0\}$, and its derivatives. Hence,

$$K(h\otimes a, 0 ; k\otimes b, 0) = \sum B_{ab,r} \int dx\, h(x) k^{(r)}(x) \qquad (5.4)$$

where the sum is finite, the B' s are constants, $k^{(0)} = k$ and the other $k^{(r)}$' s are derivatives of k. Further, it follows from (4.5)-(4.7) that the l.h.s. of (5.4) is invariant under the scale transformations $h, k \to h_p , k_p$, where

$$h_p(x) = p^{d/2} h(p x).$$

Since the term for which r = 0 is the only one on the r.h.s. of (5.4) that is invariant under these transformations, that equation must reduce to the form

$$K(h\otimes a, 0 ; k\otimes b, 0) = B_{ab} \int dx\, h(x) k(x) \qquad (5.5)$$

where B_{ab} is a constant. Moreover, as the l.h.s. of this formula is linear in each of the vectors a and b, it follows that B_{ab} is of the form (a, B b), where B is an operator in $\mathbf{R}^n$ and the brackets denote the inner product in that space. Hence, (5.5) implies the formula (5.3), and assumption (I) signifies that B is Hermitian.

□

Turning now to the time-dependent two-point functions K(f, t; g, 0), we make two further assumptions, that correspond to those at the basis of the Onsager theory [1]. The first of these is that, at the microscopic level, both the observables $\hat{q}(f)$ and the dynamics of S are invariant under time-reversals, and thus

$$(w, \hat{q}(f, t) \hat{q}(g) w) = (w, \hat{q}(f, -t) \hat{q}(g) w).$$

By (4.5)-(4.7), this implies the following time-reversibility condition at the macroscopic level.

(III) $K(f, t; g, 0) = K(f, -t; g, 0)\ \forall f, g \in S(X)\otimes \mathbf{R}^n, t\in \mathbf{R}$ (5.6)

The next assumption is the following version of Onsager' s [1] hypothesis, relating phenomenological laws to the regressions of the fluctuations of the relevant macroscopic variables.

(IV) In the limit $L\to\infty$, the regressions of the fluctuation field $\hat{q}_L$ conform to the phenomenological law (2.5), i. e.

$$K(f, t; g, 0) = K(T^*(t)f, 0; g, 0)\ \forall f, g \in S(X)\otimes\mathbf{R}^n, t\in\mathbf{R}_+ \qquad (5.7)$$

<u>Theorem 3.</u> Assuming (I)-(IV), the matrices L and B, governing the phenomenological law (2.2) and the static correlation functions of (5.3), respectively, satisfy the Onsager reciprocity condition that LB is Hermitian.

Proof. It follows from (III) and (IV), together with the continuity of $T^*(t)$, that the condition of Theorem 1(b) is fulfilled. Hence, by Th. 1(b), assumption (III) and the time-translational invariance of $\hat{W}$, and thus of K,

$$K(f, t; g, 0) = K(f, -t; g, 0) = K(f, 0; g, t) = K(g, t; f, 0) \quad \forall f, g \in S(X) \otimes R^n, t \in R \qquad (5.8)$$

Consequently, by assumption (IV),

$$K(T^*(t)f, 0; g, 0) = K(T^*(t)g, 0; f, 0) \ \forall t \in R_+$$

i. e., by Theorem 2 and equation (2.6),

$$\langle (\exp(\Delta \otimes L^*)t)f, (I \otimes B)g \rangle = \langle (\exp(\Delta \otimes L^*)t)g, (I \otimes B)f \rangle \ \forall t \in R_+$$

On taking the dervative of this formula w.r.t. t, at t = 0, we see that

$$\langle (\Delta \otimes L^*)f, (I \otimes B)g \rangle = \langle (\Delta \otimes L^*)g, (I \otimes B)f \rangle \equiv \langle (I \otimes B)f, (\Delta \otimes L^*)g \rangle,$$

as the inner products are those of a real Hilbert space. Hence, it follows that $LB = B^*L^*$, i. e. that LB is Hermitian, as required.

□

6. THE CLASSICAL MARKOV PROCESS.

We shall now investigate the general properties of the fluctuation process governed by the set of r-point functions K, on the basis of the assumption that the corresponding truncated r-point functions vanish. This is basically an assumption to the effect that the equilibrium state $\hat{W}$ has suitably short-range correlations w.r.t. both space and time translations, since the distance and time-scales on which the formulation of K is based are L and L^2, with $L \to \infty$. A version of this assumption, related to spatial translations, has already been substantiated [15] under very general conditions.

Our assumption, then, is the following.

$$(V) \qquad K(f_1 , t_1; \dots ; f_r , t_r) = \sum \prod K(f_j , t_j; f_k , t_k), \qquad (6.1)$$

the sum being taken over all products corresponding to partitions of (1, 2,. , r) into pairs (j, k), with j < k.

Theorem 4. Under the assumptions (I)-(V), the functions K correspond to a stationary, classical, Gaussian, process q_c, indexed by $(S(X)\otimes R^n) \times \mathbb{R}$, that is Markovian w.r.t. time evolution, with zero mean and covariance given by

$$E(q_c(f, t) q_c(g, 0)) = \langle (\exp(\Delta \otimes L^*) \mid t \mid)f, (I\otimes B)g \rangle \qquad (6.2)$$

Proof. Since, by (5.8) and time-translational invariance, $K(f, t; g, u)$ is invariant under $(f, t) \rightleftarrows (g, u)$, it follows from (6.1) that $K(f_1 , t_1;... ; f_r , t_r)$ is invariant under the permutations $(f_j , t_j) \rightleftarrows (f_k , t_k)$. Hence, by (V), the K' s correspond to a classical Gaussian process, q_c, indexed by $(S(X)\otimes R^n) \times \mathbb{R}$, with zero mean and covariance

$$E(q_c(f, t) q_c(g, u)) = K(f, t; g, u) \qquad (6.3)$$

Further, the invariance of the K' s under time-translations implies that the process is stationary, while it follows from (2.6), (5.3), (5.6) (5.7) and (6.3) that the covariance function is indeed given by (6.2).

To establish the Markov property, we note that, as the process is Gaussian, it follows from (2.6) and (6.3) that, if $t_1 , t_2 ,... , t_r > 0$, then

$$E(\exp i(q_c(f , 0) + \sum_{j=1}^{r} q_c(f_j , t_j))) =$$

$$\exp - \tfrac{1}{2}(\langle f, (I\otimes B)f \rangle + 2\sum_{j=1}^{r} \langle T^*(t_j)f_j , (I\otimes B)f \rangle +$$

$$\sum_{j,k=1}^{r} \langle T^*(|t_j - t_k|)f_j , (I \otimes B)f_k \rangle), \qquad (6.4)$$

from which it follows that

$$E(\exp i(\sum_{j=1}^{r} q_c(f_j , t_j) \mid q_c(t = 0))) = \exp i(\sum_{j=1}^{r} T^*(t_j)f_j) \times$$

$$\exp -\tfrac{1}{2}(\sum_{j,k=1}^{r} \langle T^*(|t_j - t_k|)f_j , (I\otimes B)f_k \rangle) \times$$

$$\exp \tfrac{1}{2}(\sum_{j,k=1}^{r} \langle T^*(t_j)f_k , (I\otimes B)T^*(t_k)f_j \rangle) \qquad (6.5)$$

If, further, $0 \geq t_{r+1} , .. , t_{r+s}$, then, we may use the formula (6.4), with $f = 0$ and j, k running from 1 to $r + s$, i. e.

$$E(\exp i(\sum_{j=1}^{r+s} q_c(f_j , t_j)))= \exp -\tfrac{1}{2}(\sum_{j,k=1}^{r+s} \langle T^*(|t_j - t_k|)f_j , (I\otimes B)f_k \rangle) \qquad (6.6)$$

It follows now from (6.5), (6.6) and the semigroup property of $T^{*}(\mathbb{R}_{+})$ that, in view of the positivity of the first r f_j' s and the non-positivity of the last s of them that

$$E(\, E(\exp i(\sum_{j=1}^{r} q_c(f_j, t_j))) \mid q_c(t=0))\, \exp i(\sum_{k=r+1}^{r+s} q_c(f_k, t_k)) = E(\exp i(\sum_{j=1}^{r+s} q_c(f_j, t_j))),$$

which proves that the process is Markovian. □

Comments. (1) The result of this theorem was effectively assumed in the the work of Onsager and Machlup [16].
(2) A simple quantum model, exhibiting a classical Markovian fluctuation process has been treated by Torres [17].

REFERENCES.

1. L. Onsager: Phys. Rev. 37, 405 (1931) ; and 38, 2265 (1931)
2. D. Ruelle: ' Statistical Mechanics ', Benjamin, New York, 1969
3. G. G. Emch: ' Algebraic Methods in Statistical Mechanics and Quantum Field Theory', Wiley, New York, London, 1972
4. W. Thirring: ' Quantum Mechanics of Large Systems ', Springer, New York, Vienna, 1980.
5. G. L. Sewell: ' Quantum Theory of Collective Phenomena ', Clarendon Press, Oxford, 1986
6. O. E. Lanford and D. Ruelle: Commun. Math. Phys. 13, 194 (1969)
7. G. L. Sewell: ' Macrostatistics and Non-Equilibrium Thermodynamics ', Preprint, to appear in the Proceedings of the 1988 Ascona Conference on ' Stochastics Processes, Physics and Geometry '
8. R. F. Streater and A. S. Wightman: ' PCT, Spin and Statistics and All That ', Benjamin, New York, 1964
9. D. A. Dubin and G. L. Sewell: J. Math. Phys. 11, 2990 (1970)
10. G. L. Sewell: Lett. Math. Phys. 6, 209 (1982)
11. M. Takesaki: ' Tomita' s Theory of Modular Hilbert Algebras and its Applications', Springer Lec. Notes in Maths. 128, New York, Berlin, 1970
12. G. G. Emch and H. J. F. Knops: J. Math. Phys. 11, 3008 (1970)
13. G. L. Sewell: J. Math. Phys. 11, 1868 (1970)
14. H. J. Borchers: Nuov. Cim. 24, 214 (1962)
15. D. Goderis, A. Verbeure and P. Vets: ' Non-Commutative Central Limits ', Preprint; and D. Goderis and P. Vets, ' Central Limit Theorem for Mixing Quantum Systems and

the CCR Algebra of Fluctuations ', Preprint.
16. L. Onsager and S. Machlup: Phys. Rev. 91, 1505 (1953); and 91, 1512 (1953)
17. P. L. Torres: J. Math. Phys. 18, 301 (1977)

This paper is in final form and no similar paper has been or is being submitted elsewhere.

CORRECTION TO THE HYDRODYNAMICAL APPROXIMATION FOR GROUPS OF BOGOLJUBOV TRANSFORMATIONS

Alexei G. Shuhov
Institute for Problems of Information Transmission
USSR Academy of Sciences
19, Yermolova str., Moscow 101447 USSR

Yuri M. Suhov
CPT CNRS and U.E.R; Luminy
Case 907 F-13288 Marseille Cedex 9 France

Abstract. The paper is devoted to the problem of deriving the equations which describe the evolution of the space profile of local equilibrium parameters of a state of CAR or CCR C^*-algebra in the course of the time evolution determined by a group the Bogoljubov transformations. The correction to the hydrodynamic limit is obtained.

1.Introduction. Preliminary facts

The paper is devoted to asymptotic analysis of the states of CAR or CCR C^*-algebras which are of the form $T^*_{\epsilon^{-1}t}Q^\epsilon$, $\epsilon > 0$, $t \in R^1$. Here T_τ, $\tau \in R^1$, is a group of the Bogoljubov transformations (BT), Q^ϵ, $\epsilon > 0$, is a "hydrodynamical" family of initial states. We investigate the evolution of the initial space profile of "local equilibrium" parameters which are given by the quadruple of operators, $M_{\alpha,\beta}(x)$, $\alpha, \beta = 1, 2$, depending upon the space point $x \in R^\nu$. We refer the reader to the papers [1 - 5] for the detailed statement of the problem. In particular, the reader is referred to [5] where the basic notations are taken from.

Parameter ϵ characterizes the ratio of the microscopic and macroscopic space-time scales. The asymptotics under consideration is $\epsilon \to 0$. The limiting formulas and limiting equation for operators $M_{\alpha,\beta}(t, x)$, $\alpha, \beta = 1, 2$, $t \in R^1$, $x \in R^\nu$, determining the space profile of local equilibrium parameters for the states $T^*_{\epsilon^{-1}t}Q^\epsilon$ as $\epsilon \to 0$ are investigated in [5]. The question of writing "corrections", of a given accuracy in ϵ, to limiting formulas was not studied so far. However, this question is interesting from both the physical and mathematical point of view. In particular, the correction of the first order in ϵ must lead, generally speaking, to Navier-Stokes type equations (limiting equations which are derived in [5] are interpreted as Euler type equations).

In this paper we write corrections of the first order for the equation derived in [5]. The question about higher order corrections remains open (see Remark 2.4 in the next section). Nevertheless, we formulate the main result, Theorem 1, as a general form assertion which gives, in principle, the possibility to find, at the moment $t \neq 0$, the correction of any order in ϵ provided that one knows the expansion up to the same order for the states Q^ϵ at the initial moment of time. As we think, the related equations must appear in analysis of the situation on "larger" time intervals when, instead of the states $T^*_{\epsilon^{-1}t}Q^\epsilon$, one considers the states $T^*_{\epsilon^{-s}t}Q^\epsilon$, $s > 1$. We notice that the idea of deriving the Navier-Stokes type equations on the time intervals $\sim \epsilon^{-2}t$ is now actively discussed in the literature (cf. [6,7]) as a rival to the previously discussed conception based on the limiting picture for time intervals $\sim \epsilon^{-1}t$.

Examples of families of states Q^ϵ satisfying conditions of Theorem 1 (for the case of the first order correction) are given in a separate paper.

Theorem 1 is formulated for the case of continuous fermion systems; very simple modifications lead to the same result for the continuous boson case (cf. [5]). On the other hand, for the lattice case some technical complications appear which make the statement more cumbersome, and we shall not consider the lattice case in this paper.

Let $V = L_2(R^\nu)$ be the physical Hilbert space of the system, $J: V \to V$ be an operator of complex conjugation, $\mathbf{H} = \exp^{\oplus}_{-} V$ be the corresponding fermion Fock space, $\mathfrak{R}$ be the CAR C*-algebra generated by the fermion creation-annihilation operators $a^+(f), a(g), f, g \in V$, acting in $\mathbf{H}$ (we suppose that $a^+(f)$ depends on f linearly and $a(f)$ antilinearly). The group of BT, $\mathrm{T} = (T_\tau, \tau \in R^1)$, of the C*-algebra $\mathfrak{R}$ is uniquely determined by the one parameter group of operator matrices, $\mathbf{T} = (\mathbf{T}_\tau, \tau \in R^1)$,

$$\mathbf{T}_\tau = \begin{pmatrix} \mathbf{T}^1_\tau & \mathbf{T}^2_\tau \\ \mathbf{T}^2_\tau & \mathbf{T}^1_\tau \end{pmatrix}, \tau \in R^1,$$

where T^1_τ is linear and T^2_τ antilinear bounded operator in V:

$$T_\tau a^+(f) = a^+(\mathbf{T}^1_\tau f) + a(\mathbf{T}^2_\tau f).$$

Let $\mathrm{U} = (U_y, y \in R^\nu)$ be the ν-parameter unitary group of space translations in V. The action of U induces the action of the group of $*$-automorphisms of $\mathfrak{R}$. The last group is called again the group of space translations and denoted by U.

We shall suppose that the group T commutes with U. Denote by $\mathbf{D}$ the generator of the group $\mathbf{T}_\tau$: $\mathbf{T}_\tau = \exp(i\tau\mathbf{D})$. $\mathbf{D}$ is an operator matrix, $(D_{\alpha\beta}, \alpha, \beta = 1, 2)$, where $D_{11} = D_{22} = B$, $D_{12} = D_{21} = C$. The operators B and C in the spectral (Fourier) representation have the form $\hat{B}\hat{f}(k) = b(k)\hat{f}(k)$, $\hat{C}\hat{f}(k) = c(k)\hat{J}\hat{f}(k)$, $\hat{f} \in \hat{V}$, where $\hat{V}$ is the dual to V, b is a real function and c is an odd function on R^ν, $\hat{J}\hat{f}(k) = \hat{f}(-k)_-$ is the operator of complex conjugation in $\hat{V}$. In the sequel we shall use the following functions:

$$b_\pm(k) = 1/2(b(k) \pm b(-k)), w = (b_+^2 + |c|^2)^{1/2},$$

$$b_1 = b_+ w^{-1}, \omega_\pm = b_- \pm w.$$

Given a state Q of the C*-algebra $\mathfrak{R}$, we define the family of states

$$T^*_\tau Q(A) = Q(T_{-t}A), A \in \mathfrak{R},$$

with $T^*_0 Q = Q$, which describes the time evolution of Q.

A convenient characteristic of a state Q is the operator matrix $\mathbf{M}^{(Q)} = (M_{\alpha\beta}, \alpha, \beta = 1, 2)$ which is defined by

$$Q(a^+(f)a^+(g)) = < f, M_{11}^{(Q)} g >, Q(a(f)a(g)) = < M_{22}^{(Q)} g, f >,$$

$$Q(a^+(f)a(g)) = < f, M_{12}^{(Q)} g >, f, g \in V,$$

$$M_{21}^{(Q)} = E - M_{12}^{(Q)*},$$

where E is the unit operator in V. Diagonal operators $M_{11}^{(Q)}$, $M_{22}^{(Q)}$ are antilinear, off-diagonal operators $M_{12}^{(Q)}, M_{21}^{(Q)}$ are linear with $\| M_{\alpha\beta}^{(Q)} \| \leq 1$, $\alpha, \beta = 1, 2$.

We shall assume that the group $\mathbf{T}$ satisfies the following non-degeneracy condition

(D) The functions $b, c \in C^\infty$, the set $\beta_2(\omega_+) \cup \beta_2(\omega_-) \cup \beta_1(w, 0)$ is of zero Lebesgue measure, and for any compact $K \subset R^\nu$ and any $h \in R^\nu$ the set $\beta_1(\omega_\kappa, h) \cap K$, $\kappa = \pm$, is finite. here and below

$$\beta_2(\omega_\pm) = [k : det \frac{\partial^2 \omega_\pm}{\partial k_j \partial k_s} = 0],$$

$$\beta_1(w, k) = [k : \mathbf{grad} w = h].$$

2. The hydrodynamical approximation for BT groups : the n-thorder correction to the limiting equation

Fix positive integer n. Suppose a family of operator matrices, $\mathbf{M}^\epsilon(x) = (M_{\alpha,\beta}^\epsilon(x), \alpha, \beta = 1, 2)$, $x \in R^\nu$, is given which commutes with space translations and satisfies the following conditions (for simplicity we shall omit the index ϵ):

(i) For any $x \in R^\nu$ the diagonal operators $M_{11}(x), M_{22}(x)$ are antilinear and off-diagonal operators are linear in V of the norm ≤ 1 and the matrix $\mathbf{M}(x)$ obeys

$$(i\mathbf{D})^* \mathbf{M}(x) + \mathbf{M}(x)(i\mathbf{D}) = 0. \tag{2.1}$$

The equality (2.1) means that the matrix $\mathbf{M}(x)$ has the property of invariance wrt the group $\mathbf{T}$: $\mathbf{T}_{-\tau}^* \mathbf{M}(x) \mathbf{T}_{-\tau} = \mathbf{M}(x)$.

(ii) The operators $M_{\alpha\beta}$, $\alpha, \beta = 1, 2$, depend on x in a smooth manner (relative to the norm topology), and the norms $\| \frac{\partial^s}{\partial x_j^s} M_{\alpha\beta}(x) \|$, $0 \leq s \leq 2\nu + n + 1$, are uniformly bounded provided x remains bounded.

(iii) The linear operators $M_{\alpha\alpha} J$, $\alpha = 1, 2$, and $M_{\alpha\beta}$, $\alpha \neq \beta$, are integral ones with kernels $m_{\alpha\beta}(x; y, y') = m_{\alpha\beta}(x, y - y')$ satisfying the condition

$$\sup_{\alpha,\beta} \sup_{\epsilon, x} | m_{\alpha\beta}(x, y) | < \phi^{(0)}(| y |) \tag{2.2}$$

where a function $\phi^{(0)} : R_+^1 \to R_+^1$ is such that $s^r \phi^{(0)}(s) \to 0$ as $s \to \infty$ for any $r > 0$.

Let us suppose now that a family $(Q^\epsilon, \epsilon > 0)$ of states of the C$*$-algebra $\Re$ is given which has the following properties.

(A$_1$) The linear operators $M_{\alpha\alpha}^{(Q^\epsilon)}J$, $\alpha = 1, 2$, and $M_{\alpha\beta}^{Q^\epsilon}$, $\alpha \neq \beta$, are integral ones with kernels $m_{\alpha\beta}^{(Q^\epsilon)}(y, y')$ satisfying the condition:

$$\sup_{\alpha,\beta} \sup_{\epsilon} | m_{\alpha\beta}^{(Q^\epsilon)}(y, y') | < \phi^{(1)}(| y - y' |) \tag{2.3}$$

where a function $\phi^{(1)} : R^1_+ \rightarrow R^1$ has $\int \phi^{(1)}(| y |)dy < \infty$.

(A$_2$) For any $y, y' \in R^\nu$

$$\sup_{\alpha,\beta} | m_{\alpha\beta}^{(Q^\epsilon)}(y, y') - m_{\alpha\beta}(\epsilon\bar{y}, y - y') | < \phi_\epsilon(| y - y' |) \tag{2.4}$$

where $\bar{y} = 1/2(y + y')$ and

$$\lim_{\epsilon \to 0} \epsilon^{-n} \int \phi_\epsilon(| y |)dy = 0.$$

Remark 2.1. The condition (A$_2$) means that the family $\mathrm{M}(x)$, $x \in R^\nu$, gives an $o(\epsilon^n)$ approximation to the lower moment functionals of the states Q^ϵ. Generally speaking, the matrix $\mathrm{M}(x)(= \mathrm{M}^\epsilon(x))$ must contain terms of order $O(\epsilon^m)$, $m < n$.

We introduce the family of operators, $\mathrm{M}_n(t; x)(= \mathrm{M}_n^\epsilon(t; x))$, $x \in R^\nu$, in the space $V \oplus V$ which are connected with $\mathrm{M}(x)$ in the following way. It is convenient to pass to the Fourier representation in the dual space $\hat{V} \oplus \hat{V}$. The operator $\hat{\mathrm{M}}_n(t; x)$ is written in the form

$$(\hat{\mathrm{M}}_n(t; x)\hat{f})(k) = 1/2[(1 + b_1(k))D_n(t, \epsilon, \omega_+)\hat{\mathrm{M}}(x, k) +$$

$$+ (1 - b_1(k))D_n(t, \epsilon, \omega_-)\hat{\mathrm{M}}(x, k)]\hat{f}(k) \tag{2.6}$$

Here $D_n(t, \epsilon, \omega_\pm)$is the linear operator :

$$D_n(t, \epsilon, \omega_\pm)\hat{m}(x, k) = \exp[-t(\mathbf{grad}_k\omega_\pm(k)_*\mathbf{grad}_x -$$

$$-\frac{1}{3!}(\frac{\epsilon}{2})^2\mathbf{grad}_k^{\otimes 3}\omega_\pm(k)_*\mathbf{grad}_x^{\otimes 3} + \ldots +$$

$$+\frac{(-1)^{\frac{\tilde{n}-1}{2}}}{\tilde{n}!}(\frac{\epsilon}{2})^{\tilde{n}-1}\mathbf{grad}_k^{\otimes\tilde{n}}\omega_\pm(k)_*\mathbf{grad}_x^{\otimes\tilde{n}})]\hat{m}(x, k) \tag{2.6a}$$

where $\tilde{n} = n$ for odd n and $\tilde{n} = n - 1$ for even n ($\tilde{n} = 2[\frac{n-1}{2}] + 1$).

Equivalently, the family $\mathrm{M}_n(t, x)$ can be defind as the solution of a differential equation. We write down this equation in the one dimensional case($\nu = 1$) where it has a particularly simple form:

$$(\frac{\partial}{\partial t} - P_n(\epsilon, \omega_+))(\frac{\partial}{\partial t} - P_n(\epsilon, \omega_-))\hat{\mathrm{M}}_n(t; x) = 0 \tag{2.7}$$

where

$$P_n(\epsilon,\omega_{+,-}) = -(\omega'_{+,-}(k)\frac{\partial}{\partial x} - \frac{1}{3!}(\frac{\epsilon}{2})^2\omega^{(3)}(k)(\frac{\partial}{\partial x})^3 + \ldots +$$

$$+\frac{(-1)^{\frac{\tilde{n}-1}{2}}}{\tilde{n}!}\omega^{(\tilde{n})}(k)(\frac{\epsilon}{2})^{\tilde{n}-1}(\frac{\partial}{\partial x})^{\tilde{n}}).$$

Denote

$$\tilde{V} = (f \in V : \hat{f} \in C_0^\infty, (\mathrm{supp}\hat{f} \cup \mathrm{supp}(\hat{J}\hat{f})) \cap (\beta_2(\omega_-) \cup$$

$$\cup \beta_2(\omega_+) \cup \beta_1(w)) = \emptyset).$$

Theorem 1. *Assume a BT group* T, *family of operator matrices* $(\mathrm{M}(x), x \in R^\nu)$ *and family of states* $(Q^\epsilon, \epsilon > 0)$ *satisfy the conditions (D), (i) - (iii) and* (A_1) - (A_2), *respectively. Then for any* $f, g \in \tilde{V}$ *and* $t^0, t^1, 0 < t^0, t^1 < \infty$,

$$\lim_{\epsilon\to 0} \epsilon^{-n} |< U_{-\epsilon^{-1}x} M^{(T^*_{\epsilon^{-1}}, Q^\epsilon)} U_{\epsilon^{-1}x} f, g > -$$

$$- \int\int dl dl' m^{(n)}_{\alpha\beta}(t, x + \epsilon\frac{(l+l')}{2}, l - l')(J^{|\alpha-\beta|+1} f)(l') g^-(l) | = 0 \qquad (2.8)$$

uniformly in $t \in (t^0, t^1)$, *where* $m_{\alpha\beta}(t, x, .)$ *is the kernel of operator* $(M_n)_{\alpha\beta}$ *for* $\alpha \neq \beta$ *and is the kernel of operator* $(M_n)_{\alpha\alpha} J$ *for* $\alpha = \beta$.

Remark 2.2. The statement of Theorem 1 remains true if one uses in condition (A_2) the function $1/2(m_{\alpha\beta}(\epsilon y, y - y') + m_{\alpha\beta}(\epsilon y', y - y'))$ instead of $m_{\alpha\beta}(\epsilon\bar{y}, y - y')$. However, in that case this condition seems unnatural from the physical point of view. One can propose "unsymmetric" versions of the condition (A_2) replacing $\bar{y} = (y + y')/2$ by $(py + (1-p)y'), p \in (0, 1)$. Then the statement of Theorem 1 takes place with replacing $\mathrm{M}_n(t; x + \epsilon(l + l')/2, l - l')$ by the matrix $\mathrm{M}_{n,p}(t; x + \epsilon pl + \epsilon(1-p)l', l - l')$ which is determined in the Fourier representation by (2.6) with replacing D_n by $D_{n,p}$ where

$$D_{n,p}(t, \epsilon, \omega_\pm)\hat{m}(x, k) = \exp(it(\mathrm{grad}_k\omega_\pm(k)_*(i\mathrm{grad}_x) +$$

$$\frac{1}{2!}\epsilon(p^2 - (p-1)^2)\mathrm{grad}_k^{\otimes 2}\omega(k)_*(i\mathrm{grad}_x)^{\otimes 2} + \ldots +$$

$$+\frac{1}{n!}\epsilon^{n-1}(p^n - (p-1)^n)\mathrm{grad}^{\otimes n} k\omega(k)_*(i\mathrm{grad}_x)^{\otimes n})\hat{m}(x, k).$$

Remark 2.3. Relating Theorem to Theorems 4.1 and 6;1 from [5] it is useful to notice the following. Suppose that operators $\mathrm{M}(x)(= \mathrm{M}^\epsilon(x)), x \in R^\nu$ have the limits, as $\epsilon \to 0$ (in the weak operator topology),

$$\mathrm{N}(x) = \lim_{\epsilon\to 0} \mathrm{M}^\epsilon(x).$$

Then, under the assumptions of Theorem 1 (where it suffices now to suppose that (2.5) is valid only for $n = 0$), for any $f, g \in V \oplus V$ and any $t^0, t^1 \in R^1$ with $0 < t^0 < t^1 < \infty$,

$$\lim_{\epsilon\to 0} < \mathbf{U}_{-\epsilon^{-1}x}\mathbf{M}^{(T^*_{\epsilon^{-1}t},Q^\epsilon)}\mathbf{U}_{\epsilon^{-1}x}f,g> = < \mathbf{N}(t;x)f,g>$$

uniformly in $t \in (t^0,t^1)$. Here $\mathbf{N}(t;x)$ is defined as the limit of (2.6) as $\epsilon \to 0$:

$$\hat{\mathbf{N}}(t;x) = \frac{1}{2}(1+\hat{B}_1)\hat{\mathbf{N}}_+(t;x) + \frac{1}{2}(1-\hat{B}_1)\hat{\mathbf{N}}_-(t;x)$$

where

$$\hat{\mathbf{N}}_\pm(t;x)\hat{f}(k) = \exp(-t\,\mathbf{grad}_k\omega(k)_*\mathbf{grad}_x)\hat{\mathbf{N}}(x,k)\hat{f}(k), f \in V \oplus V,$$

$$\hat{B}_1\hat{g}(k) = b_1(k)\hat{g}(k), g \in V \oplus V.$$

This remark allows to obtain the generalization of Theorems 4.1 and 6.1 from [5] which is related to the replacement of the scaling $n(\epsilon,x) = (\mathtt{int.part}(\epsilon^{-1}xN_\epsilon^{-1})N_\epsilon$ (cf. [5], p. 191) by the more natural scaling $\epsilon^{-1}x$. Moreover, the non-degeneracy condition (D) for the group of BT T is weaker than the corresponding conditios (5.5)-(5.7) from [5].

Remark 2.4 Examples of families (Q^ϵ) satisfying the condition (A_2) are discussed in a separate paper. We notice here that in these examples one deals with $n = 1$ and, moreover, the function $m_{\alpha\beta}(x,y)$ does not depend on ϵ. As to the verification of condition (A_2) for $n > 1$, it seems that natural physical examples do not exist. However, we think that equation (2.8) has the physical sense. Namely, there is a natural conjecture that this equation describes the hydrodynamic approximation on larger time intervals of order $\epsilon^{-n}t$ (when t is replaced by $\epsilon^{-n}t$).

Remark 2.5. In the special case of free motion where $B = -1/2\Delta$ and $C = 0$, one has

$$b_1 = 1, c = 0, \omega_\pm = \pm b_+(k) = \pm k^2,$$

and

$$(\hat{\mathbf{M}}_n(t;x)\hat{f})(k) = D_n(t,\epsilon,b_+)\hat{\mathbf{M}}(t;x).$$

Instead of (2.8), the following equation arises

$$\left(\frac{\partial}{\partial t} - P_n(\epsilon,b_+)\right)\hat{\mathbf{M}}(t;x) = 0.$$

For $n = 1$ this is merely the transfer equation

$$\frac{\partial}{\partial t}\hat{\mathbf{M}}(t;x) = - < \mathbf{grad}b_+, \mathbf{grad}_x\hat{\mathbf{M}}(t;x) > .$$

In the case of unsymmetric version of condition (A_2) (see Remark 2.2), for $n = 1$ the matrix $\mathbf{M}_{1,p}, p \in (0,1)$, is described by the equation of the second order. It has particularly simple form for $\nu = 1$

$$\left(\frac{\partial}{\partial t} + b'_+(k)\frac{\partial}{\partial x} + \frac{i\epsilon}{2}(2p-1)b''(k)\frac{\partial^2}{\partial x^2}\hat{\mathbf{M}}_{1,p}(t;x)\right) = 0.$$

Proof of Theorem 1. As in [5], we restrict ourselves to the analysis of one of four terms which give the contribution to the quantity $< U_{-\epsilon^{-1}x} M_{1,2}^{(Q^{\epsilon}_{\epsilon^{-1}t})} U_{\epsilon^{-1}x} f, g >$. Namely, we consider the addend

$$< U_{-\epsilon^{-1}x}(T^1_{-\epsilon^{-1}t} U_{\epsilon^{-1}x} f, g > \tag{2.9}$$

for $t > 0$ and $f, g \in \tilde{V}$. Using the conditions (iii), (A_1), (A_2), one passes from (2.9) to the approximation

$$\int\int dy dy' (T^1_{-\tau} f)(y') \overline{T^1_{-\tau} g)(y)} m_{12}(\epsilon \bar{y} + x, y - y') \tag{2.10}$$

where $\bar{y} = 1/2(y + y'), \tau = \epsilon^{-1} t$. The difference between (2.9) and (2.10) is of the order $o(\epsilon^n)$ and estimated uniformly in $t \in (t^0, t^1)$ due to the condition (A_2).

Now, as in [5], we pass to the Fourier transform. In the Fourier representation, $\hat{T}^1_{-\tau}$ is the multiplication by the function

$$\exp(i\tau b_-)(\cos(\tau w) + (ib_+/w)\sin(\tau w)$$

(see [9], formulas (0.7) - (0.8)). Substituting this quantity into (2.10) and writing cos(.) and sin(.) as the sums of exponents, one gets the 16 addends of the same type (cf. [5], (3.26)):

$$\frac{1}{4}(2\pi)^{\frac{-3\nu}{2}} \int du du' \int dk dk' \hat{f}(k) \overline{\hat{g}(k')} \exp[i\tau\omega_{\kappa'}(k') - i\tau\omega_{\kappa}(k) -$$

$$ik(u' + u/2) + ik'(u' - u/2)] \int dk'' \exp(ik''u) \hat{m}_{12}(\epsilon u' + x, k'') =$$

$$= \frac{1}{4}(2\pi)^{-\frac{\nu}{2}} \int du' \int dk dk' \hat{f}(k) \overline{\hat{g}(k')} \exp[i\tau\omega_{\kappa'}(k') - i\tau\omega_{\kappa}(k) -$$

$$-i(k - k')u'] \hat{m}_{12}(\epsilon u' + x, (k + k')/2), \kappa, \kappa' = \pm. \tag{2.11}$$

Given $x \in R^\nu$, we introduce the function $\hat{m}^0(z, k)$ such that

1) $\hat{m}^0(z, k) = \hat{m}_{12}(z, k) \forall z \in I(x, c\tau)$ where

$$c > 2 \sup_k (| \mathbf{grad}\omega_+(k) | + | \mathbf{grad}\omega_-(k) |)$$

and the sup is taken over $k \in (\mathbf{supp}\hat{f} \cup \mathbf{supp}\hat{g})$;

2) $\hat{m}^0(z, k) = 0 \forall z \notin I(x, ct + c_1)$ where $c_1 > 0$ is a fixed constant;

3) for any $k, \hat{m}^0(., k) \in C_0^{2\nu+n+1}$ and the norm

$$\sup_{\alpha:|\alpha| \leq 2\nu+n+1} \| \hat{m}^0(., k) \|_{C^\alpha}$$

is bounded uniformly in$k$ and ϵ.

Let us replace $\hat{m}_{12}$ in (2.11) by $\hat{m}^0$. The difference of the corresponding integrals is of order $o(\epsilon^n)$ uniformly in $t \in (t^0, t^1)$ (cf.[9], Ch. 3, Lemma 2.1 and Corollary 3.1).

After changing the variables $k = (k+k')/2, \phi = k' - k$, we get the following expression

$$\frac{1}{4}\int dk d\phi \hat{f}(k-\phi/2)\overline{\hat{g}(k+\phi/2)}\exp[i\tau\omega_{\kappa'}(k+\phi/2)-$$
$$-i\tau\omega_{\kappa}(k-\phi/2)]\epsilon^{-\nu}\hat{m}_x^{\wedge 0}(\epsilon^{-1}\phi, k) \tag{2.12}$$

where

$$\hat{m}_x^{\wedge O}(\xi, k) = (2\pi)^{-\nu/2}\int dy \exp(i\xi y)\hat{m}_x^0(y,k),$$
$$\hat{m}_x^0(y,k) = \hat{m}^0(x+y,k).$$

Denoting $z = \epsilon^{-1}\phi$ we rewrite (2.12) in the form

$$\frac{1}{4}\int dk \int dz \hat{f}(k-\epsilon z/2)\overline{\hat{g}(k+\epsilon z/2)}\exp[i\tau\omega_{\kappa'}(k+\epsilon z/2)-$$
$$-i\tau\omega_{\kappa}(k-\epsilon z/2)]\hat{m}_x^{\wedge 0}(z,k). \tag{2.13}$$

Due to the property 3) of $\hat{m}^0$, untegral (2.13) may be represented in the following form

$$\frac{1}{4}\int dk \exp(i\tau\omega_{\kappa'}(k) - i\tau_{\kappa}(k))\int dz[\sum_{j=0}^{n}(-1)^j \hat{f}^{(j)}(k)(\frac{\epsilon z}{2})^j]$$
$$[\sum_{j=0}^{n}\hat{g}^{(j)}(k)(\frac{\epsilon z}{2})^j]\exp[it\sum_{j=1}^{n+1}\omega_{\kappa'}^{(j)}(k)(\frac{z}{2})^j\frac{\epsilon^{j-1}}{j!}-$$
$$-it\sum_{j=1}^{n+1}\omega_{\kappa}^{(j)}(k)(-\frac{z}{2})^j\frac{\epsilon^{j-1}}{j!}]\hat{m}_x^{\wedge O}(z,k) + o(\epsilon^n). \tag{2.14}$$

For $\kappa \neq \kappa'$ the equality $\omega_{\kappa'}(k) - \omega_{\kappa}(k) = \pm w(k)$ holds and the integral (2.14) is of order $o(\epsilon^n)$ (see, for instance, [9], Ch. 3, 2) uniformly int $\in (t^0, t^1)$. For $\kappa = \kappa'$, using the properties 1) -3) of $\hat{m}^0$, one can rewrite (2.14) in the form

$$\int\int dl dl' m_n(t; x+\epsilon(l+l')/2, l-l')f(l')\overline{g}(l) + o(\epsilon^n)$$

where the function m_n in the Fourier representation gives $D_n(t,\epsilon,\omega_\kappa)\hat{m}_{12}(x,k)$.

The other terms in (2.10) are treated similarly. Writing the answer in invariant form we obtain (2.8).

REFERENCES

1. Boldrighini C., Dobrushin R.L., Suhov Yu.M. J. Stat. Phys., 1983, 31, 577-615.
2. Dobrushin R.L., Pellegrinotti A., Suhov Yu.M., Triolo L. J; Stat. Phys., 1986, 43, 571-607.

3. De Masi A., Ianiro N., Pellegrinotti A., Presutti E. In: Nonequilibrium Phenomena, II. From Stochastics to Hydrodynamucs (Montroll E.W. and Lebowutz J.L., eds.). North-Holland, Amsterdam, 1984.
4. Dobrushin R.L., Sinai Ya.G., Suhov Yu.M. In: Itogi Nauki i Tehniki. Sovremennye Problemy Matematiki. Fundamentalnye Napravlenija. 1985, v. 2, 235-284.
5.Suhov Yu.M., Shuhov A.G. Tr. Mosc. Mat. Ob-va, 1987, 50, 156-208.
6. Boldrighini C.,Wick D.Fluctuations in one-dimensional mechanical systems Preprint, 1988.
7. Dobrushin R.L., Pellegrinotti A., Suhov Yu.M., Triolo L. J. Stat. Phys., 1988, 52, 423-439.
8. Shuhov A.G., Suhov Yu.M. Ann. Phys., 1987, 175:2, 231-266.
9. Fedorjuk M.V. Metod Perevala. Nauka, Moskwa, 1987.

This paper is in final form and no similar paper has been or is being submitted elsewhere.

BELL'S INEQUALITIES AND QUANTUM FIELD THEORY*

STEPHEN J. SUMMERS**

Centre de Physique Théorique***
CNRS - Luminy, Case 907
F-13288 MARSEILLE CEDEX 09 (FRANCE)

Abstract. **The present state of mathematically rigorous results about Bell's inequalities in relativistic quantum field theory is reviewed. In addition, the nature of the statistical independence of algebras of observables associated to spacelike separated spacetime regions is discussed.**

I. INTRODUCTION.

Motivated by the desire to bring into the realm of testable hypotheses at least some of the important matters concerning the interpretation of quantum mechanics that were evoked in the controversy surrounding the Einstein-Podolsky-Rosen paradox [11,26], Bell discovered the first version [9,10] of a series of related inequalities that are now generally called Bell's inequalities and that have received a great deal of attention (for reviews see [16,7]). These inequalities provide an upper bound on the strength of correlations between systems that are no longer interacting but have interacted in their past.

The class of correlation experiments involved in these inequalities can be described briefly. A source provides an ensemble of identically prepared systems, one after another, and, as part of the preparation, splits each system into two subsystems, directing these to separate arms of the experiment. At one arm the arriving subsystem is subjected to a measuring device chosen from a class $\mathcal{A}$ of suitable devices, and at the other arm the incident subsystem interacts with a measuring device from a class $\mathcal{B}$. In the simplest situations, $\mathcal{A}$ and $\mathcal{B}$ each consists of two devices. For each device $A \in \mathcal{A}$ and each $B \in \mathcal{B}$ with possible outcome sets $\hat{A}$ and $\hat{B}$, the relative frequencies $p(\alpha, \beta)$ of the measurement of $\alpha \in \hat{A}$ on one arm and $\beta \in \hat{B}$ on the other arm are determined. An operational condition of independence of the two arms of the experiment is required:

$$\sum_{\beta \in \hat{B}} p(\alpha, \beta) \equiv p(\alpha)$$

must be independent of the choice of $B \in \mathcal{B}$, and

$$\sum_{\alpha \in \hat{A}} p(\alpha, \beta) \equiv p(\beta)$$

---

*Invited lecture at the 5th Workshop on Quantum Probability, Universität Heidelberg, Sept. 26-30,1988.
**Present address: Department of Mathematics, University of Florida, Gainesville, FL 32611, USA.
***Laboratoire Propre LP.7061, Centre National de la Recherche Scientifique.

must be independent of the choice of $A \in \mathcal{A}$. If there are $|\mathcal{A}|$ devices in $\mathcal{A}$ and $|\mathcal{B}|$ devices in $\mathcal{B}$, the one must carry out $|\mathcal{A}| \cdot |\mathcal{B}|$ correlation experiments to obtain the necessary data.

Bell's inequality, in the form of Clauser and Horne [15], is:

$$p(\alpha_1, \beta_1) + p(\alpha_1, \beta_2) + p(\alpha_2, \beta_1) - p(\alpha_2, \beta_2) \leq p(\alpha_1) + p(\beta_1), \tag{1.1}$$

for all $\alpha_i \in \hat{A}_i$, $\beta_j \in \hat{B}_j$, $A_i \in \mathcal{A}$, $B_j \in \mathcal{B}$. Bell's theorem (and the many generalizations that followed) is a metatheoretical theorem that states that all theories of a certain class that describe such a correlation experiment must provide predictions satisfying (1.1). Hence, if in a real experiment correlation probabilities are measured that violate (1.1), then one must conclude that there are real physical processes that cannot be described by any theory in the said class. If a theory predicts a violation of (1.1), then Bell's theorem implies that not all predictions of this theory can be reproduced by any theory in the said class.

What is the class of theories that must produce only correlation probabilities satisfying (1.1)? The details of the answer to this question depend on the particular set of hypotheses used to prove Bell's inequality, and Bell's theorem appears in many forms in the literature. Because it is not the object of this paper to review this multitude of theorems, let it suffice simply to say that most versions assume, explicitly or tacitly, that the theories in this class are "classical" and "local". Roughly speaking, this means that all the correlation probabilities are given by a single classical measure (the significance of this assumption was particularly emphasized in [52]; see also [53]) and the theory treats the two subsystems as independent of each other (independence is unfortunately a theory-dependent concept, hence it cannot be further specified here without entering into metatheoretic terrain). We refer to the review [16] for a detailed discussion of some of these "Bell's theorems" and to [44] for a very general approach to Bell's inequalities that enters more into the metatheoretic considerations that are necessary.

In this paper we are concerned with what standard theories, particularly quantum field theory, predict about the violation of (1.1). We first recall, in order to fix notation, how such theories model the experimental situation described above. There is a C*-algebra $\mathcal{C}$ (with identity 1) of observables for the system and a pair $(\mathcal{A}, \mathcal{B})$ of mutually commuting subalgebras of $\mathcal{C}$ (each containing 1) for the algebra of observables of the independent subsystems. The possible outcomes α, β are modelled by basic observables $\tilde{A} \in \mathcal{A}$, $\tilde{B} \in \mathcal{B}$ satisfying $0 \leq \tilde{A} = \tilde{A}^* \leq 1$, $0 \leq \tilde{B} = \tilde{B}^* \leq 1$ (projections are examples of such basic observables). To each device A, resp. B, there corresponds a collection $\{\tilde{A}_i\}$, resp. $\{\tilde{B}_j\}$, of such basic observables such that $\Sigma \tilde{A}_i = 1$, resp. $\Sigma \tilde{B}_j = 1$. Corresponding to the preparation of the ensemble of systems there is a state ϕ, a positive, normalized, linear functional on $\mathcal{C}$. Then the correlation probabilities $p(\alpha, \beta)$ are given by $\phi(\tilde{A}\tilde{B})$. Built into this model are the relations

$$\sum_j \phi(\tilde{A}_i\tilde{B}_j) = \phi(\tilde{A}_i) \quad , \quad \sum_i \phi(\tilde{A}_i\tilde{B}_j) = \phi(\tilde{B}_j),$$

for all devices $\{\tilde{A}_i\}$, $\{\tilde{B}_j\}$. Hence the operational condition of independence of the two subsystems is an integral part of the model.

Making the obvious substitutions into (1.1), one obtains Bell's inequality for the standard theories. For the sake of convenience, we rewrite this as:

$$-1 \leq \phi(A_1B_1) + \phi(A_1B_2) + \phi(A_2B_1) - \phi(A_2B_2) \leq 1, \tag{1.2}$$

with $-1 \leq A_i = A_i^* \leq 1$, $-1 \leq B_j = B_j^* \leq 1$, $A_i \in \mathcal{A}$, $B_j \in \mathcal{B}$. Note that $2\tilde{A} - 1 = A$ is a selfadjoint contraction if and only if $\tilde{A}$ is a basic observable. Bell's inequality for the standard theories is thus the requirement that (1.2) is satisfied for all pairs $\{A_1, A_2\} \subset \mathcal{A}$, $\{B_1, B_2\} \subset \mathcal{B}$ of selfadjoint contractions. We therefore make the following definition.

Definition 1.1: The maximal Bell correlation of the pair $(\mathcal{A}, \mathcal{B})$ of commuting subalgebras of the C*-algebra $\mathcal{C}$ in the state $\phi \in \mathcal{C}^*$ is

$$\beta(\phi, \mathcal{A}, \mathcal{B}) \equiv \sup \frac{1}{2}\phi(A_1(B_1 + B_2) + A_2(B_1 - B_2)),$$

where the supremum is taken over all selfadjoint contractions $A_i \in \mathcal{A}$, $B_j \in \mathcal{B}$. (Note that $\beta(\phi, \mathcal{A}, \mathcal{B})$ is a convex functional in ϕ and if $\mathcal{A}_1 \subset \mathcal{A}$, $\mathcal{B}_1 \subset \mathcal{B}$, then $\beta(\phi, \mathcal{A}_1, \mathcal{B}_1) \leq \beta(\phi, \mathcal{A}, \mathcal{B})$.)

Bell's inequality (1.2) can thus be expressed as

$$\beta(\phi, \mathcal{A}, \mathcal{B}) \leq 1. \tag{1.3}$$

We recall the following result.

Proposition 1.2 [14,43,45,30]: For any C*-algebra $\mathcal{C}$, commuting subalgebras $\mathcal{A}$ and $\mathcal{B}$ and state ϕ on $\mathcal{C}$,

$$\beta(\phi, \mathcal{A}, \mathcal{B}) \leq \sqrt{2}. \tag{1.4}$$

Hence, in standard theories the maximal Bell correlation is never greater than $\sqrt{2}$. A measured violation of (1.4) could serve to exclude all theories providing C*-algebras as models, just as a violation of (1.3) excludes "classical, local" theories. (For further information on this point, see [14,32].) However, here we are working within the standard theories.

Definition 1.3: The pair $(\mathcal{A}, \mathcal{B})$ of commuting subalgebras of a C*-algebra $\mathcal{C}$ maximally violates Bell's inequality in the state ϕ if $\beta(\phi, \mathcal{A}, \mathcal{B}) = \sqrt{2}$.

Of course, $\beta(\phi, \mathcal{A}, \mathcal{B}) > 1$ already entails violation of Bell's inequality. The following theorem collects a number of general situations where Bell's inequality must be satisfied in standard theories.

Theorem 1.4 [45]: Let $(\mathcal{A}, \mathcal{B})$ be a pair of commuting subalgebras of a C*-algebra $\mathcal{C}$.

a) If $\phi|\mathcal{A} \vee \mathcal{B}$, the restriction of ϕ to the C*-algebra generated by $\mathcal{A}$ and $\mathcal{B}$, is a convex sum of product states over $(\mathcal{A}, \mathcal{B})$, then $\beta(\phi, \mathcal{A}, \mathcal{B}) = 1$.

b) If $\mathcal{A}$ or $\mathcal{B}$ is abelian, then $\beta(\phi, \mathcal{A}, \mathcal{B}) = 1$, for all states ϕ.

c) If $\phi|\mathcal{A}$ or $\phi|\mathcal{B}$ is a pure state, then $\beta(\phi, \mathcal{A}, \mathcal{B}) = 1$.

Part (a) asserts that if the preparation of the subsystems is such that the state is a (sum of) product state over $(\mathcal{A}, \mathcal{B})$, i.e. $\phi(AB) = \phi(A)\phi(B)$ for all $A \in \mathcal{A}$, $B \in \mathcal{B}$, then the correlations between the observables of the subsystems in this state are weak enough to satisfy Bell's inequality. From (b) we learn that if at least one of the two subsystems is classical, i.e. all observables commute, then Bell's inequality is satisfied in every state. And in (c) the purity of the restriction to one of the subsystems entails weak Bell correlations.

It has been known for some time [9,10] that quantum mechanics, to state the matter in our language, predicts the existence of $(\mathcal{A}, \mathcal{B})$ and ϕ such that $\beta(\phi, \mathcal{A}, \mathcal{B}) = \sqrt{2}$. We state and prove this fact.

Theorem 1.5: Let $\mathcal{A}$ and $\mathcal{B}$ be mutually commuting copies of the two by two complex matrices $M_2(\mathbf{C})$ acting on the Hilbert space $\mathcal{H}$. Then there exists a normal state ϕ on $\mathcal{B}(\mathcal{H})$, the algebra of all bounded, linear operators on $\mathcal{H}$, such that $\beta(\phi, \mathcal{A}, \mathcal{B}) = \sqrt{2}$.

Proof: In $\mathcal{A}$, resp. $\mathcal{B}$, there is a copy $\{\sigma_x, \sigma_y, \sigma_z\}$, resp. $\{\sigma'_x, \sigma'_y, \sigma'_z\}$ of the Pauli spin matrices. Let $\Phi_\pm$, resp. $\Phi'_\pm$, satisfy $\sigma_z\Phi_\pm = \pm\Phi_\pm$, resp. $\sigma'_z\Phi'_\pm = \Phi'_\pm$, and let $\mathcal{K}$ be the four-dimensional Hilbert subspace of $\mathcal{H}$ generated by $\{\Phi_\pm, \Phi'_\pm\}$. Furthermore, let $\mathcal{D}$ be the C*-algebra generated by $\{\sigma_x, \sigma_y, \sigma_z\} \cup \{\sigma'_x, \sigma'_y, \sigma'_z\}$. Then $\mathcal{K}$ is unitarily equivalent to $\mathbf{C}^2 \otimes \mathbf{C}^2$ and $\mathcal{D}|\mathcal{K}$ is unitarily equivalent to $\mathcal{B}(\mathbf{C}^2) \otimes \mathcal{B}(\mathbf{C}^2) = M_2(\mathbf{C}) \otimes M_2(\mathbf{C})$. Let $\chi_+ = \binom{1}{0} \in \mathbf{C}^2, \chi_- = \binom{0}{1} \in \mathbf{C}^2$, and similarly for $\chi'_\pm$. Define $\Psi \equiv 2^{-1/2}(\chi_+ \otimes \chi'_- - \chi_- \otimes \chi'_+) \in \mathbf{C}^2 \otimes \mathbf{C}^2$. If $\hat{x}$, $\hat{z}$ are the obvious unit vectors in $\mathbf{R}^3$ and $\hat{\alpha} \equiv \cos\alpha\, \hat{z} + \sin\alpha\, \hat{x}$, then $\vec{\sigma} \cdot \hat{\alpha} = \cos\alpha\sigma_z + \sin\alpha\sigma_x$. A straightforward calculation yields

$$< \Psi, (\vec{\sigma} \cdot \hat{\alpha}) \otimes (\vec{\sigma} \cdot \hat{\beta})\Psi >= -\hat{\alpha} \cdot \hat{\beta}.$$

Using the mentioned unitary equivalence, this can be understood to obtain for some $\Psi \in \mathcal{K} \subset \mathcal{H}$. Choosing $\hat{\alpha} = \hat{z}$, $\hat{\alpha}' = \hat{x}$, $\hat{\beta} = \cos\frac{\pi}{4}\hat{z} + \sin\frac{\pi}{4}\hat{x}$, and $\hat{\beta}' = \cos\frac{3\pi}{4}\hat{z} + \sin\frac{3\pi}{4}\hat{x}$, and also $A_1 = \vec{\sigma} \cdot \hat{\alpha}'$, $A_2 = \vec{\sigma} \cdot \hat{\alpha}$, $B_1 = \vec{\sigma}' \cdot \hat{\beta}$, $B_2 = oh\sigma' \cdot \hat{\beta}'$, then one sees that if ϕ denotes the vector state on $\mathcal{B}(\mathcal{H})$ generated by Ψ, $\phi(A_1(B_1 + B_2) + A_2(B_1 - B_2)) = \sqrt{2}$.□

Hence, if two commuting algebras $\mathcal{A}$, $\mathcal{B}$ contain copies of $M_2(\mathbf{C})$ on a Hilbert space $\mathcal{H}$, they maximally violate Bell's inequalities in some normal state on $\mathcal{B}(\mathcal{H})$. Landau showed the following result.

Theorem 1.6 [30]: Let $(\mathcal{A}, \mathcal{B})$ be a pair of commuting von Neumann algebras on a Hilbert space $\mathcal{H}$ such that if $A \in \mathcal{A}$, $B \in \mathcal{B}$ and $AB = 0$, then either $A = 0$ or $B = 0$. Then if neither $\mathcal{A}$ nor $\mathcal{B}$ is abelian, there exists a normal state ϕ on $\mathcal{B}(\mathcal{H})$ such that $\beta(\phi, \mathcal{A}, \mathcal{B}) = \sqrt{2}$.

Sketch of proof: For any projection P, $2P - 1$ is a selfadjoint contraction. Let $P_i \in \mathcal{A}$, $Q_j \in \mathcal{B}$ be projections and $A_i = 2P_i - 1$, $B_j = 2Q_j - 1$. Then

$$\|A_1(B_1 + B_2) + A_2(B_1 - B_2)\| = 2\sqrt{1 + 4\,\|\,[P_1, P_2][Q_1, Q_2]\,\|}.$$

One can find a normal state ϕ on $\mathcal{B}(\mathcal{H})$ such that

$$\frac{1}{2}\,|\phi(A_1(B_1 + B_2) + A_2(B_1 - B_2))\,| = \sqrt{1 + 4\,\|\,[P_1, P_2][Q_1, Q_2]\,\|}.$$

The condition that $A \in \mathcal{A}, B \in \mathcal{B}$ and $AB = 0$ imply either $A = 0$ or $B = 0$ entails that $\|\,[P_1, P_2][Q_1, Q_2]\,\| = \|\,[P_1, P_2]\,\|\;\|\,[Q_1, Q_2]\,\|$ [38]. Since in any nonabelian von Neumann algebra $\mathcal{M}$ two projections $P_1, P_2 \in \mathcal{M}$ can be found such that $\|\,[P_1, P_2]\,\| = \frac{1}{2}$, the theorem's claim follows. □

Hence, one has only two possible situations. Either $\mathcal{A}$ or $\mathcal{B}$ is abelian, so from Theorem 1.4 (b) Bell's inequality is satisfied in all states, or both are nonabelian and (up to the additional hypothesis in Theorem 1.6) there exists a normal state in which Bell's inequality is maximally violated. The next result establishes the interesting fact that only copies of the Pauli spin matrices provide maximal violation.

Proposition 1.7 [45]: Let $(\mathcal{A}, \mathcal{B})$ be a pair of commuting subalgebras of a C*-algebra $\mathcal{C}$ and let $A_i \in \mathcal{A}$, $B_j \in \mathcal{B}$ be selfadjoint contractions such that for a state ϕ on $\mathcal{C}$ with $\phi|\mathcal{A}$ and $\phi|\mathcal{B}$ faithful,

$$\frac{1}{2}\phi(A_1(B_1 + B_2) + A_2(B_1 - B_2)) = \sqrt{2}.$$

Then $A_i^2 = 1$ and $A_1A_2 + A_2A_1 = 0$ (similarly for B_j), so A_1, A_2 and $A_3 \equiv -\frac{i}{2}[A_1, A_2]$ form a realization of the Pauli spin matrices in $\mathcal{A}$ (similarly for B_j in $\mathcal{B}$). (Moreover, A_1, A_2, A_3, resp. B_1, B_2, B_3, are contained in the centralizer of $\mathcal{A}$ in ϕ, resp. centralizer of $\mathcal{B}$ in ϕ).

Sketch of proof: Under the stated assumptions, we may identify $\mathcal{A}$ and $\mathcal{B}$ with a pair of commuting von Neumann algebras on a Hilbert space $\mathcal{H}$ with ϕ realized as a vector state by $\Phi \in \mathcal{H}$, where Φ is cyclic and separating for $\mathcal{A}$ and $\mathcal{B}$. Let $A_i \in \mathcal{A}$, $B_j \in \mathcal{B}$ be selfadjoint contractions such that

$$\frac{1}{2} < \Phi, (A_1(B_1+B_2) + A_2(B_1-B_2))\Phi >= \sqrt{2},$$

and let $\tilde{A} = \frac{1}{2}(A_1 + iA_2)$, $\tilde{B} = \frac{1}{2\sqrt{2}}(B_1 + B_2 + i(B_1 + B_2))$. Then $\tilde{A}\Phi = \tilde{B}\Phi$, $\tilde{A}^*\Phi = \tilde{B}^*\Phi$, and

$$\frac{1}{2} < \Phi, (A_1^2 + A_2^2)\Phi > = < \Phi, (\tilde{A}^*\tilde{A} + \tilde{A}\tilde{A}^*)\Phi >= 1.$$

Hence $A_i^2\Phi = \Phi$ and $(\tilde{A}^*\tilde{A} + \tilde{A}\tilde{A}^*)\Phi = \Phi$. Therefore, for any $A \in \mathcal{A}$, $\phi(AA_i^2) = < \Phi, AA_i^2\Phi >= \phi(A)$, $\phi(A(A_1+iA_2)) = 2 < \Phi, A\tilde{A}\Phi >= 2 < \Phi, A\tilde{B}\Phi >= 2 < \tilde{B}^*\Phi, A\Phi >$ $= 2 < \tilde{A}^*\Phi, A\Phi >= \phi((A_1+iA_2)A)$, and $(A_1A_2 + A_2A_1)\Phi = -2i(\tilde{A}^2 - \tilde{A}^{*2})\Phi = -2i(\tilde{B}^2 - \tilde{B}^{*2})\Phi = (B_1^2 - B_2^2)\Phi = 0$. □

Therefore, if one is designing an experiment to test violation of Bell's inequalities, one should only choose observables (like particle spins, polarizations, etc.) that can be modelled in standard theories by Pauli spin matrices. This is, in fact, what was done in the experiments carried out to date [6, 7, 8, 16], and to an extremely high accuracy, the prediction of a Bell correlation equal to $\sqrt{2}$ was verified. A natural question now is: what does quantum field theory predict about Bell's inequalities?

This question has been examined in a mathematically rigorous manner in very few publications, the first such paper appearing as late as 1985 [43]. Presently the papers that directly address the topic of Bell's inequalities and quantum field theory are [43, 44, 45, 46, 47, 30, 31, 48, 33]. The main results of these papers are reviewed in the next section. We provide a brief overview of these papers.

Paper [43] was an announcement of some of the main results from [44, 45, 46], where it was first proven that any free quantum field theory predicts that Bell's inequalities are maximally violated in the vacuum. In other words, already the vacuum fluctuations in any noninteracting quantum field model entail correlations (for spacelike separated and thus commuting observables) that maximally violate Bell's inequalities. This result indicated that maximal violation of Bell's inequalities had nothing to do with interaction or with special preparation of the system. In [30] it was emphasized that already the nonabelian character of the local algebras of observables sufficed to conclude maximal violation in some (unspecified) state, and the nonclassical nature of the vacuum state was re-established in [31] (however, not by showing that Bell's inequalities were maximally violated in the vacuum - see Theorem 2.1). In the paper [47] (the generality of which was significantly extended in [48]) it was shown that, in fact, the axioms of quantum field theory actually entailed that Bell's inequalities were maximally violated in every (normal) state in essentially every quantum field model. This is a result that is not true in nonrelativistic quantum mechanics. Finally, in paper [33] Landau exhibited, using the construction of [46], (exponentials of) quadratic expressions in free quantum field operators which violate Bell's inequality (not maximally) in thermal states for all sufficiently low temperatures and which have a physical interpretation as 'local' charges associated with symmetry transformations. In Section III we briefly discuss Bell's inequalities and quantum field theory in the more general context of statistical independence.

II. QUANTUM FIELD THEORY.

Ordinary quantum field theory on Minkowski space, formalized in the Wightman axioms [41], provides models of the type considered in the previous section. It is known that up to minor technical assumptions (see e.g. [25]) quantum field theories provide nets of C*-algebras assigning to each open region O of Minkowski space a C*-algebra $\mathcal{A}(O)$ such that the net $\{\mathcal{A}(O)\}$ satisfies certain standard axioms [3, 29] (isotony, locality, Poincaré covariance, and the existence of a Poincaré-covariant representation with positive energy satisfying the relativistic spectrum condition) that were naturally motivated by the interpretation of each $\mathcal{A}(O)$ as the algebra generated by all the observables that can be measured in the spacetime region O. In a certain technical sense [25], the quantum field operators smeared with test functions having support in O generate the algebra $\mathcal{A}(O)$. Since in this section all results refer to normal states, we may consider $\{\mathcal{A}(O)\}$ to be a net of von Neumann algebras in a Hilbert space $\mathcal{H}$ satisfying the mentioned axioms. In this section the algebra $\mathcal{C}$ is the C*-algebra generated by all the algebras in the net $\{\mathcal{A}(O)\}$.

By the locality axiom, if $O_1 \times O_2$, i.e. if all points in O_1 are spacelike separated from all points in O_2, then $\mathcal{A}(O_1) \subseteq \mathcal{A}(O_2)'$, the commutant of $\mathcal{A}(O_2)$ in $\mathcal{B}(\mathcal{H})$. Hence $(\mathcal{A}(O_1), \mathcal{A}(O_2))$ is a pair of commuting C*-algebras as in the previous section. Since at this level of generality we can only consider spacelike separated regions, we have in mind only correlation experiments where the measurements on the two arms are performed far enough apart and in a short enough time that they are spacelike separated (as in [8]).

Since local algebras $\mathcal{A}(O)$ in quantum field theories are very nonabelian, it is clear from Theorems 1.5 and 1.6 that there are going to be many states in which Bell's inequalities are maximally violated. In fact, typical local algebras contain an <u>infinite</u> product of copies of $M_2(\mathbf{C})$ [5,2,47,48], so that by Theorem 1.5 whenever $O_1 \times O_2$ there are infinitely many normal states ϕ such that $\beta(\phi, \mathcal{A}(O_1), \mathcal{A}(O_2)) = \sqrt{2}$. Landau demonstrated the following proposition using Theorem 1.6. The region O_1 is said to be strictly spacelike separated from O_2 if there exists a neighborhood $\mathcal{N}$ of the origin in $\mathbf{R}^4$ such that $O_1 + \mathcal{N} \times O_2$. (This relation is symmetric.)

<u>Theorem 2.1</u> [30,31]: Let $\{\mathcal{A}(O)\}$ be a net of local algebras in a physical representation with

a unique vacuum vector Ω.

a) For any strictly spacelike separated regions O_1, O_2 and any $A_i \in \mathcal{A}(O_1)$, $B_j \in \mathcal{A}(O_2)$ selfadjoint contractions such that $[A_1, A_2] \neq 0$ and $[B_1, B_2] \neq 0$, there exists a normal state ϕ such that

$$\frac{1}{2}\phi(A_1(B_1 + B_2) + A_2(B_1 - B_2)) > 1.$$

b) If Ω is cyclic for all $\mathcal{A}(O)$, $O \neq \emptyset$, and O_1, O_2, O_3 are any three mutually strictly separated spacetime regions, there exists a dense set $\mathcal{S}$ of normal states on $\mathcal{B}(\mathcal{H})$ (containing all states with bounded energy with respect to the vacuum) such that for any selfadjoint contractions $A_i \in \mathcal{A}(O_1)$, $B_j \in \mathcal{A}(O_2)$ satisfying $[A_1, A_2] \neq 0$ and $[B_1, B_2] \neq 0$ there is a projection $P \in \mathcal{A}(O_3)$ and a translation $x \in \mathbf{R}^4$, depending on $\phi \in \mathcal{S}$, so that the translates $A_1(x)$, $A_2(x)$, $B_1(x)$, $B_2(x)$ and $P(x)$ do not have a joint classical distribution in the state ϕ.

Remarks: (1) If the regions O_i above are not very pathological, for example if both are bounded and $O_i'' = O_i$, then they need only be spacelike separated from each other.

(2) The conclusion in (b) is weaker than that in (a), but it also illustrates an aspect of the nonclassical behavior of quantum field theory. Note that because the vacuum state is translation invariant, the assertion in part (b) simplifies somewhat for the vacuum. Since this paper is about Bell's inequalities, we shall sketch only the proof of part (a).

Proof of Theorem 2.1 (a): Under the stated assumptions, it is known [39] that $A \in \mathcal{A}(O_1)$, $B \in \mathcal{A}(O_2)$ and $AB = 0$ imply $A = 0$ or $B = 0$. Hence part (a) is a direct corollary of the proof to Theorem 1.6. Note that if $O_i'' = O_i$ and O_1 is spacelike separated from O_2 (and not necessarily strictly spacelike separated) then it follows from Theorem 3.5 in [20] that $AB = 0$ if and only if $A = 0$ or $B = 0$, so that Theorem 1.6 may be applied once again to yield the claim in Remark (1).□

Although a few natural questions remain open here, it is now clear that quantum field theory predicts the violation of Bell's inequalities in many states for any pair of algebras associated to spacelike separated spacetime regions, no matter how far apart the regions are. If, however, the spacelike separated regions are tangent, then we shall see below that the corresponding algebras of observables maximally violate Bell's inequalities in all normal states. Tangent spacetime regions are spacelike separated regions whose closures intersect, and we shall consider two classes of such regions in this section.

Let $W_R = \{x \in \mathbf{R}^4 | x_1 > |x_0|\}$ denote the "right wedge". Then $\mathcal{W}$, the collection of all "wedge" regions, is the set of all Poincaré transforms of W_R. If O is a spacetime region, O' denotes the interior of its causal complement (the set of all points spacelike separated from O). Then for any $W \in \mathcal{W}$, one has $W' \in \mathcal{W}$. Moreover, the pair (O, O') is always tangent for ordinary regions. The set $\mathcal{K}$ of double cones is described as follows. Let $x,\ y \in \mathbf{R}^4$ be timelike separated with x in y's future light cone. Then a double cone is obtained as the interior of the intersection of x's past light cone with y's future light cone. Taking all such x, y one generates all double cones. Note that double cones are bounded regions, while wedges are not. For both classes of regions, $O = O''$.

Since we are examining situations in which Bell's inequalities are maximally violated in all normal states, we make the following definition.

Definition 2.2: A pair $(\mathcal{A}, \mathcal{B})$ of commuting subalgebras of a W*-algebra $\mathcal{C}$ is called maximally correlated if for any normal state ϕ on $\mathcal{A} \vee \mathcal{B}$, one has $\beta(\phi, \mathcal{A}, \mathcal{B}) = \sqrt{2}$.

Theorem 2.3 [47]: a) In any vacuum sector, in any superselection sector of a global gauge group, in any massive particle representation, $(\mathcal{A}(W),\ \mathcal{A}(W)')$ is maximally correlated, for all $W \in \mathcal{W}$. Hence, if $\mathcal{A}(W)$ is weakly associated to a Wightman field in the sense of [25], $(\mathcal{A}(W),\ \mathcal{A}(W'))$ is maximally correlated for all $W \in \mathcal{W}$.

b) In any free field theory, in any local Fock field theory (e.g. $P(\phi)_2$ [28], Yukawa$_2$ [40], etc.) and in any dilatation-invariant theory, $(\mathcal{A}(O_1),\ \mathcal{A}(O_2))$ is maximally correlated for any pair $(O_1,\ O_2)$ of tangent double cones.

The three cases that enter into the hypothesis in part (a) above – vacuum sectors (a physical representation with at least one cyclic vacuum vector), superselection sectors [18], and massive particle representations [13] – include all physically interesting situations except for the charged sectors of a gauge theory with local gauge group and a massless particle (like quantum electrodynamics). This latter type of physical setting is not included in this theorem because it is still not known how to describe such a sector rigorously in terms of algebras of observables, not because the theorem is false in such a sector. In part (b) more restrictive conditions are assumed for technical reasons arising from limitations in the method of proof, not because the conclusion is believed to be false more generally. In fact, at the end of this section we shall describe our conjecture on the generality of the result in part (b). But first we shall sketch some aspects of the proof of this theorem in order to give the reader a sense of the ideas behind such results. However, we are obliged to refer the reader to the original papers for complete details. We begin with a discussion of some abstract structure properties.

Definition 2.4: Let $\mathcal{A}$ be a C*-algebra with unit 1. Then $N \in \mathcal{A}$ is called a I_2-generator if $N^2 = 0$ and $NN^* + N^*N = 1$.

Let $V_{\mathcal{A}}$ denote the set of I_2-generators in $\mathcal{A}$. Clearly, if N is contained in $V_{\mathcal{A}}$, then N^*N and NN^* are nonzero complementary projections, i.e. their sum is 1 and their product is 0, and the C*-algebra generated by N is isomorphic to $M_2(\mathbb{C})$ and contains the unit 1 of $\mathcal{A}$. Conversely, if $\mathcal{A}$ contains a copy of $M_2(\mathbb{C})$ containing 1, then $V_{\mathcal{A}} \neq \emptyset$. Note that if $A_i \in \mathcal{A}$ satisfies $A_i^* = A_i$, $A_i^2 = 1$ and $A_1A_2 + A_2A_1 = 0$ (which is the case if A_1, A_2 are maximal violators of Bell's inequalities in some faithful state on $\mathcal{A}$ (Prop. 1.7)), then $N \equiv \frac{1}{2}(A_1 + iA_2)$ is an element of $V_{\mathcal{A}}$. We introduce some standard definitions.

Definition 2.5: A von Neumann algebra $\mathcal{A}$ is said to have the property L_λ (resp. L'_λ) with $\lambda \in [0, 1/2]$ if for every $\epsilon > 0$ and any normal state $\phi \in \mathcal{A}^*$ (resp. finite family $\{\phi_i\}_{i=1}^{n}$ of normal states on $\mathcal{A}$), there exists an $N \in V_{\mathcal{A}}$ such that for any $A \in \mathcal{A}$,

$$|\lambda\phi(AN) - (1-\lambda)\phi(NA)| \leq \epsilon\|A\| \tag{2.1}$$

(resp. for any $A \in \mathcal{A}$ and $i = 1, \ldots, n$

$$|\lambda\phi_i(AN) - (1-\lambda)\phi_i(NA)| \leq \epsilon\|A\|.$$

Definition 2.6: The asymptotic ratio set $r_\infty(\mathcal{A})$ of a von Neumann algebra $\mathcal{A}$ is the set of all $\alpha \in [0,1]$ such that $\mathcal{A}$ is W^*-isomorphic to $\mathcal{A} \otimes \mathcal{R}_\alpha$, where $\{\mathcal{R}_\alpha\}_{\alpha\in[0,1]}$ is the family of hyperfinite factors constructed by Powers [35].

It is known that property L'_λ is strictly stronger than property L_λ [4], that property L'_λ implies property $L'_{1/2}$ [4,5], and that property L'_λ for $\mathcal{A}$ is equivalent to $\lambda/1-\lambda \in r_\infty(\mathcal{A})$ [4]. Using Prop. 1.7 one easily sees that if $A_1, A_2 \in \mathcal{A}$ are maximal violators of Bell's inequalities in the normal state ϕ on $\mathcal{A} \vee \mathcal{B}$, where $\mathcal{B} \subset \mathcal{A}'$, then $N \equiv \frac{1}{2}(A_1 + iA_2) \in V_{\mathcal{A}}$ satisfies (2.1) with $\epsilon = 0$ and $\lambda = 1/2$.

These properties are intimately related to the occurrence of $\beta(\phi, \mathcal{A}, \mathcal{A}') = \sqrt{2}$.

Theorem 2.7 [47,48]: For a von Neumann algebra $\mathcal{A}$ with a cyclic and separating vector in a separable Hilbert space $\mathcal{H}$, the following conditions are equivalent.

(a) $\mathcal{A} \approx \mathcal{A} \otimes \mathcal{R}_1$, i.e. $\mathcal{A}$ has property $L'_{1/2}$.

(b) The pair $(\mathcal{A}, \mathcal{A}')$ is maximally correlated.

(c) There exist sequences of selfadjoint contractions $\{A_{1,\alpha}\}_{\alpha\in N}$, $\{A_{2,\alpha}\}_{\alpha\in N}$ $\subset \mathcal{A}$, $\{B_{1,\alpha}\}_{\alpha\in N}$, $\{B_{2,\alpha}\}_{\alpha\in N} \subset \mathcal{A}'$ such that $T_\alpha \equiv \frac{1}{2}(A_{1,\alpha}(B_{1,\alpha} + B_{2,\alpha}) + A_{2,\alpha}(B_{1,\alpha} - B_{2,\,g\alpha}))$ converges to $\sqrt{2}\cdot 1$ in the σ-weak operator topology on $\mathcal{B}(H)$ as $\alpha \to \infty$.

Also the following conditions are equivalent.

(d) $\mathcal{A}$ has the property $L_{1/2}$.

(e) For any vector state $\omega(A) = <\Omega, A\Omega>$, $\Omega \in \mathcal{H}$, one has $\beta(\omega, \mathcal{A}, \mathcal{A}') = \sqrt{2}$.

Remark: Condition (c) means that there exists a sequence of admissible observables that in the limit maximally violate Bell's inequalities in all normal states at once. Note also that $\mathcal{R}_1$ is an infinite product of copies of $M_2(\mathbb{C})$ [5].

Contained in Theorem 2.7 is a characterization of von Neumann algebras $\mathcal{A}$ such that $(\mathcal{A}, \mathcal{A}')$ is maximally correlated. If $\mathcal{A}$ and $\mathcal{B} \subset \mathcal{A}'$ are von Neumann algebras, then $(\mathcal{A}, \mathcal{B})$ maximally correlated implies that $(\mathcal{A}, \mathcal{A}')$ and $(\mathcal{B}, \mathcal{B}')$ are both maximally correlated. The converse is false [48], so we mention a characterization of maximally correlated pairs of von Neumann algebras $(\mathcal{A}, \mathcal{B})$.

Theorem 2.8 [48]: Let $(\mathcal{A}, \mathcal{B})$ be a pair of commuting von Neumann algebras acting on a separable Hilbert space $\mathcal{H}$. Then the pair $(\mathcal{A}, \mathcal{B})$ is maximally correlated if and only if there exists a type I factor $\mathcal{M} \subset \mathcal{A} \vee \mathcal{B}$ such that $\mathcal{A} \cap \mathcal{M}$ and $\mathcal{B} \cap \mathcal{M}$ are (spatially) isomorphic to $\mathcal{R}_1$ and are relative commutants of each other in $\mathcal{M}$.

Now that the connection between maximal violation of Bell's inequalities and structure properties of the algebras is somewhat clearer, we can proceed to the situation in quantum field theory.

Theorem 2.9 [48]: Let $\{\mathcal{A}(O)\}$ be a net of observable algebras in an irreducible vacuum representation such that $[\bigcup_{O \in \mathcal{K}} \mathcal{A}(O)]\ \Omega$ is dense in the representation space $\mathcal{H}$, where Ω is the (up to a factor) unique vacuum vector. Then for each $W \in \mathcal{W}$, $\mathcal{A}(W)$ is a type III_1 factor that has property L'_λ for all $\lambda \in [0, 1/2]$.

Proof: Under the above assumptions each wedge algebra $\mathcal{A}(W)$ is nontrivial [24] and must be a type III_1 factor [20]. Let $\{V(t)\}_{t \in R}$ denote the strongly continuous unitary group on $\mathcal{H}$ implementing the velocity transformation subgroup of the Poincaré group that leaves W invariant. Then Ω is the (up to a factor) unique $V(\mathbb{R})$-invariant vector in $\mathcal{H}$ and

$$\underset{|a| \to \infty}{\text{w-lim}}\, V(a) A V(a)^{-1} = < \Omega, A\Omega > \cdot 1 \tag{2.2}$$

for every $A \in \mathcal{A}$ (Prop. I.1.6 in [22]).

By [17], because $\mathcal{A}(W)$ is a type III_1 factor, for any $\epsilon > 0$ and $\lambda \in [0, 1/2]$ there exists a I_2-generator $N \in \mathcal{A}(W)$ such that for every $A \in \mathcal{A}(W)$,

$$|\lambda < \Omega, AN\Omega > -(1-\lambda) < \Omega, NA\Omega > | \leq \epsilon \|A\|.$$

Since Ω is invariant under $V(R)$ and since

$$V(a)\ \mathcal{A}(W)\ V(a)^{-1} \equiv \alpha_a(\mathcal{A}(W)) = \mathcal{A}(W)$$

for all $a \in \mathbf{R}$, one also has

$$|\lambda < \Omega, A\alpha_a(N)\Omega > -(1-\lambda) < \Omega, \alpha_a(N)A\Omega > | \leq \epsilon\|A\|, \tag{2.3}$$

for all $A \in \mathcal{A}(W)$ and $a \in \mathbf{R}$. Let $\{\omega_i\}_{i=1}^{n}$ be a finite family of normal states on $\mathcal{A}(W)$. Again by [17] there exist unitaries $U_i, i = 1, \ldots, n$, in $\mathcal{A}(W)$ such that

$$| < \Omega, U_i A U_i^* \Omega > -\omega_i(A)| \leq \epsilon\|A\|, \tag{2.4}$$

for all $A \in \mathcal{A}(W)$, $i = 1, \ldots, n$. Choosing $b \in \mathbf{R}$ such that

$$\|[\alpha_b(N), U_i]\Omega\| \leq \epsilon \tag{2.5}$$

and

$$\|[\alpha_b(N), U_i^*]\Omega\| \leq \epsilon \tag{2.6}$$

which is possible by (2.2), locality and the cyclicity of Ω for $\mathcal{A}(W)$ (see, e.g. the proof of (A) in [21]), one has

$$\begin{aligned}
&|\lambda\omega_i(A\alpha_b(N)) - (1-\lambda)\omega_i(\alpha_b(N)A)| \\
&\leq \lambda|\omega_i(A\alpha_b(N)| - < \Omega, U_i\, A\, \alpha_b(N)\, U_i^*\Omega > | \\
&+ \lambda| < \Omega, U_i A\alpha_b(N)\, U_i^*\Omega > - < \Omega, U_i A\, U_i^*\, \alpha_b(N)\Omega > | \\
&+ |\lambda < \Omega, U_i A\, U_i^* \alpha_b(N)\Omega > -(1-\lambda) < \Omega, \alpha_b(N)U_i\, A\, U_i^*\Omega > | \\
&+ (1-\lambda)| < \Omega, \alpha_b(N)\, U_i\, A\, U_i^*\Omega > - < \Omega, U_i\alpha_b(N)A\, U_i^*\Omega > | \\
&+ (1-\lambda)| < \Omega, U_i\alpha_b(N)A\, U_i^*\Omega > -\omega_i(\alpha_b(N)A)| \\
&\leq 5\epsilon\|A\|,
\end{aligned}$$

for all $A \in \mathcal{A}(W)$ and $i = 1, \ldots, n$, using (2.3) – (2.6). Since $\alpha_b(N) \in V_{\mathcal{A}(W)}$, the theorem is proved.□

Remarks: (1) The above proof is a modification of an argument sketched by Testard in [50].

(2) By using methods of [24], the assumption that the vacuum vector is unique can be dropped and one can still conclude that for each $W \in \mathcal{W}$ $\mathcal{A}(W)$ is type III and has property L'_λ for all $\lambda \in [0, 1/2]$. (See [48])

Since the property L'_λ, $\lambda \in [0, 1/2]$, is an isomorphic invariant, the above theorem is also true for nets of local algebras in representations such as those occuring in the Doplicher, Haag, Roberts theory of superselection structure [18] and also the massive single particle representations of Buchholz and Fredenhagen [13], which include, in principle, charged sectors of theories like quantum chromodynamics. Hence by evoking Theorem 2.7 we have finished the sketch of the proof of part (a) of Theorem 2.3.

We commence the discussion of part (b) of Theorem 2.3 by proving the following result for dilatation-invariant theories. The dilatation invariance of a theory with unique vacuum is expressed by the existence of a strongly continuous, unitary representation $D(\mathbf{R}_+)$ of the dilatation group on $\mathbf{R}^d$ acting such that

$$\delta_\lambda(\mathcal{A}(O)) \equiv D(\lambda)\,\mathcal{A}(O)\,D(\lambda)^{-1} = \mathcal{A}(\lambda O), \lambda > 0.$$

where $\lambda O \equiv \{\lambda x | x \in O\}$ and $D(\lambda)\Omega = \Omega$ for any $\lambda \in \mathbf{R}_+$ (Ω is the unique vacuum vector of the theory).

Theorem 2.10 [48]: Let $\{\mathcal{A}(O)\}$ be a net of local von Neumann algebras in an irreducible vacuum representation of a dilatation-invariant theory such that the wedge algebras are locally generated [24] and $\mathcal{A}(W)' = \mathcal{A}(W')$ for each $W \in \mathcal{W}$ (both of which are true if the net is locally associated to a Wightman field in the sense of [25]). Then for any tangent double cones O_1, O_2 the pair $(\mathcal{A}(O_1), \mathcal{A}(O_2))$ is maximally correlated and thus all double cone algebras have property $L'_{1/2}$.

Proof: It is known [37] that under the given assumptions, for any $A \in \mathcal{C}$, $\delta_\lambda(A)$ converges weakly to $\phi_0(A) \cdot 1$ as $\lambda \downarrow 0$, where $\phi_0 = \phi_0 \circ \delta_\lambda$ is the vacuum state on $\mathcal{C}$. Thus, for any locally normal state $\phi \in \mathcal{B}(\mathcal{H})^*$, $\phi \circ \delta_\lambda \to \phi_0$ pointwise on $\mathcal{C}$ as $\lambda \downarrow 0$. Without loss of generality, it maybe assumed that the point of tangency for O_1 and O_2 is the origin and that $O_1 \subset W_R$, $O_2 \subset W'_R$.

$\mathcal{A}(W_R)$ is a type III_1 factor [20, 21]. Since type III_1 factors have property $L_{1/2}$ [17, 47], it follows from Theorem 2.7 that $\beta(\phi, \mathcal{A}(W_R), \mathcal{A}(W'_R)) = \sqrt{2}$ for every vector state ϕ on $\mathcal{B}(\mathcal{H})$. In particular, $\beta(\phi_0, \mathcal{A}(W_R), \mathcal{A}(W'_R)) = \sqrt{2}$. Let $\epsilon > 0$ be arbitrary and pick selfadjoint contractions $A_i \in \mathcal{A}(W_R)$, $B_j \in \mathcal{A}(W'_R)$, $i,j = 1,2$, such that with $T_\epsilon \equiv \frac{1}{2}(A_1(B_1+B_2) + A_2(B_1 - B_2))$ one has $\phi_0(T_\epsilon) \geq \sqrt{2} - \epsilon$. Let also $\delta > 0$ be arbitrary and pick two sufficiently large tangent double cones $\hat{O}_1$, $\hat{O}_2$ (with $O_1 \subset \hat{O}_1 \subset W_R$ and $O_2 \subset \hat{O}_2 \subset W'_R$) such that there exist selfadjoint contractions $\hat{A}_i \in \mathcal{A}(\hat{O}_1)$, $\hat{B}_j \in \mathcal{A}(\hat{O}_2)$ satisfying $|\phi_0(T_\epsilon - \hat{T}_{\epsilon,\delta})| < \delta$, where $\hat{T}_{\epsilon,\delta} \equiv \frac{1}{2}(\hat{A}_1(\hat{B}_1 + \hat{B}_2) + \hat{A}_2(\hat{B}_1 - \hat{B}_2))$ (this is possible by Kaplansky's density theorem and the assumption that the wedge algebras are locally generated).

Then for any locally normal state $\phi \in \mathcal{B}(\mathcal{H})^*$,

$$\phi \circ \delta_\lambda(\hat{T}_{\epsilon,\delta}) \underset{\lambda \to 0}{\to} \phi_0(\hat{T}_{\epsilon,\delta}) \geq \sqrt{2} - \epsilon - \delta.$$

But for every $\lambda \in \mathbf{R}_+$, one has $\delta_\lambda(\hat{A}_i) \in \mathcal{A}(\lambda\hat{O}_1)$ and $\delta_\lambda(\hat{B}_j) \in \mathcal{A}(\lambda\hat{O}_2)$, and there exists a $\lambda_0 > 0$ such that $\lambda\hat{O}_1 \subset O_1$ and $\lambda\hat{O}_2 \subset O_2$ for all $\lambda < \lambda_0$. Hence the assertion of the theorem follows at once.□

Because the scaling limit of the models mentioned in part (b) of Theorem 2.3 is the massless, free field, which is dilatation-invariant, one can extend the result of Theorem 2.10 to include such models as well. Let $f(x) \to f_\lambda(x) \equiv f(\lambda^{-1}x)$ be the induced action of the dilatation group on the test function space $\mathcal{S}(\mathbf{R}^d)$. It is well known that there exists a scaling function $N(\lambda)$ (monotone, nonnegative for $\lambda > 0$) such that for all $f_1, f_2 \in \mathcal{S}(\mathbf{R}^d)$,

$$\lim_{\lambda\to 0} N(\lambda)^2 W_m{}^{(2)}(f_{1,\lambda}, f_{2,\lambda}) = W_0{}^{(2)}(f_1, f_2),$$

where $W_m{}^{(2)}(.,.)$ is the two-point Wightman function of the free field with mass m.

Sufficient conditions in terms of test functions have been given in [46] that insure that Bell's inequalities are maximally violated in the vacuum state by any free field algebras containing the spectral projections of field operators smeared with test functions satisfying said conditions. It is shown in [48] that with the above scaling one can insure that for any pair of tangent double cones (O_1, O_2) one can find test functions with appropriate support satisfying the said conditions. The proof of maximal correlation is then completed by the following theorem.

Theorem 2.11 [48]: Let $(\mathcal{A}, \mathcal{B})$ be a pair of commuting von Neumann algebras acting on a separable Hilbert space. Then the following are equivalent.

(a) $(\mathcal{A}, \mathcal{B})$ is maximally correlated.

(b) There exists a faithful state $\omega \in (\mathcal{A} \vee \mathcal{B})_*$ such that $\beta(\omega, \mathcal{A}, \mathcal{B}) = \sqrt{2}$.

Proof: The implication (a) $\to$ (b) is trivial. To verify the other implication, first note that for a faithful state $\omega \in (\mathcal{A} \vee \mathcal{B})_*$ and an arbitrary normal state $\phi \in (\mathcal{A} \vee \mathcal{B})_*$, ϕ can be arbitrarily well approximated in norm by elements of the set of all states $\psi \in (\mathcal{A} \vee \mathcal{B})_*$ such that there is some $\lambda > 0$ with $\psi \le \lambda\omega$ (see the proof of Theorem 2.1 in [48]). Let $\{A_1{}^{(n)}, A_2{}^{(n)}, B_1{}^{(n)}, B_2{}^{(n)}\}_{n\in N}$ be a sequence of selfadjoint contractions with $A_i{}^{(n)} \in \mathcal{A}$, $B_j{}^{(n)} \in \mathcal{B}$, $i, j = 1, 2, n \in N$, satisfying

$$\frac{1}{2}\omega(A_1{}^{(n)}(B_1{}^{(n)} + B_2{}^{(n)}) + A_2{}^{(n)}(B_1{}^{(n)} - B_2{}^{(n)})) \to \sqrt{2}$$

as $n \to \infty$, and let $\psi \in (\mathcal{A} \vee \mathcal{B})_*$ be a state with $\psi \le \lambda\omega$ for some $\lambda > 0$. Then with

$$T_n \equiv \sqrt{2} - \frac{1}{2}(A_1{}^{(n)}(B_1{}^{(n)} + B_2{}^{(n)}) + A_2{}^{(n)}(B_1{}^{(n)} - B_2{}^{(n)})) \ge 0$$

(by Prop. 1.2) one has $\psi(T_n) \le \lambda\omega(T_n) \to 0$ as $n \to \infty$. Since such states $\psi \in (\mathcal{A} \vee \mathcal{B})_*$ are norm dense in the normal states on $(\mathcal{A} \vee \mathcal{B})$, the desired implication follows. □

Thus, to verify that $(\mathcal{A}, \mathcal{B})$ is maximally correlated, it suffices to check that $\beta(\omega, \mathcal{A}, \mathcal{B}) = \sqrt{2}$ for one, conveniently chosen faithful state $\omega \in (\mathcal{A} \vee \mathcal{B})_*$. In particular, since in quantum field theory the vacuum state ϕ_0 is typically faithful on the local algebras of observables, it follows that if already the vacuum fluctuations are such that Bell's inequalities are maximally violated, then all other preparations of the system will lead to violation of Bell's inequalities, as well.

It is worth emphasizing that the above scaling arguments show that the spacetime supports of the observables that give a Bell correlation converging to $\sqrt{2}$ converge to the point of tangency of the pair (O_1, O_2).

We expect that the following conjecture is true. Let $\{\mathcal{A}(O)\}$ be a net of local observable von Neumann algebras in a vacuum representation to which is locally associated a quantum field in the sense of [25] and for which assumption (A) below holds. Then $(\mathcal{A}(O_1), \mathcal{A}(O_2))$ is maximally correlated for any tangent double cones $O_1, O_2 \in \mathcal{K}$.

(A) There exists a scaling function $N(\lambda)$ (monotone, nonnegative for $\lambda > 0$) such that for all test functions f_i

$$\lim_{\lambda \to 0} N(\lambda)\phi_0(\varphi(f_\lambda)) = W_0{}^{(1)}(f)\ (= 0),$$

$$\lim_{\lambda \to 0} N(\lambda)^2 \phi_0(\varphi(f_{1,\lambda})\varphi(f_{2,\lambda})) = W_0{}^{(2)}(f_1, f_2),$$

and

$$\lim_{\lambda \to 0} N(\lambda)^4 \phi_0(\varphi(f_{1,\lambda})\varphi(f_{2,\lambda})\varphi(f_{3,\lambda})\varphi(f_{4,\lambda})) = W_0{}^{(4)}(f_1, f_2, f_3, f_4),$$

where $\{W_0{}^{(j)}\}_{j=1,2,4}$ are the Wightman functions corresponding to the vacuum state of the free, massless field.

Condition (A) is known to be true in most of the quantum field models that have been constructed. It is a weak, rigorous way of saying that the theory has a well-defined Gell-Mann-Low limit.

If we may briefly summarize: The results above show that in relativistic quantum field theory, how ever the field has been prepared and no matter what the particular dynamics of the field may be, there are observables associated to spacelike tangent regions that maximally violate Bell's inequalities. The maximal violation of Bell's inequalities in every normal state is a consequence of the most basic axioms of quantum field theory. The same axioms imply that though it is true that pairs of algebras of observables associated with regions that are spacelike separated by an arbitrary nonzero distance will have many states satisfying Bell's inequalities, nonetheless even in that case there are (infinitely) many states on the same pairs in which Bell's inequalities are maximally violated.

III. MAXIMAL CORRELATION, SPLIT PROPERTY AND STATISTICAL INDEPENDENCE

In this section we shall briefly contrast maximal correlation with the split property and place them both in the context of statistical independence. For a more complete discussion, see [54]. Having stated maximal correlation as a property of a pair of commuting algebras, we shall do the same for the split property (but see [12, 19]).

Definition 3.1: Let $(\mathcal{A}, \mathcal{B})$ be a pair of commuting von Neumann algebras on a Hilbert space $\mathcal{H}$. The pair $(\mathcal{A}, \mathcal{B})$ is split if there exists a type I factor $\mathcal{M}$ such that $\mathcal{A} \subset \mathcal{M} \subset \mathcal{B}'$.

As we shall recall below more formally, $(\mathcal{A}, \mathcal{B})$ is split if and only if there are many normal product states across the algebra $\mathcal{A} \vee \mathcal{B}$. Hence by Theorem 1.4(a), the split property and maximal correlation of $(\mathcal{A}, \mathcal{B})$ are mutually exclusive, indeed they are each other's opposite, in a sense that we want to indicate in this section. However, we first weave the thread of independence of algebras of observables into the discussion.

In quantum mechanics if the algebras of observables $\mathcal{A}, \mathcal{B}$ of two systems mutually commute, they are viewed as independent, insofar as all measurements on one system are compatible with all measurements on the other system. However, there are stronger conditions of independence that are also of interest.

Definition 3.2: A pair $(\mathcal{A}, \mathcal{B})$ of commuting subalgebras of a C*-algebra $\mathcal{C}$ is said to be C*-independent if for each state $\phi_1 \in \mathcal{A}^*$ and each state $\phi_2 \in \mathcal{B}^*$ there exists a state $\phi \in \mathcal{C}^*$ such that $\phi|\mathcal{A} = \phi_1$ and $\phi|\mathcal{B} = \phi_2$.

Hence two systems with associated algebras of observables $\mathcal{A}, \mathcal{B}$ that are C^*- independent can each be prepared in any state independently of the state of the other system. Roos showed [38] that in fact a pair $(\mathcal{A}, \mathcal{B})$ of commuting C*-algebras is C*-independent if and only if any pair of states $\phi_1 \in \mathcal{A}^*$, $\phi_2 \in \mathcal{B}^*$ has a common extension ϕ that is a **product** state across $\mathcal{A} \vee \mathcal{B}$. This is entailed if, for example, $\mathcal{A} \vee \mathcal{B}$ is naturally isomorphic to the tensor product of $\mathcal{A}$ with $\mathcal{B}$. One has the following result in quantum field theory.

Theorem 3.3 [39, 20, 38, 49]: In an irreducible vacuum representation, for any strictly spacelike separated regions $\mathcal{O}_1$, $\mathcal{O}_2$, $(\mathcal{A}(\mathcal{O}_1), \mathcal{A}(\mathcal{O}_2))$ is C*-independent. For any tangent double cones $\mathcal{O}_1, \mathcal{O}_2 \in \mathcal{K}$, resp. any wedge $W \in \mathcal{W}$, the pair $(\mathcal{A}(\mathcal{O}_1), \mathcal{A}(\mathcal{O}_2))$, resp. $(\mathcal{A}(W), \mathcal{A}(W'))$, is C*-independent.

So C*-independence is typical in quantum field theory. There is a yet stronger condition of independence.

Definition 3.4: Let $\mathcal{A}$ and $\mathcal{B}$ be commuting subalgebras of a W^*-algebra $\mathcal{C}$. The pair $(\mathcal{A}, \mathcal{B})$ is said to be W^*-independent (in the product sense) if for every normal state $\phi_1 \in \mathcal{A}_*$ and every normal state $\phi_2 \in \mathcal{B}_*$ there exists a product normal state $\phi \in \mathcal{C}^*$ such that $\phi|(AB) = \phi_1(A)\phi_2(B)$ for all $A \in \mathcal{A}$ and $B \in \mathcal{B}$

It is known [39, 49] that if $(\mathcal{A}, \mathcal{B})$ is W^*-independent, then $(\mathcal{A}, \mathcal{B})$ is C*-independent. We collect the following characterizations of W^*-independence.

Theorem 3.5 [12, 1, 2, 51, 49]: In an irreducible vacuum representation for which the vacuum vector is cyclic for all local algebras, the following are equivalent for any two spacelike separated double cones or wedges $\mathcal{O}_1, \mathcal{O}_2$.

(1) Local preparability of all normal states: for every normal state ϕ_0 there is a normal positive map $T : \mathcal{B}(\mathcal{H}) \to \mathcal{B}(\mathcal{H})$ such that $T(A) = \phi_0(A)T(1)$ for all $A \in \mathcal{A}(\mathcal{O}_1)$ and $T(B) = T(1)B$ for $B \in \mathcal{A}(\mathcal{O}_2)$.

(2) $(\mathcal{A}(\mathcal{O}_1), \mathcal{A}(\mathcal{O}_2))$ is W^*-independent (in the product sense).

(3) $(\mathcal{A}(\mathcal{O}_1), \mathcal{A}(\mathcal{O}_2))$ is split.

(4) $\mathcal{A}(\mathcal{O}_1) \vee \mathcal{A}(\mathcal{O}_2) \approx \mathcal{A}(\mathcal{O}_1) \otimes \mathcal{A}(\mathcal{O}_2)$.

By Theorem 1.4(a), the pair $(\mathcal{A}(\mathcal{O}_1), \mathcal{A}(\mathcal{O}_2))$ of algebras associated to the tangent spacetime regions considered in Theorem 2.3 are not W^*-independent, even though, by Theorem 3.3, they are C*-independent. On the other hand, strictly spacelike separated pairs of double cones are known in many cases [12, 42, 2] to be split, hence W^*-independent. $(\mathcal{A}(\mathcal{O}_1), \mathcal{A}(\mathcal{O}_2))$ maximally correlated implies $(\mathcal{A}(\mathcal{O}_1), \mathcal{A}(\mathcal{O}_2))$ is very badly nonsplit. Being nonsplit is a property that is strictly weaker than being maximally correlated [48]. It is known that algebras of observables associated to tangent, spacelike separated spacetime regions are not split, in general. We mention only the following result.

Theorem 3.6 [48]: Let $\mathcal{O}_1$ and $\mathcal{O}_2$ be (Poincaré transforms of) tangent spacelike separated spacetime regions for which there exists a $\lambda_0 > 0$ such that $\lambda\mathcal{O}_i \subset \mathcal{O}_i$ for all $0 < \lambda < \lambda_0$, $i = 1, 2$. And let $\{\mathcal{A}(\mathcal{O})\}$ be a net of local von Neumann algebras in an irreducible vacuum representation, to which is locally associated a quantum field in the sense of [25]. With assumption (A) at the end of the previous section, then is no (locally) normal state ϕ on $\mathcal{C}$ such that $\phi(AB) = \phi(A)\phi(B)$ for all $A \in \mathcal{A}(\mathcal{O}_1)$, $B \in \mathcal{A}(\mathcal{O}_2)$.

So we see that very generally, tangent spacetime regions have associated to them nonsplit algebras; however, it is known [49] that for a class of such regions $\mathcal{O}_1, \mathcal{O}_2$, one has $\beta(\phi, \mathcal{A}(\mathcal{O}_1), \mathcal{A}(\mathcal{O}_2) < \sqrt{2} - \epsilon$ for all normal states ϕ, where ϵ depends on geometric properties of the regions at the point of tangency. Hence they are nonsplit but not maximally correlated.

We summarize: If ϕ is a product state over $(\mathcal{A}(O_1), \mathcal{A}(O_2))$ then $\beta(\phi, \mathcal{A}(O_1), \mathcal{A}(O_2)) = 1$. Hence, if $\beta(\phi, \mathcal{A}(O_1), \mathcal{A}(O_2)) > 1$ then ϕ is not a product state over $(\mathcal{A}(O_1), \mathcal{A}(O_2))$. The larger the number $\beta(\phi, \mathcal{A}(O_1), \mathcal{A}(O_2))$ the stronger the correlations between $\mathcal{A}(O_1)$ and $\mathcal{A}(O_2)$ in the state ϕ and the less "product-like" ϕ is across $(\mathcal{A}(O_1), \mathcal{A}(O_2))$. If $\beta(\phi, \mathcal{A}(O_1), \mathcal{A}(O_2)) = \sqrt{2}$, the correlation between $\mathcal{A}(O_1)$ and $\mathcal{A}(O_2)$ is the maximum possible. In the case of some tangent spacetime regions one has $\beta(\phi, \mathcal{A}(O_1), \mathcal{A}(O_2)) = \sqrt{2}$ for all normal states, whereas in the same situation $\beta(\phi, \mathcal{A}(O_1), \mathcal{A}(O_2)) = 1$ for many nonnormal states. The ultraviolet effect that is responsible for the maximal violation of Bell's inequalities in all normal states over tangent spacetime regions entails that tangent spacetime regions are quantitatively (as measured by $\beta(\phi, \mathcal{A}(O_1), \mathcal{A}(O_2))$) and maximally far from W^*-independence while remaining C*-independent.

Acknowledgements

This review grew out of a very pleasant collaboration with Dr. Reinhard Werner, and it was written at the Centre de Physique Théorique while the author was supported by a fellowship from the French Ministry of National Education. The author wishes to express his gratitude to the French Government, to Prof. Jean Béllissard, and to the Centre de Physique Théorique, Marseille..

REFERENCES

[1] C. D'Antoni and R. Longo, Interpolation by type I factors and the flip automorphism, J. Funct. Anal. 51, 361–371 (1983).

[2] C. D'Antoni, D. Buchholz and K. Fredenhagen, The universal structure of local algebras, Commun. Math. Phys. 111, 123–135 (1987)

[3] H. Araki, Local quantum theory, I, in Local Quantum Theory, ed. R. Jos Academic Press, New York, 1969.

[4] H. Araki, Asymptotic ratio set and property L'_λ, Publ. RIMS, Kyoto Univ 6, 443–460 (1970–1971).

[5] H. Araki and E.J. Woods, A classification of factors, Publ. RIMS, Kyoto Uni 4, 51–130 (1968).

[6] A. Aspect, Experiences basées sur les inégalités de Bell, J. Phys. (Pari Suppl. 42, C2 (1981).

[7] A. Aspect, Experimental tests of Bell's inequalities in atomic physics, Atomic Physics 8, ed. A. Lindgren, A. Rosen and S. Svanberg, Plenum, Ne York, 1983.

[8] A. Aspect, J. Dalibard and G. Roger, Experimental test of Bell's inequalities using time-varying analyzers, Phys. Rev. Lett., 49, 1804–1807 (1982).

[9] J.S. Bell, On the Einstein-Podolsky-Rosen paradox, Physics (NY) 1, 195–200 (1964).

[10] J.S. Bell, On the problem of hidden variables in quantum mechanics, Rev. Mod. Phys., 38, 447-452 (1966).

[11] N. Bohr Can quantum-mechanical description of physical reality be considered complete?, Phys. Rev., 48, 696–702 (1935).

[12] D. Buchholz, Product states for local algebras, Commun. Math. Phys., 36, 287–304 (1974).

[13] D. Buchholz and K. Fredenhagen, Locality and the structure of particle states, Commun. Math. Phys., 84, 1–54 (1982).

[14] B.S. Cirel'son, Quantum generalization of Bell's inequalities, Lett. Math. Phys., 4, 93–100 (1980).

[15] J.F. Clauser and M.A. Horne, Experimental consequences of objective local theories, Phys. Rev. D. 10, 526–535 (1974).

[16] J.F. Clauser and A. Shimony, Bell's theorem: Experimental tests and implications, Rep. Prog. Phys., 41, 1881–1927 (1978).

[17] A. Connes and E. Stormer, Homogeneity of the state space of factors of type III_1, J. Funct. Anal., 28, 187–196 (1978).

[18] S. Doplicher, R. Haag and J.E. Roberts, Fields, observables and gauge transformations, I, II, Commun. Math. Phys., 13, 1–23 (1969) and 15, 173–200 (1969).

[19] S. Doplicher and R. Longo, Standard and split inclusions of von Neumann algebras, Invent. Math., 75, 493–536 (1984)

[20] W. Driessler, Comments on lightlike translations and applications in relativistic quantum field theory, Commun. Math. Phys., 44, 133–141 (1975).

[21] W. Driessler, On the type of local algebras in quantum field theory, Commun. Math. Phys., 53, 295–297 (1977).

[22] W. Driessler, On the structure of fields and algebras on null planes, I, Local algebras, Acta Phys. Aust., 46, 63-96 (1977).

[23] W. Driessler, Duality and absence of locally generated superselection sectors for CCR-type algebras, Commun. Math. Phys., 70, 213–220 (1979).

[24] W. Driessler and S.J. Summers, Central decomposition of Poincaré invariant nets of local field algebras and absence of spontaneous breaking of the Lorentz group, Ann. Inst. Henri Poincaré, A43, 147–166 (1985).

[25] W. Driessler, S.J. Summers and E.H. Wichmann, On the connection between

quantum fields and von Neumann algebras of local operators, Commun. Math. Phys., 105, 49–84 (1986).

[26] A. Einstein, B. Podolsky and N. Rosen, Can quantum-mechanical description of physical reality be considered complete?, Phys. Rev., 47, 777–780 (1935).

[27] K. Fredenhagen, On the modular structure of local algebras of observables, Commun. Math. Phys., 97, 79–89 (1985).

[28] J. Glimm and A. Jaffe, The $\lambda(\phi^4)_2$ quantum field theory without cutoff, III: The physical vacuum, Acta Math., 125, 204–267 (1970).

[29] R. Haag and D. Kastler, An algebraic approach to quantum field theory, J. Math. Phys., 5, 848–861 (1964).

[30] L.J. Landau, On the violation of Bell's inequality in quantum theory, Phys. Lett., A 120, 54–56 (1987).

[31] L.J. Landau, On the nonclassical structure of the vacuum, Phys. Lett., A 123, 115–118 (1987).

[32] L.J. Landau, Experimental tests of general quantum theories, Lett. Math. Phys., 14, 33–40 (1987).

[33] L.J. Landau, Gaussian quantum fields and stochastic electrodynamics, Phys. Rev. A, 37, 4449–4460 (1988).

[34] L.J. Landau, Empirical two-point correlation functions, Found. Phys., 18, 449–460 (1988).

[35] R.F. Powers, Representation of uniformly hyperfinite algebras and their associated von Neumann rings, Ann. Math., 86, 138–171 (1967).

[36] R.F. Powers, UHF algebras and their applications to representations of the anticommutation relations, in Cargése Lectures in Physics, ed. D. Kastler, Gordon and Breach, New York, 1970.

[37] J.E. Roberts, Some applications of dilatation invariance to structural questions in the theory of local observables, Commun. Math. Phys., 37, 273–286 (1974).

[38] H. Roos, Independence of local algebras in quantum field theory, Commun. Math. Phys., 16, 238–246 (1970).

[39] S. Schlieder, Einige Bemerkungen über Projektionsoperatoren, Commun. Math. Phys., 13, 216–225 (1969).

[40] R. Schrader, A Yukawa quantum field theory in two spacetime dimensions without cutoffs, Ann. Phys., 70, 412–457 (1972).

[41] R.J. Streater and A.S Wightman, PCT, Spin and Statistics, and All That, Benjamin, New York, 1964.

[42] S.J. Summers, Normal product states for fermions and twisted duality for

CCR and CAR-type algebras with application to the Yukawa quantum field model, Commun. Math. Phys., 86, 111–141 (1982).

[43] S.J. Summers and R. Werner, The vacuum violates Bell's inequalities, Phys. Lett., A 110, 257–259 (1985).

[44] S.J. Summers and R. Werner, Bell's inequalities and quantum field theory, I: General setting, preprint, unabridged version of [45], available from authors.

[45] S.J. Summers and R. Werner, Bell's inequalities and quantum field theory, I: General setting, J. Math. Phys., 28, 2440–2447 (1987).

[46] S.J. Summers and R. Werner, Bell's inequalities and quantum field theory, II: Bell's inequalities are maximally violated in the vacuum, J. Math. Phys., 28, 2448–2456 (1987).

[47] S.J. Summers and R. Werner, Maximal violation of Bell's inequalities is generic in quantum field theory, Commun. Math. Phys., 110, 247–259 (1987).

[48] S.J. Summers and R. Werner, Maximal violation of Bell's inequalities for algebras of observables in tangent spacetime regions, Ann. Inst. Henri Poincaré, 49, 215–243 (1988).

[49] S.J. Summers and R. Werner, unpublished.

[50] D. Testard, Asymptotic ratio set of von Neumann algebras generated by temperature states in statistical mechanics, Rep. Math. Phys., 12, 115–118 (1977).

[51] R. Werner, Local preparability of states and the split property in quantum field theory, Lett. Math. Phys., 13, 325-329 (1987).

[52] L. Accardi and A. Fedullo, On the statistical meaning of complex numbers in quantum theory, Lett. Nuovo Cim., 34, 161-172 (1982).

[53] L. Accardi, Foundations of quantum mechanics: a quantum probabilistic approach, in Quantum Paradoxes, Reidel, Amsterdam, 1987.

[54] S.J. Summers, On the independence of local algebras in quantum field theory, preprint.

LECTURE NOTES IN MATHEMATICS
Edited by A. Dold, B. Eckmann and F. Takens

Some general remarks on the publication of proceedings of congresses and symposia

Lecture Notes aim to report new developments - quickly, informally and at a high level. The following describes criteria and procedures which apply to proceedings volumes. <u>The editors of a volume are strongly advised to inform contributors about these points at an early stage.</u>

§1. One (or more) expert participant(s) of the meeting should act as the responsible editor(s) of the proceedings. They select the papers which are suitable (cf. §§ 2, 3) for inclusion in the proceedings, and have them individually refereed (as for a journal). It should not be assumed that the published proceedings must reflect conference events faithfully and in their entirety. Contributions to the meeting which are not included in the proceedings can be listed by title. The series editors will normally not interfere with the editing of a particular proceedings volume - except in fairly obvious cases, or on technical matters, such as described in §§ 2, 3. The names of the responsible editors appear on the title page of the volume.

§2. The proceedings should be reasonably homogeneous (concerned with a limited area). For instance, the proceedings of a congress on "Analysis" or "Mathematics in Wonderland" would normally not be sufficiently homogeneous.

One or two longer survey articles on recent developments in the field are often very useful additions to such proceedings - even if they do not correspond to actual lectures at the congress. An extensive introduction on the subject of the congress would be desirable.

§3. The contributions should be of a high mathematical standard and of current interest. Research articles should present new material and not duplicate other papers already published or due to be published. They should contain sufficient information and motivation and they should present proofs, or at least outlines of such, in sufficient detail to enable an expert to complete them. Thus resumes and mere announcements of papers appearing elsewhere cannot be included, although more detailed versions of a contribution may well be published in other places later.

Contributions in numerical mathematics may be acceptable without formal theoréms resp. proofs if they present new algorithms solving problems (previously unsolved or less well solved) or develop innovative qualitative methods, not yet amenable to a more formal treatment. .

Surveys, if included, should cover a sufficiently broad topic, and should in general not simply review the author's own recent research. In the case of such surveys, exceptionally, proofs of results may not be necessary.

§4. "Mathematical Reviews" and "Zentralblatt für Mathematik" recommend that papers in proceedings volumes carry an explicit statement that they are in final form and that no similar paper has been or is being submitted elsewhere, if these papers are to be considered for a review. Normally, papers that satisfy the criteria of the Lecture Notes in Mathematics series also satisfy this requirement, but we strongly recommend that the contributing authors be asked to give this guarantee explicitly at the beginning or end of their paper. There will occasionally be cases where this does not apply but where, for special reasons, the paper is still acceptable for LNM.

§5. Proceedings should appear soon after the meeeting. The publisher should, therefore, receive the complete manuscript (preferably in duplicate) within nine months of the date of the meeting at the latest.

§6. Plans or proposals for proceedings volumes should be sent to one of the editors of the series or to Springer-Verlag Heidelberg. They should give sufficient information on the conference or symposium, and on the proposed proceedings. In particular, they should contain a list of the expected contributions with their prospective length. Abstracts or early versions (drafts) of some of the contributions are helpful.

§7. Lecture Notes are printed by photo-offset from camera-ready typed copy provided by the editors. For this purpose Springer-Verlag provides editors with technical instructions for the preparation of manuscripts and these should be distributed to all contributing authors. Springer-Verlag can also, on request, supply stationery on which the prescribed typing area is outlined. Some homogeneity in the presentation of the contributions is desirable.

Careful preparation of manuscripts will help keep production time short and ensure a satisfactory appearance of the finished book. The actual production of a Lecture Notes volume normally takes 6 -8 weeks.

Manuscripts should be at least 100 pages long. The final version should include a table of contents.

§8. Editors receive a total of 50 free copies of their volume for distribution to the contributing authors, but no royalties. (Unfortunately, no reprints of individual contributions can be supplied.) They are entitled to purchase further copies of their book for their personal use at a discount of 33.3 %, other Springer mathematics books at a discount of 20 % directly from Springer-Verlag. Contributing authors may purchase the volume in which their article appears at a discount of 33.3 %.

Commitment to publish is made by letter of intent rather than by signing a formal contract. Springer-Verlag secures the copyright for each volume.

<u>Addresses:</u>

Professor A. Dold, Mathematisches Institut, Universität Heidelberg,
Im Neuenheimer Feld 288, 6900 Heidelberg, Federal Republic of Germany

Professor B. Eckmann, Mathematik, ETH-Zentrum
8092 Zürich, Switzerland

Prof. F. Takens, Mathematisch Instituut, Rijksuniversiteit Groningen,
Postbus 800, 9700 AV Groningen, The Netherlands

Springer-Verlag, Mathematics Editorial, Tiergartenstr. 17,
6900 Heidelberg, Federal Republic of Germany, Tel.: (06221) 487-410

Springer-Verlag, Mathematics Editorial, 175, Fifth Avenue,
New York, New York 10010, USA, Tel.: (212) 460-1596

Vol. 1290: G. Wüstholz (Ed.), Diophantine Approximation and Transcendence Theory. Seminar, 1985. V, 243 pages. 1987.

Vol. 1291: C. Mœglin, M.-F. Vignéras, J.-L. Waldspurger, Correspondances de Howe sur un Corps p-adique. VII, 163 pages. 1987

Vol. 1292: J.T. Baldwin (Ed.), Classification Theory. Proceedings, 1985. VI, 500 pages. 1987.

Vol. 1293: W. Ebeling, The Monodromy Groups of Isolated Singularities of Complete Intersections. XIV, 153 pages. 1987.

Vol. 1294: M. Queffélec, Substitution Dynamical Systems – Spectral Analysis. XIII, 240 pages. 1987.

Vol. 1295: P. Lelong, P. Dolbeault, H. Skoda (Réd.), Séminaire d'Analyse P. Lelong – P. Dolbeault – H. Skoda. Seminar, 1985/1986. VII, 283 pages. 1987.

Vol. 1296: M.-P. Malliavin (Ed.), Séminaire d'Algèbre Paul Dubreil et Marie-Paule Malliavin. Proceedings, 1986. IV, 324 pages. 1987.

Vol. 1297: Zhu Y.-l., Guo B.-y. (Eds.), Numerical Methods for Partial Differential Equations. Proceedings. XI, 244 pages. 1987.

Vol. 1298: J. Aguadé, R. Kane (Eds.), Algebraic Topology, Barcelona 1986. Proceedings. X, 255 pages. 1987.

Vol. 1299: S. Watanabe, Yu.V. Prokhorov (Eds.), Probability Theory and Mathematical Statistics. Proceedings, 1986. VIII, 589 pages. 1988.

Vol. 1300: G.B. Seligman, Constructions of Lie Algebras and their Modules. VI, 190 pages. 1988.

Vol. 1301: N. Schappacher, Periods of Hecke Characters. XV, 160 pages. 1988.

Vol. 1302: M. Cwikel, J. Peetre, Y. Sagher, H. Wallin (Eds.), Function Spaces and Applications. Proceedings, 1986. VI, 445 pages. 1988.

Vol. 1303: L. Accardi, W. von Waldenfels (Eds.), Quantum Probability and Applications III. Proceedings, 1987. VI, 373 pages. 1988.

Vol. 1304: F.Q. Gouvêa, Arithmetic of *p*-adic Modular Forms. VIII, 121 pages. 1988.

Vol. 1305: D.S. Lubinsky, E.B. Saff, Strong Asymptotics for Extremal Polynomials Associated with Weights on ℝ. VII, 153 pages. 1988.

Vol. 1306: S.S. Chern (Ed.), Partial Differential Equations. Proceedings, 1986. VI, 294 pages. 1988.

Vol. 1307: T. Murai, A Real Variable Method for the Cauchy Transform, and Analytic Capacity. VIII, 133 pages. 1988.

Vol. 1308: P. Imkeller, Two-Parameter Martingales and Their Quadratic Variation. IV, 177 pages. 1988.

Vol. 1309: B. Fiedler, Global Bifurcation of Periodic Solutions with Symmetry. VIII, 144 pages. 1988.

Vol. 1310: O.A. Laudal, G. Pfister, Local Moduli and Singularities. V, 117 pages. 1988.

Vol. 1311: A. Holme, R. Speiser (Eds.), Algebraic Geometry, Sundance 1986. Proceedings. VI, 320 pages. 1988.

Vol. 1312: N.A. Shirokov, Analytic Functions Smooth up to the Boundary. III, 213 pages. 1988.

Vol. 1313: F. Colonius, Optimal Periodic Control. VI, 177 pages. 1988.

Vol. 1314: A. Futaki, Kähler-Einstein Metrics and Integral Invariants. IV, 140 pages. 1988.

Vol. 1315: R.A. McCoy, I. Ntantu, Topological Properties of Spaces of Continuous Functions. IV, 124 pages. 1988.

Vol. 1316: H. Korezlioglu, A.S. Ustunel (Eds.), Stochastic Analysis and Related Topics. Proceedings, 1986. V, 371 pages. 1988.

Vol. 1317: J. Lindenstrauss, V.D. Milman (Eds.), Geometric Aspects of Functional Analysis. Seminar, 1986–87. VII, 289 pages. 1988.

Vol. 1318: Y. Felix (Ed.), Algebraic Topology – Rational Homotopy. Proceedings, 1986. VIII, 245 pages. 1988

Vol. 1319: M. Vuorinen, Conformal Geometry and Quasiregular Mappings. XIX, 209 pages. 1988.

Vol. 1320: H. Jürgensen, G. Lallement, H.J. Weinert (Eds.), Semigroups, Theory and Applications. Proceedings, 1986. X, 416 pages. 1988.

Vol. 1321: J. Azéma, P.A. Meyer, M. Yor (Eds.), Séminaire de Probabilités XXII. Proceedings. IV, 600 pages. 1988.

Vol. 1322: M. Métivier, S. Watanabe (Eds.), Stochastic Analysis. Proceedings, 1987. VII, 197 pages. 1988.

Vol. 1323: D.R. Anderson, H.J. Munkholm, Boundedly Controlled Topology. XII, 309 pages. 1988.

Vol. 1324: F. Cardoso, D.G. de Figueiredo, R. Iório, O. Lopes (Eds.), Partial Differential Equations. Proceedings, 1986. VIII, 433 pages. 1988.

Vol. 1325: A. Truman, I.M. Davies (Eds.), Stochastic Mechanics and Stochastic Processes. Proceedings, 1986. V, 220 pages. 1988.

Vol. 1326: P.S. Landweber (Ed.), Elliptic Curves and Modular Forms in Algebraic Topology. Proceedings, 1986. V, 224 pages. 1988.

Vol. 1327: W. Bruns, U. Vetter, Determinantal Rings. VII,236 pages. 1988.

Vol. 1328: J.L. Bueso, P. Jara, B. Torrecillas (Eds.), Ring Theory. Proceedings, 1986. IX, 331 pages. 1988.

Vol. 1329: M. Alfaro, J.S. Dehesa, F.J. Marcellan, J.L. Rubio de Francia, J. Vinuesa (Eds.): Orthogonal Polynomials and their Applications. Proceedings, 1986. XV, 334 pages. 1988.

Vol. 1330: A. Ambrosetti, F. Gori, R. Lucchetti (Eds.), Mathematical Economics. Montecatini Terme 1986. Seminar. VII, 137 pages. 1988.

Vol. 1331: R. Bamón, R. Labarca, J. Palis Jr. (Eds.), Dynamical Systems, Valparaiso 1986. Proceedings. VI, 250 pages. 1988.

Vol. 1332: E. Odell, H. Rosenthal (Eds.), Functional Analysis. Proceedings, 1986–87. V, 202 pages. 1988.

Vol. 1333: A.S. Kechris, D.A. Martin, J.R. Steel (Eds.), Cabal Seminar 81–85. Proceedings. 1981–85. V, 224 pages. 1988.

Vol. 1334: Yu.G. Borisovich, Yu. E. Gliklikh (Eds.), Global Analysis – Studies and Applications III. V, 331 pages. 1988.

Vol. 1335: F. Guillén, V. Navarro Aznar, P. Pascual-Gainza, F. Puerta, Hyperrésolutions cubiques et descente cohomologique. XII, 192 pages. 1988.

Vol. 1336: B. Helffer, Semi-Classical Analysis for the Schrödinger Operator and Applications. V, 107 pages. 1988.

Vol. 1337: E. Sernesi (Ed.), Theory of Moduli. Seminar, 1985. VIII, 232 pages. 1988.

Vol. 1338: A.B. Mingarelli, S.G. Halvorsen, Non-Oscillation Domains of Differential Equations with Two Parameters. XI, 109 pages. 1988.

Vol. 1339: T. Sunada (Ed.), Geometry and Analysis of Manifolds. Procedings, 1987. IX, 277 pages. 1988.

Vol. 1340: S. Hildebrandt, D.S. Kinderlehrer, M. Miranda (Eds.), Calculus of Variations and Partial Differential Equations. Proceedings, 1986. IX, 301 pages. 1988.

Vol. 1341: M. Dauge, Elliptic Boundary Value Problems on Corner Domains. VIII, 259 pages. 1988.

Vol. 1342: J.C. Alexander (Ed.), Dynamical Systems. Proceedings, 1986–87. VIII, 726 pages. 1988.

Vol. 1343: H. Ulrich, Fixed Point Theory of Parametrized Equivariant Maps. VII, 147 pages. 1988.

Vol. 1344: J. Král, J. Lukeš, J. Netuka, J. Veselý (Eds.), Potential Theory – Surveys and Problems. Proceedings, 1987. VIII, 271 pages. 1988.

Vol. 1345: X. Gomez-Mont, J. Seade, A. Verjovski (Eds.), Holomorphic Dynamics. Proceedings, 1986. VII, 321 pages. 1988.

Vol. 1346: O. Ya. Viro (Ed.), Topology and Geometry – Rohlin Seminar. XI, 581 pages. 1988.

Vol. 1347: C. Preston, Iterates of Piecewise Monotone Mappings on an Interval. V, 166 pages. 1988.

Vol. 1348: F. Borceux (Ed.), Categorical Algebra and its Applications. Proceedings, 1987. VIII, 375 pages. 1988.

Vol. 1349: E. Novak, Deterministic and Stochastic Error Bounds in Numerical Analysis. V, 113 pages. 1988.

Vol. 1350: U. Koschorke (Ed.), Differential Topology. Proceedings, 1987. VI, 269 pages. 1988.

Vol. 1351: I. Laine, S. Rickman, T. Sorvali, (Eds.), Complex Analysis, Joensuu 1987. Proceedings. XV, 378 pages. 1988.

Vol. 1352: L.L. Avramov, K.B. Tchakerian (Eds.), Algebra – Some Current Trends. Proceedings, 1986. IX, 240 Seiten. 1988.

Vol. 1353: R.S. Palais, Ch.-l. Terng, Critical Point Theory and Submanifold Geometry. X, 272 pages. 1988.

Vol. 1354: A. Gómez, F. Guerra, M.A. Jiménez, G. López (Eds.), Approximation and Optimization. Proceedings, 1987. VI, 280 pages. 1988.

Vol. 1355: J. Bokowski, B. Sturmfels, Computational Synthetic Geometry. V, 168 pages. 1989.

Vol. 1356: H. Volkmer, Multiparameter Eigenvalue Problems and Expansion Theorems. VI, 157 pages. 1988.

Vol. 1357: S. Hildebrandt, R. Leis (Eds.), Partial Differential Equations and Calculus of Variations. VI, 423 pages. 1988.

Vol. 1358: D. Mumford, The Red Book of Varieties and Schemes. V, 309 pages. 1988.

Vol. 1359: P. Eymard, J.-P. Pier (Eds.), Harmonic Analysis. Proceedings, 1987. VIII, 287 pages. 1988.

Vol. 1360: G. Anderson, C. Greengard (Eds.), Vortex Methods. Proceedings, 1987. V, 141 pages. 1988.

Vol. 1361: T. tom Dieck (Ed.), Algebraic Topology and Transformation Groups. Proceedings, 1987. VI, 298 pages. 1988.

Vol. 1362: P. Diaconis, D. Elworthy, H. Föllmer, E. Nelson, G.C. Papanicolaou, S.R.S. Varadhan. École d'Été de Probabilités de Saint-Flour XV–XVII, 1985–87. Editor: P.L. Hennequin. V, 459 pages. 1988.

Vol. 1363: P.G. Casazza, T.J. Shura. Tsirelson's Space. VIII, 204 pages. 1988.

Vol. 1364: R.R. Phelps, Convex Functions, Monotone Operators and Differentiability. IX, 115 pages. 1989.

Vol. 1365: M. Giaquinta (Ed.), Topics in Calculus of Variations. Seminar, 1987. X, 196 pages. 1989.

Vol. 1366: N. Levitt, Grassmannians and Gauss Maps in PL-Topology. V, 203 pages. 1989.

Vol. 1367: M. Knebusch, Weakly Semialgebraic Spaces. XX, 376 pages. 1989.

Vol. 1368: R. Hübl, Traces of Differential Forms and Hochschild Homology. III, 111 pages. 1989.

Vol. 1369: B. Jiang, Ch.-K. Peng, Z. Hou (Eds.), Differential Geometry and Topology. Proceedings, 1986–87. VI, 366 pages. 1989.

Vol. 1370: G. Carlsson, R.L. Cohen, H.R. Miller, D.C. Ravenel (Eds.), Algebraic Topology. Proceedings, 1986. IX, 456 pages. 1989.

Vol. 1371: S. Glaz, Commutative Coherent Rings. XI, 347 pages. 1989.

Vol. 1372: J. Azéma, P.A. Meyer, M. Yor (Eds.), Séminaire de Probabilités XXIII. Proceedings. IV, 583 pages. 1989.

Vol. 1373: G. Benkart, J.M. Osborn (Eds.), Lie Algebras, Madison 1987. Proceedings. V, 145 pages. 1989.

Vol. 1374: R.C. Kirby, The Topology of 4-Manifolds. VI, 108 pages. 1989.

Vol. 1375: K. Kawakubo (Ed.), Transformation Groups. Proceedings, 1987. VIII, 394 pages, 1989.

Vol. 1376: J. Lindenstrauss, V.D. Milman (Eds.), Geometric Aspects of Functional Analysis. Seminar (GAFA) 1987–88. VII, 288 pages. 1989.

Vol. 1377: J.F. Pierce, Singularity Theory, Rod Theory, and Symmetry-Breaking Loads. IV, 177 pages. 1989.

Vol. 1378: R.S. Rumely, Capacity Theory on Algebraic Curves. III, 437 pages. 1989.

Vol. 1379: H. Heyer (Ed.), Probability Measures on Groups IX. Proceedings, 1988. VIII, 437 pages. 1989

Vol. 1380: H.P. Schlickewei, E. Wirsing (Eds.), Number Theory, Ulm 1987. Proceedings. V, 266 pages. 1989.

Vol. 1381: J.-O. Strömberg, A. Torchinsky. Weighted Hardy Spaces. V, 193 pages. 1989.

Vol. 1382: H. Reiter, Metaplectic Groups and Segal Algebras. XI, 128 pages. 1989.

Vol. 1383: D.V. Chudnovsky, G.V. Chudnovsky, H. Cohn, M.B. Nathanson (Eds.), Number Theory, New York 1985–88. Seminar. V, 256 pages. 1989.

Vol. 1384: J. Garcia-Cuerva (Ed.), Harmonic Analysis and Partial Differential Equations. Proceedings, 1987. VII, 213 pages. 1989.

Vol. 1385: A.M. Anile, Y. Choquet-Bruhat (Eds.), Relativistic Fluid Dynamics. Seminar, 1987. V, 308 pages. 1989.

Vol. 1386: A. Bellen, C.W. Gear, E. Russo (Eds.), Numerical Methods for Ordinary Differential Equations. Proceedings, 1987. VII, 136 pages. 1989.

Vol. 1387: M. Petković, Iterative Methods for Simultaneous Inclusion of Polynomial Zeros. X, 263 pages. 1989.

Vol. 1388: J. Shinoda, T.A. Slaman, T. Tugué (Eds.), Mathematical Logic and Applications. Proceedings, 1987. V, 223 pages. 1989.

Vol. 1000: Second Edition. H. Hopf, Differential Geometry in the Large. VII, 184 pages. 1989.

Vol. 1389: E. Ballico, C. Ciliberto (Eds.), Algebraic Curves and Projective Geometry. Proceedings, 1988. V, 288 pages. 1989.

Vol. 1390: G. Da Prato, L. Tubaro (Eds.), Stochastic Partial Differential Equations and Applications II. Proceedings, 1988. VI, 258 pages. 1989.

Vol. 1391: S. Cambanis, A. Weron (Eds.), Probability Theory on Vector Spaces IV. Proceedings, 1987. VIII, 424 pages. 1989.

Vol. 1392: R. Silhol, Real Algebraic Surfaces. X, 215 pages. 1989.

Vol. 1393: N. Bouleau, D. Feyel, F. Hirsch, G. Mokobodzki (Eds.), Séminaire de Théorie du Potentiel Paris, No. 9. Proceedings. VI, 265 pages. 1989.

Vol. 1394: T.L. Gill, W.W. Zachary (Eds.), Nonlinear Semigroups, Partial Differential Equations and Attractors. Proceedings, 1987. IX, 233 pages. 1989.

Vol. 1395: K. Alladi (Ed.), Number Theory, Madras 1987. Proceedings. VII, 234 pages. 1989.

Vol. 1396: L. Accardi, W. von Waldenfels (Eds.), Quantum Probability and Applications IV. Proceedings, 1987. VI, 355 pages. 1989.

Vol. 1397: P.R. Turner (Ed.), Numerical Analysis and Parallel Processing. Seminar, 1987. VI, 264 pages. 1989.

Vol. 1398: A.C. Kim, B.H. Neumann (Eds.), Groups – Korea 1988. Proceedings. V, 189 pages. 1989.

Vol. 1399: W.-P. Barth, H. Lange (Eds.), Arithmetic of Complex Manifolds. Proceedings, 1988. V, 171 pages. 1989.

Vol. 1400: U. Jannsen. Mixed Motives and Algebraic K-Theory. XIII, 246 pages. 1990.

Vol. 1401: J. Steprāns, S. Watson (Eds.), Set Theory and its Applications. Proceedings, 1987. V, 227 pages. 1989.

Vol. 1402: C. Carasso, P. Charrier, B. Hanouzet, J.-L. Joly (Eds.), Nonlinear Hyperbolic Problems. Proceedings, 1988. V, 249 pages. 1989.

Vol. 1403: B. Simeone (Ed.), Combinatorial Optimization. Seminar, 1986. V, 314 pages. 1989.

Vol. 1404: M.-P. Malliavin (Ed.), Séminaire d'Algèbre Paul Dubreil et Marie-Paul Malliavin. Proceedings, 1987 – 1988. IV, 410 pages. 1989.

Vol. 1405: S. Dolecki (Ed.), Optimization. Proceedings, 1988. V, 223 pages. 1989.

Vol. 1406: L. Jacobsen (Ed.), Analytic Theory of Continued Fractions III. Proceedings, 1988. VI, 142 pages. 1989.

Vol. 1407: W. Pohlers, Proof Theory. VI, 213 pages. 1989.

Vol. 1408: W. Lück, Transformation Groups and Algebraic K-Theory. XII, 443 pages. 1989.

Vol. 1409: E. Hairer, Ch. Lubich, M. Roche. The Numerical Solution of Differential-Algebraic Systems by Runge-Kutta Methods. VII, 139 pages. 1989.

Vol. 1410: F.J. Carreras, O. Gil-Medrano, A.M. Naveira (Eds.), Differential Geometry. Proceedings, 1988. V, 308 pages. 1989.